Electromagnetic Scattering in Disperse Media

Inhomogeneous and Anisotropic Particles

Springer
Berlin
Heidelberg
New York
Hong Kong
London
Milan
Paris
Tokyo

V. A. Babenko, L. G. Astafyeva and V. N. Kuzmin

Electromagnetic Scattering in Disperse Media

Inhomogeneous and Anisotropic Particles

Springer

Published in association with
Praxis Publishing
Chichester, UK

PRAXIS

Dr Victor A. Babenko
Senior Research Scientist
B.I. Stepanov Institute of Physics
Belarus National Academy of Sciences
Minsk
Belarus

Dr Liudmila G. Astafyeva
Leading Research Scientist
B.I. Stepanov Institute of Physics
Belarus National Academy of Sciences
Minsk
Belarus

Dr Vladimir N. Kuzmin
Sciential Secretary
Belarus Republican Scientific & Technical
Centre for Remote Sensing (RSTC "Ecomir")
Minsk
Belarus

SPRINGER–PRAXIS BOOKS IN ENVIRONMENTAL SCIENCES
SUBJECT *ADVISORY EDITOR*: John Mason B.Sc., M.Sc., Ph.D.

ISBN 978-3-642-07820-0 Springer-Verlag Berlin Heidelberg New York

Bibliographic information published by Die Deutsche Bibliothek

Die Deutsche Bibliothek lists this publication in the Deutsche Nationalbibliografie; detailed bibliographic data are available from the Internet at http://dnb.ddb.de

Library of Congress Cataloging-in-Publication Data
Babenko, V.A. (Victor Aleksandrovich)
Electromagnetic scattering in dispersed media : inhomogenous and anisotropic particles / V.A. Babenko, L.G. Astafyeva, and V.N. Kuzmin.
p. cm. – (Springer-Praxis books in environmental sciences)
Includes bibliographical references and index.

1. Electromagnetic waves—Scattering. 2. Light–Scattering. 3. Light absorption 4. Anisotropy. 5. Inhomogeneous materials. I. Astafyeva, L.G. (Liudmila, Grigor'evna), 1945- II. Kuzmin, V.N. (Vladimir Nikolaevich) III. Title. IV. Series.

QC665.S3B33 2003
539.2–dc21 2003050474

Printed in Germany

Cover design: Jim Wilkie
Project Management: Originator Publishing Services, Gt Yarmouth, Norfolk, UK

Printed on acid-free paper

To the memory of Prof. Anatoly P. Prishivalko

Contents

Preface

The rapid development of the optics of scattering media and its practical applications continuously brings forth new problems about electromagnetic radiation interaction with scatterers of different types and their ensembles. Such problems arise in the optics of the sea and atmosphere, in astrophysics, biophysics, physical chemistry, chemistry and physics of polymers, thermal physics, and in other branches of science.

The Russian edition of this book was written by researchers from the Laboratory of Optical Diagnostics of Disperse Media in the Stepanov Institute of Physics, Belarus Academy of Sciences (Minsk, Belarus). This Laboratory together with the Laboratory of Scattering Media Optics at the Stepanov Institute formed the well known (in the former Soviet Union) Minsk Scientific School of Light Scattering.

When this book was published many years ago the only model commonly used for description and interpretation of scattering properties of dispersion media, including atmospheric aerosols was an ensemble of homogeneous spherical particles. Naturally, in many cases such a model is too simplified and idealized, as the actual scattering particles frequently are inhomogeneous, and the substances, comprising them, can have an anisotropy of optical properties. For deriving the objective quantitative data it is necessary to take into account the presence of the internal structure of such objects, as atmospheric condensation nuclei, hailstones, carbon-ice particles of cosmic dust, micro-organisms and so on. The optical properties of clouds and some other polydisperse media can depend on an anisotropy of particles within them.

The investigation of the scattering by non-uniform and anisotropic particles here hindered partly by the circumstance, that the mathematical apparatus bound with the solution of these problems was complex enough, the analytical solutions was very cumbersome and, almost as a rule, the opportunity to examine the problem in the general form was eliminated.

With the purpose of filling of the knowledge gap the authors carried out investigations, directed to develop the theory and computational methods of radiation scattering and absorption by particles with internal inhomogeneity of different types. We compared the features of light scattering by homogeneous and inhomogeneous, isotropic and anisotropic particles, and systems of such particles, and cleared up the areas and requirements of maximal responsivity of the scattering characteristics to change of optical properties and internal structure of particles and so on. On the basis of these investigations, some new practical applications of light scattering by non-uniform and anisotropic particles (radiation monochromatization, detection of the thin surface layers of the absorbed substances on a particle surface, adjustment of absorption efficiency) were designed and justified.

Although basic attention was given to optical properties of atmospheric aerosol as dispersion medium (rather important in practice), it is necessary to stress, that our purpose was not at all restricted by objects and problems of aerosol optics. The obtained solutions and deductions are applicable to a wide circle of other dispersion media.

The Russian edition was the first monograph devoted especially to the problem of light scattering and absorption by inhomogeneous (such as two-layered, three-layered, radially-inhomogeneous) and anisotropic spherical particles. The distinguishing features of the book were the Green's functions approach to solving the problems of diffraction by such particles and many results (more than 100 figures) computed for polydisperse particle systems. Regrettably, this book remains almost totally unknown to the English-speaking light scattering community.

During the years after the issue of the Russian edition, a real scientific revolution took place in light scattering science, that resulted from both the explosion-like progress of IT and the great theoretical successes. A great deal from things that twenty years ago were considered as extremely hard or principally impossible are routine tasks now. A number of people involved in light scattering theory and practical applications increased by an order of magnitude. Now very interesting investigations are being conducted not only in the traditional centers, but also in the countries where such investigations were not conducted at all before. Unfortunately, this rapid growth is not taking place in the states formed after the collapse of the Soviet Union, including Belarus where science is under severe conditions now.

In recent years, great achievements were obtained in the theory and practice of light scattering by non-spherical particles. In this regard, those interested in the subject are strongly recommended to read two newly edited books by Mishchenko *et al.* (200a, 2002a). A lot of new information can also be found in the *Proceedings of the International Conferences on Electromagnetic and Light Scattering by Nonspherical Particles* (the last one was in 2002 (Gustafson *et al.*, 2002)).

So, when it was suggested to re-edit our book in English, we had doubts. Is it appropriate to re-edit the book, parts of which have become obsolete? Is it better to write something completely new (e.g., to take the particle non-spherisity properly into account)? However, the simultaneous consideration of both inhomogeneity and/or anisotropy and non-sphericity is a very difficult problem even now. Therefore, as it is customary in this situation, we made a compromise, keeping the main ideas of the Russian edition unaltered but considerably updating and appending the text and references. We did not strike the Russian papers off the reference list (except for "grey" literature that is beyond the reach of the western reader).

In particular, the new Chapters 5–7 deal with subjects related to various aspects of light absorption inside particles, including field distributions inside homogeneous (Chapter 6) and layered (Chapter 7) spheres, MDR resonances, and absorption in restricted regions inside particles (Chapter 5). Incorporation of this issue in the book is somewhat connected with the interests of our laboratory. We were always interested not only in scattered fields, but in internal fields as well, enabling the complex thermophysical-optical problems of the particle heating and destruction to be solved. We realize that the content of Chapters 6 and 7 is somewhat concise, but the relatively comprehensive consideration of issues related to the non-linear optics of aerosols require more voluminous book. The discussion on the methodology of light scattering calculations is widely extended (Sections 1.4–1.5). The detailed algorithm to calculate the light scattering by a sphere with the arbitrary number of concentric layers (Section 3.5), consideration of scattering by a sphere with homogeneous core and radially-inhomogeneous shell (Section 8.4), application of the radially-inhomogeneous models to describe the scattering by the fractal cluster (Section 8.5) – all these themes were also absent in the Russian edition. The book is supplemented by five Appendices, where useful analytic formulae for the Riccati–Bessel functions (Appendix A), the angular functions (Appendix B) and the vector spherical wavefunctions (Appendix C) are listed. Appendix D is the guide to our light scattering codes.

In Appendix E we provide references to the websites containing many relevant computer programs developed by various research groups and made publicly available through the Internet. Although the majority of these programs (including ours mentioned by Appendix D) have been extensively tested and are expected to generate reliable results in most cases provided that they are used as instructed, it is not inconceivable that some of them contain errors or idiosyncrasies. Furthermore, input parameters can be used that are outside the range of values from which results can be computed accurately. For these reasons the authors of this book and the publisher disclaim all liability for any damage that may result from the use of these programs. In addition, although the publisher has used its best endeavours to ensure that the URLs for external websites referred to in this book are correct and active at the time of going to press, the publisher has no responsibilty for the websites and can make no guarantee that a site will remain live or that the content is or will remain appropriate.

Victor A. Babenko
Liudmila G. Astafyeva
Vladimir N. Kuzmin
Minsk, Belarus, March 2003

Acknowledgments

This book is dedicated to memory of Professor Anatoly P. Prishivalko who was a coordinator of the project on writing and publishing the Russian version of the book and coauthor of many articles from which the book was originated. We very highly appreciate his efforts to promote the Russian edition.

We thank Alexander Kokhanovsky who initialized the English edition. Without his constant reminding the writing of this book would not have been completed.

We are extremely thankful to Michael Mishchenko who took the trouble to read very raw material and made numerous valuable corrections and comments which resulted in a much improved manuscript.

Special thanks are given to present and former researchers of the Optical Diagnostics of Disperse Media Laboratory in Minsk (Stepanov Institute of Physics, Belarus), where we spent many years in studies of various aspects of light scattering.

We thank many researchers in light scattering for helping us to get papers that were not readily accessible in Belarus.

The preparation of this book was partially sponsored by the INTAS grant 99/652 and the ISTC grant B-678.

Last, but not least, we would like to underline that this book would not have been prepared without help and understanding of our families. In particular, we thank Pavel Babenko who read some of the first chapters, improving its readability.

Abbreviations

3D	three-dimensional
APs	anisotropic particles
BR theory	Bruggeman theory
EMT	effective medium theory
FED	field of experimental data
HP	homogeneous particle
HOP	hollow particles
IR	infra red
MDR	morphology dependent resonances
MG theory	Maxwell Garnett theory
MM	Mueller matrix
RIP	radially inhomogeneous particles
RI	radially inhomogeneous
SAS	surface active substances
TLPs	two-layered particles
WKB	Wentzel–Kramers–Brillouin

List of figures

List of tables

1

Coordinate and momentum representations in the theory of light scattering by disperse media

In this chapter the most general equations and relations of optics of disperse media are considered. These are based on classic electrodynamics and statistical physics. The main attention is given to the electrodynamic part of the problem; the statistical part is also discussed but in less detail. The theoretical analysis of the basic relationships will be carried out with the help of a covariant description of scattering processes; the basis for this approach is the method of Green's functions and scattering matrices. Such an approach appears to be useful not only for the investigation of light scattering by systems of particles, but also in the case of scattering by an individual particle, as it ensures the generality of a solution procedure for many problems complicated by the introduction of individual coordinate systems.

To illustrate the importance of switching from local electrodynamic radiation characteristics to the experimental photometric characteristics, a simple way of introducing quantities such as intensity is offered, and their relation to quantities recorded by light receivers is shown.

The light scattering by a homogeneous isotropic spherical particle is considered as an example of the specific application of the mathematical apparatus used in this book. This simplest case helps the reader to move on to more complex problems of light scattering by non-homogeneous and anisotropic particles.

1.1 DIFFERENTIAL AND INTEGRAL EQUATIONS OF ELECTROMAGNETIC SCATTERING

The solution of the diffraction problem as well as of any boundary-value problem consists of the solution of the relevant differential equations, which satisfy certain boundary conditions. In particular, to solve the problem of electromagnetic wave diffraction on an object of finite size it is necessary to solve the set of macroscopic Maxwell equations:

$$\nabla \times \mathbf{H} = \mathbf{J}_F + \frac{\partial \mathbf{D}}{\partial t} \quad \nabla \times \mathbf{E} = -\frac{\partial \mathbf{B}}{\partial t} \quad \nabla \cdot \mathbf{D} = \rho_F \quad \nabla \cdot \mathbf{B} = 0 \qquad (1.1.1)$$

where $\mathbf{E}$ and $\mathbf{H}$ are the electric and magnetic fields, $\mathbf{D}$ and $\mathbf{B}$ are the electric displacement and magnetic induction, $\mathbf{J}_F$ is the free current density, and ρ_F is the free charge density along with:

(1) the linear constitutive relations

$$\mathbf{D} = \varepsilon\varepsilon_0\mathbf{E} \qquad \mathbf{B} = \mu\mu_0\mathbf{H} \qquad \mathbf{J}_F = \sigma\mathbf{E} \tag{1.1.2}$$

where ε_0 is the permittivity and μ_0 is the permeability of free space, ε is the dielectric function, μ is the permeability of the medium under consideration, and σ is the specific conductivity;

(2) the boundary conditions on the object's surface

$$E_{\tau^+} = E_{\tau^-} \qquad H_{\tau^+} = H_{\tau^-} \tag{1.1.3}$$

where τ labels the tangential components of fields; and

(3) the conditions at infinity

$$\begin{aligned} \mathbf{E}(\mathbf{r}) &\overset{r\to\infty}{=} \mathbf{E}_I(\mathbf{r}) + r^{-1}\exp(\pm ik_0 r)\mathcal{E}(\mathbf{r}) \\ \mathbf{H}(\mathbf{r}) &\overset{r\to\infty}{=} \mathbf{H}_I(\mathbf{r}) + r^{-1}\exp(\pm ik_0 r)\mathcal{H}(\mathbf{r}) \end{aligned} \tag{1.1.4}$$

where $\mathbf{E}_I, \mathbf{H}_I$ are incident electric and magnetic fields.

The structure of the field $(\mathbf{E}_I, \mathbf{H}_I)$ incident on the object can be rather arbitrary. However, it is assumed that this field can be expanded in terms of harmonic plane waves practically in any case. Such representation is most convenient for the solution of different diffraction problems and with the help of an inverse transformation one can always return to the initial form. With this in mind we further consider the diffraction of a harmonic plane wave $\mathbf{E}_I(t) \sim \exp(-i\omega t + ik_0 r)$. The constitutive relations seen in Eq.(1.1.2) are assumed to be linear. Because we are not considering magneto-optical effects, the magnetic permeability is supposed to be equal to unity ($\mu = 1$) both inside and outside the body. The dielectric function inside the body is a complex quantity (in the general case it is a tensor). Outside the body it is a constant scalar that, without loss of generality, can be taken to be unity. The sign in the exponential factor in the conditions at infinity is defined by the causality principle. With the chosen time dependence for the incident field these conditions are $\mathbf{E}(\mathbf{r}) = \mathbf{E}_I(\mathbf{r}) + r^{-1}\exp(ik_0 r)\mathcal{E}(\mathbf{r})$ and $\mathbf{H}(\mathbf{r}) = \mathbf{H}_I(\mathbf{r}) + r^{-1}\exp(ik_0 r)\mathcal{H}(\mathbf{r})$.

Some authors represent a harmonic wave as $\mathbf{E}_I \sim \exp(i\omega t)$. In this case, results for linear combinations of electromagnetic fields are complex conjugates of the ones in the present work. For quadratic combinations the choice of the sign in a representation of a harmonic wave has no importance.

Based on the above it is possible to derive a wave equation for the electric field (Newton, 1982):

$$\nabla \times \nabla \times \mathbf{E}(\mathbf{r}) - k_0^2\xi^2(\mathbf{r})\mathbf{E}(\mathbf{r}) = 0 \tag{1.1.5}$$

where $k_0 = 2\pi/\lambda$ is the wave number for free space, $\xi(\mathbf{r}) = 1$ outside the particle, and $\xi(\mathbf{r}) = N(\mathbf{r})$ inside it. In the simplest scalar case the complex refractive index $N(\mathbf{r})$ relates to the complex dielectric function $\hat{\varepsilon} = \varepsilon + i\sigma/\omega$ thus:

$$N = \operatorname{Re} N + i \operatorname{Im} N \tag{1.1.6}$$

$$n = \operatorname{Re} N = \left[\frac{1}{2}\left(|\hat{\varepsilon}| + \operatorname{Re}\hat{\varepsilon}\right)\right]^{1/2} \qquad \kappa = \operatorname{Im} N = \left[\frac{1}{2}\left(|\hat{\varepsilon}| - \operatorname{Re}\hat{\varepsilon}\right)\right]^{1/2}$$

The magnetic field can always be found from:

$$\mathbf{H}(\mathbf{r}) = -i\sqrt{\frac{\varepsilon_0}{\mu_0}}k_0^{-1}\nabla \times \mathbf{E}(\mathbf{r}) \tag{1.1.7}$$

Hence, only wave equations for the electric field will only be considered in what follows.

The diffracted fields must meet the boundary conditions on the body surface and at infinity. The vector wave equation (1.1.5) together with boundary conditions (1.1.3–4) is equivalent to the integral equation (Newton, 1982; Tai, 1993):

$$\mathbf{E}(\mathbf{r}) = \mathbf{E}_I(\mathbf{r}) + k_0^2 \int\limits_V \hat{G}^0(\mathbf{r}, \mathbf{r}') \left[\xi^2(\mathbf{r}') - 1\right] \mathbf{E}(\mathbf{r}') d^3\mathbf{r}' \tag{1.1.8}$$

where $\hat{G}^0(\mathbf{r}, \mathbf{r}')$ is the tensor Green's function for free space, and the integration is performed over the particle volume V. By definition, the function $\hat{G}^0(\mathbf{r}, \mathbf{r}')$ satisfies the following equations:

$$\nabla_r \times \nabla_r \times \hat{G}^0(\mathbf{r}, \mathbf{r}') - k_0^2 \hat{G}^0(\mathbf{r}, \mathbf{r}') = \hat{I}\delta(\mathbf{r} - \mathbf{r}') \tag{1.1.9}$$

$$\nabla_r \cdot \hat{G}^0(\mathbf{r}, \mathbf{r}') = 0 \qquad \mathbf{r} \neq \mathbf{r}'$$

where $\hat{I}$ is a unit tensor and $\delta(\mathbf{r} - \mathbf{r}')$ is the Dirac delta function. It obeys the radiation conditions (1.1.4) at infinity. On this basis it is possible to reach the following relation for the tensor Green's function of free space (Newton, 1982):

$$\hat{G}^0(\mathbf{r}, \mathbf{r}') = \left(\hat{I} + k_0^{-2}\nabla \otimes \nabla\right) \frac{e^{ik_0|\mathbf{r}-\mathbf{r}'|}}{4\pi|\mathbf{r} - \mathbf{r}'|} \tag{1.1.10}$$

$$= \lim_{\varepsilon \to 0^+} \int \frac{d^3\mathbf{k}}{(2\pi)^3} \frac{\hat{I}k_0^2 - \mathbf{k} \otimes \mathbf{k}}{k_0^2\left(k^2 - k_0^2 - i\varepsilon\right)} e^{i\mathbf{k}\cdot(\mathbf{r}-\mathbf{r}')}$$

where $\otimes$ is dyadic or tensor product of vectors (Tai, 1987, 1993).

The integral equation (1.1.8) can be written for a body of any geometrical configuration. Moreover, if one knows the distribution of volume or surface field for this body, then the integral equation yields the solution to the electrodynamic scattering problem. In this case the basic difficulty is finding the correct integration over the body volume and the associated representation of Green's functions in the corresponding coordinate system. Most works in diffraction theory which use Green's functions are devoted to finding different representations of these functions and to investigate their behaviour near the singular point $\mathbf{r} = \mathbf{r}'$ (e.g., see (van Bladel, 1961; Mei, 1971; Tai, 1973, 1993; Tai and Rosenfeld, 1976)).

Let us consider the single-particle Green's function which yields the solution to the boundary diffraction problem (1.1.1–4), regardless of the form of the incident wave, through an integral relation:

$$\mathbf{E}(\mathbf{r}) = \mathbf{E}_I(\mathbf{r}) + \int\limits_V \hat{G}(\mathbf{r},\mathbf{r}')k_0^2\left[\xi^2(\mathbf{r}') - 1\right]\mathbf{E}_I(\mathbf{r}')d^3\mathbf{r} \tag{1.1.11}$$

This function is a solution of the boundary problem:

$$\begin{aligned} \nabla_r \times \nabla_r \times \hat{G}(\mathbf{r},\mathbf{r}') - k_0^2\xi^2(\mathbf{r})\hat{G}(\mathbf{r},\mathbf{r}') &= \hat{I}\delta(\mathbf{r}-\mathbf{r}') \\ \mathbf{n}_r \times \hat{G}(\mathbf{r},\mathbf{r}')\,|_{S+} &= \mathbf{n}_r \times \hat{G}(\mathbf{r},\mathbf{r}')\,|_{S-} \\ \mathbf{n}_r \times \nabla_r \times \hat{G}(\mathbf{r},\mathbf{r}')\,|_{S+} &= \mathbf{n}_r \times \nabla_r \times \hat{G}(\mathbf{r},\mathbf{r}')\,|_{S-} \end{aligned} \tag{1.1.12}$$

It also obeys the radiation condition at infinity and regularity condition at the origin. In Eq.(1.1.12) S is the surface bounding the particle volume V and $\mathbf{n}_r$ is a unit vector normal to the surface. The rigorous derivation of Eqs.(1.1.11) and (1.1.12) can be obtained applying the second Green's identity to vector $\mathbf{E}(\mathbf{r})$ and dyad $\hat{G}(\mathbf{r},\mathbf{r}')$ (Zich, 1969).

As a result of inhomogeneity of space as a whole due to presence of particle the single-particle Green's function has a different structure depending on its arguments' values $\mathbf{r}$ and $\mathbf{r}'$. This allows discussion about a set of Green's functions for the given diffraction problem. Depending on whether the internal or external diffraction problem is solved, and whether the radiation source is inside or outside the body, we need to use one or another type of Green's function. The single-particle Green's functions will be considered in full in Section 1.3 considering the example of homogeneous spherical particles.

It is customary to express the solution of the external diffraction problem (or, scattering problem) through the amplitude scattering dyad $\hat{T}$:

$$\mathbf{E}(\mathbf{r}) = \mathbf{E}_I(\mathbf{r}) + \int\limits_V \hat{G}^0(\mathbf{r},\mathbf{r}')\hat{T}(\mathbf{r}',\mathbf{r}'')\mathbf{E}_I(\mathbf{r}'')d^3\mathbf{r}'d^3\mathbf{r}'' \tag{1.1.13}$$

which is closely related to the single-particle Green's function (see Sections 1.2 and 1.3). However, in the optical range the asymptotic form of the above relationship occurs more frequently. Let us consider the case of a plane incident wave. With the proviso that $r \to \infty$ the Green's function of free space becomes simpler:

$$\hat{G}^0(\mathbf{r},\mathbf{r}') \approx (\hat{I} - \hat{\mathbf{r}} \otimes \hat{\mathbf{r}})\frac{e^{ik_0 r}}{4\pi r}e^{-i\mathbf{k}_0'\cdot\mathbf{r}'} \tag{1.1.14}$$

where $\mathbf{k}_0' = k_0\mathbf{r}/|\mathbf{r}| \equiv k_0\hat{\mathbf{r}}$. After substituting into Eq.(1.1.13) and taking into account the character of the incident wave, we obtain the following simple expression for the external field:

$$\mathbf{E}(\mathbf{r}) = \mathbf{E}_I(\mathbf{r}) + \frac{1}{4\pi r}e^{ik_0 r}\hat{A}\left(\mathbf{k}_0', \mathbf{k}_0\right)\cdot\mathbf{E}_I \tag{1.1.15}$$

This expression includes the amplitude of incident wave $\mathbf{E}_I(\mathbf{r})$ and the asymptotic scattering tensor:

$$\hat{A}\left(\mathbf{k}_0', \mathbf{k}_0\right) = \int\limits_V \left(\hat{I} - \hat{\mathbf{r}} \otimes \hat{\mathbf{r}}\right) e^{-i\mathbf{k}_0' \cdot \mathbf{r}'} \hat{T}(\mathbf{r}', \mathbf{r}'') e^{i\mathbf{k}_0' \cdot \mathbf{r}''} d^3\mathbf{r}' d^3\mathbf{r}'' \tag{1.1.16}$$

This tensor represents the Fourier transform of the tensor $\hat{T}(\mathbf{r}', \mathbf{r}'')$ up to a constant factor. Due to transversality of plane waves in homogeneous non-absorbing space it has a 2×2 dimension. The first factor in the second term of Eq.(1.1.15) is the scalar Green's function of free space:

$$G^0 = (4\pi r)^{-1} e^{ik_0 r} \tag{1.1.17}$$

Up to this point we were dealing with intensities of electromagnetic fields. Let us now concentrate on quadratic (that is, energetic) quantities. In optics these quantities are observed and registered in experiments. However, they will appear for some time in our considerations as local field characteristics.

From Eqs.(1.1.1–4) it is easy to obtain a balance relationship, which may be interpreted as the law of energy conservation (Born and Wolf, 1999):

$$\nabla \cdot \mathbf{S} + \frac{\partial W}{\partial t} - W' = 0 \tag{1.1.18}$$

where

$$\mathbf{S} = \mathrm{Re}\left(\mathbf{E} \times \mathbf{H}^*\right) \qquad W = \frac{1}{2}\left(\mathbf{E} \cdot \mathrm{Re}\,\varepsilon \cdot \mathbf{E}^* + \mathbf{H} \cdot \mathbf{H}^*\right)$$
$$W' = \mathrm{Re}\left(\mathbf{E} \cdot \mathrm{Im}\,\varepsilon \cdot \frac{\partial \mathbf{E}^*}{\partial t}\right)$$

are, respectively, the energy flux density, the density of field energy due to the real part of the dielectric tensor, and the density of field energy due to the imaginary part of the dielectric tensor. Apparently, the value of W' represents the energy absorbed by a unit volume of a substance. The law of energy conservation in the case of light scattering by a particle will be considered in some details in Section 5.1.

In an optical range the most frequently used representation of energy quantities is a representation of harmonic plane waves or the spectral–angular representation, which is referred to as the momentum–frequency representation in quantum mechanics. The basic quantities here are the Stokes parameters which in the case of a plane wave in the homogeneous non-absorbing media are written as:

$$S_\lambda(\omega, \mathbf{k}) = n\mathbf{E}(\omega, \mathbf{k}) \cdot \hat{\boldsymbol{\sigma}}^\lambda \cdot \mathbf{E}^*(\omega, \mathbf{k}) \tag{1.1.19}$$

where $\hat{\sigma}^\lambda$ are the Pauli matrices (Rozenberg, 1982):

$$\hat{\sigma}^1 = \begin{pmatrix} 1 & 0 \\ 0 & 1 \end{pmatrix} \quad \hat{\sigma}^2 = \begin{pmatrix} 1 & 0 \\ 0 & -1 \end{pmatrix} \quad \hat{\sigma}^3 = \begin{pmatrix} 0 & 1 \\ 1 & 0 \end{pmatrix} \quad \hat{\sigma}^4 = \begin{pmatrix} 0 & -i \\ i & 0 \end{pmatrix}$$

The dimension of electric field vectors is equal to two and they oscillate in the plane perpendicular to wave vector **k**. The set of Stokes parameters forms a complete collection of optical quantities which occur during spectral treatment of phenomena. The quantity W can be expressed through the first Stokes parameter S_1, indicating luminosity:

$$\int W(\mathbf{r})d^3\mathbf{r} = \int d\omega \int d^3\mathbf{k}\, n(\omega)S_1(\omega,\mathbf{k}), \quad \mathbf{S} = \int d\omega \int d^2\hat{k}\, \hat{k}S_1(\omega,\mathbf{k})$$

where $\int d^3\mathbf{k}...$ is the integration over the wave vector.

Let us concentrate in greater detail on definition (1.1.19). The electric field vector $\mathbf{E}(\omega,\mathbf{k})$ represents the Fourier transform of the corresponding vector in the coordinate–time representation:

$$\mathbf{E}(\omega,\mathbf{k}) = \int dt \int d^3\mathbf{r}\, \mathbf{E}(t,\mathbf{r})e^{-i\omega t}e^{i\mathbf{k}\cdot\mathbf{r}} \tag{1.1.20}$$

The inverse transform looks like:

$$\mathbf{E}(t,\mathbf{r}) = \frac{1}{(2\pi)^4}\int d\omega \int d^3\mathbf{k}\, \mathbf{E}(\omega,\mathbf{k})e^{i\omega t}e^{-i\mathbf{k}\cdot\mathbf{r}} \tag{1.1.21}$$

Wave vectors and circular frequency occur in Eqs.(1.1.20) and (1.1.21) as independent quantities. However, the dispersion relation:

$$\mathbf{k}\times\mathbf{k}\times\mathbf{E}(\omega,\mathbf{k}) - k_0^2\mathbf{E}(\omega,\mathbf{k}) = 0$$

follows from Maxwell's equations for free space that imposes some restrictions on $\mathbf{E}(\omega,\mathbf{k})$ and the wave vector **k**.

In fact, after this relation is multiplied by the wave vector **k** we obtain $\mathbf{k}\cdot\mathbf{E}(\omega,\mathbf{k}) = 0$. Therefore, it follows that the vector of electric field oscillations in momentum representation lies in the plane perpendicular to **k** and has only two components.

After opening the double vector product we arrive at the relationship:

$$\left(k^2 - k_0^2\right)\mathbf{E}(\mathbf{r}) = 0 \tag{1.1.22}$$

from which it follows that for the inverse transformation (1.1.21) to exist, the function $\mathbf{E}(\omega,\mathbf{k})$ must be defined in space of infinite generalized functions. Let us define it with the help of the following limit:

$$\lim_{\varepsilon\to 0}\left[\left(k^2 - k_0^2\right)^2 + \varepsilon^2\right]\mathbf{E}(\omega,\mathbf{k}) = \lim_{\varepsilon\to 0}\varepsilon\mathbf{E}'(\omega,\mathbf{k}_0) \tag{1.1.23}$$

where $\mathbf{E}'(\omega,\mathbf{k}_0)$ is the usual finite function. Apparently, Eq.(1.1.22) is satisfied in the limit. It allows, without loss of generality, to present $\mathbf{E}(\omega,\mathbf{k})$ as:

$$\mathbf{E}(\omega,\mathbf{k}) = \lim_{\varepsilon\to 0}\frac{\varepsilon}{\left(k^2 - k_0^2\right)^2 + \varepsilon^2}\mathbf{E}'(\omega,\mathbf{k}_0)$$

that gives in the limit:

$$\mathbf{E}(\omega,\mathbf{k}) = 2k_0\pi\delta\left(k^2 - k_0^2\right)\mathbf{E}'(\omega,\mathbf{k}_0) \tag{1.1.24}$$

Integrating Eq.(1.1.24) over the length of vector **k**, we obtain the relationship:

$$\mathbf{E}'(\omega,\mathbf{k}_0) = \int \mathbf{E}(\omega,\mathbf{k})k^2\, dk \tag{1.1.25}$$

in view of this, the transformation (1.1.21) takes the form:

$$\mathbf{E}(t,\mathbf{r}) = \frac{1}{(2\pi)^3}\int d\omega \int d^2\mathbf{k}_0\, \mathbf{E}'(\omega,\mathbf{k}_0)e^{i\omega t}e^{-i\mathbf{k}_0\cdot\mathbf{r}} \tag{1.1.26}$$

where $d^2\mathbf{k}_0$ denotes integration over the surface of a sphere with radius k_0. Thus, Eq.(1.1.26) with the help of Eq.(1.1.25) may be determined in the space of usual finite functions.

Let us consider now the scattering problem in momentum representation. Assume that the plane harmonic wave incidents on a particle. Then, applying the transformations (1.1.20) to Eq.(1.1.10) it is easy to obtain the scattered field in a far-field region:

$$\mathbf{E}_{\text{sc}}(\omega,\mathbf{k}) = \hat{G}_0(\mathbf{k})\hat{A}(\mathbf{k},\mathbf{k}_0)\mathbf{E}_I \tag{1.1.27}$$

where $\hat{G}_0(\mathbf{k})$ is the momentum representation of the scalar Green's function for free space (1.1.17):

$$\hat{G}_0(\mathbf{k}) = \frac{\hat{I}}{(k^2 - k_0^2) - i\varepsilon} \tag{1.1.28}$$

To simplify Eq.(1.1.27), we make the usual assumption for linear optics that the radiation frequency remains unchanged during scattering:

$$\hat{A}(\mathbf{k},\mathbf{k}_0) = \begin{pmatrix} \hat{A}(\mathbf{k}_0',\mathbf{k}_0) & k = k_0 \\ 0 & k \neq k_0 \end{pmatrix} \tag{1.1.29}$$

This expression may be rewritten using the limiting relation:

$$\hat{A}(\mathbf{k},\mathbf{k}_0) = \lim_{\varepsilon \to 0} \frac{i\varepsilon\hat{A}(\mathbf{k}_0',\mathbf{k}_0)}{(k^2 - k_0^2) + i\varepsilon} \tag{1.1.30}$$

Substitution of Eqs.(1.1.30) and (1.1.28) into Eq.(2.1.27) gives:

$$\mathbf{E}_{\text{sc}}(\omega,\mathbf{k}) = \frac{i\varepsilon}{(k^2 - k_0^2) + \varepsilon^2}\hat{A}(\mathbf{k}_0',\mathbf{k}_0)\mathbf{E}_I$$

or

$$\mathbf{E}_{\text{sc}}(\omega,\mathbf{k}) = 2\pi k_0 i\delta\left(k^2 - k_0^2\right)\hat{A}(\mathbf{k}_0',\mathbf{k}_0)\mathbf{E}_I$$

that yields, as in the case of Eq.(1.1.24):

$$\mathbf{E}_{\text{sc}}'(\omega,\mathbf{k}_0) = \hat{A}(\mathbf{k}_0',\mathbf{k}_0)\mathbf{E}_I$$

After forming the Stokes parameters from the vector components of a scattered field $\mathbf{E}'_{\text{sc}}(\omega, \mathbf{k}_0)$, we obtain:

$$S_\lambda(\omega, \mathbf{k}'_0) = \sum_{\nu=1}^{4} D_{\lambda\nu}(\omega, \mathbf{k}'_0, \mathbf{k}_0)\, \mathbf{S}_{I\nu}(\omega, \mathbf{k}'_0)$$

where the Stokes parameters transformation matrix is expressed in components of matrixes $\hat{\sigma}^\lambda$ and $\hat{A}(\mathbf{k}'_0, \mathbf{k}_0)$ in the form:

$$D_{\lambda\nu}(\omega, \mathbf{k}'_0, \mathbf{k}_0) = \frac{1}{2} \sum_{ijkl=1}^{2} \sigma^\lambda_{ik} A_{ij}(\mathbf{k}'_0, \mathbf{k}_0)\, A^*_{kl}(\mathbf{k}'_0, \mathbf{k}_0)\, \sigma^\nu_{jl} \tag{1.1.31}$$

In investigations of scattering (both theoretical and experimental) it is necessary to deal with the transformation matrix of the Stokes parameters. Note that in the literature this quantity is ordinarily referred to as the scattering matrix (Bohren and Huffman, 1983; Mishchenko *et al.*, 2000*b;* Mishchenko *et al.*, 2002*a*) but we shall hold to our notation to avoid confusion. As was already noted, the quantities under discussion are the pure local field characteristics. In course of their measurement there is an interaction of the field with an optical instrument. We shall look at this in the following section.

1.2 LIGHT SCATTERING BY PARTICLE COLLECTIONS

1.2.1 Equations for multiple scattering of waves

In optics, in the majority of practical situations, we are dealing with light scattering by many particles rather than by a single particle. Therefore, to describe the process of radiation propagation it is necessary to consider a boundary-value problem with the boundary conditions on the surface of each particle. The most simple and clear way to do it is based on the apparatus of integral equations. To avoid the cumbersome mathematics, we use the operator formalism. With its help the basic integral equation of diffraction may be written as:

$$\mathbf{E} = \mathbf{E}_I + G^0 \bar{V} \mathbf{E} \tag{1.2.1}$$

where $G^0 = -\left(k_0^2 - \nabla \times \nabla \times \ + i\varepsilon\right)^{-1}$ is the operator Green's function and $\bar{V} = k_0^2\left(\xi^2 - 1\right)$ is the potential of electromagnetic wave interaction with the particle. The solution of the operator equation (1.2.1) is written as:

$$\mathbf{E} = \mathbf{E}_I + G\bar{V}\mathbf{E}_I \tag{1.2.2}$$

where $G = -\left(k_0^2 - \nabla \times \nabla \times \ + \bar{V} + i\varepsilon\right)^{-1}$ is the operator Green's function for the problem of diffraction on a single particle. Equation (1.1.8) and relation (1.1.11) are the coordinate representation of Eq.(1.2.1) and its solution (1.2.2).

For external diffraction problems it is convenient to use the operator scattering tensor T, which is defined by the relation:

$$\mathbf{E} = \mathbf{E}_I + G^0 T \mathbf{E}_I \tag{1.2.3}$$

It is seen from Eq.(1.2.3) that the interaction of radiation with the particle is described by operator T, and radiation propagation outside the particle by the free space Green's function G^0. Comparing Eqs.(1.2.2) and (1.2.3) we derive the relationship between the operator of scattering tensor and the operator Green's function:

$$G^0 T = G\bar{V} \tag{1.2.4}$$

Upon applying the identity (Goldberger and Watson, 1973) $A^{-1} - B^{-1} = A^{-1}(B - A)B^{-1} = B^{-1}(B - A)A^{-1}$ to the operators $A^{-1} = G$ and $B^{-1} = G^0$ we have:

$$T = \bar{V} + \bar{V} G \bar{V} \tag{1.2.5}$$

Comparing it with Eq.(1.2.4), we formulate the equation for the operator of scattering tensor:

$$T = \bar{V} + \bar{V} G^0 T \tag{1.2.6}$$

Equation (1.2.6) is called the Lippmann–Schwinger equation and is widely used in the non-relativistic quantum theory of scattering (Goldberger and Watson, 1973).

Let us consider now the scattering of an electromagnetic wave by N particles. In this case the integral equation for the electric field is as follows:

$$\mathbf{E}(\mathbf{r}) = \mathbf{E}_I(\mathbf{r}) + \sum_{\alpha=1}^{N} \int_V G^0(\mathbf{r}, \mathbf{r}') k_0^2 \left[\xi_\alpha^2(\mathbf{r}') - 1\right] \mathbf{E}(\mathbf{r}') d^3\mathbf{r}' \tag{1.2.7}$$

where $\xi_\alpha(\mathbf{r}') = N_\alpha$, if $\mathbf{r}'$ belongs to the part of space occupied by the particle with the number α, and $\xi(\mathbf{r}') = 1$, if $\mathbf{r}'$ does not belong to this part. The operator form of Eq.(1.2.7) is the same as that of Eq.(1.2.1) but the interaction potential is different:

$$\bar{V} = \sum_{\alpha=1}^{N} \bar{V}_\alpha = \sum_{\alpha=1}^{N} k_0^2 \left[\xi_\alpha^2(\mathbf{r}') - 1\right] \tag{1.2.8}$$

From the formal point of view, the solution of Eq.(2.2.1) with potential (2.2.8) is ensured with a set of the bound operational equations (Goldberger and Watson, 1973):

$$\mathbf{E} = \mathbf{E}_I + \sum_{\alpha=1}^{N} G^0 T_\alpha \mathbf{E}_\alpha \qquad \mathbf{E}_\alpha = \mathbf{E}_I + \sum_{\beta(\neq\alpha)=1}^{N} G^0 T_\beta \mathbf{E}_\beta \tag{1.2.9}$$

where the operator T_β of a scattering tensor satisfies the equation:

$$T_\beta = \bar{V}_\beta + \bar{V}_\beta G^0 T_\beta \tag{1.2.10}$$

This is the equation for the scattering tensor operator for the isolated particle with number β. In fact, by virtue of Eq.(1.2.8) the relation:

$$\bar{V}G^0T_\alpha = \bar{V}_\alpha G^0 T_\alpha + \sum_{\beta(\neq\alpha)=1}^{N} \bar{V}_\beta G^0 T_\alpha$$

occurs or, with account of Eq.(1.2.10):

$$\bar{V}G^0T_\alpha = T_\alpha - \bar{V}_\alpha + \sum_{\beta(\neq\alpha)=1}^{N} \bar{V}_\beta G^0 T_\alpha \tag{1.2.11}$$

Now substitute the first relation of Eqs.(1.2.9) into Eq.(1.2.1):

$$\mathbf{E} = \mathbf{E}_I + G^0\bar{V}\mathbf{E}_I + G^0\bar{V}\sum_{\alpha=1}^{N} G^0 T_\alpha \mathbf{E}_\alpha$$

Taking into account Eq.(1.2.11), we obtain:

$$\mathbf{E} = \mathbf{E}_I + G^0\bar{V}\mathbf{E}_I + G^0\sum_{\beta=1}^{N}\bar{V}_\beta \sum_{\alpha(\neq\beta)=1}^{N} G^0T_\alpha\mathbf{E}_\alpha + G^0\sum_{\alpha=1}^{N} T_\alpha\mathbf{E}_\alpha - G^0\sum_{\alpha=1}^{N}\bar{V}_\alpha\mathbf{E}_\alpha \tag{1.2.12}$$

Grouping of terms in Eq.(1.2.12) gives:

$$\begin{aligned}\mathbf{E} = {} & \left(\mathbf{E}_I + G^0\sum_{\alpha=1}^{N} T_\alpha\mathbf{E}_\alpha\right) + G^0\bar{V}\mathbf{E}_I \\ & + G^0\sum_{\beta=1}^{N}\bar{V}_\beta \sum_{\alpha(\neq\beta)=1}^{N} G^0T_\alpha\mathbf{E}_\alpha - G^0\sum_{\alpha=1}^{N}\bar{V}_\alpha\mathbf{E}_\alpha \\ = {} & \left(\mathbf{E}_I + G^0\sum_{\alpha=1}^{N} T_\alpha\mathbf{E}_\alpha\right) + G^0\bar{V}\mathbf{E}_I + G^0\sum_{\alpha=1}^{N}\bar{V}_\alpha\left(\sum_{\alpha(\neq\beta)=1}^{N} G^0T_\alpha\mathbf{E}_\alpha - \mathbf{E}_\alpha\right)\end{aligned} \tag{1.2.13}$$

To transform the last term in Eq.(1.2.13) let us employ the second relation in Eq.(2.2.9), that is:

$$\mathbf{E} = \mathbf{E}_I + G^0\sum_{\alpha=1}^{N} T_\alpha\mathbf{E}_\alpha + G^0\bar{V}\mathbf{E}_I - G^0\sum_{\alpha=1}^{N}\bar{V}_\alpha\mathbf{E}_I \equiv \mathbf{E}$$

Thus, the system of the meshing equations (1.2.9) converts Eq.(1.2.1) into an identity. It is a set of Lippmann–Schwinger equations for the case of electromagnetic radiation scattering by N particles. Note that T_α is the operator of scattering tensor for a particle with number α in the absence of all other particles, and the condition necessary to satisfy this statement is the additivity of the interaction potential $\bar{V}$.

Using the relation (1.2.4), the set of Eqs.(2.2.9) can be written in terms of single-particle Green's functions:

$$\mathbf{E} = \mathbf{E}_I + \sum_{\alpha=1}^{N} G_\alpha \bar{V}_\alpha \mathbf{E}_\alpha \qquad \mathbf{E}_\alpha = \mathbf{E}_I + \sum_{\alpha(\neq\beta)=1}^{N} G_\beta \bar{V}_\beta \mathbf{E}_\beta \tag{1.2.14}$$

This set of operator equations, as well as Eq.(1.2.14), is exact and transforms into a set of integral equations when switching to one or another representation. For example, the problem of plane wave scattering on N bodies located at given points in space is formulated on the basis of the following set of equations:

$$\mathbf{E}(\mathbf{r}) = \mathbf{E}_I(\mathbf{r}) + \sum_{\alpha=1}^{N} \int_V \hat{G}_\alpha(\mathbf{r},\mathbf{r}') k_0^2 \left[\xi_\alpha^2(\mathbf{r}') - 1\right] \mathbf{E}_\alpha(\mathbf{r}')\, d^3\mathbf{r}' \tag{1.2.15}$$

$$\mathbf{E}_\alpha(\mathbf{r}) = \mathbf{E}_I(\mathbf{r}) + \sum_{\beta(\neq\alpha)=1}^{N} \int_V \hat{G}_\beta(\mathbf{r},\mathbf{r}') k_0^2 \left[\xi_\beta^2(\mathbf{r}') - 1\right] \mathbf{E}_\beta(\mathbf{r}')\, d^3\mathbf{r}'$$

where $\hat{G}_\alpha(\mathbf{r},\mathbf{r}')$ and $\hat{G}_\beta(\mathbf{r},\mathbf{r}')$ are the Green's functions for particles α and β with the relevant refractive indexes.

The solution of the set (1.2.15), even for two particles, represents a problem that can only be solved numerically. However, a large class of scattering systems can be described with a simpler set of equations than Eq.(1.2.15). Let us make the following assumptions:

$$k_0 l \gg 1, \qquad l \gg a \tag{1.2.16}$$

where the former means that re-scattering happens in the far-field region and the latter means that the re-scattering happens at distances which are much greater than the typical size of a scattering particle, l being the distance between the particles.

Because the scattering matrix concept is, as a rule, used in the scattering theory, the set of equations (1.2.9) will be considered further which in the coordinate representation takes the form:

$$\mathbf{E}(\mathbf{r}) = \mathbf{E}_I(\mathbf{r}) + \sum_{\alpha=1}^{N} \int_V \hat{G}^0(\mathbf{r},\mathbf{r}') \hat{T}_\alpha(\mathbf{r}',\mathbf{r}'') \mathbf{E}_\alpha(\mathbf{r}'')\, d^3\mathbf{r}' d^3\mathbf{r}'' \tag{1.2.17}$$

$$\mathbf{E}_\alpha(\mathbf{r}) = \mathbf{E}_I(\mathbf{r}) + \sum_{\beta(\neq\alpha)=1}^{N} \int_V \hat{G}^0(\mathbf{r},\mathbf{r}') \hat{T}_\beta(\mathbf{r}',\mathbf{r}'') \mathbf{E}_\beta(\mathbf{r}'')\, d^3\mathbf{r}' d^3\mathbf{r}''$$

After applying the asymptotic Green's function of free space (1.1.17) under conditions (1.2.16) and performing integration in the set (1.2.17) it is possible to obtain the set of algebraic equations:

$$\mathbf{E}(\mathbf{r}_\alpha) = \mathbf{E}_I(\mathbf{r}_\alpha) + \sum_{\beta(\neq\alpha)=1}^{N} \mathbf{E}_{\alpha\beta}(\mathbf{r}_\alpha) \tag{1.2.18}$$

$$\mathbf{E}_{\alpha\beta}(\mathbf{r}_\alpha) = G^0(R_{\alpha\beta})\hat{A}(\mathbf{k}_{\alpha\beta}, \mathbf{k}_I)\mathbf{E}_I + \sum_{\sigma(\neq\beta)=1}^{N} G^0(R_{\alpha\beta})\hat{A}(\mathbf{k}_{\alpha\beta}, \mathbf{k}_{\beta\sigma})\mathbf{E}_{\beta\sigma}(\mathbf{r}_\beta)$$

where $\mathbf{r}_\alpha$ is the radius-vector of a particle with number α, $\mathbf{R}_{\alpha\beta} = \mathbf{r}_\alpha - \mathbf{r}_\beta$, $\mathbf{k}_I$ is the wave vector of the incident radiation (the incident wave will be assumed to be plane here), $R_{\alpha\beta} = |\mathbf{R}_{\alpha\beta}|$, $\hat{\mathbf{R}}_{\alpha\beta} = \mathbf{R}_{\alpha\beta}/R_{\alpha\beta}$,and $\mathbf{k}_{\alpha\beta} = k_0\hat{\mathbf{R}}_{\alpha\beta}$. The quantity $\hat{A}(\mathbf{k}', \mathbf{k})$ is the asymptotic amplitude scattering tensor (1.1.16) and $G^0(R_{\alpha\beta})$ is the scalar Green's function of free space (see Eq.(1.1.17)) which satisfies the equation:

$$\left(\nabla^2_{\mathbf{r}_\alpha} + k_0^2\right) G^0(R_{\alpha\beta}) = -4\pi\delta\left(\mathbf{R}_{\alpha\beta}\right) \tag{1.2.19}$$

Notice that the first equation in the set (1.2.18) is written in the coordinate system associated with a particle α, and the second equation is associated with particle β. The dimension of the field vector is equal to two, and the dimension of the scattering tensor is 2×2.

The physical interpretation of the set (1.2.18) is simple enough. Indeed, the field in point $\mathbf{r}_\alpha$ consists of three parts. The first one (the first term in the set (1.2.18)) describes the incident field, the second one describes the field obtained by scattering of the incident wave on particles, and the third one describes the field obtained via repeated scattering of already scattered waves.

The set of Eq.(2.2.18) is much simpler than Eq.(2.2.17) because it is an algebraic one. Given a small number of particles, it could be numerically solved. With a particular amplitude scattering tensor the solution gives the value of the electric field at point $\mathbf{r}_\alpha$. However, in actual media the number of particles is very large, so the direct solution of the problem is practically impossible. It is necessary to invoke the methods of statistical physics. Indeed, the dispersion medium, consisting in practice of a great number of particles, represents a dynamic system whose variety of states forms a statistical ensemble. In this case the system state can be described by the N-dimensional distribution function $P_N(\sigma_{ij}) = P(\sigma_{i1}, \sigma_{i2}, ..., \sigma_{iN})$ which yields the joint probability of particles in states described by the corresponding set of parameters σ_i. It includes the spatial arrangement of particles, their size, shape, orientation, and so on.

1.2.2 Statistical characteristics of scattered waves

At propagation of electromagnetic waves the ensemble of states of a particle system generates the state ensemble of the scattered field. How to acquire distribution functions of this ensemble is the subject of statistical electrodynamics, whereas the distribution function of the particle system is supposed to be known. Obtaining distribution functions is, as a rule, connected to obtaining the field statistical moments which are experimentally measured. Knowing moments of all orders is equivalent to knowing the distribution function.

However, to describe a wide range of problems it is quite sufficient to know the moments of only the first and second order:

$$\langle \mathbf{E}(\mathbf{r}) \rangle = \int \mathbf{E}(\mathbf{r}) P_N(\sigma_i) \prod_{i,\, j=1}^{N} d\sigma_{ij} \tag{1.2.20}$$

$$\Gamma_{kl}(\mathbf{r}, \mathbf{r}') = \int E_k(\mathbf{r}) E_l^*(\mathbf{r}') P_N(\sigma_i) \prod_{i,\, j=1}^{N} d\sigma_{ij}$$

Here the second relation is written in the component form $k, l = x, y, z = 1, 2, 3$ rather than in the vector one. The phenomenological theory of radiative transfer in media deals with quantities such as those defined by Eq.(1.2.20).

The theory of electromagnetic wave propagation in such statistically non-uniform media began to develop rather recently. To formulate the equations for average quantities, this theory widely uses methods developed in the quantum theory of multiple scattering, in particular the Dyson equation for the average Green's function and the Bethe–Salpeter equation for the average value of the Green's function covariance. A variety of methods for solving the radiative transfer problems in statistically non-uniform media and analysis of the range of their applicability are outlined in Barabanenkov *et al.* (1970), Klyatskin and Tatarskiy (1972, 1973), Klyatskin (1973), Barabanenkov (1975), Kravtsov *et al.* (1975), Kravtsov (1993), and Lagendijk and van Tiggelen (1996). The radiative transfer equation received justification in the framework of this theory (Barabanenkov *et al.*, 1975). The same equation was formulated on the basis of the energy conservation law in the phenomenological theory.

In this book we do not pursue the goal of considering in detail the problems connected to radiative transfer in the randomly non-uniform media. Therefore, we present below only those results of this theory which will be required further. More detailed information on this subject can be obtained from the monographs of Tatarskiy (1967), Chernov (1975), Klyatskin (1975), Kravtsov and Tatarskiy (1975), Rytov *et al.* (1978), and Ishimaru (1981).

To facilitate further presentation we assume that the scattering happens only on particles of disperse medium with neglect to scattering on inhomogeneities such as density fluctuations. Let us also assume that the propagation of radiation in the given disperse medium can be described within the framework of geometrical optics. The last assumption allows us to apply the radiative transfer equation.

The random field scattered by particles may be represented in terms of the following sum:

$$\mathbf{E}(\mathbf{r}) = \langle \mathbf{E}(\mathbf{r}) \rangle + \mathbf{E}_f(\mathbf{r}) \tag{1.2.21}$$

where $\langle \mathbf{E}(\mathbf{r}) \rangle$ is the average (or, coherent) field and $\mathbf{E}_f(\mathbf{r})$ is the fluctuating field vanishing upon averaging. To find the equation for the coherent field it is necessary to average the set (1.2.18) over the ensemble of states of the particle system. This operation held under the conditions stated above gives the Helmholtz equation with an effective dielectric tensor (Watson, 1969; Mishchenko, 2002*a*):

$$\left(\nabla^2 + k_0^2 \hat{\varepsilon}_{\text{eff}}\right) \mathbf{E}_c(\mathbf{r}) = 0 \tag{1.2.22}$$

where

$$\mathbf{E}_c(\mathbf{r}) = \langle \mathbf{E}(\mathbf{r}) \rangle \qquad \hat{\varepsilon}_{\text{eff}} = 1 + \frac{4\pi}{k_0^2} \left\langle \hat{A}(\hat{k}_0) \right\rangle C_N \tag{1.2.23}$$

In the last expression the quantity $\left\langle \hat{A}(\hat{k}_0) \right\rangle$ is the amplitude forward-scattering tensor, averaged over possible parameters (size, orientation), given that the wave incident on the particle system is plane. C_N is the number density of scattering particles.

In the general case the effective dielectric function is a tensor, that is, the medium is anisotropic as a whole. This can be the result of a preferred orientation of non-spherical particles or the anisotropy of their properties. For more complex statistical properties of the medium, than those considered here, the anisotropy can also arise due to formation of crystal lattices. Notice that the effective dielectric function (1.2.23) is a 2×2 tensor because, firstly, the longitudinal components can be neglected owing to Eq.(1.2.16), and, secondly, this function is defined in the coordinate system connected with the propagation direction of incident radiation. In the simplest case such a geometrical situation corresponds to the Huygens construction in crystal optics which yields the refractive indexes for ordinary and non-ordinary waves. For example, in transition to another coordinate system rotated through some angle with regard to the initial one, the dielectric function tensor becomes a three-dimensional quantity.

In general, the amplitude forward-scattering tensor is an asymmetrical complex tensor which becomes three-dimensional in an arbitrarily chosen coordinate system. Hence, the tensor inverse to the dielectric function tensor can be written in the axial representation (Fedorov, 1958) as:

$$\hat{\varepsilon}_{\text{eff}}^{-1} = a + b\left(\mathbf{c} \otimes \mathbf{c}' + \mathbf{c} \otimes \mathbf{c}'\right) + \mathbf{g}^{\times}$$

where a and b are complex scalar functions, $\mathbf{c}$ and $\mathbf{c}'$ are eigenvalue complex vectors of the symmetric part of the dielectric tensor, and $\mathbf{g}^{\times}$ is a vector dual to the antisymmetric part. The expression (1.2.23) differs from similar expressions for continuous anisotropic crystals in that all quantities in the right-hand term depend on the incident radiation wave vector. In the general case, such anisotropic disperse medium has anisotropy and gyrotropy of the general form and, hence, it is possible to describe the processes of coherent field propagation within the framework of the covariant approach developed by Fedorov (1958).

Having the theorem of Wiener and Khinchin (Perina, 1974), it is not difficult to switch to the representation of harmonic plane waves. This is given by formula (1.1.20) where the coherent or average field is contained in an integrand. In this representation Eq.(1.2.22) for the electric field yields the dispersion equation, on which basis the eigenvalue waves propagating in disperse medium, optic axes, and similar quantities are determined.

When radiation propagates in such a disperse medium the wave vector of space is determined by the effective refractive index $k_{\text{eff}}^2 = k_0^2 \varepsilon_{\text{eff}} = k_0^2 N_{\text{eff}}^2$. For simplicity, we consider here the case of an isotropic medium. In general, this vector is a complex quantity and, hence, the harmonic plane wave representation is yielded by a Laplace instead of a Fourier transform which is defined in space of real wave vectors. Of course, it is possible to operate with complex wave vectors. However, it is desirable that they be real and yield the direction of propagation of phase wavefronts that is adequate to geometric optics representations on which basis any optical instrument is created.

Let us consider how to define the Stokes parameters on the basis of the above considerations. The coherent part of the correlation function of the second order (or, simply, the correlation function (1.2.20)), as follows from Eq.(1.2.22), obeys both of the two following equations:

$$2\nabla_R \cdot \nabla_R \,\hat{\Gamma}_c(\mathbf{R},\boldsymbol{\rho}) + \frac{1}{2}\nabla_\rho \cdot \nabla_\rho \,\hat{\Gamma}_c(\mathbf{R},\boldsymbol{\rho}) + 2\,\mathrm{Re}\,\left(k_{\text{eff}}^2\right)\hat{\Gamma}_c(\mathbf{R},\boldsymbol{\rho}) = 0 \tag{1.2.24}$$

$$4\nabla_R \cdot \nabla_\rho \,\hat{\Gamma}_c(\mathbf{R},\boldsymbol{\rho}) + 2i\,\mathrm{Im}\,\left(k_{\text{eff}}^2\right)\hat{\Gamma}_c(\mathbf{R},\boldsymbol{\rho}) = 0 \tag{1.2.25}$$

where the new variables are introduced:

$$\mathbf{R} = (\mathbf{r} + \mathbf{r}') \qquad \rho = \frac{1}{2}(\mathbf{r} - \mathbf{r}') \tag{1.2.26}$$

Let us define the Fourier transform of the coherent part of the second field moment through Fourier transformation over the difference variable ρ:

$$\hat{\Gamma}_c(\mathbf{R},\mathbf{k}) = \int \hat{\Gamma}_c(\mathbf{R},\boldsymbol{\rho})e^{i\mathbf{k}\cdot\mathbf{r}}d^3\boldsymbol{\rho} \tag{1.2.27}$$

Then Eq.(1.2.24) taking account of Eq.(1.2.27) takes the form of the somewhat unusual dispersion equation:

$$-4k^2\hat{\Gamma}_c(\mathbf{R},\mathbf{k}) + \nabla_R^2\hat{\Gamma}_c(\mathbf{R},\mathbf{k})\hat{\Gamma}_c(\mathbf{R},\mathbf{k}) = 0 \tag{1.2.28}$$

whose solution can be found by using the Laplace transformation on variable R (it is possible to satisfy the Eq.(1.2.25) only in this case). In doing so we assume that the inverse Laplace transformation also exists. In view of the above the dispersion equation (1.2.28) takes the form:

$$\left[-2k^2 + \frac{1}{2}s^2 + 2\,\mathrm{Re}\,\left(k_{\text{eff}}^2\right)\right]\hat{\Gamma}_c(\mathbf{s},\mathbf{k}) = 0$$

where $\mathbf{s}$ is the parameter of the Laplace transformation. The solution of this equation for analogous variables can be written for homogeneous waves with the help of Eq.(1.2.25) as:

$$k^2 = 4k_0^2 n_{\text{eff}}^2 \qquad s^2 = k_0^2\kappa_{\text{eff}}^2$$

Using the Fourier representation only, we have:

$$\hat{\Gamma}_c(\mathbf{R},\boldsymbol{\rho}) = \frac{1}{(2\pi)^3}\int d^3\mathbf{k}\,\hat{\Gamma}'(\mathbf{R},\mathbf{k}_{\text{eff}})\exp\left(-2i\mathbf{k}_{\text{eff}}\cdot\boldsymbol{\rho}\right) \tag{1.2.29}$$

where $\mathbf{k}_{\text{eff}} = k_0 n_{\text{eff}}\hat{\mathbf{k}}_{\text{eff}}$ ($\hat{\mathbf{k}}_{\text{eff}}$ is the unit vector along $\mathbf{k}_{\text{eff}}$).

Let us now define the Stokes parameters (see Eq.(1.1.19)):

$$S_{\lambda c} = nSp\left[\sigma^{\lambda}\hat{\Gamma}'_{c}(\mathbf{R},\mathbf{k})\right] \tag{1.2.30}$$

Such definition of the Stokes parameters corresponds, firstly, to the case of plane waves in a non-absorbing medium. Secondly, the Stokes parameters defined, obey the classic transfer equation. To show this, we switch to the momentum representation in Eq.(1.2.25). With account of Eqs.(1.2.29) and (1.2.30) we obtain:

$$\hat{\mathbf{k}}_{\text{eff}} \cdot \nabla_R\, S_{\lambda c}(\mathbf{R},\mathbf{k}_{\text{eff}}) + \kappa k_0 S_{\lambda c}(\mathbf{R},\mathbf{k}_{\text{eff}}) = 0 \tag{1.2.31}$$

The most important constituent of Eq.(1.2.31) is the differential translation operator $\hat{\mathbf{k}}_{\text{eff}} \cdot \nabla + \kappa k_0$, whose form immediately follows from Maxwell's equations. The construction process of Stokes parameters is directly related to specific properties of scattering medium and can be strongly complicated if we complicate the dispersion equations (e.g., in the case of the regular spatial dependence of the effective refractive index, the integral dispersion equations are obtained, the anisotropy of space gives the whole system of the dispersion equations, existence of inhomogeneous waves results in the incompleteness of Stokes parameter system and so on).

We do not consider all these problems here. We only give the transfer equation for the coherent part of the second field moment in the case of the disperse anisotropic medium, which will be required further. It can be obtained from procedure described above. It takes the form:

$$\hat{\mathbf{k}} \cdot \nabla_R\, S_{\lambda c}(\mathbf{R},\mathbf{k}) + \sum_{\nu=1}^{4} \alpha_{\lambda\nu} S_{\nu c}(\mathbf{R},\mathbf{k}) = 0 \tag{1.2.32}$$

where $\alpha_{\lambda\nu}$ is the extinction matrix. It relates to the amplitude of the forward-scattering tensor by the following expression:

$$\alpha = \frac{2\pi C_N}{k_0} \times \begin{bmatrix} \operatorname{Im}\left(\left\langle A_{11}(\hat{k}_0)\right\rangle + \left\langle A_{22}(\hat{k}_0)\right\rangle\right) & \operatorname{Im}\left(\left\langle A_{11}(\hat{k}_0)\right\rangle - \left\langle A_{22}(\hat{k}_0)\right\rangle\right) & \operatorname{Im}\left(\left\langle A_{12}(\hat{k}_0)\right\rangle + \left\langle A_{21}(\hat{k}_0)\right\rangle\right) & \operatorname{Re}\left(\left\langle A_{12}(\hat{k}_0)\right\rangle - \left\langle A_{21}(\hat{k}_0)\right\rangle\right) \\ \operatorname{Im}\left(\left\langle A_{11}(\hat{k}_0)\right\rangle - \left\langle A_{22}(\hat{k}_0)\right\rangle\right) & \operatorname{Im}\left(\left\langle A_{11}(\hat{k}_0)\right\rangle + \left\langle A_{22}(\hat{k}_0)\right\rangle\right) & \operatorname{Im}\left(\left\langle A_{12}(\hat{k}_0)\right\rangle - \left\langle A_{21}(\hat{k}_0)\right\rangle\right) & \operatorname{Re}\left(\left\langle A_{12}(\hat{k}_0)\right\rangle + \left\langle A_{21}(\hat{k}_0)\right\rangle\right) \\ \operatorname{Im}\left(\left\langle A_{12}(\hat{k}_0)\right\rangle + \left\langle A_{21}(\hat{k}_0)\right\rangle\right) & -\operatorname{Im}\left(\left\langle A_{12}(\hat{k}_0)\right\rangle - \left\langle A_{21}(\hat{k}_0)\right\rangle\right) & \operatorname{Im}\left(\left\langle A_{11}(\hat{k}_0)\right\rangle + \left\langle A_{22}(\hat{k}_0)\right\rangle\right) & \operatorname{Re}\left(\left\langle A_{22}(\hat{k}_0)\right\rangle - \left\langle A_{11}(\hat{k}_0)\right\rangle\right) \\ \operatorname{Re}\left(\left\langle A_{12}(\hat{k}_0)\right\rangle - \left\langle A_{21}(\hat{k}_0)\right\rangle\right) & -\operatorname{Re}\left(\left\langle A_{12}(\hat{k}_0)\right\rangle + \left\langle A_{21}(\hat{k}_0)\right\rangle\right) & \operatorname{Re}\left(\left\langle A_{11}(\hat{k}_0)\right\rangle - \left\langle A_{22}(\hat{k}_0)\right\rangle\right) & \operatorname{Im}\left(\left\langle A_{11}(\hat{k}_0)\right\rangle + \left\langle A_{22}(\hat{k}_0)\right\rangle\right) \end{bmatrix} \tag{1.2.33}$$

The concept of intensity as the field characteristic implies to some degree a concretization of the medium, in which the radiation propagates or, in other words, the availability of specific information on radiation interaction with the substance. Therefore, it is not such an universal characteristic as, for example, the electric field $\mathbf{E}(\mathbf{r})$. In this respect it is useful and even necessary to introduce the concept of generalized intensity, defined in space of generalized functions as a Fourier transform from the sum of diagonal elements of the correlation matrix. Just this functional interconnection reflects the most essential part of the relationship between the electrodynamic and optical characteristics of the electromagnetic field. It is the most simple and general characteristic for all processes of optical radiation propagation.

The generalized intensity was introduced within the framework of theory of multiple scattering of radiation by statistically non-uniform media, with the help of direct comparison of some quantities entering into the phenomenological equation of radiative transfer with quantities entering into the equation of transfer obtained from the asymptote of the Bethe–Salpeter equation for the correlation matrix (Barabanenkov *et al.*, 1975). This allowed going into detailed analysis of the intensity concept used in classical optics, as well as going into simpler methods of intensity definition within the electrodynamics framework, as, for example, is done in this book.

It is essential to statistical electrodynamics that the fluctuating part of the correlation matrix (the incoherent part in optics) which, as seen from Eq.(1.2.21), is defined by the relation:

$$\left\langle \mathbf{E}(\mathbf{r}) \otimes \mathbf{E}(\mathbf{r}')\right\rangle_f = \left\langle \mathbf{E}(\mathbf{r}) \otimes \mathbf{E}(\mathbf{r}')\right\rangle - \mathbf{E}_c(\mathbf{r}) \cdot \mathbf{E}_c(\mathbf{r}')$$

is not small in contrast to thermodynamic fluctuations. It plays a decisive role in many electrodynamic processes including radiative transfer in disperse media. Until now, we only considered the coherent part of the scattered field and, therefore, it is necessary to briefly expand on the incoherent component in the process of radiative transfer to complete our description.

Let us introduce the average Green's function. This function is similar to the free space Green's function and it satisfies the equation:

$$\left(\nabla^2 + k_0^2 \hat{\varepsilon}_{\text{eff}}\right) \hat{G}_c(\mathbf{r}, \mathbf{r}') = -4\pi\delta\left(\mathbf{r} - \mathbf{r}'\right) \tag{1.2.34}$$

Taking into consideration Eq.(1.2.22) for the average field, Eq.(1.2.34) can be written in integral form:

$$\begin{aligned} \mathbf{E}_c\left(\mathbf{r}\right) &= \mathbf{E}_I\left(\mathbf{r}\right) + \int \hat{G}^0(\mathbf{r}, \mathbf{r}') k_0^2 \left(\hat{\varepsilon}_{\text{eff}} - 1\right) \mathbf{E}_c(\mathbf{r}') d^3\mathbf{r}' \\ \hat{G}_c(\mathbf{r}, \mathbf{r}') &= \hat{G}^0(\mathbf{r}, \mathbf{r}') + \int \hat{G}^0(\mathbf{r}, \mathbf{r}'') k_0^2 \left(\hat{\varepsilon}_{\text{eff}} - 1\right) \hat{G}_c(\mathbf{r}'', \mathbf{r}') d^3\mathbf{r}'' \end{aligned} \tag{1.2.35}$$

It is not difficult now to show by direct substitution that the equation:

$$\mathbf{E}\left(\mathbf{r}\right) = \mathbf{E}_c\left(\mathbf{r}\right) + \sum_{\alpha=1}^{N} \int \hat{G}_c(\mathbf{r}, \mathbf{r}') k_0^2 \left[\hat{\varepsilon}_\alpha\left(\mathbf{r}'\right) - \hat{\varepsilon}_{\text{eff}}\left(\mathbf{r}\right)\right] \mathbf{E}(\mathbf{r}')\, d^3\mathbf{r}' \tag{1.2.36}$$

is equivalent to Eq.(1.2.7), and, hence, in place of system (1.2.18) it is possible to write the set of equations where the coherent field is separated:

$$\begin{aligned} \mathbf{E}(\mathbf{r}_\alpha) &= \mathbf{E}_c(\mathbf{r}_\alpha) + \sum_{\alpha(\neq\beta)=1}^{N} \mathbf{E}'_{\alpha\beta}(\mathbf{r}_\alpha) \\ \mathbf{E}'_{\alpha\beta}(\mathbf{r}_\alpha) &= \hat{G}_c(R_{\alpha\beta})\hat{A}'(\mathbf{k}'_{\alpha\beta}, \mathbf{k}_I)\mathbf{E}_c(\mathbf{r}_\beta) \\ &+ \sum_{\sigma(\neq\beta)=1}^{N} \hat{G}_c(R_{\alpha\beta})\hat{A}'(\mathbf{k}'_{\alpha\beta}, \mathbf{k}_{\beta\sigma})\mathbf{E}'_{\beta\sigma}(\mathbf{r}_\beta) \end{aligned} \tag{1.2.37}$$

Comparing the first equation from set (1.2.37) with the first equation in set (1.2.18), we find that $\mathbf{E}'_{\alpha\beta}(\mathbf{r}_\alpha)$ is the fluctuating part of the field

$$\mathbf{E}'_{\alpha\beta}(\mathbf{r}_\alpha) = \mathbf{E}_{\alpha\beta}(\mathbf{r}_\alpha) - \langle \mathbf{E}_{\alpha\beta}(\mathbf{r}_\alpha) \rangle$$

The amplitude light scattering tensor for particles embedded in "effective medium" enters into the second equation of system (1.2.37). It means that the radiation is scattered by particles put into a medium with an effective dielectric function $\hat{\varepsilon}_{\text{eff}}$.

The most notable thing in the modified system (1.2.37) is that it divides the field into coherent and incoherent parts in clear way. Upon averaging this system over the ensemble of states of scattering particles and simplifying it on the basis of the above assumptions in the "ladder" approximation (Mishchenko, 2002*a*) it is possible to obtain the radiative transfer integral equation for the abovementioned Stokes parameters. Applying the differential translation operator $\left(\hat{\mathbf{k}} \cdot \nabla + \alpha\right)$ to it, we exclude the explicit dependence on the coherent field and come to the standard transfer equation:

$$\hat{\mathbf{k}} \cdot \nabla_R\, S_\lambda(R,\Omega) = -\sum_{\nu=1}^{4} \alpha_{\lambda\nu} S_\nu(R,\Omega) + \sum_{\nu=1}^{4} \int d\Omega' D'_{\lambda\nu}(\Omega, \Omega')\, S_\nu(R,\Omega') \tag{1.2.38}$$

where $D'_{\lambda\nu}(\Omega, \Omega')$ is the matrix of Stokes parameters transformation whose explicit form is obtained from this equation (cf. Eq.(1.1.31):

$$D'_{\lambda\nu}(\Omega, \Omega') = \frac{C_N}{2} \sum_{ijlk=1}^{2} \sigma^\lambda_{ik} \left\langle A'_{ij}(\Omega, \Omega')\, A'^*_{kl}(\Omega, \Omega') \right\rangle \sigma^\nu_{jl} \tag{1.2.39}$$

where < ... >, as before, designates the averaging over possible statistical parameters.

It is necessary to mention here the following circumstance. In transition from the set of integral equations (1.2.17) to the system of the algebraic equations (1.2.18) we used the so-called one-group approximation (Barabanenkov, 1975), that is, we neglected the terms which describe the part of radiation that is small in comparison with the basic flux. Of course, it can lead to violation of the energy conservation law which is formulated in the multiple scattering theory as the optical theorem (Barabanenkov and Finkelberg, 1968). In particular, according to this theorem, the equality:

$$\alpha_{11} = \frac{1}{4\pi} \int D'_{11}(\Omega, \Omega')\, d\Omega\, d\Omega'$$

should be satisfied for non-absorbing media. This relation, being expressed through the amplitude scattering matrixes, coincides with the optical theorem, by form, for the case of single scattering:

$$\frac{4\pi}{k_0} \operatorname{Im} \left[\left\langle A_{11}(\hat{k}_0) \right\rangle + \left\langle A_{22}(\hat{k}_0) \right\rangle \right] = \frac{1}{4\pi} \sum_{ij=1}^{2} \int A'_{ji}(\Omega, \Omega')\, A'^{*}_{ij}(\Omega, \Omega')\, d\Omega\, d\Omega' \tag{1.2.40}$$

Apparently, this equality will be satisfied if amplitude matrixes refer to the same scattering potential. It is clear that the matrix $\hat{A}'(\Omega, \Omega')$ is a more precise quantity with regards to the one-group approximation, because it turns into the matrix $\hat{A}(\Omega, \Omega')$ at $\hat{\varepsilon}_{\mathrm{eff}} \to 1$. Thus, it is possible to satisfy equality (1.2.40) for small concentrations by neglecting small terms in $\hat{A}'(\Omega, \Omega')$. Hence, we obtain the following expression for Stokes parameters transformation matrix instead of Eq.(1.2.39):

$$D'_{11}(\Omega, \Omega') = \frac{C_N}{2} \sum_{ijlk=1}^{2} \sigma^{\lambda}_{ik} \left\langle A_{ij}(\Omega, \Omega')\, A^{*}_{kl}(\Omega, \Omega') \right\rangle \sigma^{\nu}_{jl} \tag{1.2.41}$$

These characteristics are widely used in the optics of scattering media (Chandrasekhar, 1960; Ivanov, 1969). They can be obtained from experimental data or calculated on the basis of different theoretical models of scattering particles. The results presented in this book generally deal with these characteristics. It should be mentioned that more general than the extinction matrix characteristic of elementary volume is an effective dielectric tensor appearing in the equation for the average field, as the extinction matrix is formed on the basis of this tensor. Moreover, in the presence of sharp boundaries between scattering media it is definitely impossible to correctly describe the propagation of the coherent part of radiation on the basis of the extinction matrix only.

1.3 ELECTROMAGNETIC SCATTERING BY A HOMOGENEOUS SPHERICAL PARTICLE IN COORDINATE AND MOMENTUM REPRESENTATIONS

The general characteristics of light scattering which were introduced in previous sections, are concretized here using an example of homogeneous isotropic spherical particles. It pursues two purposes. The first one is to present a procedure on which diffraction problems are considered in this book. It is, more specifically, the covariant approach to scattering processes with use of the integral equation method and its related Green's functions and scattering matrixes. The second purpose is to define and consider in brief the mathematical functions and concepts used.

Let us build $\hat{G}$-function (1.1.12) with the help of eigenfunctions of wave equation (1.1.5) which, as shown by Stratton (1941), are the vector spherical wave functions $\mathbf{M}, \mathbf{N}$, and $\mathbf{L}$ (see Appendix C):

$$\begin{aligned} \mathbf{M}^{(i)}_{\sigma ml}(k\mathbf{r}) &= \frac{\psi_l^{(i)}(kr)}{kr}\mathbf{m}_{\sigma ml} \quad m = 0,1,2,... \quad l = m, m+1, ... \quad \sigma = e, o \\ \mathbf{N}^{(i)}_{\sigma ml}(k\mathbf{r}) &= \frac{\psi_l^{(i)}(kr)}{k^2r^2}\mathbf{l}_{\sigma ml} + \frac{\psi_l^{(i)\prime}(kr)}{kr}\mathbf{n}_{\sigma ml} \\ \mathbf{L}^{(i)}_{\sigma ml}(k\mathbf{r}) &= \frac{1}{krl(l+1)}\left(\psi_l^{(i)\prime}(kr) - \frac{\psi_l^{(i)}(kr)}{kr}\right)\mathbf{l}_{\sigma ml} + \frac{\psi_l^{(i)}(kr)}{k^2r^2}\mathbf{n}_{\sigma ml} \end{aligned} \tag{1.3.1}$$

where the angular vector wave functions $\mathbf{m}$, $\mathbf{n}$, and $\mathbf{l}$ are defined as:

$$\begin{aligned} \mathbf{m}_{\substack{e\\o} ml} &= \mp \mathbf{e}_\theta m\pi_l^{(m)}(\mu)\begin{pmatrix}\sin m\varphi\\ \cos m\varphi\end{pmatrix} - \mathbf{e}_\varphi \tau_l^{(m)}(\mu)\begin{pmatrix}\cos m\varphi\\ \sin m\varphi\end{pmatrix} \\ \mathbf{n}_{\substack{e\\o} ml} &= \left[\mathbf{e}_r \times \mathbf{m}_{\substack{e\\o} ml}\right] \\ &= \mathbf{e}_\theta \tau_l^{(m)}(\mu)\begin{pmatrix}\cos m\varphi\\ \sin m\varphi\end{pmatrix} \mp \mathbf{e}_\varphi m\pi_l^{(m)}(\mu)\begin{pmatrix}\sin m\varphi\\ \cos m\varphi\end{pmatrix} \\ \mathbf{l}_{\substack{e\\o} ml} &= \mathbf{e}_r l(l+1)\pi_l^{(m)}(\mu)\sin\theta\begin{pmatrix}\cos m\varphi\\ \sin m\varphi\end{pmatrix}, \quad \mu = \cos\theta \end{aligned} \tag{1.3.2}$$

In Eqs.(1.3.1) and (1.3.2) the following notation is used: $\mathbf{e}_r$, $\mathbf{e}_\theta$, and $\mathbf{e}_\varphi$ are the unit basis vectors of the spherical coordinate system (r, θ, φ) with origin at the sphere centre; $k = k_0\xi(\mathbf{r})$; indexes e and o describe even and odd functions, respectively; $\psi_l^{(i)}(kr)$ $(i = 1, 2, 3)$ are the Riccati–Bessel functions (see Appendix A):

$$\begin{aligned} \psi_l^{(1)}(kr) &\equiv \psi_l(kr) = \sqrt{\frac{\pi kr}{2}}J_{l+1/2}(kr) \\ \psi_l^{(2)}(kr) &\equiv \chi_l(kr) = -\sqrt{\frac{\pi kr}{2}}Y_{l+1/2}(kr) \\ \psi_l^{(3)}(kr) &\equiv \xi_l(kr) = \sqrt{\frac{\pi kr}{2}}H^{(1)}_{l+1/2}(kr) \end{aligned} \tag{1.3.3}$$

where $J_{l+1/2}(kr)$, $Y_{l+1/2}(kr)$, and $H^{(1)}_{l+1/2}(kr)$ are the Bessel, Neimann, and Hankel functions with the half-integer order), and $\pi_l^{(m)}(\mu)$ and $\tau_l^{(m)}(\mu)$ are so-called angular functions (see Appendix B) related to the associated Legendre polynomials $P_l^{(m)}(\mu)$:

$$\begin{aligned} \pi_l^{(m)}(\mu) &= \frac{P_l^{(m)}(\cos\theta)}{\sin\theta} = \frac{P_l^{(m)}(\mu)}{(1-\mu^2)^{1/2}} \\ \tau_l^{(m)}(\mu) &= \frac{dP_l^{(m)}(\cos\theta)}{d\theta} = -\frac{dP_l^{(m)}(\mu)}{d\mu}(1-\mu^2)^{1/2} \end{aligned} \tag{1.3.4}$$

In contrast to many other books on light scattering, the Riccati–Bessel functions are used throughout this book instead of the spherical Bessel functions $z_l(kr)$ because use of the latter sometimes results in unjustifiably cumbersome expressions (e.g., see Barber and Hill (1990)).

The relationship between the **M** and **N** functions is:

$$\mathbf{M}_{\substack{e\\o}ml}(k\mathbf{r}) = \frac{1}{k}\nabla \times \mathbf{N}_{\substack{e\\o}ml}(k\mathbf{r}) \qquad \mathbf{N}_{\substack{e\\o}ml}(k\mathbf{r}) = \frac{1}{k}\nabla \times \mathbf{M}_{\substack{e\\o}ml}(k\mathbf{r}) \tag{1.3.5}$$

The angular vector wave functions obey the orthogonality relations:

$$\mathbf{m}_{\sigma ml}\cdot\mathbf{l}_{\sigma ml} = \hat{\mathbf{r}}\cdot\mathbf{m}_{\sigma ml} = 0 \qquad \sigma = e, o$$

$$\int\limits_{(4\pi)} \mathbf{m}_{\sigma ml}(\theta,\varphi)\cdot\mathbf{m}_{\sigma' m' l'}(\theta,\varphi)\sin\theta\, d\theta\, d\varphi = \delta_{ll'}\delta_{mm'}\delta_{\sigma\sigma'}\lambda_{ml}^{-1} \tag{1.3.6}$$

$$\int\limits_{(4\pi)} \mathbf{l}_{\sigma ml}(\theta,\varphi)\cdot\mathbf{l}_{\sigma' m' l'}(\theta,\varphi)\sin\theta\, d\theta\, d\varphi = \delta_{ll'}\delta_{mm'}\delta_{\sigma\sigma'}l(l+1)\lambda_{ml}^{-1}$$

where

$$\lambda_{ml} = \frac{(2l+1)(l-m)!}{2\pi l(l+1)(l+m)!(1+\delta_{0m})}$$

δ_{ij} is the Kronecker delta and the integration is done over the entire solid angle.

Let us go on to the particular construction of $\hat{G}$-functions. According to Eq.(1.1.9) and symmetry property:

$$\hat{G}(\mathbf{r}',\mathbf{r}'') = \left[\hat{G}(\mathbf{r}'',\mathbf{r}')\right]^T \tag{1.3.7}$$

(where $[...]^T$ denotes a transposed dyad) the expansion of Green's function in terms of the vector wave functions should be a superposition of linearly independent terms such as $\mathbf{M}\otimes\mathbf{M}$ and $\mathbf{N}\otimes\mathbf{N}$. The **L**-functions do not appear in this superposition because, in contrast to **M** and **N** functions, they are not solenoidal ($\nabla\cdot\mathbf{L}\neq\mathbf{0}$) and they describe the longitudinal part of the wave.

It is also evident that of the three modifications (1.3.3) of the Riccati–Bessel function the one should be chosen that physically corresponds to range of argument. If this range contains the origin then it is necessary to choose the function $\psi_l^{(1)}(kr)\equiv\psi_l(kr)$ due to its regularity in the origin; if the range of argument could contain an infinite point then it is necessary to choose $\psi_l^{(3)}(kr)\equiv\xi_l(kr)$, because, given a chosen sign of the time factor $(e^{-i\omega t})$, only the Hankel function of the first kind describes an outgoing spherical wave at infinite distance from a particle.

For these reasons, the Green's function for infinite homogeneous space (usually termed fundamental function) is represented in the form:

$$\hat{G}^{(k)}(\mathbf{r},\mathbf{r}') = \sum_{l=0}^{\infty}\sum_{m=0}^{l}\sum_{\sigma=e,o}\sum_{l'=0}^{\infty}\sum_{m'=0}^{l'}\sum_{\sigma'=e,o}\left\{\nu_{\sigma ml}^{\sigma'm'l'}\begin{bmatrix}\mathbf{M}_{\sigma'm'l'}^{(3)}(k\mathbf{r})\otimes\mathbf{M}_{\sigma ml}^{(1)}(k\mathbf{r}')\\ \mathbf{M}_{\sigma'm'l'}^{(1)}(k\mathbf{r})\otimes\mathbf{M}_{\sigma ml}^{(3)}(k\mathbf{r}')\end{bmatrix}\right.$$
$$\left.+\mu_{\sigma ml}^{\sigma'm'l'}\begin{bmatrix}\mathbf{N}_{\sigma'm'l'}^{(3)}(k\mathbf{r})\otimes\mathbf{N}_{\sigma ml}^{(1)}(k\mathbf{r}')\\ \mathbf{N}_{\sigma'm'l'}^{(1)}(k\mathbf{r})\otimes\mathbf{N}_{\sigma ml}^{(3)}(k\mathbf{r}')\end{bmatrix}\right\}\begin{matrix} r>r'\\ r<r'\end{matrix} \tag{1.3.8}$$

where $\nu_{\sigma ml}^{\sigma'm'l'}$ and $\mu_{\sigma ml}^{\sigma'm'l'}$ are superposition coefficients, which can be found by applying the second Green identity to $\hat{G}^{(k)}(\mathbf{r},\mathbf{r}')$ and to the wave equation eigenfunctions $\mathbf{M}_{\sigma ml}^{(1)}(k\mathbf{r})$ and $\mathbf{N}_{\sigma ml}^{(1)}(k\mathbf{r}')$ for the spherical area that includes the point $\mathbf{r} = \mathbf{r}'$. After transformations with the help of relations (1.3.1)–(1.3.6) we have:

$$\nu_{\sigma ml}^{\sigma'm'l'} = \mu_{\sigma ml}^{\sigma'm'l'} = \delta_{ll'}\delta_{mm'}\delta_{\sigma\sigma'}ik\lambda_{ml} \tag{1.3.9}$$

Relation (1.3.9) allows us to rewrite Eq.(1.3.8) as:

$$\hat{G}^{(k)}(\mathbf{r},\mathbf{r}') = ik\sum_{\sigma ml}\lambda_{ml}\begin{bmatrix}\mathbf{M}_{\sigma ml}^{(3)}(k\mathbf{r})\otimes\mathbf{M}_{\sigma ml}^{(1)}(k\mathbf{r}')\\ \mathbf{M}_{\sigma ml}^{(1)}(k\mathbf{r})\otimes\mathbf{M}_{\sigma ml}^{(3)}(k\mathbf{r}')\end{bmatrix.$$
$$\left.+\begin{matrix}\mathbf{N}_{\sigma ml}^{(3)}(k\mathbf{r})\otimes\mathbf{N}_{\sigma ml}^{(1)}(k\mathbf{r}')\\ \mathbf{N}_{\sigma ml}^{(1)}(k\mathbf{r})\otimes\mathbf{N}_{\sigma ml}^{(3)}(k\mathbf{r}')\end{matrix}\right]\begin{matrix} r>r'\\ r<r'\end{matrix} \tag{1.3.10}$$

From Eq.(1.3.10) onwards, we will use a compact notation for the sum over m, l, and σ. The function $\hat{G}^{(0)}(\mathbf{r},\mathbf{r}')$ introduced earlier is a special case of Eq.(1.3.10) at $k = k_0$.

The space inhomogeneity forces us to introduce different Green's functions for different ranges of variables $\mathbf{r}$ and $\mathbf{r}'$. This may be clearly illustrated on the diagram in a phase space (Figure 1.1). On the diagram the following Green's functions are introduced:

$$\begin{aligned}
\hat{G}_{11} &= ik_0\sum_{\sigma ml}\lambda_{ml}\left[\alpha_{ml}^{11}\mathbf{M}_{\sigma ml}^{(1)}(k\mathbf{r})\otimes\mathbf{M}_{\sigma ml}^{(1)}(k\mathbf{r}')\right.\\
&\qquad\left.+\beta_{ml}^{11}\mathbf{N}_{\sigma ml}^{(1)}(k\mathbf{r})\otimes\mathbf{N}_{\sigma ml}^{(1)}(k\mathbf{r}')\right]\\
\hat{G}_{33} &= ik_0\sum_{\sigma ml}\lambda_{ml}\left[\alpha_{ml}^{33}\mathbf{M}_{\sigma ml}^{(3)}(k\mathbf{r})\otimes\mathbf{M}_{\sigma ml}^{(3)}(k\mathbf{r}')\right.\\
&\qquad\left.+\beta_{ml}^{33}\mathbf{N}_{\sigma ml}^{(3)}(k\mathbf{r})\otimes\mathbf{N}_{\sigma ml}^{(3)}(k\mathbf{r}')\right]\\
\hat{G}_{13} &= ik_0\sum_{\sigma ml}\lambda_{ml}\left[\alpha_{ml}^{13}\mathbf{M}_{\sigma ml}^{(1)}(k\mathbf{r})\otimes\mathbf{M}_{\sigma ml}^{(3)}(k\mathbf{r}')\right.\\
&\qquad\left.+\beta_{ml}^{13}\mathbf{N}_{\sigma ml}^{(1)}(k\mathbf{r})\otimes\mathbf{N}_{\sigma ml}^{(3)}(k\mathbf{r}')\right]\\
\hat{G}_{31} &= ik_0\sum_{\sigma ml}\lambda_{ml}\left[\alpha_{ml}^{31}\mathbf{M}_{\sigma ml}^{(3)}(k\mathbf{r})\otimes\mathbf{M}_{\sigma ml}^{(1)}(k\mathbf{r}')\right.\\
&\qquad\left.+\beta_{ml}^{31}\mathbf{N}_{\sigma ml}^{(3)}(k\mathbf{r})\otimes\mathbf{N}_{\sigma ml}^{(1)}(k\mathbf{r}')\right]
\end{aligned} \tag{1.3.11}$$

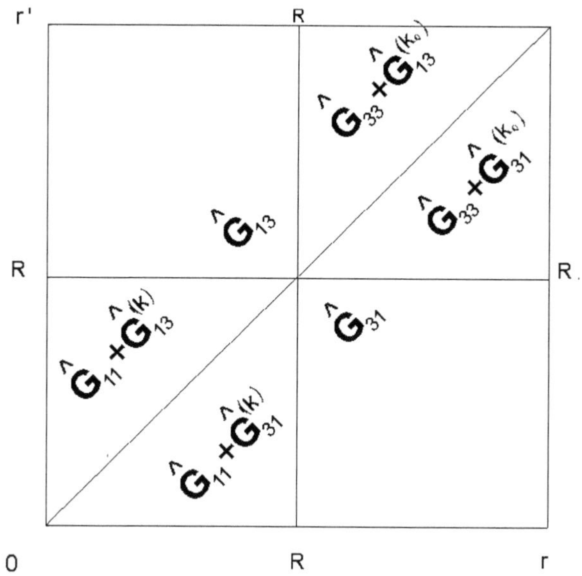

Figure. 1.1. The phase diagram of Green's function in accordance with the range of its arguments (R is the radius of the scattering particle).

$\hat{G}_{13}^{(k)}, \hat{G}_{31}^{(k)}, \hat{G}_{13}^{(k_0)}$, and $\hat{G}_{31}^{(k_0)}$ are the fundamental Green's functions (the purpose of the indexes is clear from Eqs.(1.3.8) and (1.3.11)). The superposition coefficients can be found from boundary conditions on lines $r = R$ and $r' = R$ (where R is the radius of spherical particle):

$$
\begin{aligned}
\mathbf{n}_r \times \left(\hat{G}_{13}^{(k_0)} + \hat{G}_{33}\right)_{R+} &= \mathbf{n}_r \times \hat{G}_{13}\Big|_{R-} \\
\mathbf{n}_r \times \nabla_r \times \left(\hat{G}_{31}^{(k)} + \hat{G}_{11}\right)_{R-} &= \mathbf{n}_r \times \nabla_r \times \hat{G}_{31}\Big|_{R+} \\
\mathbf{n}_r \times \nabla_r \times \left(\hat{G}_{13}^{(k_0)} + \hat{G}_{33}\right)_{R+} &= \mathbf{n}_r \times \nabla_r \times \hat{G}_{13}\Big|_{R-} \\
\mathbf{n}_r \times \left(\hat{G}_{31}^{(k)} + \hat{G}_{11}\right)_{R-} &= \mathbf{n}_r \times \hat{G}_{31}\Big|_{R+}
\end{aligned}
\tag{1.3.12}
$$

$$
\begin{aligned}
\mathbf{n}_{r'} \times \left(\hat{G}_{13}^{(k)} + \hat{G}_{11}\right)_{R-} &= \mathbf{n}_{r'} \times \hat{G}_{13}\Big|_{R+} \\
\mathbf{n}_{r'} \times \nabla_{r'} \times \left(\hat{G}_{13}^{(k)} + \hat{G}_{11}\right)_{R-} &= \mathbf{n}_{r'} \times \nabla_{r'} \times \hat{G}_{13}\Big|_{R+} \\
\mathbf{n}_{r'} \times \left(\hat{G}_{31}^{(k_0)} + \hat{G}_{33}\right)_{R+} &= \mathbf{n}_{r'} \times \hat{G}_{31}\Big|_{R-} \\
\mathbf{n}_{r'} \times \nabla_{r'} \times \left(\hat{G}_{31}^{(k_0)} + \hat{G}_{33}\right)_{R+} &= \mathbf{n}_{r'} \times \nabla_{r'} \times \hat{G}_{31}\Big|_{R-}
\end{aligned}
\tag{1.3.13}
$$

The systems (1.3.12) and (1.3.13) are not independent and only one of them (for example Eq.(1.3.13)) is sufficient to obtain the coefficients. The substitution of expressions (1.3.11) into Eq.(1.3.13) gives:

$$\alpha_{ml}^{33} = \frac{\psi_l'(x)\,\psi_l(Nx) - N\psi_l(x)\,\psi_l'(Nx)}{N\xi_l(x)\,\psi_l'(Nx) - \xi_l'(x)\,\psi_l(Nx)}$$
$$\beta_{ml}^{33} = \frac{\psi_l(x)\,\psi_l'(Nx) - N\psi_l'(x)\,\psi_l(Nx)}{N\xi_l'(x)\,\psi_l(Nx) - \xi_l(x)\,\psi_l'(Nx)} \tag{1.3.14}$$

$$\alpha_{ml}^{11} = \frac{\xi_l'(x)\,\xi_l(Nx) - N\xi_l(x)\,\xi_l'(Nx)}{N\xi_l(x)\,\psi_l'(Nx) - \xi_l'(x)\,\psi_l(Nx)}$$
$$\beta_{ml}^{11} = \frac{\xi_l(x)\,\xi_l'(Nx) - N\xi_l'(x)\,\xi_l(Nx)}{N\xi_l'(x)\,\psi_l(Nx) - \xi_l(x)\,\psi_l'(Nx)} \tag{1.3.15}$$

$$\alpha_{ml}^{31} = N\alpha_{ml}^{13} = iN\,\left[\xi_l'(x)\,\psi_l(Nx) - N\xi_l(x)\,\psi_l'(Nx)\right]^{-1}$$
$$\beta_{ml}^{31} = N\beta_{ml}^{13} = -iN\,\left[\xi_l(x)\,\psi_l'(Nx) - N\xi_l'(x)\,\psi_l(Nx)\right]^{-1} \tag{1.3.16}$$

where N is the complex refractive index of particle $(N = n + i\kappa)$, x is a particle size parameter $(x = k_0R)$, the prime denotes differentiation over the function argument, and index m is further omitted for the coefficients (1.3.14)–(1.3.16) do not depend on it. If we introduce the following notations for logarithmic derivations of the Riccati–Bessel functions:

$$D_l(z) = \frac{\psi_l'(z)}{\psi_l(z)} \qquad G_l(z) = \frac{\xi_l'(z)}{\xi_l(z)} \tag{1.3.17}$$

then Eqs.(1.3.14–16) can be written in the form more suitable for numerical calculations (see Section 1.5):

$$\alpha_l^{33} = \frac{\psi_l(x)}{\xi_l(x)} \cdot \frac{D_l(x) - ND_l(Nx)}{ND_l(Nx) - G_l(x)}$$
$$\beta_l^{33} = \frac{\psi_l(x)}{\xi_l(x)} \cdot \frac{D_l(Nx) - ND_l(x)}{NG_l(x) - D_l(Nx)} \tag{1.3.18}$$

$$\alpha_l^{11} = \frac{\xi_l(Nx)}{\psi_l(Nx)} \cdot \frac{G_l(x) - NG_l(Nx)}{ND_l(Nx) - G_l(x)}$$
$$\beta_l^{11} = \frac{\xi_l(Nx)}{\psi_l(Nx)} \cdot \frac{G_l(Nx) - NG_l(x)}{NG_l(x) - D_l(Nx)} \tag{1.3.19}$$

$$\alpha_{ml}^{31} = N\alpha_{ml}^{13} = -\frac{1}{\xi_l(x)\psi_l(Nx)}\frac{iN}{ND_l(Nx) - G_l(x)}$$
$$\beta_{ml}^{31} = N\beta_{ml}^{13} = \frac{1}{\xi_l(x)\psi_l(Nx)}\frac{iN}{NG_l(x) - D_l(Nx)} \tag{1.3.20}$$

The Green's function enables us to write the solution for the diffraction problem both in the external and the internal regions. It is sufficient just to know the function $\hat{G}_{31}(\mathbf{r},\mathbf{r}')$ to find the external field:

$$\mathbf{E}(\mathbf{r}) = \mathbf{E}_I(\mathbf{r}) + \int\limits_V \hat{G}_{31}(\mathbf{r},\mathbf{r}') k_0^2 (N^2 - 1)\mathbf{E}_I(\mathbf{r}') d^3\mathbf{r}' \tag{1.3.21}$$

and the functions $\hat{G}_{11}(\mathbf{r},\mathbf{r}')$ and $\hat{G}^{(k)}(\mathbf{r},\mathbf{r}')$ to find the internal field:

$$\mathbf{E}_{\text{int}}(\mathbf{r}) = \mathbf{E}_I(\mathbf{r}) + \int\limits_V \left[\hat{G}_{11}(\mathbf{r},\mathbf{r}') + \hat{G}^{(k)}(\mathbf{r},\mathbf{r}')\right] k_0^2 (N^2 - 1)\mathbf{E}_I(\mathbf{r}') d^3\mathbf{r}' \tag{1.3.22}$$

The expression (1.3.22) is inconvenient in practice, as the Green's functions in integrands have singularities at $\mathbf{r} = \mathbf{r}'$. The structure of the singularities (van Bladel, 1961; Tai, 1973, 1993; Tai and Rosenfeld, 1976), is such that Eq.(1.3.22) can be replaced with the equivalent relation:

$$\mathbf{E}_{\text{int}}(\mathbf{r}) = \int\limits_{(R)} \left[\hat{G}_{11}(\mathbf{r},\mathbf{r}') + \hat{G}^{(k)}(\mathbf{r},\mathbf{r}')\right] k_0^2 (N^2 - 1)\mathbf{E}_I(\mathbf{r}') d^3\mathbf{r}' \tag{1.3.23}$$

where $\int\limits_{(R)}$ shows that the value of the integral is taken only at the upper boundary (at $r' = R$).

In Eqs.(1.3.21) and (1.3.23) the incident field is arbitrary, provided it can be expanded in terms of wave equation eigenfunctions:

$$\mathbf{E}_I(\mathbf{r}) = E_I \sum_{\sigma m l} \left[\gamma_{\sigma m l} \mathbf{M}^{(j)}_{\sigma m l}(k_p \mathbf{r}) + \zeta_{\sigma m l} \mathbf{N}^{(j)}_{\sigma m l}(k_p \mathbf{r})\right] \tag{1.3.24}$$

where $j = 1, 3$ (the case $j = 1$ corresponds to the incoming wave and $j = 3$ to the outgoing wave), if $k_p = k_0$ then the source is outside the particle, if $k_p = k$ then the source is inside the particle, and E_I is the amplitude of the incident field. The expansion coefficients $\gamma_{\sigma m l}$ and $\zeta_{\sigma m l}$ are determined by the properties of the source and its location relative to the scatterer. It should be mentioned that though the deduction of analytical expressions for these coefficients is complicated enough even in rather simple situations (for example, for a spherical outgoing wave), their numerical calculation does not represent the special difficulties.

Substituting Eq.(1.3.24) into Eqs.(1.3.21) and (1.3.23), we acquire (taking into account Eq.(1.3.6)):

$$\mathbf{E}(\mathbf{r}) = \mathbf{E}_I(\mathbf{r}) + ik_0^3(N^2-1)E_I \sum_{\sigma ml} \left[\alpha_l^{31} \gamma_{\sigma ml} \, {}^{(1)}I_l^{1j}(k,k_p)\Big|_0^R \mathbf{M}_{\sigma ml}^{(3)}(k_0\mathbf{r}) \right.$$
$$\left. + \beta_l^{31} (kk_p)^{-1} \zeta_{\sigma ml} \, {}^{(2)}I_l^{1j}(k,k_p)\Big|_0^R \mathbf{N}_{\sigma ml}^{(3)}(k_0\mathbf{r}) \right] \tag{1.3.25}$$

$$\mathbf{E}_{\text{int}}(\mathbf{r}) = ikk_0^2(N^2-1)E_I \sum_{\sigma ml} \left\{ \gamma_{\sigma ml} \left[\alpha_l^{11} \, {}^{(1)}I_l^{1j}(k,k_p) \right.\right.$$
$$\left. + {}^{(1)}I_l^{3j}(k,k_p) \right] \Big|^R \mathbf{M}_{\sigma ml}^{(1)}(k\mathbf{r}) \tag{1.3.26}$$
$$\left. + (kk_p)^{-1} \zeta_{\sigma ml} \left[\beta_l^{11} \, {}^{(2)}I_l^{1j}(k,k_p) + {}^{(2)}I_l^{3j}(k,k_p) \right] \Big|^R \mathbf{N}_{\sigma ml}^{(1)}(k\mathbf{r}) \right\}$$

where for simplicity the following compact notations of indefinite integrals (see Appendix A) are introduced:

$${}^{(1)}I_l^{ij}(\alpha,\beta) \equiv \frac{1}{\alpha\beta} \int \psi_l^{(i)}(\alpha r) \psi_l^{(j)}(\beta r) dr \tag{1.3.27}$$
$${}^{(2)}I_l^{ij}(\alpha,\beta) \equiv \int \left[\psi_l^{(i)\prime}(\alpha r) \psi_l^{(j)\prime}(\beta r) + \frac{l(l+1)}{r^2\alpha\beta} \psi_l^{(i)}(\alpha r) \psi_l^{(j)}(\beta r) \right] dr$$

The integrals (1.3.27) may be taken in the closed form (Gradshteyn and Ryzhik, 1963; Malushkov, 1974):

$${}^{(1)}I_l^{ij}(\alpha,\beta) = \frac{1}{(\alpha^2-\beta^2)\,\alpha\beta} \left[\beta\psi_l^{(i)}(\alpha r) \psi_l^{(j)\prime}(\beta r) - \alpha\psi_l^{(i)\prime}(\alpha r) \psi_l^{(j)}(\beta r) \right]$$
$${}^{(2)}I_l^{ij}(\alpha,\beta) = \frac{1}{\alpha^2-\beta^2} \left[\alpha\psi_l^{(i)}(\alpha r) \psi_l^{(j)\prime}(\beta r) - \beta\psi_l^{(i)\prime}(\alpha r) \psi_l^{(j)}(\beta r) \right] \tag{1.3.28}$$

The expressions (1.3.25), (1.3.26), and (1.3.28) yield the complete solution to the problem of the arbitrary harmonic wave diffraction on a spherical particle.

In the most simple situation when the homogeneous plane wave incidents on a particle (let us also assume that the wave with electric vector along the $\mathbf{e}_x$ axis propagates in the positive direction of the $\mathbf{e}_z$ axis) expansion coefficients in Eq.(1.3.24) equal:

$$k_\nu = k \quad k_p = k_0$$
$$\gamma_{o1l} = i^l\,(2l+1)\,/\,[l(l+1)] \qquad \zeta_{e1l} = -i^{l+1}\,(2l+1)\,/\,[l(l+1)]$$

the remaining coefficients are equal to zero (Stratton, 1941). Then we have for the scattered field:

$$\mathbf{E}(\mathbf{r}) = \mathbf{E}_I(\mathbf{r}) + ik_0^3(N^2-1)E_I \sum_{l=1}^{\infty} \alpha_l^{31} \gamma_{o1l} I_l^{11}(k,k_0)\big|_0^R \mathbf{M}_{o1l}^{(3)}(k_0\mathbf{r})$$
$$+ (kk_0)^{-1} \zeta_{e1l} \beta_l^{31} \, \bar{I}_l^{11}(k,k_0)\big|_0^R \mathbf{N}_{\sigma ml}^{(3)}(k_0\mathbf{r}) \Big\} \tag{1.3.29}$$

and for the internal field:

$$\begin{aligned}\mathbf{E}_{\text{int}}(\mathbf{r}) = ikk_0^2(N^2-1)E_I\sum_{l=1}^{\infty}\{\gamma_{o1l}\left[\alpha_l^{11}I_l^{11}(k,k_0)\right.\\ \left.+I_l^{31}(k,k_0)\right]\Big|^R\mathbf{M}_{o1l}^{(1)}(k\mathbf{r})\\ +(kk_p)^{-1}\zeta_{e1l}\left[\beta_l^{11}\bar{I}_l^{1j}(k,k_0)+\bar{I}_l^{3j}(k,k_0)\right]\Big|^R\mathbf{N}_{e1l}^{(1)}(k\mathbf{r})\Big\}\end{aligned} \tag{1.3.30}$$

The simple calculations give:

$$\begin{aligned}ik_0^3(N^2-1)\alpha_l^{31}\,I_l^{11}(k,k_0)\big|_0^R &= \alpha_l^{33}\\ ik_0(N^2-1)\beta_l^{31}\,\bar{I}_l^{11}(k,k_0)\big|_0^R &= k\beta_l^{33}\\ ikk_0^2(N^2-1)\left[\alpha_l^{11}I_l^{11}(k,k_0)+I_l^{31}(k,k_0)\right]\Big|^R &= \alpha_l^{31}\\ ik_0(N^2-1)\left[\beta_l^{11}\bar{I}_l^{1j}(k,k_0)+\bar{I}_l^{3j}(k,k_0)\right]\Big|^R &= \beta_l^{31}\end{aligned} \tag{1.3.31}$$

So, as follows from Eqs.(1.3.29)–(1.3.31), when the homogeneous plane wave incidents on particle, expressions for fields outside and inside the particle take the form of:

$$\begin{aligned}\mathbf{E}(\mathbf{r}) &= \mathbf{E}_I(\mathbf{r})+E_I\sum_{l=1}^{\infty}i^l\frac{2l+1}{l(l+1)}\left[\alpha_l^{33}\mathbf{M}_{o1l}^{(3)}(k_0\mathbf{r})-i\beta_l^{33}\mathbf{N}_{e1l}^{(3)}(k_0\mathbf{r})\right]\\ \mathbf{E}_{\text{int}}(\mathbf{r}) &= E_I\sum_{l=1}^{\infty}i^l\frac{2l+1}{l(l+1)}\left[\alpha_l^{31}\mathbf{M}_{o1l}^{(1)}(k\mathbf{r})-i\beta_l^{31}\mathbf{N}_{e1l}^{(1)}(k\mathbf{r})\right]\end{aligned} \tag{1.3.32}$$

These expressions completely coincide with the formulae for fields obtained by usual methods of Mie theory (Shifrin, 1951; Bohren and Huffman, 1983).

Let us now derive the scattering tensor, as Kuzmin (1979) derived it. We write down the equation for the scattering tensor elements or the scattering tensor in coordinate representation. Having in mind the functional relation between the Green's function operator and the scattering tensor operator (1.2.4) we arrive at:

$$\hat{G}(\mathbf{r},\mathbf{r}') = \left[k_0^2(N^2-1)\right]^{-1}\int_V \hat{G}^0(\mathbf{r},\mathbf{r}'')\hat{T}(\mathbf{r}'',\mathbf{r}')d^3\mathbf{r}'' \tag{1.3.33}$$

If the Green's function $\hat{G}(\mathbf{r},\mathbf{r}')$ is known, then it is possible to find from Eq.(1.3.33) the scattering tensor in coordinate representation. In the case of spherical particles with a complex refractive index $\hat{G}(\mathbf{r},\mathbf{r}') = \hat{G}_{31}(\mathbf{r},\mathbf{r}')$ (see Eq.(1.3.11)). We will search for this scattering tensor according to Eq.(1.3.33) and take into account the complex character of Green's function dependence on its arguments.

It can be shown that expression:

$$\hat{B}(\mathbf{r},\mathbf{r}') = k_0^2(N^2-1)\delta(\mathbf{r}-\mathbf{r}') + \left[k_0^2(N^2-1)\right]^2\left[\hat{G}_{11}(\mathbf{r},\mathbf{r}')+\hat{G}^{(k_0)}(\mathbf{r},\mathbf{r}')\right]$$

satisfies Eq.(1.3.33). If we take into account the form of Green's functions in singular points $\mathbf{r} = \mathbf{r}'$ (van Bladel, 1961; Tai, 1973, 1993) then it is possible to write expression for $\hat{B}(\mathbf{r}, \mathbf{r}')$ as:

$$\hat{B}(\mathbf{r}, \mathbf{r}') = \left[k_0^2(N^2 - 1)\right]^2 \text{ v.p. } \left[\hat{G}_{11}(\mathbf{r}, \mathbf{r}') + \hat{G}^{(k_0)}(\mathbf{r}, \mathbf{r}')\right] \tag{1.3.34}$$

where v.p. denotes that the integral in Eq.(1.3.34) is taken in the sense of the principal value. Writing Eq.(1.3.33) in the expanded form and substituting Eq.(1.3.34) we have:

$$\begin{aligned}\hat{G}_{31}(\mathbf{r}, \mathbf{r}') &= -kk_0^3(N^2 - 1) \sum_{\sigma ml} \sum_{\sigma' m' l'} \lambda_{ml} \lambda_{m'l'} \\ &\times \left\{ \mathbf{M}^{(3)}_{\sigma ml}(k_0 \mathbf{r}) \otimes \mathbf{M}^{(1)}_{\sigma' m' l'}(k\mathbf{r}') \left[\alpha^{11}_{l'} \, {}^{(1)}I^{\,11}_{\,mlm'l'}(k_0, k) + {}^{(1)}I^{\,13}_{\,mlm'l'}(k_0, k) \right] \Big|^R \right. \\ &\left. + \mathbf{N}^{(3)}_{\sigma ml}(k_0 \mathbf{r}) \otimes \mathbf{N}^{(1)}_{\sigma' m' l'}(k\mathbf{r}') \left[\beta^{11}_{l'} \, {}^{(2)}I^{\,11}_{\,mlm'l'}(k_0, k) + {}^{(2)}I^{\,13}_{\,mlm'l'}(k_0, k) \right] \Big|^R \right\}\end{aligned} \tag{1.3.35}$$

where the following notation for integrals (see Appendix C) is introduced:

$$\begin{aligned}{}^{(1)}I^{\,ij}_{\,mlm'l'}(\alpha, \beta) &= \int_V \mathbf{M}^{(i)}_{\sigma ml}(\alpha \mathbf{r}) \cdot \mathbf{M}^{(j)}_{\sigma' m' l'}(\beta \mathbf{r}) d^3\mathbf{r} \\ {}^{(2)}I^{\,ij}_{\,mlm'l'}(\alpha, \beta) &= \int_V \mathbf{N}^{(i)}_{\sigma ml}(\alpha \mathbf{r}) \cdot \mathbf{N}^{(j)}_{\sigma' m' l'}(\beta \mathbf{r}) d^3\mathbf{r}\end{aligned} \tag{1.3.36}$$

Having the expressions (1.3.1)–(1.3.3) and (1.3.27) it is easy to show that these integrals are reduced to the earlier defined integrals (1.3.26):

$$\begin{aligned}{}^{(1)}I^{\,ij}_{\,mlm'l'}(\alpha, \beta) &= \delta_{ll'} \delta_{mm'} \delta_{\sigma\sigma'} \lambda^{-1}_{ml} \, {}^{(1)}I^{\,ij}_{\,l}(\alpha, \beta) \\ {}^{(2)}I^{\,ij}_{\,mlm'l'}(\alpha, \beta) &= \delta_{ll'} \delta_{mm'} \delta_{\sigma\sigma'} \lambda^{-1}_{ml} (\alpha\beta)^{-1} \, {}^{(2)}I^{\,ij}_{\,l}(\alpha, \beta)\end{aligned} \tag{1.3.37}$$

Substituting the third and fourth relations from Eqs.(1.3.31) into Eq.(1.3.35), we have:

$$\hat{G}_{31}(\mathbf{r}, \mathbf{r}') = ik_0 \sum_{\sigma ml} \lambda_{ml} \left[\alpha^{31}_l \mathbf{M}^{(3)}_{\sigma ml}(k_0 \mathbf{r}) \otimes \mathbf{M}^{(1)}_{\sigma ml}(k\mathbf{r}') + \beta^{31}_l \mathbf{N}^{(3)}_{\sigma ml}(k_0 \mathbf{r}) \otimes \mathbf{N}^{(1)}_{\sigma ml}(k\mathbf{r}') \right]$$

that coincides with the definition of the function $\hat{G}_{31}(\mathbf{r}, \mathbf{r}')$. Thus, after substitution of expression (1.3.30) into Eq.(1.3.29) the latter is satisfied identically and, therefore, Eq.(1.3.34) defines the required scattering tensor in the coordinate representation.

Let us find now the scattering tensor in momentum representation, that is, the Fourier transform (within a constant value) of the scattering tensor in coordinate representation:

$$\hat{T}(\mathbf{k}, \mathbf{k}') = \int_V e^{-i\mathbf{k} \cdot \mathbf{r}''} \hat{T}(\mathbf{r}'', \mathbf{r}') e^{-i\mathbf{k}' \cdot \mathbf{r}'} d^3\mathbf{r}'' d^3\mathbf{r}'$$

or in the equivalent form:

$$\hat{T}(\mathbf{k},\mathbf{k}') = \int\limits_V \hat{I} e^{-i\mathbf{k}\cdot\mathbf{r}''} \hat{T}(\mathbf{r}'',\mathbf{r}') e^{-i\mathbf{k}'\cdot\mathbf{r}'} \hat{I} d^3\mathbf{r}'' d^3\mathbf{r}' \tag{1.3.38}$$

where $\hat{I}$ is a unit matrix. The scattering tensor (1.3.38) can be calculated for any values of $\mathbf{k}$ and $\mathbf{k}'$. However, for elastic scattering it only has a meaning when $\mathbf{k} = \mathbf{k}_0$, $\mathbf{k}' = \mathbf{k}_0'$ ($|\mathbf{k}_0| = |\mathbf{k}_0'|$). For such values of wave vectors the further calculation is done:

$$\hat{T}(\mathbf{k}_0,\mathbf{k}_0') = \int\limits_V \hat{I} e^{-i\mathbf{k}_0\cdot\mathbf{r}''} \hat{T}(\mathbf{r}'',\mathbf{r}') e^{-i\mathbf{k}_0'\cdot\mathbf{r}'} \hat{I}\, d^3\mathbf{r}'' d^3\mathbf{r}' \tag{1.3.39}$$

To calculate Eq.(1.3.39) we need to expand the matrix function $\hat{I}e^{\pm i\mathbf{k}_0\cdot\mathbf{r}}$ in terms of vector spherical functions.This expansion looks like (Morse and Feshbach, 1953):

$$\begin{aligned}\hat{I}\, e^{\pm i\mathbf{k}_0\cdot\mathbf{r}} = & \sum_{\sigma ml} 4\pi\lambda_{ml} (\pm i)^l \Big[\mp i\mathbf{l}_{\sigma ml}(u,v)\otimes \mathbf{L}^{(1)}_{\sigma ml}(k_0\mathbf{r}) \\ & + \mathbf{m}_{\sigma ml}(u,v)\otimes \mathbf{M}^{(1)}_{\sigma ml}(k_0\mathbf{r}) \mp i\mathbf{n}_{\sigma ml}(u,v)\otimes \mathbf{N}^{(1)}_{\sigma ml}(k_0\mathbf{r})\Big]\end{aligned} \tag{1.3.40}$$

where u and v are angular components of the vector $\mathbf{k}_0$ along the spherical coordinate axes.

In the specific case when the wave vector $\mathbf{k}_0$ is parallel to the axis $\mathbf{e}_z$ of the introduced coordinate system the relation (1.3.40) becomes (Morse and Feshbach, 1953):

$$\begin{aligned}\hat{I}\, e^{\pm ik_0 z} = & \sum_{l=0}^{\infty} (\pm i)^l \frac{2l+1}{l(l+1)} \Big\{\mathbf{e}_x \otimes \Big[\mathbf{M}^{(1)}_{o1l}(k_0\mathbf{r}) \mp i\mathbf{N}^{(1)}_{e1l}(k_0\mathbf{r})\Big] \\ & -\mathbf{e}_y \otimes \Big[\mathbf{M}^{(1)}_{e1l}(k_0\mathbf{r}) \pm i\mathbf{N}^{(1)}_{o1l}(k_0\mathbf{r})\Big] \mp il(l+1)\mathbf{e}_z \otimes \mathbf{L}^{(1)}_{e0l}(k_0\mathbf{r})\Big\}\end{aligned} \tag{1.3.41}$$

where $\mathbf{e}_x$, $\mathbf{e}_y$ and $\mathbf{e}_z$ are the basis vectors of the Cartesian coordinate system. Then, upon substituting expansions (1.3.40), (1.3.34), (1.3.10), and (1.3.11) into Eq.(1.3.39), integrating over $\mathbf{r}'$ and $\mathbf{r}''$ ($\mathbf{r}'' > \mathbf{r}'$) with respect to relations (1.3.28), (1.3.34), and (1.3.37) and using the integrals (1.3.42) and (1.3.43) calculated by Malushkov (1974) (see Appendix C):

$$\int\limits_V \mathbf{N}^{(i)}_{\sigma ml}(k_\nu \mathbf{r})\cdot \mathbf{L}^{(j)}_{\sigma' m' l'}(k_p\mathbf{r})\, d^3r = \lambda^{-1}_{ml}\delta_{\sigma\sigma'}\delta_{mm'}\delta_{ll'} \frac{1}{k_\nu k_p}\, {}^{(3)}I^{ij}_l(k_\nu,k_p) \tag{1.3.42}$$

$$\begin{aligned}{}^{(3)}I^{ij}_l(\alpha,\beta) = & \int \Big[\frac{1}{\alpha r}\psi^{(i)}_l(\alpha r)\psi^{(j)\,\prime}_l(\beta r) + \frac{1}{\beta r}\psi^{(i)\,\prime}_l(\alpha r)\psi^{(j)}_l(\beta r) \\ & - \frac{1}{\alpha\beta r^2}\psi^{(i)}_l(\alpha r)\psi^{(j)}_l(\beta r)\Big] dr = \frac{1}{\alpha\beta r}\psi^{(i)}_l(\alpha r)\psi^{(j)}_l(\beta r)\end{aligned} \tag{1.3.43}$$

after some transformations we obtain:

$$\hat{T}(\mathbf{k}_0, \mathbf{k}'_0) = 16\pi^2 i \sum_{\sigma ml} \lambda_{ml} \left\{ -k_0^{-1} \alpha_l^{33} \mathbf{m}_{\sigma ml}\left(\hat{\mathbf{k}}_0\right) \otimes \mathbf{m}_{\sigma ml}\left(\hat{\mathbf{k}}'_0\right) \right. \tag{1.3.44}$$

$$- k_0^{-1} \beta_l^{33} \mathbf{n}_{\sigma ml}\left(\hat{\mathbf{k}}_0\right) \otimes \mathbf{n}_{\sigma ml}\left(\hat{\mathbf{k}}'_0\right) + \chi_l \mathbf{l}_{\sigma ml}\left(\hat{\mathbf{k}}_0\right) \otimes \mathbf{l}_{\sigma ml}\left(\hat{\mathbf{k}}'_0\right)$$

$$\left. +\gamma_l \mathbf{l}_{\sigma ml}\left(\hat{\mathbf{k}}_0\right) \otimes \mathbf{n}_{\sigma ml}\left(\hat{\mathbf{k}}'_0\right) + \varphi_l \mathbf{n}_{\sigma ml}\left(\hat{\mathbf{k}}_0\right) \otimes \mathbf{l}_{\sigma ml}\left(\hat{\mathbf{k}}'_0\right) \right\}$$

where $\hat{\mathbf{k}}_0$ and $\hat{\mathbf{k}}'_0$ are the unit orths of vectors $\mathbf{k}_0$ and $\mathbf{k}'_0$, respectively,

$$\chi_l = \frac{(N^2-1)^2}{k_0 N^2 x} \psi_l(x) \psi_l(Nx) \eta_l(x) \beta_l^{13} \tag{1.3.45}$$

$$\gamma_l = -i \frac{N^2-1}{k_0 N x} \psi_l(x) \psi_l(Nx) \beta_l^{33}, \qquad \varphi_l = -i \frac{N^2-1}{k_0 N x} \psi_l(x) \psi_l(Nx) \beta_l^{31}$$

Let us define notations for combined angular functions S, Q, and T:

$$\begin{aligned} S_{\substack{e\\o}ml}\left(\hat{\mathbf{k}}_0\right) &= -\pi_l^{(m)}(\theta) \begin{matrix}\cos\\ \sin\end{matrix} m\phi \\ Q_{\substack{e\\o}ml}\left(\hat{\mathbf{k}}_0\right) &= \mp \tau_l^{(m)}(\theta) m \begin{matrix}\sin\\ \cos\end{matrix} m\phi \\ T_{\substack{e\\o}ml}\left(\hat{\mathbf{k}}_0\right) &= l(l+1) \pi_l^{(m)}(\theta) \sin\theta \begin{matrix}\cos\\ \sin\end{matrix} m\phi \end{aligned} \tag{1.3.46}$$

Then, the components of the scattering tensor in the momentum representation may be written as:

$$\begin{aligned} \hat{T}_{rr}(\mathbf{k}_0, \mathbf{k}'_0) &= 16\pi^2 i \sum_{\sigma ml} \lambda_{ml} \chi_l T_{\sigma ml}\left(\hat{\mathbf{k}}_0\right) T_{\sigma ml}\left(\hat{\mathbf{k}}'_0\right) \\ \hat{T}_{r\theta}(\mathbf{k}_0, \mathbf{k}'_0) &= -16\pi^2 i \sum_{\sigma ml} \lambda_{ml} \gamma_l T_{\sigma ml}\left(\hat{\mathbf{k}}_0\right) S_{\sigma ml}\left(\hat{\mathbf{k}}'_0\right) \\ \hat{T}_{r\varphi}(\mathbf{k}_0, \mathbf{k}'_0) &= 16\pi^2 i \sum_{\sigma ml} \lambda_{ml} \gamma_l T_{\sigma ml}\left(\hat{\mathbf{k}}_0\right) Q_{\sigma ml}\left(\hat{\mathbf{k}}'_0\right) \\ \hat{T}_{\theta r}(\mathbf{k}_0, \mathbf{k}'_0) &= -16\pi^2 i \sum_{\sigma ml} \lambda_{ml} \varphi_l S_{\sigma ml}\left(\hat{\mathbf{k}}_0\right) T_{\sigma ml}\left(\hat{\mathbf{k}}'_0\right) \\ \hat{T}_{\varphi r}(\mathbf{k}_0, \mathbf{k}'_0) &= 16\pi^2 i \sum_{\sigma ml} \lambda_{ml} \varphi_l Q_{\sigma ml}\left(\hat{\mathbf{k}}_0\right) T_{\sigma ml}\left(\hat{\mathbf{k}}'_0\right) \end{aligned} \tag{1.3.47}$$

and

$$
\begin{aligned}
\hat{T}_{\varphi\varphi}(\mathbf{k}_0, \mathbf{k}_0') &= -16\pi^2 i \sum_{\sigma ml} \lambda_{ml} k_0^{-1} \left[\alpha_l^{33} S_{\sigma ml}\left(\hat{\mathbf{k}}_0\right) S_{\sigma ml}\left(\hat{\mathbf{k}}_0'\right)\right. \\
&\quad \left. +\beta_l^{33} Q_{\sigma ml}\left(\hat{\mathbf{k}}_0\right) Q_{\sigma ml}\left(\hat{\mathbf{k}}_0'\right)\right] \\
\hat{T}_{\theta\theta}(\mathbf{k}_0, \mathbf{k}_0') &= -16\pi^2 i \sum_{\sigma ml} \lambda_{ml} k_0^{-1} \left[\alpha_l^{33} Q_{\sigma ml}\left(\hat{\mathbf{k}}_0\right) Q_{\sigma ml}\left(\hat{\mathbf{k}}_0'\right)\right. \\
&\quad \left. +\beta_l^{33} S_{\sigma ml}\left(\hat{\mathbf{k}}_0\right) S_{\sigma ml}\left(\hat{\mathbf{k}}_0'\right)\right] \qquad (1.3.48) \\
\hat{T}_{\theta\varphi}(\mathbf{k}_0, \mathbf{k}_0') &= -16\pi^2 i \sum_{\sigma ml} \lambda_{ml} k_0^{-1} \left[\alpha_l^{33} Q_{\sigma ml}\left(\hat{\mathbf{k}}_0\right) S_{\sigma ml}\left(\hat{\mathbf{k}}_0'\right)\right. \\
&\quad \left. -\beta_l^{33} S_{\sigma ml}\left(\hat{\mathbf{k}}_0\right) Q_{\sigma ml}\left(\hat{\mathbf{k}}_0'\right)\right] \\
\hat{T}_{\varphi\theta}(\mathbf{k}_0, \mathbf{k}_0') &= -16\pi^2 i \sum_{\sigma ml} \lambda_{ml} k_0^{-1} \left[\alpha_l^{33} S_{\sigma ml}\left(\hat{\mathbf{k}}_0\right) Q_{\sigma ml}\left(\hat{\mathbf{k}}_0'\right)\right. \\
&\quad \left. -\beta_l^{33} Q_{\sigma ml}\left(\hat{\mathbf{k}}_0\right) S_{\sigma ml}\left(\hat{\mathbf{k}}_0'\right)\right]
\end{aligned}
$$

Thus, the scattering tensor is obtained, which completely describes scattering of the plane wave with an arbitrary direction of the wave vector $\mathbf{k}_0'$ with respect to the spherical coordinate system.

For the coordinate system with the axis $\mathbf{e}_z$ parallel to the wave vector of incident wave $\mathbf{k}_0'$, the analytical expression for scattering tensor becomes considerably simpler because in this case:

$$
\begin{aligned}
S_{\sigma ml}\left(\hat{\mathbf{k}}_0' \upuparrows \mathbf{e}_z\right) &= S_{\sigma ml}(0,0) = -\frac{1}{2} l(l+1)\delta_{1m}\delta_{\sigma e} \\
Q_{\sigma ml}\left(\hat{\mathbf{k}}_0' \upuparrows \mathbf{e}_z\right) &= Q_{\sigma ml}(0,0) = \frac{1}{2} l(l+1)\delta_{1m}\delta_{\sigma o}
\end{aligned}
$$

Taking account of this equation and in the absence of the radial part, the scattering matrix components (1.3.47) and (1.3.48) may be written as:

$$
\begin{aligned}
\hat{T}_{\theta\theta}\left(\mathbf{k}_0\right) &= 4\pi A_{\theta\theta}\left(\mathbf{k}_0\right) \\
&= 4\pi i k_0^{-1} \sum_{l=1}^{\infty} \frac{2l+1}{l(l+1)} \left[-\alpha_l^{33} Q_{o1l}\left(\hat{\mathbf{k}}_0\right) + \beta_l^{33} S_{e1l}\left(\hat{\mathbf{k}}_0\right)\right] \\
\hat{T}_{\theta\varphi}\left(\mathbf{k}_0\right) &= 4\pi A_{\theta\varphi}\left(\mathbf{k}_0\right) \qquad (1.3.49) \\
&= 4\pi i k_0^{-1} \sum_{l=1}^{\infty} \frac{2l+1}{l(l+1)} \left[\alpha_l^{33} Q_{e1l}\left(\hat{\mathbf{k}}_0\right) + \beta_l^{33} S_{o1l}\left(\hat{\mathbf{k}}_0\right)\right]
\end{aligned}
$$

and

$$\hat{T}_{\varphi\theta}(\mathbf{k}_0) = 4\pi A_{\varphi\theta}(\mathbf{k}_0)$$
$$= 4\pi i k_0^{-1} \sum_{l=1}^{\infty} \frac{2l+1}{l(l+1)} \left[-\alpha_l^{33} S_{o1l}\left(\hat{\mathbf{k}}_0\right) - \beta_l^{33} Q_{e1l}\left(\hat{\mathbf{k}}_0\right)\right] \tag{1.3.50}$$
$$\hat{T}_{\varphi\varphi}(\mathbf{k}_0) = 4\pi A_{\varphi\varphi}(\mathbf{k}_0)$$
$$= 4\pi i k_0^{-1} \sum_{l=1}^{\infty} \frac{2l+1}{l(l+1)} \left[\alpha_l^{33} S_{e1l}\left(\hat{\mathbf{k}}_0\right) - \beta_l^{33} Q_{o1l}\left(\hat{\mathbf{k}}_0\right)\right]$$

From Eqs.(1.1.15) and (1.3.49–50) expressions for transverse scattered electric field components in the far-field region ($k_0 r \gg 1$) are easily obtained:
for $\mathbf{E}_I(\mathbf{r}) \upuparrows \mathbf{e}_x$

$$E_\theta^{\text{sc}} = E_I \frac{e^{ik_0 r}}{r} A_{\theta\theta}$$
$$= E_I \frac{e^{ik_0 r}}{k_0 r} \sum_{l=1}^{\infty} i \frac{2l+1}{l(l+1)} \left[-\alpha_l^{33} Q_{o1l}\left(\hat{\mathbf{k}}_0\right) + \beta_l^{33} S_{e1l}\left(\hat{\mathbf{k}}_0\right)\right] \tag{1.3.51}$$
$$E_\varphi^{\text{sc}} = E_I \frac{e^{ik_0 r}}{r} A_{\varphi\theta}$$
$$= E_I \frac{e^{ik_0 r}}{k_0 r} \sum_{l=1}^{\infty} i \frac{2l+1}{l(l+1)} \left[-\alpha_l^{33} S_{o1l}\left(\hat{\mathbf{k}}_0\right) - \beta_l^{33} Q_{e1l}\left(\hat{\mathbf{k}}_0\right)\right]$$

for $\mathbf{E}_I(\mathbf{r}) \upuparrows \mathbf{e}_y$

$$E_\theta^{\text{sc}} = E_I \frac{e^{ik_0 r}}{r} A_{\theta\theta}$$
$$= E_I \frac{e^{ik_0 r}}{k_0 r} \sum_{l=1}^{\infty} i \frac{2l+1}{l(l+1)} \left[\alpha_l^{33} Q_{e1l}\left(\hat{\mathbf{k}}_0\right) + \beta_l^{33} S_{o1l}\left(\hat{\mathbf{k}}_0\right)\right] \tag{1.3.52}$$
$$E_\varphi^{\text{sc}} = E_I \frac{e^{ik_0 r}}{r} A_{\theta\theta}$$
$$= E_I \frac{e^{ik_0 r}}{k_0 r} \sum_{l=1}^{\infty} i \frac{2l+1}{l(l+1)} \left[\alpha_l^{33} S_{e1l}\left(\hat{\mathbf{k}}_0\right) - \beta_l^{33} Q_{o1l}\left(\hat{\mathbf{k}}_0\right)\right]$$

It is easy to obtain also the expressions of scattered fields for elliptically polarized and partially polarized incident beams (Kuzmin, 1979). It can be seen that the relation (1.3.51) and the relation obtained from Eq.(1.3.32) completely coincide in the case of external fields in the far-field region. Notice that the calculation of elements of the scattering matrix in momentum representation is carried out on the energy surface $|\mathbf{k}_0| = |\mathbf{k}'_0|$. However, it can be carried out in the more general case too, that is, outside the energy surface. This approach may be used to solve the problems of inelastic scattering, for example, the problem of scattering by fluorescent particles (Chew *et al.*, 1978).

Before obtaining expressions of integral characteristics of light scattering, we should consider one important property of angular vector functions (1.3.2). After comparing components of these functions with components of the vector spherical harmonics $\mathbf{Y}_{lm}^{(\lambda)}(\theta,\varphi)$ (Varshalovich *et al.*, 1988) we can write their relationship as:

$$\begin{aligned}
\mathbf{Y}_{lm}^{(1)}(\theta,\varphi) &= \lambda_{ml}^{1/2}\frac{1}{\sqrt{2}}\left[\mathbf{n}_{eml}(\theta,\varphi)+i\mathbf{n}_{oml}(\theta,\varphi)\right] \\
\mathbf{Y}_{lm}^{(0)}(\theta,\varphi) &= \lambda_{ml}^{1/2}\frac{1}{\sqrt{2}}\left[\mathbf{m}_{eml}(\theta,\varphi)+i\mathbf{m}_{oml}(\theta,\varphi)\right] \qquad (1.3.53)\\
\mathbf{Y}_{lm}^{(-1)}(\theta,\varphi) &= \lambda_{ml}^{1/2}\frac{1}{\sqrt{2}}\left[\mathbf{l}_{eml}(\theta,\varphi)+i\mathbf{l}_{oml}(\theta,\varphi)\right]
\end{aligned}$$

The vector spherical harmonics are of great use in the quantum theory of angular momentum. Their properties are well investigated. They are also frequently used in the theory of scattering on macroscopic particles (for example, see Newton, 1982). Having in mind the well known relationship for vector spherical harmonics:

$$\sum_{m=-l}^{l}\mathbf{Y}_{lm}^{(\lambda')*}(\theta,\varphi)\cdot\mathbf{Y}_{lm}^{(\lambda)}(\theta,\varphi)=\frac{2l+1}{4\pi}\delta_{\lambda\lambda'}$$

and Eq.(1.3.53), we obtain the relations for angular functions:

$$\begin{aligned}
\sum_{\sigma=e,o}\sum_{m=0}^{l}\lambda_{ml}\mathbf{m}_{\sigma ml}(\theta,\varphi)\cdot\mathbf{m}_{\sigma ml}(\theta,\varphi) &= \frac{2l+1}{4\pi}\\
\sum_{\sigma=e,o}\sum_{m=0}^{l}\lambda_{ml}\mathbf{n}_{\sigma ml}(\theta,\varphi)\cdot\mathbf{n}_{\sigma ml}(\theta,\varphi) &= \frac{2l+1}{4\pi} \qquad (1.3.54)\\
\sum_{\sigma=e,o}\sum_{m=0}^{l}\lambda_{ml}\mathbf{l}_{\sigma ml}(\theta,\varphi)\cdot\mathbf{l}_{\sigma ml}(\theta,\varphi) &= \frac{2l+1}{4\pi}
\end{aligned}$$

Let us substitute the amplitude tensor of scattering (1.3.44) into the expression for the extinction cross section σ_{ext} (Newton, 1982):

$$\sigma_{\text{ext}}=\frac{2\pi}{k_0}\operatorname{Im}\left[Sp\hat{A}(\mathbf{k}_0,\mathbf{k}_0')\right] \qquad (1.3.55)$$

Taking into the account transverse components only, we obtain

$$\begin{aligned}
\sigma_{\text{ext}} &= \frac{2\pi}{k_0} \qquad (1.3.56)\\
&\times\operatorname{Im}\left\{\frac{4\pi i}{k_0}\sum_{\sigma ml}\lambda_{ml}\left[-\alpha_l^{33}\mathbf{m}_{\sigma ml}(\theta,\varphi)\cdot\mathbf{m}_{\sigma ml}(\theta,\varphi)-\beta_l^{33}\mathbf{n}_{\sigma ml}(\theta,\varphi)\cdot\mathbf{n}_{\sigma ml}(\theta,\varphi)\right]\right\}
\end{aligned}$$

As α_l^{33} and β_l^{33} do not depend on index m, it is possible to use Eqs.(1.3.54) here, which eventually results in an expression for the extinction cross section

$$\sigma_{\text{ext}} = \frac{2\pi}{k_0^2} \operatorname{Re} \sum_{l=1}^{\infty} (2l+1) \left[\alpha_l^{33} + \beta_l^{33}\right] \tag{1.3.57}$$

To obtain expressions for the scattering cross section σ_{sc} it is necessary to employ the orthogonality property of angular functions (1.3.6) as well as Eqs. (1.3.54):

$$\sigma_{\text{sc}} = \frac{2\pi}{k_0^2} \sum_{l=1}^{\infty} (2l+1) \left[\left|\alpha_l^{33}\right|^2 + \left|\beta_l^{33}\right|^2\right] \tag{1.3.58}$$

where $|...|$ denotes the absolute value.

The absorption cross section is usually calculated by subtraction:

$$\sigma_{\text{abs}} = \sigma_{\text{ext}} - \sigma_{\text{sc}} \tag{1.3.59}$$

However, the formula (1.3.59) can sometimes give big computational errors. In this case, use of another expression makes sense, that is, the so-called formula of Kattawar and Eisner (1970) which can be obtained from Eq.(1.1.16) by integration over the particle volume (see Section 5.2).

1.4 CALCULATED CHARACTERISTICS OF SCATTERED LIGHT

Some characteristics of the field scattered by individual particles were defined in previous sections. Now we complete this subject by defining other characteristics to be used in the following sections and we introduce their notations. These characteristics will also be defined for the polydisperse ensemble of particles.

1. Let us begin with the *integral characteristics* (i.e., ones which do not depend on the scattering angle). The cross sections of scattering σ_{sc}, extinction σ_{ext}, and absorption σ_{abs} were defined above. The ratio of scattering to extinction cross section is called the *particle albedo* (or, single scattering albedo):

$$\Lambda = \frac{\sigma_{\text{sc}}}{\sigma_{\text{ext}}} = \frac{\sigma_{\text{sc}}}{\sigma_{\text{sc}} + \sigma_{\text{abs}}} \tag{1.4.1}$$

which varies in limits $0 < \Lambda < 1$. The quantity $(1 - \Lambda)$ is called the *quantum absorption probability*.

Radiation transfers a momentum, so it is frequently important to consider the *radiation pressure cross section:*

$$\sigma_{\text{pr}} = \sigma_{\text{ext}} - \langle\cos\theta\rangle\, \sigma_{\text{sc}} = \sigma_{\text{abs}} + (1 - \langle\cos\theta\rangle)\, \sigma_{\text{sc}} \tag{1.4.2}$$

where σ_{ext} is proportional to momentum deleted from the beam and $\langle\cos\theta\rangle\,\sigma_{\text{sc}}$ is the recoil momentum of scattered radiation ($\langle\cos\theta\rangle$ is the so-called average cosine of the phase function or the *asymmetry parameter*).

All cross sections σ are connected to the corresponding dimensionless *efficiency factors* Q via the relation $\sigma = GQ$, where G is the geometrical area of the particle's shadow. The efficiency factors equal the ratio of energy removed from the beam by the particle to energy removed by a screen with area G (G is equal to πR^2 for sphere).

The volume *extinction coefficient* for the unit volume containing C_N particles (C_N is a *number density* with dimensionality L^{-3}) with the same radius R is defined as a sum of cross sections of all particles in this volume (dimensionality L^{-1}):

$$K_{\text{ext}} = C_N\sigma_{\text{ext}} = C_N\pi R^2 Q_{\text{ext}}(R) \tag{1.4.3}$$

Values of K_{sc}, K_{abs}, and K_{pr} are defined similarly. If we introduce the *volume concentration* C_V (C_V is the non-dimensional fraction of the volume occupied by particles) as $C_V = 4C_N\pi R^3/3$ and the *mass concentration* C_M as $C_M = 4C_N\rho\pi R^3/3$ (C_M is the mass of all particles in the unit volume and ρ is the density of the particle substance) then Eq.(1.4.3) takes a form:

$$K_{\text{ext}} = \frac{3}{4}C_V\frac{Q_{\text{ext}}}{R} = \frac{3\pi}{2\lambda}C_M\frac{Q_{\text{ext}}}{x} \tag{1.4.4}$$

where λ is the wavelength of the incident radiation.

Let us now consider a polydisperse system of particles. There is no orientation parameter for a system of spherical particles and, therefore, to find characteristics for polydisperse system it is sufficient to multiply characteristics of an individual particle by number density C_N, and to average over a distribution function of particles sizes $f(R)$. We define $\overline{f(R)}$ to be a function normalized to unity. Note that such simple one-parameter functional dependence exists only for the spherical homogeneous particles. In more complex cases the distribution function has more parameters (e.g. the radii of the core and the shell for a case of layered spherical particles). In the simplest case:

$$K_{\text{ext}} = \pi C_N \int_{R_1}^{R_2} R^2 Q_{\text{ext}}(R)\overline{f(R)}dR \tag{1.4.5}$$

where R_1 and R_2 are the minimal and maximum particle sizes in the distribution. Note that the determination of computationally reasonable values of R_1 and R_2 is not simple in practice. Since:

$$C_N = \frac{3C_V}{4\pi}\int_{R_1}^{R_2} R^3\overline{f(R)}dR \equiv \frac{C_V}{\pi}E^{-1} \tag{1.4.6}$$

Equation (1.4.5) may be transformed into:

$$K_{\text{ext}} = \frac{C_V}{E}\int_{R_1}^{R_2} R^2 Q_{\text{ext}}(R)\overline{f(R)}dR \equiv C_V K'_{\text{ext}} \tag{1.4.7}$$

where K'_{ext} is the so-called normalized extinction coefficient. The unity-normalized distribution function may be replaced in Eq.(1.4.7) by an arbitrarily normalized distribution function since the normalization factor is cancelled.

2. Now let us turn to the quantities characterizing the angular structure of scattered radiation. The matrix of Stokes parameters transformation $D_{\lambda\mu}(\Omega, \Omega')$ and the amplitude scattering matrix $A_{ij}(\Omega, \Omega')$ are defined above. The net monochromatic energy flux, scattered by particle in the direction (θ, φ) is $E_\theta E_\theta^* + E_\varphi E_\varphi^*$. Therefore, taking into account Eq.(1.3.51), intensity of light scattered by the elementary volume V of monodisperse medium is:

$$I(\theta, \varphi) = \frac{C_N V I_0}{k_0^2 r^2} i = \frac{3 C_M V I_0}{2\rho x^3 \lambda r^2} i \tag{1.4.8}$$

where r is the distance to observation point, $i = i_1 + i_2$,

$$\begin{aligned} i_1(\theta) &= k_0^2 \left|A_{\theta\varphi}\right|^2 \\ &= \left|\frac{2l+1}{l(l+1)} \left[\alpha_l^{33} \tau_l(\cos\theta) + \beta_l^{33} \pi_l(\cos\theta)\right]\right|^2 \sin^2\varphi \end{aligned} \tag{1.4.9}$$

$$\begin{aligned} i_2(\theta) &= k_0^2 \left|A_{\theta\theta}\right|^2 \\ &= \left|\frac{2l+1}{l(l+1)} \left[\alpha_l^{33} \pi_l(\cos\theta) + \beta_l^{33} \tau_l(\cos\theta)\right]\right|^2 \cos^2\varphi \end{aligned} \tag{1.4.10}$$

where i_1 and i_2 are the non-dimensional quantities sometimes referred to as *the Mie intensities* and I_0 is the intensity of the incident plane wave:

$$I_0 = \frac{1}{2} E_I^2 \sqrt{\frac{\varepsilon_0}{\mu_0}} \tag{1.4.11}$$

It is evident that $i_1 = D_{11} - D_{21}$ and $i_2 = D_{11} + D_{21}$. Equation (1.4.8) defines the *phase function* of elementary volume. The phase function normalized to unity has the form of:

$$F(\theta, \varphi) = \frac{i}{\pi x^2 Q_{\text{sc}}} \tag{1.4.12}$$

As for the unit volume of the polydisperse medium, the intensity of radiation scattered in the direction (θ, φ) is:

$$\langle I \rangle = \frac{I_0 C_V}{k_0^2 r^2 \pi E} \int_{R_1}^{R_2} i f(R) dR \tag{1.4.13}$$

Normalized elements of the scattering matrix are:

$$f_{\lambda\mu} = D_{\lambda\mu} \Big/ \int_{R_1}^{R_2} i f(R) dR, \qquad \lambda, \mu = 1, 2, 3, 4 \tag{1.4.14}$$

3. We use two additional parameters to describe the form of the phase function. These are the *average cosine* of the phase function $\langle \cos\theta \rangle$:

$$\langle \cos\theta \rangle = \frac{\int\limits_{(4\pi)} i(\theta,\varphi)\cos\theta d\Omega}{\int\limits_{(4\pi)} i(\theta,\varphi) d\Omega} \tag{1.4.15}$$

and the *asymmetry factor* η, which is the ratio of flux scattered into the forward hemisphere $(0 \leq \theta \leq \pi/2)$ to flux scattered into the back hemisphere $(\pi/2 \leq \theta \leq \pi)$:

$$\eta = \frac{\int\limits_0^{\pi/2}\int\limits_0^{2\pi} i(\theta,\varphi) d\theta d\varphi}{\int\limits_{\pi/2}^{\pi}\int\limits_0^{2\pi} i(\theta,\varphi) d\theta d\varphi} \tag{1.4.16}$$

The integrals in Eqs.(1.4.15)–(1.4.16) can be evaluated analytically (Chýlek, 1973; Bohren and Huffman, 1983):

$$\langle \cos\theta \rangle = \frac{4}{x^2 Q_{\text{sc}}}\sum_{l=1}^{\infty}\left[\frac{l(l+2)}{l+1}\,\mathrm{Re}\left(\alpha_l^{33}\alpha_l^{33\,*}+\beta_l^{33}\beta_l^{33\,*}\right)+\frac{2l+1}{l(l+1)}\,\mathrm{Re}\left(\beta_l^{33}\alpha_l^{33\,*}\right)\right] \tag{1.4.17}$$

$$\eta = \frac{x^2 Q_{\text{sc}} - 4A}{x^2 Q_{\text{sc}} + 4A} \tag{1.4.18}$$

where

$$A = \sum_{j=1}^{\infty}{}' \frac{(2j+1)\; j!!}{j(j+1)(j-1)!!}\sum_{l=1}^{\infty}{}'(-1)^{(j+l)/2}\frac{(2l+1)\; l!!\,\mathrm{Re}\left(\beta_l^{33}\alpha_l^{33\,*}\right)}{l(l+1)(l-1)!!} + \sum_{j=1}^{\infty}{}'' \frac{(j-1)!!}{j!!}\sum_{l=1}^{\infty}{}''(-1)^{(j+l-1)/2}\frac{(2l+1)\; l!!\,\mathrm{Re}\left(\alpha_l^{33}\alpha_l^{33\,*}+\beta_l^{33}\beta_l^{33\,*}\right)}{(j-l)\,(j+l+1)(l-1)!!} \tag{1.4.19}$$

where $\sum'$ and $\sum''$ denote a sum over odd and even indexes, respectively. It is easy to see that in the case of the polydisperse system:

$$\langle \eta \rangle = \frac{\int\limits_{R_1}^{R_2} B_1(R)\, f(R) dR}{\int\limits_{R_1}^{R_2} B_2(R)\, f(R) dR} \qquad \langle\langle \cos\theta \rangle\rangle = \frac{4\int\limits_{R_1}^{R_2} B_3(R)\, f(R) dR}{k_0^2 \int\limits_{R_1}^{R_2} Q_{\text{sc}}(R)\, R^2 f(R) dR} \tag{1.4.20}$$

where $B_1(R)$ and $B_2(R)$ are the numerator and the denominator in Eq.(1.4.18), respectively, and $B_3(R)$ is the sum from Eq.(1.4.17).

1.5 COMPUTATIONAL ASPECTS OF THE MIE PROBLEM

1.5.1 General considerations

1. The computational expressions for integral characteristics of light scattering by a sphere (cf. Eqs.(1.3.32)) are the infinite series ($l = 1, 2, ...$) over amplitude coefficients α_l^{33} and β_l^{33} for the external field (Eq.(1.3.14)). In practice these series must be truncated at some number L after reaching the predetermined accuracy of calculations. At first glance it seems very simple. If the desired accuracy is not reached at some l, this value must be incremented, with the next step of corresponding upward recursions performed for necessary functions, new terms added to sums, convergence checked, and so on. The very first attempts to implement Mie theory on a computer used precisely this approach. However, it was discovered very quickly that the simplest approach frequently yields incorrect results (for example, $Q_{\rm sc} > Q_{\rm ext}$ for absorbing spheres). It has been found long ago that use of these two ideas vastly improves the stability of Mie computations:

1) Calculation of *logarithmic derivatives* of Riccati–Bessel and Riccati–Hankel functions instead of the functions itself. If we introduce the notations for log-derivatives

$$D_l(z) = \frac{\psi_l'(z)}{\psi_l(z)} \qquad G_l(z) = \frac{\xi_l'(z)}{\xi_l(z)} \tag{1.5.1}$$

then we can write the amplitude coefficients β_l^{33} and α_l^{33} in the form:

$$\beta_l^{33} = \frac{\psi_l(x)}{\xi_l(x)} \cdot \frac{D_l(x) - ND_l(Nx)}{ND_l(Nx) - G_l(x)} \qquad \alpha_l^{33} = \frac{\psi_l(x)}{\xi_l(x)} \cdot \frac{D_l(Nx) - ND_l(x)}{NG_l(x) - D_l(Nx)} \tag{1.5.2}$$

Hence, we must calculate three arrays ($l = 1, 2, ..., L$) of log-derivatives: two ($D_l(x)$ and $G_l(x)$) with the real arguments x and one ($D_l(Nx)$) with the complex argument Nx, as well as the array of ratios $\psi_l(x)/\xi_l(x)$ with real arguments (the details on calculation of this ratio can be found in Section 1.5.2.2). Note that the recursion for log-derivatives:

$$D_{l+1}(z) = -\frac{l+1}{z} - \left[D_l(z) - \frac{l+1}{z}\right]^{-1} \tag{1.5.3}$$

$$G_{l+1}(z) = -\frac{l+1}{z} - \left[G_l(z) - \frac{l+1}{z}\right]^{-1} \tag{1.5.4}$$

is a two-term one, in contrast to the three-term recursions for Bessel functions.

2) Calculation of some log-derivatives with *downward recursion* instead of upward recursion.

In upward recursion (Eq.(1.5.4)) for the log-derivatives $G_l(z)$ the relative error does not accumulate so we use this recursion with an initial value $G_0(z) = i$. At the same time a large number of numerical experiments showed that the evaluation of $D_l(z)$ with the recursion (1.5.3) in the upward direction results in numerical instability at $l > |z|$ (this instability is most frequent for larger $\operatorname{Im} z$), however, the recursion (1.5.4) is numerically stable for downward calculations. Therefore, our software calculates $D_l(z)$ with downward recursion. However, if any function entering into the amplitude coefficients is to be calculated with the downward recursion, the above simplest scheme of calculation is not sufficient, because we must know the number L beforehand, where the series must be truncated and/or number $L^* > L$, where the downward recursion begins.

There are two principally different approaches to organize the downward recursion. In the first approach, basing on the asymptotic behavior of a log-derivative ($D_l(z) \to (l+1)/z$ at $l \to \infty$) we assume $D_{L^*} = (L^*+1)/z$ for some index value $L^* \gg z$ and, after having substituted it into Eq.(1.5.3), calculate $D_l(z)$ with indexes $L^*-1, L^*-2, ...,$ and so on. Values computed thus converge rapidly to exact values of log-derivatives and at some great L^* the computational error becomes insensitive to a choice of an initial number.

In the second approach, which is widely used in contemporary Mie codes, the $D_L(z)$ is directly calculated using specially devised methods (see Section 1.5.2.1). Notice once again that this approach requires knowledge of the value L in advance. This value is selected on the base of physical reasoning and computational experience. A great popularity was gained by Wiscombe's (1980) estimate of L:

$$L = x + 4.05x^{1/3} + 2 \tag{1.5.5}$$

and its variants. It is significant that this estimate does not contain the refractive index of the particle. Numerous calculations of many authors demonstrated the reliability of this estimate with regard to the scattered field with the possible exception of very narrow morphology-dependent resonances (Roll *et al.*, 1998*a*). The calculation of the internal and near fields requires increasing this estimate.

Some additional computational difficulties arise in the case of the layered and multilayered spheres (and, also, when we calculate the internal fields). In particular, we need to calculate some combinations of Riccati–Bessel and Riccati–Hankel functions of different complex arguments (see Section 1.5.2.3).

If we want to study the angular characteristics of scattering, in addition to Bessel functions we must calculate the so-called angular functions (see Section 1.3 and Appendix B). The algorithms of their generation is described in Section Ch.1.5.3.

1.5.2 Bessel functions

1.5.2.1 Generating the logarithmic derivative of Bessel functions

1. The most popular method used to calculate the logarithmic derivatives of Bessel functions $D_l(z) = \psi_l'(z)/\psi_l(z)$ ($l = 1, 2, ...,$ up to some number L that truncates a Mie series) lies in the use of *continued fractions* for a maximum number L (Lentz, 1976), with ensuing computation of functions with numbers $l = L-1, ..., 2, 1$ through the downward recursion.

Based on work by Lentz (1976), let us consider the method of generating logarithmic derivatives of Bessel functions $\Delta_\nu(z) = J_\nu'(z)/J_\nu(z)$ with the use of continued fractions.

To generalize, we do not restrict our consideration to half-integer orders $\nu = l + 1/2$, however, we assume ν and z to be the arbitrary complex values. To avoid confusion, we retain the notation $D_l(z)$ only for the half-integer orders. In this case it is evident that:

$$D_l(z) = \Delta_{l+1/2}(z) + \frac{1}{2z} \tag{1.5.6}$$

It follows from a well known relationship $zJ'_\nu(z) + \nu J_\nu(z) = zJ_{\nu-1}(z)$ that:

$$\Delta_\nu(z) = -\frac{\nu}{z} + \frac{J_{\nu-1}(z)}{J_\nu(z)} \tag{1.5.7}$$

The continued fraction is constructed for the ratio $J_{\nu-1}(z)/J_\nu(z)$ rather than $\Delta_\nu(z)$. This fraction is easily obtained from recursion $J_{\nu-1} + J_{\nu+1} = 2\nu J_\nu/z$, if divided term by term by J_ν:

$$\frac{J_{\nu-1}}{J_\nu} = \frac{2\nu}{z} - \frac{J_{\nu+1}}{J_\nu} = \frac{2\nu}{z} - \frac{1}{J_\nu/J_{\nu+1}}$$

Further:

$$\frac{J_\nu}{J_{\nu+1}} = \frac{2(\nu+1)}{z} - \frac{J_{\nu+2}}{J_{\nu+1}}$$

that is:

$$\frac{J_{\nu-1}}{J_\nu} = \frac{2\nu}{z} - \frac{1}{2(\nu+1)/z+} \frac{1}{2(\nu+2)/z - ...} \tag{1.5.8}$$

This continued fraction must converge for all complex ν.

Let us introduce some short notations for the infinite continued fraction:

$$a_1 + \frac{1}{a_2+} \frac{1}{a_3+} \frac{1}{a_4 + ...} = [a_1, a_2, a_3, a_4...]$$

and for the l-partial fraction:

$$f_l(z) = a_1 + \frac{1}{a_2+} \frac{1}{a_3+} ... \frac{1}{+a_l} = [a_1, a_2, a_3, ..., a_l]$$

The main idea of the Lentz's method consists of representation of the n-partial fraction in the form:

$$f_l = \frac{A_1 A_2 ... A_l}{B_1 B_2 ... B_l} \tag{1.5.9}$$

where

$$\begin{aligned} A_1 &= a_1 \\ A_l &= [a_l, a_{l-1}, ..., a_1] \qquad l = 2, 3, ... \\ B_1 &= 1, \quad B_2 = a_2 \\ B_l &= [a_l, a_{l-1}, ..., a_2] \qquad l = 3, 4, ... \end{aligned} \tag{1.5.10}$$

The accuracy of Eqs.(1.5.9–10) may be verified by direct substitution after some cumbersome transformations. The advantage of such a representation lies in reverse order of terms (that is, one can begin from the initial term $l = 1$ rather than from a relatively big l, then check convergence at $l+1, l+2, ...$, and recalculate the continued fraction at every step).

It is evident that:

$$a_l = (-1)^{l-1}\frac{2}{z}(\nu + l - 1) \tag{1.5.11}$$

from which it follows that the coefficients a_l $(l = 1, 2, ...)$ may be calculated with a multiplicative recursion:

$$a_1 = \frac{2\nu}{z} \qquad a_{l+1} = -a_l\left(1 + \frac{1}{\nu + l - 1}\right) \tag{1.5.12}$$

From the definition of the continued fraction:

$$A_l = a_l + \frac{1}{A_{l-1}}\ (l = 2, 3, ...) \qquad B_l = a_l + \frac{1}{B_{l-1}}\ (l = 3, 4, ...) \tag{1.5.13}$$

It is also evident that:

$$f_l = f_{l-1}\frac{A_l}{B_l} \tag{1.5.14}$$

For each l the convergence is checked. If, for some $l = L,$ the condition:

$$\left|\frac{f_L}{f_{L-1}} - 1\right| \le 10^{-\varepsilon}$$

(where ε may be 15, for instance) is satisfied then we assume that the fraction has converged and:

$$f_L = \frac{J_{\nu-1}(z)}{J_\nu(z)} \qquad \Delta_\nu(z) = -\frac{\nu}{z} + f_L \tag{1.5.15}$$

$$D_L(z) = -\frac{L}{z} + f_L \quad \text{(for half-integer orders)} \tag{1.5.16}$$

If the condition is not satisfied, n is incremented.

The remaining elements of the array of log-derivatives are calculated by ordinary downward recursion:

$$\Delta_l(z) = \frac{l}{z} - \frac{1}{\Delta_{l+1}(z) + (l+1)/z} \tag{1.5.17}$$

$$D_l(z) = \frac{l+1}{z} - \frac{1}{D_{l+1}(z) + (l+1)/z} \qquad l = L-1, L-2, ..., 1$$

2. This algorithm meets a minor problem with real refractive indexes (Jaaskelainen and Ruuskanen, 1981; Seltman, 1985). To illustrate the cause of this problem we write down the following explicit expressions:

$$
\begin{aligned}
A_2 &= \frac{z^2-(2L+1)(2L+3)}{z\,(2L+1)} \qquad A_3 = \frac{2L+3}{z}\,\frac{2z^2-(2L+1)(2L+5)}{z^2-(2L+1)(2L+3)} \\
B_3 &= -\frac{z^2-(2L+3)(2L+5)}{z\,(2L+3)} \\
B_4 &= -\frac{2L+5}{z}\,\frac{2z^2-(2L+3)(2L+7)}{z^2-(2L+3)(2L+5)}
\end{aligned}
\tag{1.5.18}
$$

It is clear that these coefficients may be zero for the following real values of the z-variable:

$$
\begin{aligned}
A_2 &= 0 \text{ at } z = [(2L+1)(2L+3)]^{1/2} \\
A_3 &= 0 \text{ at } z = [(2L+1)(2L+5)/2]^{1/2} \\
B_3 &= 0 \text{ at } z = [(2L+3)(2L+5)]^{1/2} \\
B_4 &= 0 \text{ at } z = [(2L+3)(2L+7)/2]^{1/2}
\end{aligned}
$$

On the next step of the continued fraction calculation the numerator turns to the denominator (this is seen from Eq.(1.5.18) and the analogous expressions for A and B coefficients with subsequent indexes), so, an overflow occurs for the above values of z. Therefore, the infinite value of A_l (or B_l) is always preceded by a zero value for A_{l-1} (or B_{l-1}). In addition, it can be shown that simultaneous zeroing of A_l and B_l is impossible.

For example, let $A_{m-1} = \alpha$ and $A_m \equiv \beta \approx 0$. Then $A_{m+1} \equiv \gamma = a_{m+1} + 1/\beta \to \infty$. However, the product $A_m A_{m+1} \equiv \chi = 1 + \beta a_{m+1} \to 1$ and the next term $A_{m+2} = (a_{m+2}\gamma + 1)/\gamma = (a_{m+2}\chi + \beta)/\chi \to a_{m+2}$. This provides a means of negociating this overflow in a simple way. If $|A_l| \le 10^{-\varepsilon}$, then we can assume that $A_l = 1$, $A_{l+1} = 1$, and $A_{l+2} = a_{l+2}$ (and so for B_l) and the following terms are calculated with the ordinary formulae. Evidently, convergence is not checked at the step where A_l (or B_l) is set to 1.

This calculation algorithm of an array of log-derivatives D_l is implemented in subroutine DLD (see Appendix D).

3. Another method to calculate the ratio $J_{\nu-1}(z)/J_\nu(z)$ was devised by Cachorro *et al.* (1989) and Cachorro and Salsedo (1991). Let us consider this approach for the half-integer orders $\nu = l + 1/2$. Any sequence of Riccati-cylindrical functions $U_l(z)$ as well as any linear combination of such functions obey the known recursion:

$$
U_{l+1}(z) = \frac{2l+1}{z} U_l(z) - U_{l+1}(z) \tag{1.5.19}
$$

Let $U_l(z)$ and $V_l(z)$ be two arbitrary sequences that satisfy recursion (1.5.19) for some constant z-value. Then, for any l the Wronskian is independent of l:

$$
U_l(z)V_{l+1}(z) - U_{l+1}(z)V_l(z) = C(z) \tag{1.5.20}
$$

Let us reformulate this relationship as a difference equation:

$$V_l V_{l+1}\left(\frac{U_l}{V_l}-\frac{U_{l+1}}{V_{l+1}}\right)=C \tag{1.5.21}$$

It is easy to check that:

$$\frac{U_l}{V_l}=\alpha+C\sum_{j=l}^{\infty}(V_j V_{j+1})^{-1} \text{ (where } \alpha \text{ is constant)} \tag{1.5.22}$$

is a solution of Eq.(1.5.21) in U_l/V_l provided V_l is a sequence that goes to infinity for large l. When α equals zero, U_l tends to zero as $l \to \infty$ and, as a consequence, U_l must be proportional to the Riccati–Bessel function $\psi_l(z)$:

$$\psi_l(z)=C_1(z)V_l\sum_{j=l}^{\infty}(V_j V_{j+1})^{-1} \tag{1.5.23}$$

If we rewrite Eq.(1.5.23) for the ratio $\psi_{l-1}(z)/\psi_l(z)$ the normalization constant $C_1(z)$ is dropped:

$$\frac{\psi_{l-1}(z)}{\psi_l(z)}=V_l^{-1}\left\{V_{l-1}+V_l^{-1}\left[\sum_{j=l}^{\infty}(V_j V_{j+1})^{-1}\right]^{-1}\right\} \tag{1.5.24}$$

Because of arbitrariness of the V_l sequence we can set $V_{L-1}=0$ and $V_L=1$ for simplicity. Then, finally, for the log-derivative $D_L(z)$ with the truncating number L we have:

$$D_L(z)=-\frac{L}{z}+\left[\sum_{j=L}^{\infty}(V_j V_{j+1})^{-1}\right]^{-1} \tag{1.5.25}$$

Cachorro and Salsedo (1991) pointed out that the convergence of the series in Eq.(1.5.25) is very fast when $L > \operatorname{Re} z$ (in the so-called exponential region). Truncation of this series at some $j = L^*$ is defined by criterion $\left|(V_M(z)V_{M+1})^{-1}\right| \leq \varepsilon$, where ε is some small quantity. It can be shown that if this criterion is met then the relative error in $\psi_{l-1}(z)/\psi_l(z)$ is less than $\varepsilon/Q_{\text{ext}}$. Consequently, the method of Cachorro and Salsedo (1991), contrary to Lentz's method, enables us to control the required precision of $D_L(z)$.

Similarly to Lentz's approach, this method meets a minor difficulty when V_j goes too near to zero for some j, but this difficulty is easily negociated. From recursion (1.5.19) it follows that:

$$\frac{1}{V_{j-1}V_j}+\frac{1}{V_j V_{j+1}}=\frac{(2j+1)}{zV_{j-1}V_{j+1}} \tag{1.5.26}$$

and, thus, if V_j is too small then two consequent terms of the sum in Eq.(1.5.25) are calculated with the aid of Eq.(1.5.26).

Both approaches described above follow the same principle and, therefore, have similar convergence properties.

1.5.2.2 Generating the ratio $\psi_l(x)/\xi_l(x)$

An almost obligatory element of the expressions for amplitude coefficients of a scattered field is the ratio of the Bessel function of real argument x to the Hankel function of the same argument:

$$\gamma_l = \psi_l(x)/\xi_l(x) \tag{1.5.27}$$

A frequently used method to compute the ratio γ_l consists of upward recursions for functions $\psi_l(x)$ and $\chi_l(x)$:

$$\begin{aligned} \psi_{l+1}(x) &= \frac{2l+1}{x}\psi_l(x) - \psi_{l-1}(x) \\ \chi_{l+1}(x) &= \frac{2l+1}{x}\chi_l(x) - \chi_{l-1}(x) \qquad l = 1, 2, ..., L-1 \end{aligned} \tag{1.5.28}$$

that begin with elementary functions:

$$\begin{aligned} \psi_0(x) &= \sin x \qquad \psi_1(x) = \frac{\sin x}{x} - \cos x \\ \chi_0(x) &= \cos x \qquad \chi_1(x) = \frac{\cos x}{x} + \sin x \end{aligned} \tag{1.5.29}$$

However, this approach faces two computational problems:

(1) While the upward recursion for $\chi_l(x)$ is completely correct, accuracy of analogous recursions for $\psi_l(x)$ is doubtful. This situation is similar to the situation with log-derivatives described in Section 1.5.2.1. Although the main difficulties of upward recursion arise for complex arguments, it is not safe to say that this recursion direction provides correct $\psi_l(x)$ for real arguments too.
(2) At rather great orders $l > x$, a sharp increase of $\chi_l(x)$ occurs (it is the so-called exponential regime). As a result, an overflow may occur while attempting to divide by the complex denominator in Eq.(1.5.27). This minor difficulty may be solved with some normalization, for example:

$$\frac{\psi_l(x)}{\xi_l(x)} = \frac{\psi_l(x)/\chi_l(x)}{\psi_l(x)/\chi_l(x) - i}$$

As already stated in Section 1.5.2.1, the ultimate way to suppress these instabilities lies in use of the log-derivatives:

$$D_l(x) = \frac{\psi_l'(x)}{\psi_l(x)}, \qquad G_l(x) = \frac{\xi_l'(x)}{\xi_l(x)} \tag{1.5.30}$$

Two different methods using log-derivatives to generate γ_l are discussed below. In the first, the modification of the Bessel functions that circumvents the first difficulty is introduced. The latter uses the upward recursion for the ratio γ_l itself.

Let us assume that we need to calculate the array of functions $\psi_l(z)$ for orders $l = 1, 2, ...L$. The argument z is complex for generality. On the first step we calculate $D_L(z)$ using the method of Section 1.5.2.1. The downward recursion for ratios $f_l(z) = \psi_{l-1}(z)/\psi_l(z)$ with initial term $f_L(z) = D_L(z) + L/z$:

$$f_l(z) = \frac{2l+1}{z} - \frac{1}{f_{l+1}(z)} \tag{1.5.31}$$

immediately follows from Eqs.(1.5.27–28). If all entries of array $D_l(z)$ ($l = L, L-1, ..., 1$) are known beforehand (as expected because these log-derivatives enter the expressions for amplitude coefficients) then we can directly put:

$$f_l(z) = D_l(z) + \frac{l}{z} \quad (l = L, L-1, ..., 1) \tag{1.5.32}$$

instead of using downward recursion (1.5.31). The value of $\psi_0(z)$ is known in advance. Thus, we find $\psi_1(z) = \psi_0(z)/f_1(z)$ using this value and go further via upward recursion:

$$\psi_l(z) = \frac{\psi_{l-1}(z)}{f_l(z)} \quad (l = 2, 3, ..., L) \tag{1.5.33}$$

Dividing ordinary expressions for the derivatives of Riccati–Bessel and Riccati–Hankel functions:

$$z\psi_{l+1}(z) = (l+1)\,\psi_l(z) - z\psi_l'(z) \tag{1.5.34}$$
$$z\eta_{l+1}(z) = (l+1)\,\eta_l(z) - z\eta_l'(z) \qquad z\xi_{l+1}(z) = (l+1)\,\xi_l(z) - z\xi_l'(z)$$

term by term, we obtain the upward recursion for ratio:

$$\gamma_{l+1} = \gamma_l \frac{zD_l(z) - (l+1)}{zG_l(z) - (l+1)} \tag{1.5.35}$$

The recursion (1.5.35) has a minor drawback. For the asymptotic value of $D_l(z)$ at $l \geq z$ equals $(\,l+1)\,/z$, the subtraction of close values takes place in the numerator of this recursion at sufficiently large orders l, which could cause some loss of precision. This difficulty is eliminated using the recursions for log-derivatives. Equations (1.5.35) may be rewritten in a form more suitable for calculations:

$$\gamma_{l+1} = \gamma_l \frac{zG_{l+1}(z) + (l+1)}{zD_{l+1}(z) + (l+1)} \tag{1.5.36}$$

The initial term for this recursion is:

$$\gamma_0 = \frac{\sin z}{\sin z \mp i \cos z} \tag{1.5.37}$$

The arrays of functions $D_l(z)$ and $G_l(z)$ entering Eq.(1.5.36) are calculated in an ordinary way:

a) the algorithm of generating $D_l(z)$ is described in Section 1.5.2.1;

b) the array $G_l(z)$ is calculated via upward recursion:

$$G_{l+1}(z) = -\frac{l+1}{z} - \frac{1}{G_l(z) - (l+1)/z} \tag{1.5.38}$$

with initial term $G_0(z) = i$.

1.5.2.3 Generating the combinations of Riccati–Bessel functions of different complex arguments

Many light scattering problems (in particular, problems connected to calculation of internal fields and scattering by multi-layered particles) require computation of the following arrays of cylindrical function combinations (for simplicity we use arguments as the upper indexes in this subsection):

$$R_l^{yz} = \frac{\psi_l(y)}{\psi_l(z)} \qquad S_l^{yz} = \xi_l(z)\psi_l(y) \qquad T_l^{yy} = \xi_l(y)\psi_l(y) \tag{1.5.39}$$
$$l = 1, 2, \ldots \qquad y = Nx_1 \qquad z = Nx_2 \qquad x_2 \geq x_1$$

These combinations can be computed through the relatively simple upward recursions (Toon and Ackerman, 1981).

1. Begin with array R_l^{yz}. Since $\psi_l^u = l\psi_{l-1}^u/u - \psi_{l-1}^{\prime\,u}$ we obtain:

$$R_l^{yz} = \frac{\psi_{l-1}^y}{\psi_{l-1}^z}\frac{l - yD_{l-1}^y}{l - zD_{l-1}^z} \tag{1.5.40}$$

As in Section 1.5.2.2, such an expression is inconvenient due to the subtraction of close values. As $D_{l-1}^U = l/U - U/\left(l + UD_l^U\right)$, we have:

$$R_l^{yz} = R_{l-1}^{yz} \cdot \frac{l + zD_l^z}{l + zD_l^y} \cdot \frac{x_1}{x_2} \tag{1.5.41}$$

in place of (1.5.40). This recursion may be computed from bottom to top since relative error does not accumulate in the process. Assuming that calculation of R_{l-1}^{yz} brings in the roundoff error ε_{l-1} we have:

$$\varepsilon_l + R_l^{yz} = \left[\varepsilon_{l-1} + R_{l-1}^{yz}\right] \cdot \frac{l + zD_l^z}{l + zD_l^y} \cdot \frac{x_1}{x_2}$$

Using Eq.(1.5.41), we have for ε_l:

$$\varepsilon_l = \varepsilon_{l-1}\frac{l + zD_l^z}{l + zD_l^y} \cdot \frac{x_1}{x_2} \tag{1.5.42}$$

That is, the error is increased proportionally to the function itself and, consequently, the relative error is not changed during the recursion (1.5.41). The initial term of the recursion is:

$$R_0^{yz} = \frac{\sin(Nx_1)}{\sin(Nx_2)} \tag{1.5.43}$$

Despite its simplicity this relation contains the computational difficulty. It is able to underflow at rather large $\kappa(x_2 - x_1)$. This underflow (due to the multiplicative nature of recursion (1.5.41)) may cause zeroing of all other terms R_l^{yz}. Therefore, there is a reason to normalize:

$$\tilde{R}_l^{yz} = R_l^{yz} \exp\left[\kappa(x_2 - x_1)\right] \tag{1.5.44}$$

Now $\tilde{R}_0^{yz}$ can be written as:

$$\tilde{R}_0^{yz} = \frac{B \sin\alpha - iC\cos\alpha}{U\sin\beta - iQ\cos\beta} \tag{1.5.45}$$

where:

$$\left.\begin{matrix} B \\ C \end{matrix}\right\} = \exp(-2\kappa x_1) \pm 1 \qquad \left.\begin{matrix} U \\ Q \end{matrix}\right\} = \exp(-2\kappa x_2) \pm 1 \qquad \alpha = nx_1 \qquad \beta = nx_2$$

Evidently, the recursion (1.5.41) is the same for normalized functions.

2. Consider functions S_l^{yz} now.

Multiply $\psi_{l-1}^y = (\psi_l^y)' + l\psi_l^y/y$ by $\xi_{l-1}^z = (\xi_l^z)' + l\xi_l^z/z$. Then:

$$\psi_{l-1}^y \xi_{l-1}^z = \psi_l^y \xi_l^{z\,(1)} \left[{}^{(1)}G_l^z + \frac{l}{z}\right]\left[D_l^y + \frac{l}{y}\right]$$

that is

$$S_{l-1}^{yz} = S_l^{yz} \cdot \frac{yz}{\left[G_l^z z + l\right]\left[D_l^y y + l\right]} \tag{1.5.46}$$

with initial term:

$$\begin{aligned} S_0^{yz} = \frac{1}{2}\exp\left[-\kappa(x_2 - x_1)\right]\left[(\cos\alpha - f\cos\beta)\right. \\ \left. + i\left(\sin\alpha - f\sin\beta\right)\right] \end{aligned} \tag{1.5.47}$$

where

$$\alpha = n(x_2 - x_1) \qquad \beta = n(x_2 + x_1) \qquad f = \exp(-2\kappa x_1)$$

For the same reasons that were outlined above for R_l^{yz} the relative error is not accumulated in the course of upward recursion (1.5.46) for functions S_l^{yz}, and their normalization should be used:

$$\tilde{S}_l^{yz} = S_l^{yz} \exp\left[\kappa(x_2 - x_1)\right] \tag{1.5.48}$$

Then:

$$\tilde{S}_0^{yz} = \frac{1}{2}\left[(\cos\alpha - f\cos\beta) + i\,(\sin\alpha - f\sin\beta)\right] \tag{1.5.49}$$

and the form of recursion (1.5.46) is completely retained.

3. Finally, consider the array of functions T_l^{yy}. Upward recursion for T_l^{yy} is constructed similarly to the recursion for S_l^{yz}:

$$T_{l-1}^{yy} = T_l^{yy} \cdot \frac{y^2}{[G_l^y y + l]\,[D_l^y y + l]} \tag{1.5.50}$$

$$T_0^{yy} = \frac{1}{2} + f\left[\sin^2\gamma - 0.5 - \frac{i}{2}\sin 2\gamma\right] \quad \text{where } \gamma = nx_1$$

1.5.3 Calculation of angular functions

To calculate angular dependencies of light scattering it is necessary to generate the arrays of angular functions of the spherical angle θ:

$$\pi_l^{(m)}(\cos\theta) = \frac{P_l^{(m)}(\cos\theta)}{\sin\theta} = P_l^{(m)}(\mu)(1-\mu^2)^{-1/2} \qquad \mu = \cos\theta$$

$$\tau_l^{(m)}(\cos\theta) = \frac{dP_l^{(m)}(\cos\theta)}{d\theta} = -\frac{dP_l^{(m)}(\mu)}{d\mu}(1-\mu^2)^{1/2} \tag{1.5.51}$$

$$m = 0, 1, 2, \ldots \qquad l = m, m+1, \ldots$$

The useful analytical relations for these functions are listed in detail in Appendix B.

In problems related to spherical symmetry, angular functions with $m = 1$ only appear (these functions are designated as τ_l and π_l). In this case there are no computational difficulties. The upward recursion:

$$\pi_l(\mu) = \frac{1}{l-1}\left[\mu(2l-1)\pi_{l-1}(\mu) - l\pi_{l-2}(\mu)\right] \tag{1.5.52}$$

with initial values $\pi_1 = 1$ and $\pi_2 = 3\mu$ is very stable. The τ-functions are evaluated using already calculated π-functions:

$$\tau_l(\mu) = \frac{l^2}{2l+1}\pi_{l+1}(\mu) - \frac{(l+1)^2}{2l+1}\pi_{l-1}(\mu) \tag{1.5.53}$$

Here is a slightly modified version of these recursions (Wiscombe, 1980)

$$q_l = \mu\pi_l \qquad t_l = q_l - \pi_{l-1} \tag{1.5.54}$$

$$\tau_l = lt_l - \pi_{l-1} \qquad \pi_{l+1} = q_l + \frac{l+1}{l}t_l$$

which is used in our software.

However, the angular functions with azimuthal modes $m = 1$ are sufficient only to those scattering problems with spherical symmetry. Although for the most part this book deals with scattering by spherically symmetrical objects, the anisotropic particles (Chapters 9 and 10) do not have this symmetry. For these cases, as for many other light scattering problems, the associated angular functions $\pi_l^{(m)}(\mu)$ and $\tau_l^{(m)}(\mu)$ $(m = 0, 1, 2, ...)$ must be calculated.

These functions increase sharply with l and m. For example, from Eq.(B.16) it follows that $\pi_l^{(l)}(0) = (2l-1)!!$. The normalized functions:

$$\tilde{\pi}_l^{(m)}(\mu) = \frac{1}{\alpha_{lm}} \pi_l^{(m)}(\mu) \quad \tilde{\tau}_l^{(m)}(\mu) = \frac{1}{\alpha_{lm}} \tau_l^{(m)}(\mu) \tag{1.5.55}$$

where the normalizing coefficient $\alpha_{lm} = [(2l+1)(l+m)!/(l-m)!]^{1/2}$, would suffice in this situation. These functions increase with l and m is rather slow. Clearly:

$$\tilde{\pi}_l^{(l)}(0) = (2l-1)!! \left[\frac{(2l+1)}{(2l)!} \right]^{1/2} < \left(l + \frac{1}{2} \right)^{1/2}$$

that is, the normalized function $\tilde{\pi}_l^{(l)}(0)$ is roughly proportional to $l^{1/2}$.

Calculation of $\tilde{\pi}_l^{(m)}(\mu)$ may be done with the help of an upward scheme. The initial values $\tilde{\pi}_1^{(0)}(\mu) = \mu\sqrt{3}/A$ and $\tilde{\pi}_2^{(0)}(\mu) = \sqrt{5}(3\mu^2 - 1)/2A$, where $A = (1-\mu^2)^{1/2}$ are used in the recursion for $m = 0$:

$$\tilde{\pi}_l^{(0)}(\mu) = \frac{\sqrt{4l^2 - 1}}{l} \mu \tilde{\pi}_{l-1}^{(0)}(\mu) - \frac{l-1}{l} \sqrt{\frac{2l+1}{2l-3}} \tilde{\pi}_{l-2}^{(0)}(\mu) \quad l = 3, 4, ... \tag{1.5.56}$$

while initial values $\tilde{\pi}_1^{(1)}(\mu) = \sqrt{3/2}$, $\tilde{\pi}_2^{(1)}(\mu) = \mu\sqrt{15/2}$, $\tilde{\pi}_2^{(2)}(\mu) = A\sqrt{15/8}$, and values calculated by Eq.(1.5.56) are used in the recursion for $m = 1, 2, 3, ...$ and $l = 3, 4, ...$:

$$\tilde{\pi}_l^{(m)}(\mu) = \frac{1}{A\sqrt{l+m}} \sqrt{\frac{(2l+1)(l+m-1)}{2l-1}} \tilde{\pi}_{l-1}^{(m-1)}(\mu) - \mu\sqrt{l-m+1}\, \tilde{\pi}_l^{(m-1)}(\mu) \tag{1.5.57}$$

The functions $\tilde{\tau}_l^{(m)}(\mu)$ are calculated using the values of $\tilde{\pi}_l^{(m)}(\mu)$ and initial values $\tilde{\tau}_1^{(0)}(\mu) = -A\sqrt{3}$ and $\tilde{\tau}_1^{(1)}(\mu) = \mu\sqrt{3/2}$ via the relation:

$$\tilde{\tau}_l^{(m)}(\mu) = l\mu\tilde{\pi}_l^{(m)}(\mu) - \sqrt{\frac{(2l+1)(l+m)(l-m)}{2l-1}} \tilde{\pi}_{l-1}^{(m)}(\mu) \tag{1.5.58}$$

The factor A in the denominator of Eq.(1.5.57) prohibits of using this scheme at the ends of the variable interval $\mu = \pm 1$. In this case the relations (B.12–13) may be used which after the normalization take the form:

$$\begin{aligned}\tilde{\pi}_l^{(m)}(1) = \tilde{\tau}_l^{(m)}(1) &= \frac{\delta_{m1}}{2}\sqrt{(2l+1)l(l+1)} \\ \tilde{\pi}_l^{(m)}(-1) &= \frac{\delta_{m1}}{2}(-1)^{l+1}\sqrt{(2l+1)l(l+1)} \\ \tilde{\tau}_l^{(m)}(-1) &= \frac{\delta_{m1}}{2}(-1)^{l}\sqrt{(2l+1)l(l+1)}\end{aligned} \tag{1.5.59}$$

Note that Eq.(1.5.59) assumes that $m \neq 0$. If $m = 0$ then $\tilde{\pi}_l^{(0)}(\pm 1) \to \infty$ and $\tilde{\tau}_l^{(0)}(\pm 1) = 0$. The singularity of $\tilde{\pi}_l^{(0)}(\pm 1)$ is usually of no concern because this function appears in formulae in the non-singular combination $A\tilde{\pi}_l^{(0)}(\mu)$.

This was a brief overview of the algorithms for calculation of Bessel and angular functions. The specific details of calculations are considered in Chapters 3–8 devoted to scattering on concrete types of particles.

In summary we shall stress that literature on procedures for calculation of special functions related to light scattering problems is rather extensive. It can be explained, on the one hand, by great practical significance of these problems. On the other hand, the experience revealed that the numerical implementation of even elementary problems of this sort contains many "underwater stones" that do not allow creation of algorithms that work effectively for all possible values of parameters. Therefore, authors consider it to be expedient to attract readers' attention to some works on numerical realization of light scattering problems for homogeneous spheres (Baibulatov *et al.*, 1971; Ross, 1972; Naumenko *et al.*, 1975; Greehan and Gouesbet, 1979; Kopilovich, 1980; Felske *et al.*, 1983; Jones, 1983; Meeten, 1984; Akulinin, 1988; Shigapov, 1990; Barber and Hill, 1990; Belov, 1992; Ludlow and Everitt, 1996; Shybanov and Haltrin, 2002) and coated spheres (Krekov and Rakhimov, 1973, 1975*a*; Kaiser and Schweiger, 1993).

2

Model concepts in aerosol optics

McCartney wrote in his well known monograph (1977), "As earth dwellers we are immersed in an ocean of atmosphere several hundred miles deep, and we spend our lives at these depths. The light bringing us information from a distance, illuminating our surroundings, and bearing energy to make our world habitable, performs these functions only by travelling in this atmospheric ocean." Having penetrated into the atmosphere, the light participates in many processes and undergoes different changes. Study, interpretation, and practical use of these processes and changes are frequently impossible without the decomposition of them into basic constituents and applying to a greater or lesser extent simplified idealized models to study them. The mathematical modeling of physical and many other processes is one of powerful means to conceive the world surrounding us.

To study the processes of the electromagnetic radiation propagation in the atmosphere, the radiation balance of our planet's atmosphere, and requirements of weather and climate atmospheric models are being developed. The great progress of optical and laser systems for location, communication, aerohydro- and space-navigation requires the development of models of radiation interaction with gas and aerosol media. The presence of vertical stratification of atmospheric aerosol optical parameters demands elaboration of the vertical aerosol model. The essential distinctions of physical, physico-chemical and optical properties of aerosol masses of natural and anthropogenetic origin, and soilerosive and sea salt aerosols enforce carrying out complex international experiments aimed at creation of a global aerosol model. The different models are being developed to describe the physical properties and size distribution of aerosol particles, to describe and interpret spectral dependencies of aerosol substance optical constants, etc.

To investigate elementary acts of electromagnetic radiation interaction with single particles of aerosol, models describing particle shape, typical size, chemical composition, internal structure, and other properties are also used.

We begin this chapter with a brief analysis of the basic physical and physico-chemical characteristics of atmospheric aerosols, the processes that define their microstructure, and properties of particles comprising it.

2.1 GENERAL INFORMATION ON ATMOSPHERIC AEROSOLS

There are many articles and books on aerosols in general and on atmospheric aerosol in particular published in recent years. The physical and optical properties of aerosols are considered from different points of view and for different purposes in some reviews and books. In the abovementioned monograph (McCartney, 1977) there is a reference list of 23 books where problems of light scattering in the atmosphere are considered. In the majority of them rather extended sections are devoted to atmospheric aerosols. While not pretending to complete this list of literature on atmospheric aerosol, we woulf add works (Selezneva, 1966; Rozenberg, 1968; Green and Lane, 1972; Sedunov, 1972; Kondratiev *et al.,* 1973; Cadle, 1976; Hanel, 1976; Kondratiev, 1978; Emilenko and Tolstobrov, 1981; Kondratiev and Pozdnyakov, 1981; Krekov and Rakhimov, 1982; Ivlev and Andreev, 1986; d'Almeida *et al.*, 1991; Liou, 1992; Stephens, 1994; Kaufman *et al.*, 1997; Raes and Bates, 2000; Kokhanovsky, 2001; Mishchenko *et al.*, 2002*b*). These works contain a considerable amount of data about physical, physico-chemical, and optical properties of atmospheric aerosols that is necessary for further presentation.

Generally speaking, the concept of an atmospheric aerosol is quite broad and includes the diversity of its forms and varieties. It is possible to classify aerosol in different ways by chemical composition, origin, particle size, etc. Classification of aerosol forms based on types of optical weather (Rozenberg, 1968; Zuev, 1970) has received recognition. If we leave aside different types of precipitation (snow, rain and the like) then, according to the classification of Rozenberg (1968), it is possible to distinguish five qualitatively different types of optical weather:

- *smog* results from introduction of alien impurities into the atmosphere: dust clouds, smoke from wood fires or from industrial objects, products of volcanic activity, and so on;
- *haze* exists as the so-called pre-condensation state of air and represents the result of growth of the smallest solid particles of the disperse phase with radii of $R < 0.1\mu m$ (Aitken cores) in the presence of atmospheric moisture;
- *foggy haze* is formed in the equilibrium solution state of the atmosphere by condensation of moisture on huge ($0.1\mu m < R < 1\mu m$) and giant ($R > 1\mu m$) particles of soluble salts;
- *fog and clouds* emerge in a condensation state of the atmosphere as a result of transition of excessive vaporous moisture to a liquid-droplet phase; and
- *haze with drizzle* is the heterogeneous formation representing the usual haze being pierced by drizzle, which falls out from a layer above.

In the literature the term "atmospheric aerosol" is very frequently identified with concept of "atmospheric haze". This term will be used further in this book only in this sense. As the size and composition of particles vary essentially during condensation, sublimation, or evaporation of atmospheric water, it is necessary to distinguish the part of the solid fraction of the disperse phase, capable of presenting itself as condensation cores (particles from water-soluble substances and so-called mixed particles that contain both non-soluble and soluble substances), from the part which under usual conditions of average relative.

humidities could not be represented as condensation cores (insoluble substances) (Rozenberg, 1968; Sedunov, 1972).

The sources of particles in atmospheric aerosols are quite different. The problem of an origin of aerosol particles is comprehensively studied in works (Selezneva, 1966; Kondratiev *et al.*, 1973; Cadle, 1976; McCartney, 1977). Here we mention only two basic components of continental aerosols in the low atmospheric layer. One of these components has its origin in soil with all the consequences regarding its chemical composition, physical and optical properties to follow. The second component is the hygroscopic, so-called sulphate aerosol of natural and anthropogenic origin. A role is also played by aerosols of organic origin but information on this is still scarce and fragmentary (Kondratiev *et al.*, 1973).

In sea aerosols, condensation cores of sea salt predominate. In aerosols above continents, travelling a long way above the land, we find that the cores are mixing, interacting, and participating in different processes and reactions. As a result, even the smallest particles are composite conglomerates comprised of sets of substances. The physico-chemical properties of these conglomerates define a particular hierarchy of particles, according to which particles sequentially become condensation cores and participate in formation of the liquid-droplet phase in the atmosphere (Sedunov, 1972).

The world of small particles is characterized by very small masses and relatively large surfaces. The greater the ratio of surface area to mass, the greater the influence of environment on a particle substance. Small hygroscopic particles (condensation cores) quickly react to relative humidity changes, absorbing and losing moisture nearly as quickly as the velocity relative humidity changes (McCartney, 1977). Depending on changes in relative humidity the particles can become larger or smaller in size according to the quantity of water absorbed from the air.

Along with the processes that stipulate the creation of atmospheric particles, processes of coagulation, dropout, and washing away are also taking place. As a result of common activity of all these processes aerosol particle systems in the majority of situations are in a state of dynamic equilibrium with the surrounding medium. This explains the rather significant variations in general characteristics of this system in space and time. The detailed description of problems mentioned in this section can be found in works (Sedunov, 1972; Kondratiev *et al.*, 1973; Cadle, 1976; McCartney, 1977; Kondratiev and Pozdnyakov, 1981; Krekov and Rahimov, 1982).

2.2 MICROPHYSICAL MODEL OF AEROSOLS

The atmospheric aerosol represents a multitude of solid or liquid particles with sizes from aggregates of molecules ($R \sim 10^{-3}$ μm) up to particles with radii close to $R \sim 100$ μm (Selezneva, 1966; Zuev, 1970; McCartney, 1977). A rather important characteristic of an aerosol is particle size distribution. The experimental data on particle size is commonly represented as histograms showing how many particles there are in each chosen size class. If size classes are narrow and the number of particles is great it is possible to construct a smooth distribution function based on the histogram. Usually, such a function characterizes a specific situation.

To solve theoretical and applied problems it is necessary to have some analytical expression which approximates the experimental curves of aerosol particle size distribution. Because of the great complexity of aerosol formation processes obtaining such analytical distribution functions on the basis of the theoretical backgrounds is impossible and, therefore, the empirical formulae are used.

In many works (Junge, 1955, 1963; Bullrich, 1964; Pasceri and Fridlander, 1965; Clark and Whitby, 1967; Pueschel and Noll, 1967; Blifford and Ringer, 1969; Yamamoto and Tanaka, 1969; Gilette and Blifford, 1972) the negative power distribution was obtained in the processing of observed data that was called the Junge type distribution:

$$f(R) = dp(R)/dR \sim R^{-\alpha} \tag{2.2.1}$$

where $dp(R)$ is the number of particles with radii in the interval R to $R + dR$ per unit volume of medium and α is the distribution parameter. For practical purposes it is convenient to represent the distribution (2.2.1) in a logarithmic form:

$$f(R) = dp(R)/d(\lg R) = cR^{-\nu} \tag{2.2.1a}$$

It is easy to show that the parameter $\nu = \alpha - 1$ characterizes a ratio of quantities of large to small particles and that the normalizing factor c is expressed as:

$$c = \nu(R_{\text{min}} R_{\text{max}})^{\nu} \left(R_{\text{max}}^{\nu} - R_{\text{min}}^{\nu}\right)^{-1} \tag{2.2.2}$$

where R_{min} and R_{max} are the minimal and maximal radii of particles in the system, the particles are assumed spherical. In a series of later works it was established that specific values of the exponent ν could not always satisfactorily describe all spectra of observed particles, especially for higher R values. So, the distribution functions written in the form of some combinations of power distributions were offered (Reagan *et al.*, 1972; Whitby *et al.* 1972; Davies, 1974; Jaenicke and Davies, 1976).

To describe the particle size distribution of atmospheric aerosols many authors use the generalized or, modified, four-parametric gamma-distribution:

$$f(R) = aR^{\mu} \exp\left(-bR^{\gamma}\right) \tag{2.2.3}$$

where a, b, μ, and γ are distribution parameters. If $\gamma = 1$ then $b = \mu/R_0$, where R_0 is modal or, the most probable, radius of particle, μ is the parameter that defines the distribution half-width. The greatest popularity was gained by the haze models of Deirmendjian (1969) (parameters are given for particle concentration $C_N = 100\ \text{cm}^{-3}$):

- haze M ($a = 5.3333 \cdot 10^4$; $b = 8.9443$; $\mu = 1.0$; $\gamma = 0.5$) that describes the particle size distribution in sea or coastal aerosols;
- haze L ($a = 4.9757 \cdot 10^6$; $b = 15.1186$; $\mu = 2.0$; $\gamma = 0.5$) that represents a particle size spectrum in continental type aerosols; and
- haze H ($a = 4 \cdot 10^5$; $b = 20.0$; $\mu = 2.0$; $\gamma = 1$) that is used to describe a particle size distribution for high-altitude and stratospheric aerosols as well as submicron particles in dust layers.

Smirnov (1973) suggested using distribution (2.2.3) with negative μ for the case of coarse aerosols. At great R such a distribution is adequate to power one (2.2.1) and satisfactorily describes the distribution for the fraction of large hygroscopic condensation cores.

Many authors (Blifford and Gilette, 1971; Whitby *et al.*, 1972; Davies, 1974; Hidy *et al.*, 1974; Patterson and Gilette, 1977; Whitby, 1981) claim that in the presence of local aerosol sources it is necessary to use a normal distribution:

$$f(R) = \frac{C_N}{\sqrt{2\pi} R \ln \sigma} \exp\left[-\frac{1}{2} \left(\frac{\ln(R/R_m)}{\ln \sigma} \right)^2 \right] \tag{2.2.4}$$

where R_m is the distribution median radius and $\ln \sigma$ is the standard deviation. More detailed considerations of this problem of possible types of distributions of atmospheric aerosol particle sizes are included in books (McCartney, 1977; Krekov and Rakhimov, 1982).

The measurements carried out under different conditions of dust content in the lower atmospheric layer show that the continental atmospheric aerosol can be presented as consisting of fractions, each being characterized by normal particle size distribution (Patterson, 1977; Whitby, 1978; Kondratiev and Pozdnyakov, 1981; Krekov and Rakhimov, 1982). It is most clearly seen if we build volume $dV/d(\ln R)$ or surface $dS/d(\ln R)$ distributions instead of number distributions (Whitby, 1978). For high dust content in the atmosphere there are two following fractions or, modes: A $(1\ \mu m < R < 10\ \mu m)$; B $(10\ \mu m < R < 50\ \mu m)$. For low dust content mode A and mode C $(0.02\ \mu m < R < 0.5\ \mu m)$ (Patterson, 1977; Kondratiev and Pozdnyakov, 1981) were detected. Modes A and B have a soil origin. The mode A aerosols consist mainly of particles of clay composition. The mode B particles are mainly quartz grains with surfaces covered by small particles of clay. Mode C is formed as result of different physico-chemical processes in the atmosphere, in which the sulfur compounds play a considerable role. For this reason the mode C is sometimes called a sulphate aerosol.

A presence of mode B characterizes rather higher dust content that usually found near an aerosol source (Patterson, 1977; Kondratiev and Pozdnyakov, 1981), whereas mode A is observed under any dust content. The C mode is present at high transparencies of the atmosphere.

The study of physico-chemical processes of forming aerosols *in situ* shows that submicron particles have a bimodal distribution of surface sizes (Whitby, 1978; Kondratiev and Pozdnyakov, 1981). The so-called Aitken cores with radii from 0.015 μm up to 0.04 μm match the first mode here and the particles of so-called accumulative fraction with radii from 0.04 μm up to 1–1.5 μm match the second one.

The analysis of numerous experimental tropospheric aerosol distributions of different origins demonstrates that submicron particles exist (and behave) independently of large particles (with diameters more than 1.5–2 μm).

In some real situations the number particle distribution for accumulative and coarse dispersion modes can be satisfactorily approximated by the power Junge type distribution. The statistical processing of experimental data shows that on average the number particle size distribution of Junge type is representative enough for continental hazes of different origin and composition, though in each particular case the particle size distribution can somehow differ from the power law. The best match is observed for dry- and low-humidified hazes (at relative humidities $f < 0.8$ and particle mean refractive index $n \geq 1.4$) (McCartney, 1977).

It follows from analytical considerations of Davies (1974) that the power law distribution can be used with adequate accuracy in the interval $0.1\ \mu m \leq R \leq 1\ \mu m$. Clark and Whitby (1967), on the basis of analysis of large experimental material came to a deduction that the Junge type distribution can be applied to the interval $0.03\ \mu m \leq R \leq 3\ \mu m$.

It is shown by Foitzic (1965) that the power law distribution of particle sizes can be represented, if necessary, by a sum of several logarithmically normal distributions. This problem is studied in detail in the book by Emilenko and Tolstobrov (1981).

Our main goal is to reveal some changes in optical properties of atmospheric aerosol pertaining to the presence of non-uniform or anisotropic particles rather than to investigate in detail the optical properties of aerosols. Because of this, the power Junge type distribution, which is simple and contains small number of parameters, will be used further in this book as a typical distribution for atmospheric hazes. Its results are easily worked upon. In studies of polydisperse systems of dust particles (the coarse fraction with predominance of crystalline anisotropic particles) the gamma particle size distribution will be used.

2.3 COMPLEX REFRACTIVE INDEX OF AEROSOL PARTICLES

The atmospheric aerosol is a collection of particles with complex mineralogical and microelement composition. Therefore, the aerosol substance optical constants can be determined by averaging the values of real (n) and imaginary (κ) parts of the complex refractive index obtained in different places or, otherwise, by synthesizing dependencies $n(\lambda)$ and $\kappa(\lambda)$ on the basis of studying spectral dependencies of relevant characteristics of primary components that comprise an aerosol of given type, with regard to percentage concentration of each component. However, as it was pointed out by Kondratiev and Pozdnyakov (1981), both methods need a rather cautious approach in order to describe adequately the actual aerosol. This caution is necessary, first, because aerosol optical constants are not always linear functions of optical constants of their components and, second, because a retrieval of optical constants should be done taking into account size distributions of chemical components, instead of with simple addition of percentage averaged chemical composition.

Krekov and Rakhimov (1982) give a list of aerosol particle optical constants obtained by different methods at different levels of relative humidity in different regions of the globe. It can be concluded on the basis of the works (Pollack *et al.*, 1973; Grams *et al.*, 1974; Kondratiev and Ter-Markaryants, 1976; Patterson *et al.*, 1977) that the refractive index value for soil-erosive aerosol in the visible spectral range varies between the limits $n = 1.50$–1.56. The analysis of optical properties of basic components of atmospheric aerosols allowed Bullrich (1964) to assume that the most probable value of n is 1.55. Other authors give similar values.

Keep in mind, however, that the determination of optical constants of atmospheric aerosol substances was done in most cases at some non-zero value of air relative humidity ($f \sim 0.4$–0.8). The relative humidity affects both the particle size and structure, as well as its refractive index. Hänel (1968) measured the real part of the mean refractive index and mean density of collected probes of atmospheric aerosol particles at $T = 296K$, $\lambda = 0.589\mu m$, and air relative humidity $f = 0.40 \pm 0.05$. He received values of $n = 1.54$–1.58 for most samples, density varied from $\rho = (1,80 \pm 0.4)\ g \cdot cm^{-3}$ up to $\rho = (3.4 \pm 1)\ g \cdot cm^{-3}$. For two other samples, $n = 1.62 \pm 0.035$, $\rho = (3.3 \pm 0.8)\ g \cdot cm^{-3}$ and $n = 1.63 \pm 0.02$, $\rho = (3.4 \pm 0.7)\ g \cdot cm^{-3}$. Because of the influence of relative humidity, the absolutely dry ($f = 0$) samples should have had values of $n = 1.71$–1.72 and $\rho \approx 4\ g \cdot cm^{-3}$ in these cases. In another of Hänel's papers (1971) there is the dependence $n(f)$ which gives $n \approx 1.57$ at $f = 0.4$, and $n \approx 1.62$ at $f = 0$. In his review (1976) Hänel gives equations that permit us to find n for dry particle substances (at $f = 0$); in his judgement, n can accept values from 1.5 up to 1.7. In the work of Hänel and Bullrich (1978) a table of averaged data on spectral dependence of the complex refractive index $n + i\kappa$ for dry particles of four aerosol mass types can be found. These types are: sea aerosol (model *A*), mixed sea particles with substances from deserted areas (model *B*), aerosol particles from industrial regions (model *C*) and particles of continental background aerosol (model *D*). For the above models, the real and imaginary parts of the dry particle complex refractive index equal: $n_A = 1.55$, $\kappa_A = 0.003$; $n_B = 1.56$, $\kappa_B = 0.011$; $n_C = 1.55$, $\kappa_C = 0.05$; $n_D = 1.51$, $\kappa_D = 0.015$ for wavelength $\lambda = 0.55\ \mu m$. For all four cases n does not exceed 1.56. Hoidale and Smith (1968) showed that n lays in the range from $n = 1.5$ to 1.6 for a series of compounds detected in aerosol particles.

Many authors (Grams *et al.*, 1972; Volz, 1972; Ward *et al.*, 1973; Bhardwaja *et al.*, 1974; Weiser *et al.*, 1977; Binenko 1980; Mathai and Harrison, 1980; Reagan *et al.*, 1980) recommended using $n = 1.50$–1.55, $\kappa = 0,005$ (soluble particles), and $\kappa = 0.03$ (dust) for the visible spectral range. Note that although n doesn't vary much, remaining close to mean value $\bar{n} = 1.55$ in most cases, the imaginary part depends rather essentially on the composition of the soil-derived aerosol which is shown both by rather considerable variations of spectral dependency $\kappa(\lambda)$ and by dependence of κ on particle size. This was obviously demonstrated by the experimental data of Lindberg and Gillespie (1977). The composition and the dependency $\kappa(\lambda)$ for eight fractions of aerosol particles obtained with the help of a cascade impactor was analysed in this work. The results of the analysis are given in Tables 2.1 and 2.2. Densities of samples varied from 2 g·cm^{-3} for the 6-th and 7-th fractions up to 2.4 g·cm^{-3} for the first fraction. The numbers in the bottom line of Table 2.2 specify the ranges of particle diameters for each fraction in micrometers. As follows from Table 2.1, the submicron particles consist basically of ammonium sulphate and strongly absorbing carbon, whereas the particles with a size more than several micrometers consist primarily of low-absorbing minerals of soil origin. This is confirmed by the data in Table 2.2.

The other way to create the model of atmospheric aerosol optical constants mentioned in the beginning of this section was developed in works (Ivlev and Popova, 1972, 1973; Kondratiev *et al.*, 1973). It is founded on a guess that elements and minerals in a particle (including water) are mixed ideally. Given aerosol chemical composition and optical con-

Table 2.1. Optically significant components of fractionated dust samples (Lindberg and Gillespie, 1977). The "+" sign specifies substance presence, the "-"sign specifies, that the substance was found, but only in small concentrations.

	Fraction number							
Component	7	6	5	4	3	2	1	0
Clay minerals (montmorillonite, illite, kaolinites)	–	–	+	+	+	+	+	+
Quartz		–	+	+	+	+	+	+
Calcite		–	–	+	+	+	+	+
Gypsum			+	+	–			
Ammonium sulfate	+	+	–					
Carbon	+	+	–					

stant spectral dependencies for particle components, it is possible to find the corresponding optical characteristics of a mixture with calculation. The effective values $n(\lambda)$ and $\kappa(\lambda)$, calculated by the Kramers–Kronig method for the continental aerosol of medial latitudes, are shown in Table 2.3 (Ivlev and Popova, 1973).

As noted in Krekov and Rakhimov (1982), obtained values of n for visual and near-IR spectral ranges seem to be somewhat overestimated. It may be explained by the considerable part played in this model of the aerosol particles with the enhanced content of haematite(Fe_2O_3) coming from deserts and semi-deserts (Kondratiev *et al.*, 1973; Kondratiev and Ter-Markariants, 1976). Haematite has $n \approx 3.014$ and $0.0135 \leq \kappa \leq 0.0165$

Table 2.2. An imaginary part of the refractive index for fractioned samples of a dust (Lindberg and Gillespie, 1977). Asterisk specifies $\kappa < 0.005$.

	Fraction number							
$\lambda(\mu m)$	7	6	5	4	3	2	1	0
0.3	0.11	0.18	0.25	0.18	0.14	0.14	0.13	0.06
0.4	0.11	0.15	0.15	0.10	0.09	0.09	0.08	0.05
0.5	0.11	0.14	0.09	0.06	0.05	0.05	0.04	0.02
0.6	0.12	0.14	0.07	0.03	0.02	0.02	0.02	0.01
0.7	0.12	0.13	0.05	0.01	*	0.01	*	*
0.8	0.12	0.13	0.04	0.01	*	*	*	*
0.9	0.13	0.12	0.05	0.01	*	*	*	*
1.0	0.14	0.12	0.04	*	*	0.01	*	*
1.1	0.15	0.13	0.04	*	*	*	*	*
1.2	0.16	0.13	0.05	*	*	*	*	*
1.3	0.17	0.14	0.05	*	*	*	*	*
1.4	0.18	0.14	0.06	0.01	*	*	*	*
1.5	0.20	0.15	0.07	0.01	*	*	*	*
1.6	0.22	0.15	0.09	0.01	*	*	*	*

Table 2.3. Optical constants of "dry substance" for the atmospheric aerosol model (Ivlev and Popova, 1973).

$\lambda(\mu m)$	n	κ	$\lambda(\mu m)$	n	κ	$\lambda(\mu m)$	n	κ
0.4	1.65	0.005	2.5	1.640	0.015	6.5	1.402	0.042
0.5	1.65	0.005	2.8	1.608	0.044	6.7	1.270	0.135
0.6	1.65	0.005	3.0	1.646	0.076	7.0	1.528	0.434
0.7	1.65	0.005	3.1	1.673	0.068	7.5	1.560	0.071
0.8	1.65	0.005	3.3	1.689	0.040	8.0	1.269	0.178
0.9	1.648	0.005	3.5	1.655	0.020	8.5	1.186	0.600
1.0	1.647	0.0051	4.0	1.637	0.018	9.0	1.650	1.240
1.2	1.646	0.0052	4.5	1.620	0.018	9.5	2.342	0.600
1.5	1.645	0.0070	4.8	1.600	0.018	10.0	2.140	0.126
1.7	1.644	0.0082	5.0	1.592	0.018	10.5	1.904	0.078
2.0	1.642	0.0100	5.5	1.558	0.018	11.0	1.750	0.118
2.2	1.641	0.0120	6.0	1.515	0.045	11.5	1.794	0.331

in the visible range. The data for the visible spectral range shown in Table 2.3 is close to that obtained in works (Hänel, 1968, 1971, 1976); Hoidale and Smith, 1968).

The calculations by such a "synthetic" model of aerosol particle optical properties leads to correct results in some instances, but it is good to have in mind that the assumption about homogeneous water distribution over all particle volume is valid only with considerable restrictions, especially when air relative humidity exceeds $f \approx 0.65$–0.70.

2.4 CONTINENTAL AEROSOLS AND HOMOGENEOUS AND INHOMOGENEOUS PARTICLE MODELS

The model of homogeneous spherical particles is very frequently used to explain optical properties of aerosols. However, it seems that in some situations this model is not capable of explaining experimentally observed facts. In such cases there is a need for another, more adequate model.

As it is known (Bullrich, 1964; Sedunov, 1972), a presence of condensation cores is necessary for the liquid-droplet phase in the atmosphere to be formed. Among particles behaving as condensation cores we should mention first those consisting of water soluble substances (mostly of inorganic matter such as $(NH_4)_2S0_4$, $NaNO_3$, H_2S0_4, and so on). The condensation is also possible on insoluble but wettable cores (particles of soil and mountain rock, smoke, organic matters, micro-organisms, and the like). These cores are widely diversified: the particles have different chemical composition and each individual particle contains a great range of chemical substances. This diversity results from different origins of condensation cores as well as processes of coagulation, gas adsorption, chemical and photochemical reactions, and repeated cycles of condensation, and evaporation.

Physico-chemical and, consequently, optical properties of atmospheric aerosols depend significantly on the ability of its particles to absorb water from ambient air and on the air relative humidity. With respect to this, the basic characteristic of a particle is its share of soluble matter. While particles of sea origin almost completely consist of soluble salts, particles of continental origin contain mostly insoluble matter (70–80%), therefore being mixed. The predominance of mixed cores above continents was proved by observations with electron microscopes as early as 1952–1957 (Jakobi *et al.*, 1952; Junge 1957). It was repeatedly confirmed afterwards.

It is considered certain, that the influence of air relative humidity f on physico-chemical and optical properties of atmospheric particles begins when $f \approx 0.2$–0.3. Altering of particle properties begins by the filling micro-capillaries and surface roughness in particles, then the dissolution of water-soluble components of particles follows and, finally, given that a level of relative humidity is high enough the particles are covered by a condensed water shell. So, at relative humidities $f \geq 0.6$ the majority of particles of continental aerosols are composed of solid cores consisting of insoluble or the incompletely dissolved remainder of condensation cores and a solution of inorganic salt (Rozenberg, 1967; Georgievskiy and Rozenberg, 1973; Ivlev *et al.*, 1980). The solution concentration decreases with humidity and the solution volume increases. The increase in humidity is accompanied not only by growth of the exterior particle size and reduction in core size down to some certain value, but also by a change of optical constants of solution shell.

There were attempts to take into account the influence of relative humidity on optical properties of atmospheric aerosols; in particular, it was tried to introduce some effective optical constants of aerosol particle that depend on free water content. General acknowledgment was received for the synthetic model of aerosol particles (Ivlev and Popova 1972, 1973; Kondratiev *et al.*, 1973), that was mentioned in the previous subsection. It is based on the assumption that all dry substances and water in particles are mixed uniformly. However, as for continental aerosol (where the mixed condensation cores constitute the major portion) the assumption that water is distributed uniformly in particle volume is applicable only with significant reservations, especially when relative humidities exceed 0.65–0.70.

It was stressed by Zuev *et al.* (1973) that the disagreement in results given by different authors is caused by both great diversity of atmospheric aerosols and from lack of understanding of the physical processes taking place. Due to this, it is reasonable to analyse the influence of possible inhomogeneity of aerosol particles above continents in the relative humidity range $0.65 \leq f \leq 0.95$.

It follows from calculations (Shifrin, 1952; Prishivalko and Astafyeva, 1974, 1975; Prishivalko *et al.*, 1975) that, as for small solid cores, their impact on the efficiency of light scattering by single water-covered particles can be neglected only when the particle exterior radius is three times larger than the core radius. As for cores with radius 0.5–1.0 μm, their influence is of no importance when their ratio is 7–10 and greater. For the angular scattering characteristics this requirement is even more strong. As follows from works (Kasten, 1969; Volz, 1972; Winkler, 1973) and some experimental data (Bullrich *et al.*, 1972; Hänel, 1976), due to the water-covering of particles there is no more than an increase of twice the particle radii at the relative humidity $f \approx 0.95$. Hence, even at high humidity the influence of the insoluble part should still be significant enough (conditions close to saturation are not considered here).

At the initial stage of condensation the amount of adsorbed water can sometimes be not enough to form a salt solution; in this case water assimilation happens that only in part humidifies condensation cores. In these conditions there could exist particles with a smooth refractive index variation from the particle surface to the particle centre.

In real situations one or other type of particles predominates. Therefore, it is interesting to compare the light scattering properties of homogeneous particles and particles with different types of internal inhomogeneity, as well as to compare the atmospheric aerosol systems consisting of particles of different types (given identical size distributions). When comparing particle systems properties (with a special regard to their transformations related to relative humidity changes), we need to take into account such factors as particle concentration and shape, size distribution, content of soluble matter in condensation cores, and parameters of the condensation process. As tropospheric particles have mainly terrestrial origin and the formation mechanism of soil-erosive and sea aerosols has irregular character, the spatial and time variations of a particle content can differ greatly. The detailed analysis of a large body of data on this subject has been conducted in books of Kondratiev and Pozdnyakov (1981) and Krekov and Rakhimov (1982). Here we are interested not with absolute values of numerical particle concentration, but with its change in aerosol transformations.

Based on experimental data, Rozenberg (1967) expressed the assumption that in condensation transformations of atmospheric haze the number of particles remains almost constant if relative humidity changes within the interval 0.3–0.9. Georgievskiy (1969) established, with the help of impactor traps, that the number of solid aerosol particles does not increase with haze compression as a rule. Andreev *et al.* (1974) concluded on the basis of experimental data that at $0.4 \leq f \leq 0.65$ the coagulation of humidified small particles with radii $R < 0.3\ \mu m$ is the main physical process that occurs in atmospheric hazes when relative humidity increases. At $f > 0.65$ the condensation particle growth prevails. (Apparently, it does not mean that when humidity increases the new condensation nuclei cannot be drawn into the condensation process.) The increase of particle number with f is found by experiment (Laktionov and Bogomolov, 1972). Some inconsistency of the above data indicates that, most probably, the process is complex and depends on many factors. As the first approximation, we shall assume, the particle number per unit volume to be constant within the relative humidity range 0.65–0.95. The change of numerical concentration can easily be taken into account in calculations, if necessary.

Certainly, solid particles in the atmosphere are not spherical. However, it is necessary to say that the particles that serve as condensation nuclei are usually aggregates and, consequently, have no great asymmetry (Bullrich, 1964; Sedunov, 1972). As we will consider them at relative humidity $f > 0.65$ in most cases, then, because of humidifying and partial dissolution the departures of their shape from spherical will be exhibited to a much lesser degree. In conditions of low relative humidity the particle's non-sphericity, nevertheless, influences its light scattering ability. However, the large lidar depolarization ratios systematically measured for dust-like aerosols (see, e.g., Sassen, 2000) are indicative of non-sphericity of the non-hygroscopic dust particles even when relative humidity is high. Due to random orientation of small non-spherical particles and also to the variety of their shapes and sizes, the integral characteristics of light scattering by the ensemble of non-spherical particles and their phase function, differ relatively little from the similar characteristics for ensemble of spherical particles (Mishchenko *et al.*, 1997). Only when scattering angles are large is there a considerable disagreement in scattering phase functions for particles of the spherical and non-spherical shapes (Holland and Draper, 1967; Holland and Gagne, 1970, 1971; Plass and Kattawar, 1971; Chylek *et al.*, 1976; Volten *et al.*, 2001).

From great variety of possible particle shapes the non-spherical particles with some symmetry can be selected, for which the numerical solution of Maxwell equations with the relevant boundary conditions is possible. This problem is beyond the scope of this book. The reviews of present state of theory of scattering and absorption of radiation by particles with such specific shapes are given in new books (Mishchenko *et al.*, 2000*a*, 2000*c*, 2002*a*).

A brief note should be given to particle density. It depends on chemical composition, porosity degree, humidification degree, and some other factors. You can find the most representative experimental data on this account in the book of Kondratiev and Pozdnyakov (1981). The measurements of particle density for aerosols of mainly soil-erosive origin showed that its value for the dry remainder oscillates between 2.4 and 2.7 g·cm^{-3} on average; its greatest recorded density is 3.32 g·cm^{-3}. In actual atmospheric conditions, if $f \gtrsim 0.35$–0.40 then aerosol particles interact with water vapor which causes growth of their geometrical sizes and also specific density decline. Hänel (1976) offered a simple relation for particle density vs. the air relative humidity, valid if it is assumed that the humidified particle volume is equal to the sum of the dry substance and condensed water volumes:

$$\rho(f) = \rho_a + (\rho_0 - \rho_w)\,[R(f)/R_d]^{-3} \tag{2.4.1}$$

where ρ_0 and R_d are the mean density and equivalent radius of the aerosol particle dry substance, ρ_w is the water density, and $R(f)$ is the equivalent radius of the humidified particle (equivalent or, effective, radius is a radius of a sphere whose volume equals the actual particle volume).

From published data (Bullrich, 1964; Sedunov, 1972; Hänel, 1976; Kondratiev and Pozdnyakov, 1981; Krekov and Rakhimov, 1982) we could conclude that the relative content G of soluble substances in a typical condensation core of continental aerosol equals 0.2–0.3.

The analysis of extensive experimental data on aerosol composition in continental air, carried out by Whitby (1978) showed that approximately 95% of ammonium sulphate and other sulfates contained in aerosols account for the accumulative mode. The formation mechanism for continental aerosol accumulative mode is such that this mode is characterized by intermixing of chemical elements inside the particle (this is the so-called internal mixing [Prishivalko *et al.*, 1975; Whitby, 1978; Kondratiev and Pozdnyakov, 1981]), as opposed to exterior mixing peculiar to coarse-dispersion particles. This property of the accumulative mode allows us to assert that all its particles can be characterized by equal values of optical constants.

Thus, it follows from the above that the continental aerosol accumulative fraction with particle radii intervals of $0.4\ \mu m \le R \le 1.5\ \mu m$ has the highest content of soluble matter (20–30%). In Aitken cores ($0.01\ \mu m \le R \le 0.04\ \mu m$) this content is at the level of approximately 20% (Meszaroš, 1966, 1968). The particles with a radius more than 1 μm

usually consist of insoluble silicates, carbonates, and organic remains. Their soluble matter content declines to 5% and less with radius increase (Brown *et al.*, 1968; Hoidale and Blanko, 1968; Meszaroš, 1968; Sedunov, 1972).

It is known (see, e.g. Rozenberg (1968)) that the fraction of particles with radii $0.1\mu m \leq R \leq 1.0\mu m$ is the most optically active (in the visible spectral range). It is clear from the above that the same continental aerosol fraction participates most actively in condensation processes. Thus, it is primarily responsible for changes in atmospheric optical properties caused by variations in air temperature and relative humidity.

As it was said above, it is necessary to take into account the conditions of condensation process while comparing light scattering properties of particle systems, especially when their transformations are stipulated by relative humidity change. These conditions govern the absorbed water content and, hence, the character of particle inhomogeneity and ratio of solid core to whole particle radius. The condensation process in most cases may be considered occurring on particles that are in thermodynamic equilibrium with the surrounding medium. However, we should make a point that experimental data on particle size with air relative humidity change is scarce and contradictory in some instances (Kasten, 1969; Winkler, 1969; Hänel, 1970, 1971; Meszaroš, 1971; Bullrich *et al.*, 1972; Sedunov, 1972). Taking into account the importance of this question for the correct setting of optical investigations as well as having the purpose to construct a model that permits us to explain the inconsistent (at first glance) facts, we will consider it in more detail.

2.5 EQUILIBRIUM RADIUS OF WATER-CONTAINING PARTICLES

Data of Meszaroš (1971) shows that the ratio of water-covered particle radius to the dry core radius goes down (especially in the summer period) with the increase of particle size. Kasten (1969) building on Winkler's paper (1969) had come to the conclusion that particle condensation growth does not depend on its initial size. Hänel (1971) on the basis of experimental data, inferred that such assumption is valid for $R \geq 0.1\ \mu m$. In some works (Hänel, 1970; Bullrich *et al., 1972;* Sedunov, 1972) the noticeably greater growth of giant condensation cores in comparison with small ones was recorded at high relative humidity.

It follows from the theory of condensation processes in the atmosphere, that, if air temperature and pressure vary slow enough, and if the droplet growth rate (i.e. the rate of core water-covering) is small, then, the fields of vapor density and temperature in a fluid succeed in following the changing conditions. Then, the process of condensation may be considered as quasi-stationary. The estimates show (Sedunov, 1972) that in usual conditions the characteristic time needed to establish equilibrium between a droplet and the surrounding medium (a fraction of second) is much less than the characteristic time of change in external conditions (about several seconds). This supports the validity of the assumption that the process is quasi-stationary.

The thermodynamic equilibrium of a core composed of soluble and non-soluble parts with a surrounding medium was studied in some works (Khrgian, 1969; Sedunov, 1972; Meszaroš, 1975). For a completely wetted core the pressure of saturated water vapor e_k

above the particle surface is defined by:

$$e_k = e_0(T)\left[1 - C/\left(R^3 - R_T^3\right)\right]\exp(B/R) \tag{2.5.1}$$

where $e_0(T)$ is the saturated vapor pressure above a flat water surface at temperature T, C is a quantity that describes activity of condensation cores, R is an equilibrium radius of water-covered particles, R_T is an effective radius of the non-soluble core part, $B = 2\sigma_s/\rho_s R_{vap}T$, σ_s and ρ_s are the coefficient of surface tension and density of the solution obtained in water-covering, respectively, and R_{vap} is the gas constant for vapor.

At thermodynamic equilibrium the pressure of saturated water vapor above the droplet surface e_k is equal to the vapor pressure at infinity e_∞. Then, the ratio $e_k/e_0(T)$ is equal to the air relative humidity f. Mason (1957) suggested that $C = 8.6 m_s M^{-1}$ for weak solutions, where m_s is the mass of core soluble part and M is the relative molecular mass of the soluble substance. In the book of Sedunov (1972) it was supposed for generality that:

$$C = bR_d^{2(1+\alpha)} \tag{2.5.2}$$

where R_d is the dry core radius and the parameters b and α are determined from composition and physico-chemical properties of the core. In the simplest case $\alpha = 0.5$ and the amount of soluble substance is proportional to core volume.

After comparing Mason's expression for C with Eq.(2.5.2), we have:

$$m_s = 0.116MbR_d^{2(1+\alpha)} \tag{2.5.3}$$

Masses of soluble and insoluble substances in the condensation core are equal to $m_s = \frac{4}{3}\pi\left(R_d^3 - R_T^3\right)\rho_s$ and $m_T = \frac{4}{3}\pi R_T^3\rho_T$, where ρ_s and ρ_T are densities of soluble and insoluble core parts. Then:

$$b = \frac{8.6\cdot\frac{4}{3}\pi}{MR_d^{2(1+\alpha)}}G\left[R_T^3\rho_T + \left(R_d^3 - R_T^3\right)\rho_s\right] \tag{2.5.4}$$

It is easy to see from here that:

$$b = \frac{11.47\pi}{MR_d^{2(1+\alpha)}}R_d^3\left[(1-G)\rho_T + G\rho_s\right]G \tag{2.5.5}$$

Some uncertainty pertains to data of different authors on aerosol particle substance density. As was mentioned in Section 2.4 the density of the dry remains of aerosol substances varies from 2.4 to 2.7 g·cm^{-3}, and in some cases it may be as great as 3.3 g·cm^{-3} (Kondratiev and Pozdnyakov, 1981). In this view the assertion of Volz (1972), that mean particle density lies in range 1.76–1.96 g·cm^{-3}, apparently relates to strongly humidified particles. Similarly, Winkler's assertion (1973) that aerosol particle density can vary from $\rho = 1$ to 2.5 g·cm^{-3}, most likely stems from particles' wetting and porosity.

Because in the following text we will consider the water-covering of arbitrary condensation cores rather than that of cores with particular chemical compositions, we for simplicity put that the densities of soluble and insoluble parts of cores are approximately equal: $\rho_s \approx \rho_T$.

The influence of soluble matter on the saturated vapor pressure works primarily as a result of its mass and secondarily through the ratio of soluble and insoluble parts and through the change of solution's density and surface tension. As for diluted solutions the variations of σ_p and ρ_p are insignificant, the dependence of B on solution concentration can be neglected. Then, it is possible to estimate that $B = 0.0012\ \mu m$ (Sedunov, 1972). In the case for $R \gg 0.0012\ \mu m$, after expanding the exponent in Eq.(2.5.1) in a series, we arrive at the equation of fourth degree in R:

$$R^4(1-f) + R^3 B - RR_d^3\,[b + (1-G)(1-f)] - BR_d^3\,[b + (1-G)] = 0 \qquad (2.5.6)$$

Assuming the surface tension influence to be small, it is possible to reduce Eq.(2.5.6):

$$R = R_d \sqrt[3]{1 - G + b/(1-f)} \qquad (2.5.7)$$

To estimate the Eq.(2.5.6) applicability we need to compare its results (these are given in Table 2.4) with experimental data on the ratio R/R_d (Kasten, 1969; Bullrich *et al.*, 1972) (Table 2.5). Kasten (1969) gives values for the ratio of particle radius at some relative humidity f to its radius at minimum humidity $f = 0.1$ for continental, mean, and normal aerosols. "Normal" implies an aerosol observed in the neighborhood of Mainz (Germany), and "mean" is the average of 28 measurements for continental, normal, and sea aerosols. With the purpose of comparing data of Table 2.4 and data in the paper of Bullrich *et al.* (1972), experimental data of Kasten (1969), in Table 2.5, has been converted to zero humidity.

It follows from Table 2.4 that the calculations by Eq.(2.5.6) give a little greater R/R_d values for cores with $R_d \geq 1\ \mu m$ than for cores with $R_d < 1\ \mu m$. This difference is increased with relative humidity growth; it reaches 8% at $f = 0.95$. The approximate formula (2.5.7) gives R/R_d values rather close to values obtained from Eq.(2.5.6) for $R_d > 1\ \mu m$. Results of Eq.(2.5.6) calculations and experimental data of Bullrich *et al.* (1972) both show that for cores with $R_d \geq 1\ \mu m$ at $f < 0.95$ the R/R_d values do not depend on the dry core radius.

It is seen from the curves in Figure 2.1 that R/R_d values calculated at $G = 0.2$ are rather close to the experimental data of Kasten (1969) for continental aerosols and, when calculated at $G = 0.3$ to the experimental data of Bullrich *et al.* (1972) and Kasten (1969) for normal aerosols. The maximum difference in experimental and calculated values does not exceed 4%. This suggests that Eq.(2.5.6) can be used to define the equilibrium particle radius at different relative humidities.

Given experimental data for nine substances that could act as the principal components of soluble part of atmospheric condensation cores, Fitzgerald (1975) obtained the approximate formulae (with a precision of 5%) that can be used to calculate the equilibrium radius of particles, providing more than 60% of the particle consists of these substances at relative humidities between 0.81 to 0.993. Such calculations show that the equilibrium radius of ammonium sulphate particles coincides with values obtained from Eq.(2.5.7) within 5–7%, if $\rho_T = 2.5$–$2.65\ \mathrm{g{\cdot}cm^{-3}}$. For other substances the difference

Table 2.4. Values of the R/R_d ratio calculated with Eq.(2.5.6) (NaCl is the soluble substance, $b = 4G/3$).

	$G = 0.2$				$G = 0.3$			
	R_d (μm)							
f	0.4	0.1	1.0	10	0.4	0.1	1.0	10
0.60	1.129	1.133	1.135	1.136	1.184	1.189	1.1193	1.193
0.65	1.151	1.156	1.159	1.160	1.215	1.221	1.225	1.225
0.70	1.179	1.186	1.190	1.190	1.252	1.260	1.266	1.266
0.75	1.215	1.224	1.230	1.231	1.300	1.311	1.318	1.319
0.80	1.264	1.277	1.286	1.287	1.364	1.380	1.390	1.392
0.85	1.334	1.355	1.369	1.370	1.456	1.480	1.496	1.498
0.90	1.450	1.486	1.510	1.512	1.604	1.644	1.671	1.674
0.92	1.521	1.568	1.600	1.603	1.693	1.746	1.791	1.785
0.95	1.686	1.767	1.823	1.823	1.900	1.989	2.048	2.054
0.975	1.957	2.120	2.239	2.252	2.239	2.415	2.539	2.553
0.99	2.324	2.680	2.977	3.011	2.707	3.092	3.399	3.433

does not exceed 8–10%. These results confirm the validity of relations (2.5.6) and (2.5.7) also.

The Eq.(2.5.6) also allows explanation of possible reasons for contradictions in the available experimental data on the condensation cores growth rate. Growth coefficients $A = R/R_d$ obtained from Eq.(2.5.6) are given in Table 2.6. The notation of $A(10)/A(0.1)$ implies the ratio of R/R_d for particles with initial radius 10 μm to the corresponding quantity for particles with initial radius 0.1 μm. It is seen that if the difference between $A(10)$ and $A(0.04)$ is about 8% at $f = 0.95$, then the difference between $A(10)$ and $A(0.1)$ is only 3.5%; between $A(10)$ and $A(1)$ it is as little as 0.3%.

Kasten (1969) has carried out the analysis of condensation particle growth on the basis of Winkler's experiment (1969) who studied only particles with $R_d \geq 0.1$ μm. Therefore, it is clear that Kasten could not find a 3–4% difference in A. It was found later (Bullrich *et al.*, 1972).

Table 2.5. Experimentally obtained values of the R/R_d ratio.

	Recalculation of Kasten (1969)			Bullrich *et al.* (1972)			
	Aerosol type			R_d (μm)			
f	Continental	Normal	Mean	0.04	0.1	1.0	10
0.60	1.171	1.229	1.242	1.153	1.169	1.165	1.165
0.65	1.192	1.260	1.294	–	–	–	–
0.70	1.22	1.301	1.355	–	–	–	–
0.75	1.253	1.362	1.417	–	–	–	–
0.80	1.304	1.424	1.489	1.400	1.414	1.426	1.426
0.85	1.375	1.536	1.602	–	–	–	–
0.90	1.487	1.700	1.766	1.656	1.695	1.720	1.720
0.95	1.731	1.997	2.105	1.93	2.01	2.05	2.06

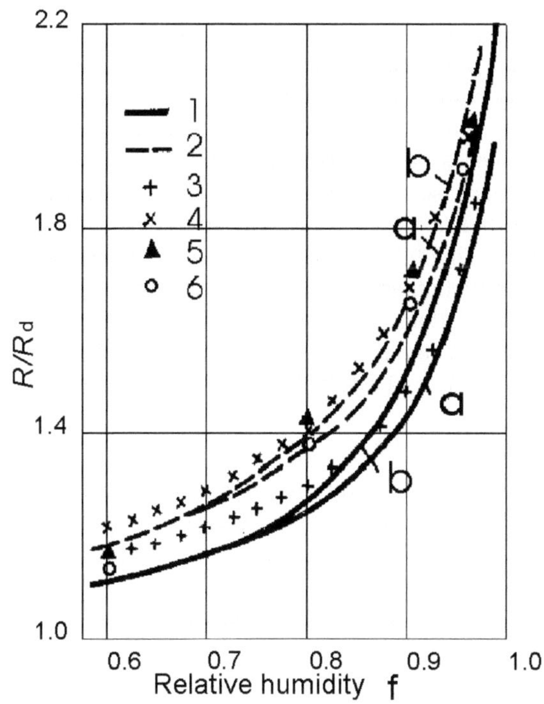

Figure 2.1. Ratio R/R_d versus the air relative humidity f: lines marked 1 and 2 represent the calculation with Eq.(2.5.6) for $G = 0.2$ (1) and 0.3 (2) at $R_d = 0.04$ (a) and 1 μm (b); line 3 represents the experiment (Kasten, 1969) after handling of Hanel (1971) for continental aerosols; line 4 represents the same, for urban aerosols; lines 5 and 6 represent the experiment (Bullrich *et al.*, 1972) at $R_d = 0.04$ (6) and 1 μm (5).

As statistics suggests, the soluble matter content decreases with increased particle size. In huge condensation cores it usually decreases down to 5% (Brown *et al.*, 1968; Hoidale and Blanko, 1968; Meszaroš, 1968; Sedunov, 1972). This circumstance, apparently, greatly affected experimental results of Meszaroš (1971). For cores with the initial radius $R_d = 0.5$ μm Meszaroš obtained R/R_d values close to Kasten's values (1969) and, in comparison with smaller particles, for groups of cores with radii $R_d = 1, 5$, and 10 μm found a noticeable decline in relative growth with increase in humidity.

Table 2.6. The growth coefficient ratios for cores of different sizes.

f	$G = 0.2$			$G = 0.3$		
	$\frac{A(10)}{A(0.04)}$	$\frac{A(10)}{A(0.1)}$	$\frac{A(10)}{A(1)}$	$\frac{A(10)}{A(0.04)}$	$\frac{A(10)}{A(0.1)}$	$\frac{A(10)}{A(1)}$
0.85	1.027	1.011	1.001	1.029	1.012	1.001
0.90	1.043	1.017	1.001	1.044	1.018	1.002
0.95	1.084	1.035	1.003	1.081	1.033	1.003
0.975	1.151	1.062	1.006	1.140	1.057	1.006

It must be noted that in natural conditions, as pointed out by Laktionov and Bogomolov (1972), the change of aerosol microstructure with humidity is masked by considerable variations of dry particle composition and properties that are attributed to change of air masses and meteorological conditions. This circumstance was clearly illustrated by experimental data of Kapustin and Lyubovtseva (1975).

Hänel (1970), on the basis of the analysis of Winkler's experimental data (1969) developed a model of dependence of the equivalent radius of aerosol particles on the ambient air relative humidity. To simplify he assumed that all particles regardless their size have identical chemical composition and structure in a dry state. However, it was underlined that speculations and calculations given are valid only when the relative humidity increases from $f = 0$ (to avoid taking into account the hysteresis effects capable of occuring when relative humidity decreases). As a result, the approximate formula was obtained:

$$R(f)/R_d = [1 + \varphi(f)f\ /(1 - f)]^{1/3} \tag{2.5.8}$$

that is valid when the relative humidity lies within the interval $f \leq 0.95$ and the dry particle radius $R_d \geq 0.1\ \mu m$. It follows from Eq.(2.5.8) that the ratio $R(f)/R_d$ does not depend on the initial particle size R_d. This is confirmed by the data of Bullrich *et al.* (1972) and Table 2.5. The quantity $\varphi(f)$ is referred to as the aerosol particle volume growth coefficient. Kasten (1969) showed that the same dependence can be represented as the semi-empirical formula:

$$R(f)/R_d = (1 - f)^{-\varepsilon} \tag{2.5.9}$$

where ε equals 0.17, 0.23, 0.26, and 0.25 for continental, normal, sea, and mean aerosols, respectively.

Hänel (1971) paid attention to the dependence of the degree ε on relative humidity and pointed out that for mean aerosols the $\varepsilon(f)$ value varies from 0.21 to 0.33 when the relative humidity changes between the interval $0.3 \leq f \leq 0.95$. During 24 hours the ε value can vary from 0.22 in the morning to 0.17 by night. Fett (1967) found that in a clear atmosphere $\varepsilon = 0.17$ at $0.3 \leq f \leq 0.98$, but in a dusty atmosphere $\varepsilon = 0.30$ at $0.6 \leq f \leq 0.9$. Laktionov and Bogomolov (1972) found that ε varies between the limits $0.08 \leq \varepsilon \leq 0.30$.

The analysis of Eq.(2.5.6) solutions shows that for $R_d = 1\ \mu m$ the corresponding values of the ratio R/R_d can be received from Eq.(2.5.9) with no more than 0.5% error in an interval $0.6 \leq f \leq 0.98$ if we substitute ε into it

$$\varepsilon = P \lg(1 - f) + Q \tag{2.5.10}$$

where $P = 0.0664$ and $Q = 0.113$ at $G = 0.2$ and $P = 0.0498$ and $Q = 0.173$ at $G = 0.3$. In applying Eq.(2.5.10), the ε values vary from 0.14 to 0.22 for $G = 0.2$ and from 0.19 to 0.26 for $G = 0.3$, which is quite close to values obtained in the experimental works already mentioned. This implies that to calculate approximate particle equilibrium radius besides using Eq.(2.5.7) it is possible to use the empirical formula (2.5.9) with the parameter ε value obtained with Eq.(2.5.10). However, use of Eq.(2.5.7) is more preferable, as it takes into account in explicit form not only of the dependence on relative humidity, but also of the influence of the soluble matter percentage and its activity.

The discussion in this section gives means to define the input data for practical calculations. In particular, it is possible, with the purposes to simplify calculations, to assume that at relative humidity $f < 0.95$ all particles are increasing to the same extent, because the tendency of quicker growth of huge cores is compensated by the decreasing of soluble matter content in them. If we take the value of the ratio R/R_d at $R_d = 0.5\ \mu m$ as a measure of growth, then we will obtain a noticeable error only for the smallest cores ($R_d = 0.04\ \mu m$), which contribute insignificantly to the overall scattering pattern.

Having considered in detail the basic question of this section, we can now decide on two other important issues directly connected with it, whose solutions are obligatory when analysing specific situations. These are questions of how the atmospheric particle optical constants change with a particle's water-covering and how the particle size distribution function is changed in this process.

Considerable attention to the problem of humidity influence on physico-chemical and optical properties of aerosol particles is given in Hänel's works (Hänel, 1968, 1970, 1972, 1975, 1976; Hänel and Bullrich, 1978; Hänel and Zankl, 1979; Hänel and Lehman, 1980, 1981). He offered the following relations for mean values of real and imaginary parts of the complex refractive index of aerosol samples (Hänel, 1976):

$$\bar{n} = n_w + (n_0 - n_w)(1-a)^{-1} \qquad \bar{\kappa} = \kappa_w + (\kappa_0 - \kappa_w)(1-a)^{-1} \tag{2.5.11}$$

where:

$$a = \frac{\rho_0}{\rho_w}\frac{m_w}{m_0} = \frac{\rho_0}{\rho_w}\bar{\mu}\frac{f}{1-f}$$

n_w, κ_w, ρ_w, and m_w are optical constants, density, and mass of pure water respectively; n_0, κ_0, ρ_0, and m_0 are the same for dry aerosol substances, and $\bar{\mu}$ is a coefficient of mean linear increase in sample mass. Note that the use of Eqs.(2.5.11) to calculate dependencies $n(f)$ and $\kappa(f)$ is possible only with the condition that $\bar{\mu}$ and the aerosol substance density in the dry state are known (if we consider n_0 and κ_0 to be known also). In the work of Hänel (1976) there is a table of $\bar{\mu}(f)$ values for typical aerosol forms. The set of $\bar{\mu}(f)$ values recommended for continental aerosol is duplicated in Table 2.7. These are approximate values.

In our opinion, the way to define $\bar{n}$ and $\bar{\kappa}$ of particles based on the use of information on particle size change vs. the relative humidity (see Tables 2.4 and 2.5) is more preferable

Table 2.7. Mean values of $\bar{\mu}(f)$ (Hänel, 1976).

f	$\bar{\mu}(f)$	f	$\bar{\mu}(f)$
0.1	0.050	0.7	0.160
0.2	0.065	0.8	0.156
0.3	0.080	0.9	0.146
0.4	0.095	0.95	0.140
0.5	0.110	0.975	0.146
0.6	0.127		

and simple. In this case the effective mean values of the real and imaginary parts of the refractive index can be approximated with:

$$\bar{n} = n_w + (n_0 - n_w)(R_d/R)^3 \qquad \bar{\kappa} = \kappa_w + (\kappa_0 - \kappa_w)(R_d/R)^3 \tag{2.5.12}$$

We will further determine the effective $\bar{n}$ and $\bar{\kappa}$ values for conditionally homogeneous particles which would have been formed, providing the condition of ideal mixing of all components inside a particle, including water is met, exactly in this way.

For two-layered particles consisting of an insoluble core and a salt solution shell, it is possible to write expressions for mean-effective n and κ values in the shell with the same degree of approximation:

$$\begin{aligned} n_{\text{shell}} &= \frac{n_w\left[(R/R_d)^3 - 1\right] - n_0\left[1 - (R_T/R_d)^3\right]}{(R_d/R)^3 - (R_T/R_d)^3} \\ \kappa_{\text{shell}} &= \frac{\kappa_w\left[(R/R_d)^3 - 1\right] - \kappa_0\left[1 - (R_T/R_d)^3\right]}{(R_d/R)^3 - (R_T/R_d)^3} \end{aligned} \tag{2.5.13}$$

where R_T is an equivalent radius of the insoluable. If we assume that $\rho_T = \rho_s$ as in Section 2.4, then the Eqs.(2.5.13) can be reduced with the help of the relation $1-(R_T/R_d)^3 = G$ (Prishivalko, 1977, 1978).

The particle size change with air relative humidity can result in deformation of the size distribution function. If we use the approximate relation (2.5.8) offered by Hänel (1970, 1972) or a relation similar to it, as:

$$R(f) \approx R_d\left(1 + \frac{\rho_0}{\rho_w}\bar{\mu}\frac{f}{1-f}\right)^{1/3} \tag{2.5.8a}$$

that are valid for relative humidity $f \lesssim 0.90$–0.95 and particles with equivalent radius above $0.1\ \mu m$ (in dry state), then it is possible to describe the change in particle size spectrum in general terms. The examples of such changes are given by Hänel (1976) for different aerosol types. Three of six reported situations could be classified as some deformation of the Junge-type distribution. Meszaroš (1971, 1977) obtained experimental data on the change in particle size spectrum in the relative humidity interval 0.35–0.98. It was found that the deformation of the Junge-type distribution (with coefficients ν that are different for different particle size intervals: $0.15 \leq R \leq 1.0$; $1.0 \leq R \leq 20$; and $20 \leq R \leq 100\ \mu m$) takes place both in summer and in winter.

Some authors in their works (e.g., Tuomi (1976)), tried to calculate the influence of the distribution function deformation on light scattering. They represented the function as a sum of two log-normal distributions with different weights.

Sedunov (1972) clearly showed that the dry core distributions, written as power functions, in a process of humidifying are being rebuilt in power distributions of water-covered cores with smaller power exponents. Such reorganization can be easily accounted for in theoretical consideration. The analysis of experimental data of Hänel (1970) shows, that if the Junge-type particle size distribution is preserved in water-covering, then its exponent

Table 2.8. The ratios of mean values $\bar{\nu}(f)/\bar{\nu}(0)$ according to Hänel (1970).

f	$R_{d1} = 0.04\ \mu m$ $R_{d2} = 0.1\ \mu m$	$R_{d1} = 0.1\ \mu m$ $R_{d2} = 1\ \mu m$	$R_{d1} = 1\ \mu m$ $R_{d2} = 10\ \mu m$
0.30	1	1	1
0.60	1	1	1
0.80	0.99	1	1
0.90	0.98	0.99	1
0.95	0.96	0.99	1
0.975	0.92	0.98	1
0.99	0.88	0.97	1
0.995	0.82	0.95	1

varies according to the formula:

$$\frac{\nu|_f}{\nu|_{f=0}} = \frac{\ln(R_{d2}/R_{d1})}{\ln(R_2/R_1)} \tag{2.5.14}$$

where R_{d1}, R_{d2}, R_1 and R_2 (see Table 2.8) are limiting radii of dry cores and particles respectively.

It is seen from this table that the exponent ν does not vary at all between the interval $1 \leq R_d \leq 10\ \mu m$, and it varies no more than 4% between the interval $1 \leq R_d \leq 10\ \mu m$ if the relative humidity $f < 0.95$. This implies that it is possible to consider ν as a constant if $f \lesssim 0.95$.

3

Light scattering and absorption by two- and multilayered spheres

The choice of a particle's optical model is the deciding factor in theoretical investigation of problems related to light scattering by different particles. This model, on the one hand, should describe a structure of actual particles more or less adequately, and on the other hand, the calculation of light scattering characteristics for such particles and their systems should not exceed the abilities of modern computers. A rather unpleasant feature of rigorous theory of light scattering is that even the insignificant complication of the model (e.g., going from sphere to spheroid) results in method and calculation difficulties, which exceeds many times the complications in the model itself. Therefore, two requirements to this model conflict frequently so it is necessary to turn to a compromised approach, where some details of internal structure and shape of particle are not considered. For atmospheric aerosols the situation is simplified to some extent by the fact that our main interest is not linked to a single particle, but rather to a system, which contains particles of different sizes, shapes, and structures.

Over many years it was suggested that due to averaging over the ensemble the individual features of particles are smoothed so greatly that very simple models can give sufficient degrees of accuracy. The investigations of the last years revealed that this statement is somewhat wrong. In particular, the model of spherically symmetric particles can never produce non-zero backscattering depolarization observed for the dust fraction of aerosols (Mishchenko *et al.*, 1997, 2002*b*). But to describe the integral scattering characteristic of dust-like aerosols and all scattering properties of other fractions of atmospheric aerosols it is quite allowable, as the first approximation, to use a simple sphere model with spherically symmetric internal structure. There are two modifications of this model. In the first variant the spherical particle consists of concentric layers with a constant refractive index inside each layer (see Section 3.5). The simplest case is a coated sphere with a homogeneous core and shell. The light scattering characteristics of such atmospheric two-layered particles (TLPs) and their polydisperse systems are considered here as well as in the Chapter 4. The second variant of the spherically symmetric model is a sphere with a smooth variation of optical properties in the radial direction. The particles of this type, which we call radially inhomogeneous, are considered in Chapter 8. In particular situations it is appropriate to introduce combined models, for example a coated sphere with a homogeneous core and radially inhomogeneous shell (see Section 8.4).

3.1 LIGHT SCATTERING BY A TWO-LAYERED SPHERICAL PARTICLE

From the methodological point of view, the theory of scattering by a coated sphere is a modification of the theory of scattering by an homogeneous sphere (so-called Mie theory), but it is much more cumbersome. However, the Mie theory was developed at the end of 19th century (e.g., see Logan, 1965), the first work on radiation diffraction by TLP appeared only in 1951 (Aden and Kerker, 1951). Then, the solution to this problem reappeared in works of Güttler (1952) and Shifrin (1952). In this section within the framework of the one-particle Green's functions formalism we give a short derivation of the basic expressions of the theory of light scattering by spherical TLPs when the plane harmonic wave incidents on it. It is assumed that the particle consists of a spherical homogeneous core of radius R_1 with the absolute complex refractive index N_1, enveloped by the spherically symmetric homogeneous shell of radius R_2 with the absolute complex refractive index N_2. The particle is located in the homogeneous non-absorbing medium with a refractive index N_2, which we can amount to unity (Figure 3.1).

Let us choose the scattering potential V as the sum:

$$V = \sum_{\alpha=1}^{2} k_0^2 \left[\varepsilon_\alpha(\mathbf{r}) - N_2^2\right] \qquad k_0 = 2\pi/\lambda \tag{3.1.1}$$

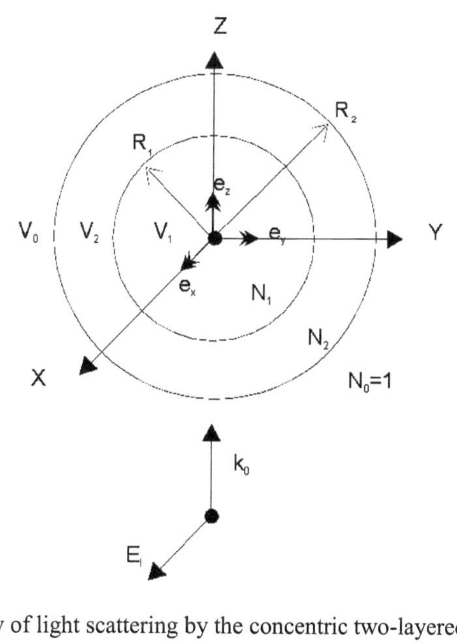

Figure 3.1. Geometry of light scattering by the concentric two-layered particle.

where $\varepsilon_\alpha(\mathbf{r}) = N_1^2$, if a point characterized by vector $\mathbf{r}$ belongs to the particle's core, and $\varepsilon_\alpha(\mathbf{r}) = 1$ if this point belongs to space outside the particle. With such a potential choice the field inside the shell satisfies the operator equation system (1.2.9). The concretization of these equations results in the simple system:

$$
\begin{aligned}
\mathbf{E}_1(\mathbf{r}) &= \int\limits_{V_0} \hat{G}_{13}(k_2\mathbf{r}, k_0\mathbf{r}'')k_0^2(1 - N_2^2)\mathbf{E}_2(\mathbf{r}'')\, d^3\mathbf{r}'' + \mathbf{E}_{\text{sh}}^{\text{hom}}(\mathbf{r}) \\
\mathbf{E}_2(\mathbf{r}) &= \int\limits_{V_1} \hat{G}_{31}(k_2\mathbf{r}, k_0\mathbf{r}')k_0^2(N_1^2 - N_2^2)\mathbf{E}_1(\mathbf{r}')\, d^3\mathbf{r}' \qquad (3.1.2) \\
\mathbf{E}_{\text{sh}}^{\text{lay}}(\mathbf{r}) &= \mathbf{E}_1(\mathbf{r}) + \mathbf{E}_2(\mathbf{r})
\end{aligned}
$$

where $\mathbf{E}_{\text{sh}}^{\text{hom}}(\mathbf{r})$ is the field inside the homogeneous particle with the refractive index equal to the shell refractive index and the size equal to the external particle radius, V_1 is the core volume, V_0 is the volume external to the particle, $k_1 = k_0 N_1$, and $k_2 = k_0 N_2$. Using the expansion (1.3.24) and integrals (1.3.27), we can find the field in the shell. The continuity of the field tangential components on the shell boundaries gives the expansion coefficients for the external (i.e., scattered) field and field inside the core.

Let the plane monochromatic wave $\mathbf{E}_I$ having a wave vector along the axis $\mathbf{e}_z$ and a vector of the velectric field oscillations along the axis $\mathbf{e}_x$ (see Fig. 3.1) be incident upon the TLP:

$$
\mathbf{E}_I(\mathbf{r}) = E_I \sum_{l=1}^{\infty} \mu_l \left[\mathbf{M}_{o1l}^{(1)}(k_0\mathbf{r}) - i\mathbf{N}_{e1l}^{(1)}(k_0\mathbf{r})\right] \qquad (3.1.3)
$$

where $\mu_l = i^l\,(2l+1)\,/\,[l(l+1)]$. Then, the scattered field $\mathbf{E}_{\text{sc}}^{\text{lay}}(\mathbf{r})$ and the fields inside the core $\mathbf{E}_{\text{cor}}^{\text{lay}}(\mathbf{r})$ and inside the shell $\mathbf{E}_{\text{sh}}^{\text{lay}}(\mathbf{r})$ be given by:

$$
\mathbf{E}_{\text{sc}}^{\text{lay}}(\mathbf{r}) = E_I \sum_{l=1}^{\infty} \mu_l \left[a_l\mathbf{M}_{o1l}^{(3)}(k_0\mathbf{r}) - ib_l\mathbf{N}_{e1l}^{(3)}(k_0\mathbf{r})\right] \qquad (3.1.4)
$$

$$
\begin{aligned}
\mathbf{E}_{\text{sh}}^{\text{lay}}(\mathbf{r}) = E_I \sum_{l=1}^{\infty} \mu_l &\left[\alpha_{2l}\mathbf{M}_{o1l}^{(1)}(k_2\mathbf{r}) + \alpha_{3l}\mathbf{M}_{o1l}^{(3)}(k_2\mathbf{r})\right. \\
&\left. - i\beta_{2l}\mathbf{N}_{e1l}^{(1)}(k_2\mathbf{r}) - i\beta_{3l}\mathbf{N}_{e1l}^{(3)}(k_2\mathbf{r})\right] \qquad (3.1.5)
\end{aligned}
$$

$$
\mathbf{E}_{\text{cor}}^{\text{lay}}(\mathbf{r}) = E_I \sum_{l=1}^{\infty} \mu_l \left[\alpha_{1l}\mathbf{M}_{o1l}^{(1)}(k_1\mathbf{r}) - i\beta_{1l}\mathbf{N}_{e1l}^{(1)}(k_1\mathbf{r})\right] \qquad (3.1.6)
$$

The expansion coefficients are given by the formulae:

$$
\begin{aligned}
&a_l = (A_2A_6 - A_1A_5)B \quad \alpha_{1l} = N_1 B \quad \alpha_{2l} = -iN_2A_2B \quad \alpha_{3l} = -iN_2A_1B \\
&b_l = (\bar{A}_2\bar{A}_6 - \bar{A}_1\bar{A}_5)\bar{B} \quad \beta_{1l} = \bar{B} \quad \beta_{2l} = -i_2\bar{A}_2\bar{B} \quad \beta_{3l} = i\bar{A}_1\bar{B} \qquad (3.1.7) \\
&B = (A_1A_4 - A_2A_3)^{-1} \quad \bar{B} = (\bar{A}_1\bar{A}_4 - \bar{A}_2\bar{A}_3)^{-1} \qquad (3.1.8)
\end{aligned}
$$

In Eqs.(3.1.7–8) the following notation is used:

$$\begin{aligned}
A_1 &= \psi_l(w)\psi_l'(y) - \frac{N_1}{N_2}\psi_l'(w)\psi_l(y) \\
A_2 &= \psi_l(w)\xi_l'(y) - \frac{N_1}{N_2}\psi_l'(w)\xi_l(y) \\
A_3 &= \psi_l(z)\xi_l'(v) - N_2\psi_l'(z)\xi_l(v) \qquad A_4 = \xi_l(z)\xi_l'(v) - N_2\xi_l'(z)\xi_l(v) \\
A_5 &= \xi_l(z)\psi_l'(v) - N_2\xi_l'(z)\psi_l(v) \qquad A_6 = \psi_l(z)\psi_l'(v) - N_2\psi_l'(z)\psi_l(v)
\end{aligned} \tag{3.1.9}$$

where ψ_l and ξ_l are the Riccati–Bessel functions and the Riccati–Hankel functions of the first kind defined in Chapter 1, the prime sign specifies the derivative with respect to the function argument, and the arguments are $w = N_1x_1$, $y = N_2x_1$, $z = N_2x_2$, and $v = x_2$, where $x_1 = 2\pi r_1/\lambda$ and $x_2 = 2\pi r_2/\lambda$. Expressions for $\bar{A}_i$ are obtained from A_i by substituting the $1/N_j$ values for the N_j values not entering the function argument.

The expressions for amplitude coefficients (3.1.7)–(3.1.9) may be expressed in a more suitable form for calculations if we introduce the log-derivatives D_l and G_l of functions ψ_l and ξ_l and combinations R_l, S_l, and T_l (see Section 1.5.2.3). For scattered fields these expressions are (Krekov and Rakhimov, 1973; Kattawar and Hood, 1976; Toon and Ackerman, 1981; Kutuzov and Klenin, 1983):

$$\begin{aligned}
a_l &= -\frac{\psi_l(v)}{\xi_l(v)}\frac{u_{2l}u_{4l} - u_{1l}(N_1 + u_{2l}u_{3l})}{u_{2l}u_{4l} - u_{5l}(N_1 + u_{2l}u_{3l})} \\
b_l &= -\frac{\psi_l(v)}{\xi_l(v)}\frac{N_2u_{7l}u_{4l} - u_{6l}(N_2 + u_{7l}u_{3l})}{N_2u_{7l}u_{4l} - u_{8l}(N_2 + u_{7l}u_{3l})}
\end{aligned} \tag{3.1.10}$$

where:

$$\begin{aligned}
u_{1l} &= D_l(z) - N_2D_l(v) & u_{5l} &= D_l(z) - N_2G_l(v) \\
u_{2l} &= N_1D_l(y) - N_2D_l(w) & u_{6l} &= N_2D_l(z) - D_l(v) \\
u_{3l} &= -i\left(T_l^{yy} - F\tilde{S}_l^{yz}r_l^{yz}\right) & u_{7l} &= N_2D_l(y) - N_1D_l(w) \\
u_{4l} &= \left(\tilde{R}_l^{yz}\right)^2 F & u_{8l} &= N_2D_l(z) - G_l(v)
\end{aligned} \tag{3.1.11}$$

where the normalized functions $\tilde{S}_l^{yz}$ and $\tilde{R}_l^{yz}$ are defined with expressions (1.5.48) and (1.5.44), respectively, and the combination T_l^{yy} is defined with Eq.(1.5.50) and the normalization constant $F = \exp\left[2\kappa_2(x_1 - x_2)\right]$. These functions can be successfully calculated through algorithms described in Section 1.5.2.3. The ratio $\psi_l(v)/\xi_l(v)$ may be calculated using procedures from Sections 1.5.2.2. It is worth noting that Fuller (1993) developed another surprisingly simple expression for scattered fields based on consideration of multiple reflections of spherical partial waves, with amplitudes given by the coefficients for coupling of plane waves to spherical partial waves.

If you compare Eqs.(3.1.4) and (1.3.32), you will notice that the structure of these expressions is identical. They differ in expansion coefficients only. It is obvious that in such cases the amplitude scattering tensor can be written similar to Eq.(1.3.44):

$$\hat{A}(\mathbf{k}_0, \mathbf{k}_0') = -\frac{4\pi}{k_0} i \sum_{\sigma ml} \lambda_{ml} \left[a_l \mathbf{m}_{\sigma ml}(\mathbf{k}_0) \otimes \mathbf{m}_{\sigma ml}(\mathbf{k}_0') + b_l \mathbf{n}_{\sigma ml}(\mathbf{k}_0) \otimes \mathbf{n}_{\sigma ml}(\mathbf{k}_0') \right] \tag{3.1.12}$$

If we know the amplitude tensor, we could find the integral characteristics of light scattering:

$$\sigma_{\text{ext}} = -\frac{2\pi}{k_0^2} \operatorname{Re} \sum_{l=1}^{\infty} (2l+1)(a_l + b_l) \qquad \sigma_{\text{sc}} = \frac{2\pi}{k_0^2} \sum_{l=1}^{\infty} (2l+1)(|a_l|^2 + |b_l|^2) \tag{3.1.13}$$

and, also, the matrix of Stokes parameters transformation through the formula (1.2.41).

Expressions for the characteristics of light scattering by a sphere with a number of layers greater than two are derived in a similar way (Mikulski and Murphy, 1963; Wait, 1963; Ferdinandov, 1967; Pontikis, 1968; Tuomi, 1980; Lopatin and Syd'ko, 1988). However, solutions are very cumbersome. In this view it is necessary to point out the iterative approach that permits calculations of scattering characteristics for a sphere with an almost unlimited number of layers (see Section 3.5 for details).

The specific difficulty arises in the calculation of scattering properties of polydisperse systems of TLPs. In the most general case, sizes of cores and sizes of shells vary independently from each other. The calculations of light scattering in view of such two-dimensional distribution of radii of cores and shells are rather labor-intensive and, furthermore, there are some methodological difficulties. Though such calculations are sometimes conducted, only small particle properties are being calculated in them. For example, Dotsenko and Zakharov (1974) investigated the nanoparticles (radius 2.5–15 nm), the centres of coloring in photochromic glasses. Taking this into account, calculation of scattering properties of polydisperse systems of TLPs is usually abide by some simplifying assumptions: a) the core size is constant, the shell size varies with some law; or, b) the core size obeys a given distribution, the shell radius is related to core radius in some way (e.g., in linear one); or, c) the shell width is a constant. As shown in Chapter 2, it is possible with a sufficient degree of reliability to apply variant b) for TLPs of atmospheric aerosols.

There are a wide variety of scattering objects that exist, which should be viewed as two-layered. The theory of light scattering by TLPs is used in radio-location, radar meteorology, physical chemistry, astrophysics, and biology. This once again underlines the role of scattering methods as powerful non-destroying tools for investigation of objects of the most different origins. We give a brief review of such works below. The authors' works are not mentioned here. Their results will be discussed in detail later.

Originally the model of a spherical TLP was used in radio-location and radar meteorology. Interest in such developments arose when experimental measurements and radio-location practice had shown, that the radar echo reflected from metal objects can be increased by an order or decreased almost to zero in the presence of a thin coat on the surface of the object. The majority of theoretical works of this type considered backscattering only, and, as a rule, they used different approximations defined by the peculiarity of the problem rather than the exact theory of diffraction on the coated sphere itself (detailed reviews can be found in books of Ruck (1970) and Kokhahovsky (2001)). Works in which the exact theory was used, were concerned in general with calculations of extinction and backscattering of radiowaves by metal (i.e., ideally reflecting) spheres with dielectric (Scharfman, 1954; Rheinstein, 1964; Arsaev, 1982; Bishop, 1986), ferrite (Richmond, 1987), chiral (Uslenghi, 1990), or magnetic (Klyuchnik *et al.*, 1986) coatings and by the watered hailstones (Herman and Battan, 1961; Abashev and Rozenberg, 1969; Rozenberg, 1970, 1972). The problem of backscattering from metal objects with plasma shells (Swarner and Peters, 1963) has also drawn attention. It is necessary to note that the model of the coated sphere with a homogeneous core and a radially inhomogeneous shell describes such objects more adequately. The investigations employing this model are rather diversified. Some of them will be mentioned in Section 8.4.

The works that use a model of the coated sphere to explain the optical and thermophysical properties of grains of interstellar and interplanetary dust are numerous. The detailed experimental investigations of spectral dependencies of interstellar extinction and discovery of the interstellar polarization at the end of 1940s resulted in the belief that the interstellar grains became the essential element of the theories of galactic structure. However, astronomers met considerable difficulties while attempting to explain observed spectral dependencies of interstellar absorption and polarization with the help of simpler models of homogeneous spherical and cylindrical particles of ice, graphite, silicates, and some metals (e.g., see (Greenberg, 1968)). There was a need to develop more complicated models, in particular, the model of a coated sphere. The paper of Güttler (1952) was the first work in this direction. In 1960s Wickramasinghe (1965, 1967) proposed a justified optical model of interstellar grains as graphite cores with mantles from the polluted ice. He also undertook the extensive calculations of scattering properties of such particles in visual, IR, and UV spectral ranges (Wickramasinghe *et al.*, 1966*a*, 1966*b*). The calculations for graphite particles with dielectric (in particular, ice) mantles were also completed in the works of Swamy and O'Dell (1967), Svatoš and Sekanina (1968), Wickramasinghe and Nandy (1968), Wickramasinghe and Swamy (1968), and Donn and Swamy (1969). Other models that were used rather widely in astrophysics were the coated sphere with silicate core and ice mantle (Wickramasinghe and Swamy, 1968; Aannestad, 1975) and metal particles covered by dielectric (Greenberg, 1968) and graphite hollow grains (Jones, 1988; Inagaki *et al.*, 1991; Jazbi and Wickramasinghe, 1991). The core–mantle model still exists now in a form of the so-called Oort–van de Hulst–Greenberg model (Li and Greenberg, 1997). However, it is necessary to note that new spectroscopic observations showed that abundances of heavy elements (including carbon, which is main component of many dust models) are much less than considered before. The inability to explain the observed extinction using the amount of carbon available in the solid phase resulted in so-called "carbon crisis" which has not been resolved up to now (Voshchinnikov, 2002). In this situation, it is appropriate to consider more complex models of grains, including the models of the multilayered sphere (Voshchinnikov and Mathis, 1999), the homogeneous spheroid (Voshchinnikov and Farafonov, 1993), the multilayered spheroid (Farafonov *et al.*, 1996), and the aggregate (see review of Lumme (2000)).

The investigations on absorption scattering of light by unicellular microorganisms were carried out with regard to the internal structure of cells. The purpose of which was the development of the express photometer methods of identification of microbe populations and obtaining the information about their major characteristics. The three-layered sphere with a thin surface layer imitating the cell shell, core, and the intermediate layer of cytoplasm was used as a model of actual unicellular organisms of the spherical shape (Lopatin and Syd'ko, 1977, 1978, 1988; Terskov *et al.*, 1977; Syd'ko *et al.*, 1990). In some cases, it is sufficient to use the more simple two-layered model ignoring the thin cell shell (Brunsting and Mullaney, 1972*a,b*, 1974; Wyatt and Philips, 1972; Fercher, 1975; Meyer and Brunsting, 1975; Mullaney and Fiel, 1976).

The list of works on light scattering by TLPs of chemical origin is rather extensive. In the 1960s the Kerker's research group had fulfilled a series of experiments on studying scattering properties of suspensions of artificially manufactured two-layered microspheres (Matijevič *et al.*, 1960, 1962; Kerker *et al.*, 1962, 1966; Espenscheid *et al.*, 1965; Kerker, 1969). The attendant calculations have shown very good agreement between the experimental and theoretical results. Later the same group detected the effect of clearing up of suspensions of pigment particles with air shells (Kerker, 1975; Kerker *et al.*, 1975; Chew and Kerker, 1976). Also, it is necessary to note the investigations of the influence of a thin water layer at a sphere surface on the phase function (Mehta *et al.*, 1971; Kerker and Cooke, 1972), of scattering by gold spherical particles with silver shell (Morriss and Collins, 1964) and by the oxide-coated silver particles (Lee *et al.*, 1986), of the scattering properties of color centres in photochromic glasses (Dotsenko and Zakharov, 1974; Dotsenko *et al.*, 1984), of the color properties of pigment TLPs (Dotsenko *et al.*, 1985, 1988), of optical influence of solvate shells on the water–oil and other emulsions (Eselson *et al.*, 1982; Oshchepkov and Sorokina, 1986; Ricka *et al.*, 1991), of optical model of marine suspension (Oshchepkov and Sorokina, 1989), and of the surface enhanced Raman scattering in TLPs (Kerker *et al.*, 1978; Kerker and Blatchford, 1982).

The special variety of TLPs are the hollow ones. Whereas the scattering by such particles have some specific features, we shall consider them separately in Section 3.4.

It was revealed, that the rather thin surface layers can in some situations result in very narrow absorption resonances in spherical particles (Gaunaurd *et al.*, 1980; Gaunaurd and Überall, 1981; Moser *et al.*, 1981; Pluchino, 1981; Hightower and Richardson, 1988; Lock, 1990; Kaizer *et al.*, 1994; Kai and D'Alessio, 1995; Bambino and Guimaraes, 1996). As such layers are frequently present on actual particles, the study of such effects presents an actual problem in the optics of disperse media.

Worthy of mention are also some papers of general purpose (Shah, 1985; Bhandari, 1986*a*, 1986*b*; Borghese *et al.*, 1987; Korostina and Rozanov, 1988; Helaly and Shafai, 1989; Vorontsov and Mirovitskaya, 1993; Choi *et al.*, 1996; Choi, 2001; Rysakov and Ston, 2001).

In almost all investigations mentioned above, the theory of scattering by a single TLP is used. The first works on applications of the coated sphere model for describing the scattering properties of atmospheric particles were restricted by the case of monodisperse systems. For example, in the paper of Fenn and Oser (1965) the monodisperse carbon particles covered with a water shell were considered with the purpose of estimation of these particles influence on the energy balance in the atmosphere. The evaluations of the basic optical characteristics of light scattering were carried for a very wide range of particle sizes ($0.1 < x_2 < 250$). Unfortunately, the authors of this paper selected the erroneous value of the carbon refractive index, that considerably lowered the value of the results. The particles with an absorbing core were also considered in the work of Ryym (1974) (for details see Section 3.3). In the paper of Pilat, (1967) the integral characteristics of scattering by atmospheric particles consisting of a salt core, covered by a coal shell were calculated. The calculations were carried out for different ratios of the core size to shell (at constant particle external diameter). The size parameter corresponding to the external particle size varied between the interval $0 < x_2 < 10$. It is necessary to note, that the carbon-containing particles are not the typical representatives of atmospheric aerosols, although they play a very important role in the thermal balance of atmosphere, especially in industrial regions (see, e.g., Chylek *et al.* (1984, 1988)).

The number of works devoted to optical properties of polydisperse systems of TLPs of atmospheric aerosols are few. Apparently, Witt's paper (1968) was the first. It was an attempt to construct the optical model of particles in the mesosphere (silvery clouds). Witt assumed that these particles were coated spheres with a dielectric or metal core and an ice shell of constant width (0.05 μm), and that the variations of external particle radius satisfy to the Junge distribution. On these assumptions the calculation of the phase function and angular dependence of the polarization degree for two wavelengths in the visual range was carried out. On the basis of other data on mesospheric aerosol microphysics Krekov and Rakhimov (1975*b*, 1982) have carried out numerical investigations of optical properties of silvery clouds with assumptions which, apparently, more adequately meet the actual nature of this aerosol. Their purpose was setting up the numerical experiments on laser sensing of silvery clouds from the cosmos. The model of the coated sphere with a metal (iron) core and contaminated ice shell was used. The core radius did not depend on the particle radius and was equal 0.05 μm, and the spectrum of the particles size was described by a generalized gamma distribution (Deirmenjian H-haze with the maximal radius of 1 μm). The calculation of the scattering matrix elements was carried out for a series of wavelengths used at laser remote sensing.

The rather specific case of a polydisperse distribution of three-layered particles was considered by Harrison *et al.* (1972). The authors made an attempt to describe the scattering properties of an ensemble of randomly oriented atmospheric aerosol particles of irregular form by a polydisperse ensemble of spherically symmetric three-layered particles.

Kattawar and Hood (1976) calculated the phase functions and polarization degree for a polydisperse ensemble of TLPs with carbon cores and water shells at wavelength $\lambda = 0.49$ μm. They viewed two cases: a) the width of water shell is constant, and the integration is over the core radius; and b) the ratio of core radius to external particle radius is constant, and integration is over the external radius. The modal radius was 1 μm, and the low and upper integration limits were selected at 0.1 and 4 μm accordingly. On the basis of calculations, the authors concluded that using the volume-averaged refractive index for describing the scattering properties of non-uniform particles of the above size range is incorrect. The interference effects caused by the presence of a thin water shell, strongly changed the scattering matrix. However, the results of this paper are only qualitative, as the authors used the particle size distribution distant from reality.

Of interest are the complex theoretical–experimental investigations of influence of condensation cores on optical properties of liquid aerosols described in the works of Savel'ev and Fadeev (1975), Krekov *et al.* (1976), Krekov and Rakhimov (1982), Ippolitov *et al.* (1987), and Rakhimov (1998). In atmospheric conditions the process of water droplet formation goes, as a rule, in the presence of condensation cores. The presence of these cores results in droplet inhomogeneity, that changes its optical properties. The complex experiments on investigation of optical properties of polydisperse water aerosols were carried out in an aerosol chamber and included measurements of the basic elements of the scattering matrix and microstructural parameters of the medium. The measurements were accompanied by calculation of the relevant scattering characteristics for models consisting of homogeneous droplets and droplets with cores at their centre. Based on the analysis of microstructure data, the authors deduced, that in most cases the droplet size distribution is quite satisfactorily described by the gamma-distribution with arithmetic mean radius R varying between 1.75–7.5 μm, and with the half-width parameter μ being 1.8–6.4. The measurements of phase function and the angular dependence of polarization of scattered radiation were carried out between a wide range of scattering angles θ=3–176°. In so doing the particle concentration in the chamber was such, that the regime of single scattering was realized. The error of obtained experimental results did not exceed 10%. The calculations were carried out for the gamma-distribution with parameters $R_0 = 3.0(1.0)7.0\ \mu m$ and $\mu = 2(2)8$ and integration limits 0.1 and 15 μm. The measured and calculated characteristics of the water aerosol consisting of TLPs were compared at $\lambda = 0.69\ \mu m$; the models with constant core radius $R_1 = 0.1$ and 0.5 μm, and also the model with core radius equal a quarter of the droplet radius, were considered. The contaminated dust with refractive index $N = 1.5 + i0.2$ was used as the matter of the condensation core.

The analysis of experimental data has shown, that the shape of the phase function depends on duration of measurement. In the initial period of fog existence the experimental phase function has three distinctly expressed small maxima at θ=22, 60, and 97°. These maxima are absent in the phase functions calculated for the ensemble of homogeneous particles; but do exist for droplets with cores a quarter of a droplet radius. With increasing duration of measurement the indicated maxima disappear and the distinction of the experimental phase function from that calculated assuming droplet homogeneity does not exceed the level of measurement error. The authors suppose, that at an initial period of fog existence in the chamber the particles have the two-layered structure, and then the particles become homogeneous due to partial dissolution and washing away of hygroscopic cores. The presence of the core in the droplet results in the change of the shape of the calculated phase function, that agrees half-qualitatively with experimental results. The same deduction was obtained on the basis of dependencies of polarization degree versus angle.

Krekov and Rakhimov (1982) have also carried out the numerical investigation of the polydisperse system of TLPs, whose core has the refractive index $N = 1.5 + i0.2$ (dry heels), with the shell representing the 50% solution of sulfuric acid. For these calculations the experimental histogram of the particle size distribution measured at a height of 5 km was used. The authors came to the conclusion, that for estimation of the basic energetic characteristics it is possible to use the model of homogeneous particles with an accuracy of 1–2% (i.e., the calculations may be carried out by Mie theory with the effective refractive index estimated according to the volume content of components). However at calculation of backscattering and, generally, scattering under angles exceeding $\theta > 165°$ they failed to select an effective refractive index. That is, the particles with the two-layered structure have qualitative features exhibiting at remote sensing measurements. The same authors carried out a series of other calculations on the light scattering characteristics for different types of TLP ensembles. The results of these investigations are given in the book of Krekov and Rakhimov (1982).

Tuomi (1980) has considered light scattering by an ensemble of model particles consisting in a dry state of a non-soluble core with a soluble shell. As the relative humidity increases the third layer (saturated solution of the shell substance) forms on the particle surface. At some critical humidity the particles become two-layered again, and at even greater humidity the solution, being in this case a shell of a non-soluble core, becomes dilute. Tuomi has traced changes of the optical characteristics of this particle ensemble at its sequential transformation. As the results obtained in this work coincide with ours in many respects, we shall look at them in more detail in Section 3.3.

As the results of some microphysical investigations show, the core of a two-layered atmospheric aerosol particle is frequently shifted with regards to the sphere centre. If we take into account this shift, it is necessary to solve the problem of diffraction by the sphere with a non-concentric spherical core. Although the formal mathematical solution of this problem has long been known (see, e.g., Ivanov, 1968; Rozenberg, 1972; Fikioris and Usunoglu, 1979), its numerical realization (Borghese *et al.*, 1994; Ngo *et al.*, 1996; Chýlek *et al.*, 1988, 1998; Fuller and Mackowski, 2000; Ioannidou and Chrissoulidus, 2002) is not so easy. On our opinion, taking into account the similar shift for atmospheric optics problems is hardly expedient, as, on the one hand, introduction of this model improvement strongly complicates the calculations, and on the other hand, in actual aerosols the influence of this core shift will be "covered" due to polydispersity, different shifts in particles of the ensemble and so on.

3.2 EFFECTS OF CORE AND SHELL RADII AND OPTICAL CONSTANTS ON ABSORPTION OF RADIATION BY TWO-LAYERED PARTICLES

In this section the size interval of cores and shells of atmospheric TLPs was selected on the basis of works (Junge, 1965; Khrgian, 1969; Green and Lane, 1972; Cadle, 1976; McCartney, 1979) that were analyzed in Section 2.2. The intervals of optical constants of cores $1.1 \leq n_1 \leq 1.7$; $0 \leq \kappa_1 \leq 0.1$, and shells $n_2 = 1.33$; $0 \leq \kappa_2 \leq 0.1$ were large enough to completely involve the variants considered in Section 1.3. Such broad ranges were chosen both for the greater generality of obtained results, and for revealing the tendencies of the light scattering characteristics by TLPs under change of optical constants of core or shell. The results for $n_1 \leq 1.4$ can are applied to other objects (e.g., particles of biological origin, particles of latexes with a surface film, and so on.

Consideration of light absorption by TLPs begins with the treatment of an effect, which represents both the theoretical, and the practical interest. Noted for the first time in the work of Fenn and Oser (1965), this effect is that if the particle core has even small absorption, then occurrence of a non-absorbing shell, and increase of its width, results in the absorption cross section of the obtained TLP increasing to begin with and then oscillating about an average value $\sigma_{\text{abs}}^{\text{asym}}$, which for brevity we shall term an asymptotic one (Prishivalko and Astafyeva, 1974). The oscillation amplitude depends on the optical constants and particle core radius (Figures 3.2 and 3.3).

By way of illustration the values of $\sigma_{\text{abs}}^{\text{asym}}$ and maximal deviations from these values $\Delta\sigma_{\text{abs}}^{\text{asym}}$ for TLPs with $R_1 = 0.1\ \mu m$ are given in Table 3.1. The increase of κ_1 at constant n_1 is accompanied by growth of oscillation amplitude. For low values of n_1 its ratio to $\sigma_{\text{abs}}^{\text{asym}}$ decreases from 10–12% at $\kappa_1 = 0.005$–0.01 to 7–8% at $\kappa_1 = 0.15$–0.20. For higher n_1 values the relative amount of deviation remains approximately in the range 4–6%. By comparison, at $R_1 = 1\ \mu m$ and fixed n_1 the oscillation amplitude varies little with the increase of κ_1, and its ratio to $\sigma_{\text{abs}}^{\text{asym}}$ rapidly decreases. It is seen from Table 3.1 also, that at $R_1 = 0.1\ \mu m$ and fixed core absorption the oscillations decrease with growth of n_1 both absolutely and relatively, if $\kappa_1 = 0.005$–0.05, and at $\kappa_1 > 0.05$ have the minimal values, when $n_1 \approx 1.5$.

In a case of $R_1 = 1\ \mu m$ and small $\kappa_1 \approx 0.01$ the oscillations of σ_{abs} increase in absolute value with n_1, and at $n_1 = 1.7$ make approximately 4% of $\sigma_{\text{abs}}^{\text{asym}}$. At higher κ_1 their amplitude in relation to $\sigma_{\text{abs}}^{\text{asym}}$ decreases.

The shell width, at which the asymptotic value of the absorption cross section is achieved, depends largely on the core size and to a lesser extent on its optical constants. The smaller the core, the greater the shell width that is necessary for reaching $\sigma_{\text{abs}}^{\text{asym}}$. For TLPs with small cores ($x_1 < 2$) the $\sigma_{\text{abs}}^{\text{asym}}$ is reached at $R_2/R_1 \approx 2.8$–3.0 for $n_1 = 1.1$; at $R_2/R_1 \approx 2.5$ for $n_1 = 1.3$–1.5; and at $R_2/R_1 \approx 1.9$–2.1 for $n_1 = 1.7$. In the case of TLPs with larger cores ($x_1 > 3$) the achievement of an asymptotic value depends to a lesser extent on core optical constants and size. This value is reached at a shell width of 40–60% the core radius. With an increase of core absorption the shell width, at which $\sigma_{\text{abs}}^{\text{asym}}$ is reached, increases to a some extent (Prishivalko and Astafyeva, 1974; 1975).

The greatest relative increase of σ_{abs} while water shell formation occurs for cores with $x_1 \lesssim 1$ throughout the entire interval of values κ_1 considered, and at $\kappa_1 > 0.05$ for $x_1 \lesssim 8$. In these cases it increases by more than 60% in comparison with the core absorption section (see Figure 3.3).

For TLPs having cores with $x_1 < 3.5$, the relative increase of the absorption section depends weakly on the κ_1 value. From $x_1 \approx 4$–5 and on, the influence of κ_1 becomes more essential. For particles with large cores ($x_1 > 10$) the quantity σ_{abs} at $\kappa_1 > 0.05$ again weakly depends on κ_1 (see Figure 3.3). In the case of large cores with small κ_1, oscillating values of $\sigma_{\text{abs}}^{\text{lay}}$ can be below $\sigma_{\text{abs}}^{\text{core}}$ at the initial stage of water-covering.

It is possible to explain the observed increase of absorption section for TLPs in the course of their water-covering in that the non-absorbing water shell focuses on absorbing core an energy flow which penetrates inside the particle, thus increasing the efficiency of its absorption.

The dependencies of the $\sigma_{\text{abs}}^{\text{asym}}$ for the water-covered particles versus the optical constants and sizes of cores can be found in Table 3.2. The relations approximating these results are obtained in the work of Prishivalko and Astafyeva (1975).

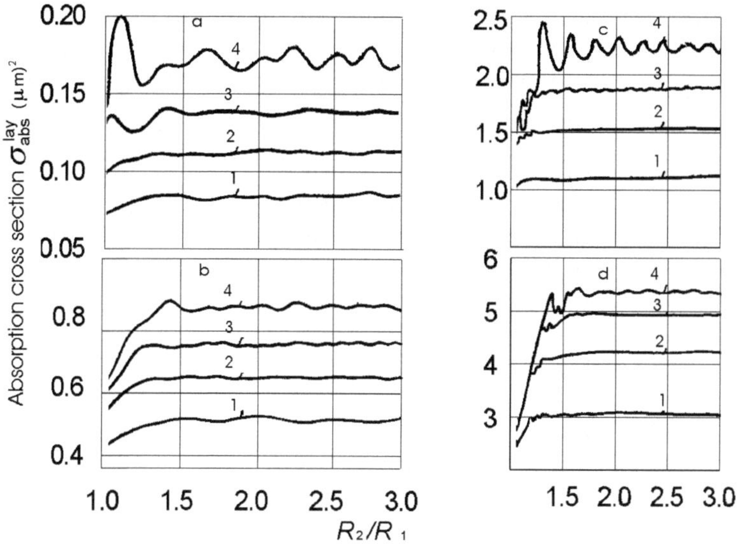

Figure 3.2. Dependencies of asymptotic value σ_{abs}^{lay} versus the ratio R_2/R_1 for $R_1 = 0.4$ (*a,b*) and 1 μm (*c,d*) at $\lambda = 0.5\ \mu m$, $n_2 = 1.33$, $\kappa_2 = 0$, and $\kappa_1 = 0.01$ (*a,c*) and 0.1 (*b,d*), $n_1 = 1.1$ (*1*), 1.33 (*2*), 1.5 (*3*), and 1.7 (*4*).

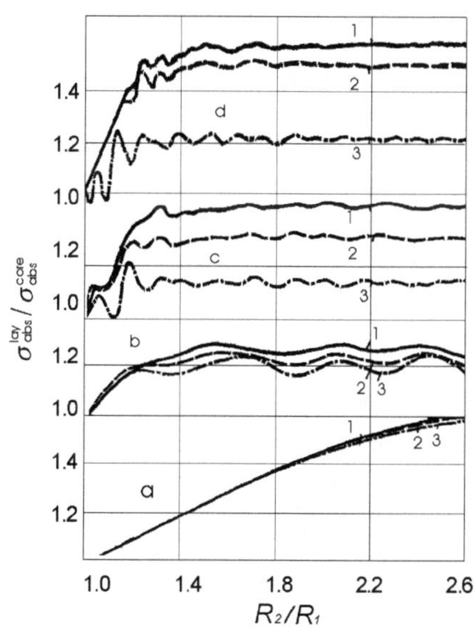

Figure 3.3. Dependencies of the ratio of absorption cross section of TLP to absorption cross section of corresponding dry core versus R_2/R_1 at $\lambda = 0.5\ \mu m$, $n_1 = 1.5$, $n_2 = 1.33$, $\kappa_2 = 0$, $\kappa_1 = 0.1(1), 0.05(2), 0.01(3)$, and $R_1 = 0.05(a), 0.25(b), 0.7(c)$, and $1\ \mu m\ (d)$.

Table 3.1. Values of $\sigma_{\text{abs}}^{\text{asym}}$ and $\Delta\sigma_{\text{abs}}^{\text{asym}}(\mu m^2 \cdot 10^2)$ for particles with $R_1 = 0.1\ \mu m$, $\lambda = 0.5\ \mu m$, $n_2 = 1.33$, and $\kappa_2 = 0$.

	$n_1 = 1.2$		$n_1 = 1.5$		$n_1 = 1.7$	
κ_1	$\sigma_{\text{abs}}^{\text{asym}}$	$\pm\Delta\sigma_{\text{abs}}^{\text{asym}}$	$\sigma_{\text{abs}}^{\text{asym}}$	$\pm\Delta\sigma_{\text{abs}}^{\text{asym}}$	$\sigma_{\text{abs}}^{\text{asym}}$	$\pm\Delta\sigma_{\text{abs}}^{\text{asym}}$
0.05	0.0864	0.009	0.103	0.06	0.112	0.005
0.010	0.172	0.017	0.203	0.014	0.221	0.014
0.015	0.256	0.026	0.304	0.018	0.331	0.015
0.025	0.421	0.041	0.499	0.028	0.543	0.024
0.050	0.819	0.072	0.965	0.45	1.05	0.04
0.075	1.19	0.11	1.40	0.06	1.52	0.07
0.100	1.54	0.12	1.82	0.06	1.98	0.10
0.125	1.86	0.14	2.18	0.08	2.35	0.11
0.150	2.17	0.16	2.54	0.09	2.73	0.13
0.175	2.45	0.18	2.88	0.10	3.08	0.15
0.200	2.74	0.18	3.19	0.10	3.40	0.17

In particular, it was established that at fixed values of R_1 and κ_1 in the intervals $1.1 \leqslant n_1 \leqslant 1.7$ and at $x_1 < 15$ the following approximate dependence takes place:

$$\sigma_{\text{abs}}^{\text{asym}} = \alpha\kappa_1^2 + \beta\kappa_1 + \varepsilon\kappa_1^{\delta} n_1 + \gamma \tag{3.2.1}$$

With values of coefficients from Table 3.3, the formula (3.2.1) gives asymptotic values of absorption sections of TLPs with a relative error not exceeding 5% at $R_1 = 0.1\ \mu m$ and $0 \leqslant \kappa_1 \leqslant 0.2$, no more than 10% for $R_1 = 0.4\ \mu m$ and $0 \leqslant \kappa_1 \leqslant 0.1$, and no more than 20% for $R_1 = 1.0\ \mu m$ and $0 \leqslant \kappa_1 \leqslant 0.05$. At fixed κ_1 values the linear dependence of $\sigma_{\text{abs}}^{\text{asym}}$ versus the core refractive index takes place. The practical interest represents the dependence of $\sigma_{\text{abs}}^{\text{asym}}$ versus the core size of water-covered particles expressed by the exponential function:

$$\sigma_{\text{abs}}^{\text{asym}} = A(n_1, \kappa_1) x_1^{B(n_1, \kappa_1)} \tag{3.2.2}$$

The values of the coefficients A and B are given in Table 3.3 also. For λ=0.5 μm, $0.1\ \mu m \leqslant R_1 \leqslant 1\ \mu m$ and $n_1 = 1.2$ the error of Eq.(3.2.2) does not exceed 1.5% at $\kappa_1 = 0.01$, 7% at $\kappa_1 = 0.05$, and 10% at $\kappa_1 = 0.1$. If $n_1 = 1.7$, then at the same κ_1 the error is no more than 9, 15 and 18% accordingly. The error of Eq.(3.2.2) grows with an increase of $n_1$and κ_1. It should not be applied in the case of TLPs having small cores with $x_1 < 1$, at $n_1 > 1.5$ and $\kappa_1 > 0.1$. For particles with large cores ($x_1 < 15$) it can give only a very approximate estimation of $\sigma_{\text{abs}}^{\text{asym}}$.

If the relative humidity is such that the absorption section asymptote is achieved, then for known n_1, κ_1, and λ and for a given size distribution function of condensation cores it is possible with the help of Eq.(3.2.2) to find an approximate value of the absorption coefficient K_{abs} for a single volume of scattering medium. For the Junge type distribution we have:

$$K_{\text{abs}} = C_N A \left(\frac{2\pi}{\lambda}\right)^B \left(\frac{\nu - 1}{B - \nu + 1}\right) R_{d_1}^{\nu-1} \left(R_{d_2}^{B-\nu+1} - R_{d_1}^{B-\nu+1}\right) \tag{3.2.3}$$

Table 3.2. Values of $\sigma_{abs}^{asym}(\mu m^2)$ at $\lambda = 0.5\ \mu m$, $n_2 = 1.33$, and $\kappa_2 = 0$.

	κ_1								
	0.01	0.05	0.1	0.01	0.05	0.1	0.01	0.05	0.1
n_1	$R_1 = 0.1\ \mu m$			$R_1 = 0.4\ \mu m$			$R_1 = 1\ \mu m$		
1.1	1.60–3	7.62–3	1.44–2	0.0830	0.327	0.513	1.10	3.20	4.09
1.2	1.72–3	8.19–3	1.54–2	0.0944	0.369	0.570	1.29	3.67	4.58
1.33	1.87–3	8.87–3	1.66–2	0.111	0.426	0.648	1.55	4.30	5.21
1.4	1.95–3	9.22–3	1.73–2	0.121	0.459	0.691	1.70	4.64	5.53
1.5	2.03–3	9.65–3	1.82–2	0.136	0.522	0.757	1.92	5.11	5.93
1.6	2.11–3	1.00–2	1.91–2	0.150	0.561	0.816	2.13	5.48	6.21
1.7	2.21–3	1.05–2	2.04–2	0.169	0.612	0.875	2.30	5.70	6.34

	n_1								
	1.2	1.5	1.7	1.2	1.5	1.7	1.2	1.5	1.7
κ_1	$R_1 = 0.1\ \mu m$			$R_1 = 0.4\ \mu m$			$R_1 = 1\ \mu m$		
0.005	8.96–3	1.03–3	1.12–3	0.049	0.070	0.085	0.70	1.06	1.27
0.015	2.56–3	3.04–3	3.31–3	0.137	0.196	0.243	1.78	2.62	3.10
0.025	4.21–3	4.99–3	5.43–3	0.214	0.302	0.372	2.54	3.69	4.26
0.075	1.19–2	1.40–2	1.52–2	0.484	0.652	0.769	4.24	5.69	6.18
0.125	1.86–2	2.18–2	2.35–2	0.937	0.827	0.947	4.79	6.03	6.37
0.150	2.17–2	2.54–2	2.73–2	0.690	0.880	0.996	4.94	6.07	6.36
0.175	2.45–2	2.88–2	3.08–2	0.733	0.918	1.03	5.05	6.09	6.33
0.200	2.74–2	3.19–2	3.40–2	0.767	0.948	1.05	5.13	6.08	6.30

where $R_{d1}$and R_{d2} are the minimal and maximal core radii for TLPs.

Notice that in all considered cases the TLP core was solid, and the shell was water. In this case the condition $n_2 < n_1$ is satisfied in the visual wavelength range. The opposite relation for refractive indexes can take place (e.g., the formation on hydrometeors (water droplets with radii $R \geq 50\ \mu m$) of films from another fluid). If this film has a refractive index smaller than that of water then some absorption efficiency increase can be observed. But the considerable part of fluids capable of forming films on the water surface, have a higher refractive index $(n_2 > n_1)$. The formation of such films on the surface of hydrometeors causes a decrease of absorption efficiency (Prishivalko, 1983, 1984).

As seen from Figures 3.4–3.5, the absorption efficiency factor Q_{abs} increases almost linearly with the radius of the homogeneous water droplet of the size range covered. During formation of the thin film the oscillating dependence $Q_{abs}(R_2)$ runs for all $n_2 > n_1$ below the analogous dependence for the homogeneous water droplet. In the first case (see Figure 3.4) the location of the first extremum of $Q_{abs}(R_2)$ is described with reasonable accuracy by the interference condition for the non-absorbing shell:

$$2\,(R_2 - R_1)\,n = j\lambda/2 \tag{3.2.4}$$

which holds rigorously for plane films (Rozenberg, 1958). In the case of Figure 3.5 this condition determines the extremum locations as a rough approximation: the extrema are located at the R_2 values that are somewhat greater than those calculated by Eq.(3.2.4).

Table 3.3. Coefficients of the Eqs.(3.2.1) and (3.2.2) at $\lambda = 0.5\ \mu m$, $n_2 = 1.33$ and $\kappa_2 = 0$.

	$\kappa_1 = 0.01$		$\kappa_1 = 0.05$		$\kappa_1 = 0.1$	
n_1	$A \cdot 10^3\ \mu m^2$	B	$A \cdot 10^3\ \mu m^2$	B	$A \cdot 10^2\ \mu m^2$	B
1.2	0.895	2.88	4.68	2.66	0.935	2.48
1.5	1.13	2.95	5.53	2.67	1.40	2.42
1.7	1.18	3.026	6.35	2.74	1.24	2.52

$R_1, \mu m$	$\alpha, \mu m^2$	$\beta, \mu m^2$	$\gamma, \mu m^2$	$\varepsilon, \mu m^2$	δ
0.1	0	0.0485	-0.56·10^{-4}	0.0639	0.878
0.4	21.1	-3.40	-0.0379	3.17	0.676
1.0	702.8	-37.1	-1.10	22.1	0.500

This shift is greater, the smaller the difference $\Delta n = n_2 - n_1$. At $j \geq 3n_2$ the form of dependence $Q_{\text{abs}}(R_2)$ gradually ceases its regularity. Absorption in the shells of TLPs of atmospheric aerosols may be somewhat distinct from absorption in pure water due to presence of soluble impurities. In normal natural conditions this difference is small. But it can grow when the high-absorbing soluble products of industrial activity discharge into the atmosphere or when substances of such type pulverize artificially. As the purpose of the present work is not only to consider daily implemented situations, but also to detect the common tendencies of the changes of TLP optical properties, thus, the analyzed interval of κ_2 values is selected with a wide enough range. First, we shall consider a non-absorbing particle core.

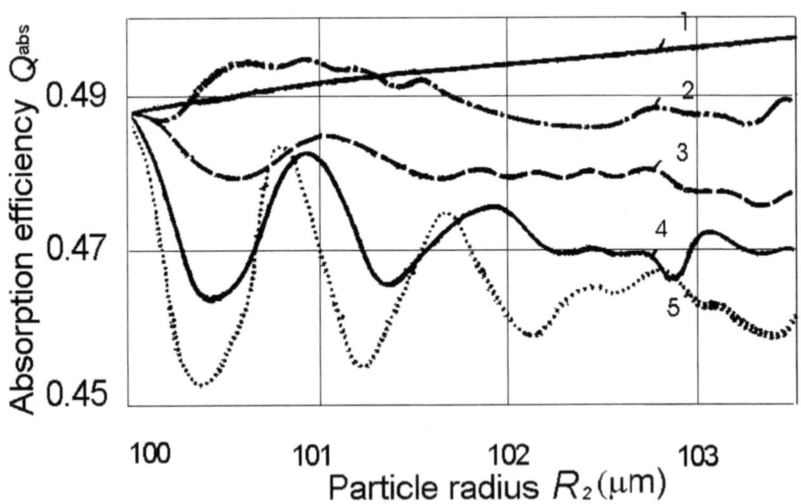

Figure 3.4. Dependencies of the absorption efficiency factor for homogeneous water droplets (*1*) and droplets with shells (*2–5*) versus the external particle radius at $\lambda = 2.36\ \mu m$, $R_1 = 100\ \mu m$, $n_1 = 1.274$, $\kappa_1 = 7.6 \cdot 10^{-4}$, $\kappa_2 = 0$, and $n_2 = 1.15\,(2)\,, 1.35(3), 1.5(4)$, and $1.6(5)$.

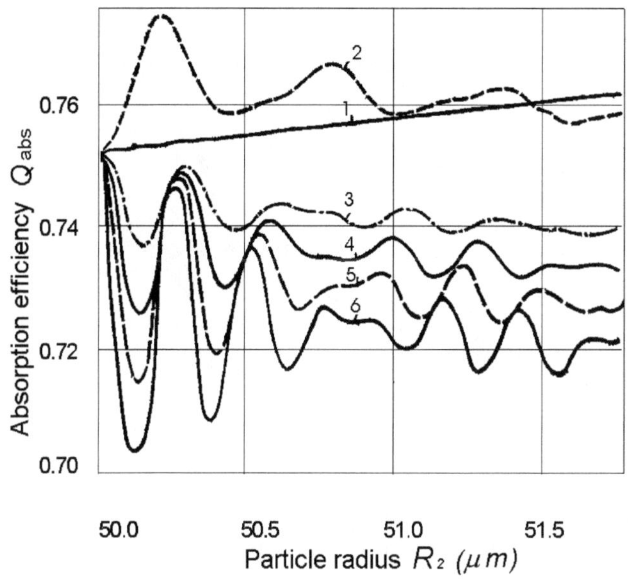

Figure 3.5. Dependencies of the absorption efficiency factor for colored homogeneous water droplets (*1*) and droplets with shells (*2–6*) versus the external particle radius at $\lambda = 0.6943\ \mu m$, $R_1 = 50\ \mu m$, $n_1 = 1.33$, $\kappa_1 = 10^{-3}$, $\kappa_2 = 0$, and $n_2 = 1.2\,(2)\,, 1.4(3),\ 1.45(4), 1.5(5)$, and $1.56(6)$.

The calculations have shown, that with growth of the absorbing shell σ_{abs}^{lay} increasing, the greater are values of R_2 and κ_2. In the case of larger cores the absorption section has only a weak dependence on κ_2 in the range $\kappa_2 = 0.05$–0.1 (Figure 3.6*a,b*).

To clarify the core influence on absorbing properties of TLPs with absorbing shells, we compared the absorption sections of these particles and absorption sections of homogeneous ones with the same external size and optical constants coincident with shell constants (Figure 3.6*c,d*). At a small shell absorption index ($\kappa_2 \leq 0.01$) the ratio $\sigma_{abs}^{lay}/\sigma_{abs}^{hom}$ is approximately constant for TLPs with cores of the different sizes (Prishivalko and Astafyeva, 1975). At $R_2/R_1 \approx 1.15$ the absorption section of the TLP is approximately three times lower, and at $R_2/R_1 \approx 1.25$ twice lower, than absorption sections of the relevant homogeneous particle (HP). In the interval $1.1 \leq R_2/R_1 \leq 1.6$ the curves corresponding to larger cores ($x_1 > 3$) oscillate slightly relative to monotonically increasing curves for particles with small cores ($x_1 < 3$). At $R_2/R_1 > 1.6$ the ratio of absorption sections is independent of the core radius. With growth of R_2/R_1 it smoothly increases, asymptotically approaching unity. For values $R_2/R_1 > 2.2$ the difference of σ_{abs} for homogeneous and TLP comprises no more than 10%. The influence of the non-absorbing core becomes practically insignificant.

In the case of more stronger shell absorption the value of σ_{abs} for TLPs depends to a greater extent on the initial core size (Figure 3.6*d*). Increasing the ratio R_2/R_1, the value of σ_{abs} for TLP approaches the value for HP considerably faster than at $\kappa_2 \approx 0.01$. The ratio $/\sigma_{abs}^{hom} = 0.9$ is achieved for $R_1 = 0.25\ \mu m$ at $R_2/R_1 = 1.76$ and for $R_1 = 1\ \mu m$

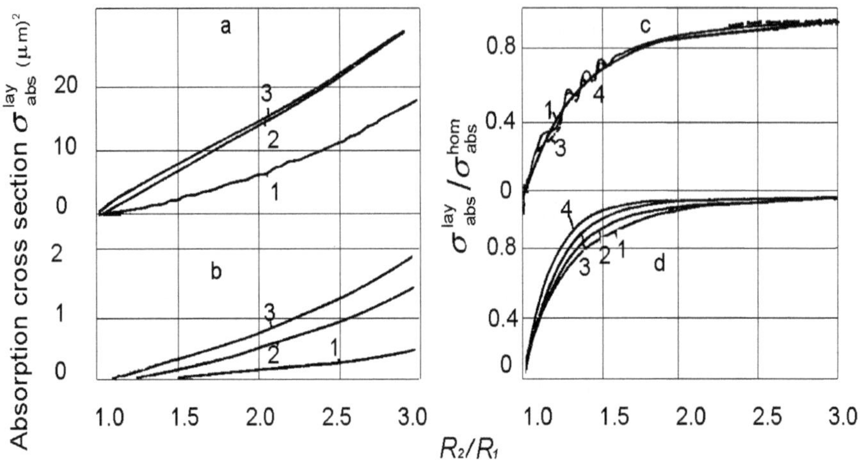

Figure 3.6. Dependencies of $\sigma_{\text{abs}}^{\text{lay}}$ (*a,b*) and $\sigma_{\text{abs}}^{\text{lay}}/\sigma_{\text{abs}}^{\text{hom}}$ (*c,d*) versus R_2/R_1 at $R_1 = 1\ \mu m$ (*a*) and $0.25\ \mu m$ (*b*), $\kappa_2 = 0.01$ (*c*) and 0.1 (*d*) at $\lambda = 0.6943\ \mu m$, $n_1 = 1.5, \kappa_1 = 0, n_2 = 1.33$; *a, b*: $\kappa_2 = 0.01(1), 0.05(2), 0.1(3)$; *c, d*: $R_1 = 0.25(1), 0.4(2), 0.7(3), 1\ \mu m(4)$.

at $R_2/R_1 = 1.4$. At the small shell absorption index the distinction of optical properties of TLP and HP depends mainly on the ratio R_2/R_1, whereas at greater κ_2 the influence of the core radius manifests itself essentially as well.

With the help of curves in Figure 3.7 it is possible to find the absorbing shell width $\Delta R = R_2 - R_1$, at which the absorption section of TLP with a given core radius coincides (with the error of 3, 5, or 10%) with the absorption section of the HP of the same size.

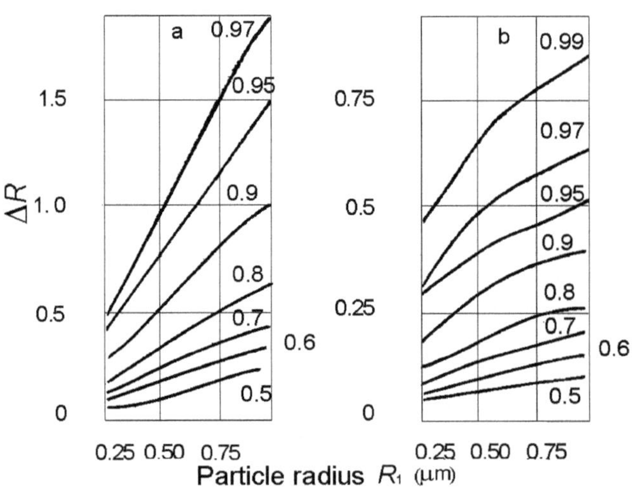

Figure 3.7. Nomographs to define the shell thickness which result in the particular value (indicated near curves) of the ratio $\sigma_{\text{abs}}^{\text{lay}}/\sigma_{\text{abs}}^{\text{hom}}$ for the TLPs with given radius R_1 at λ=0.6943 μm, $\kappa_2 = 0.01$(*a*) and 0.1 (*b*).

Table 3.4. Values of in Eq.(3.2.5) for TLPs at $\lambda = 0.5$ μm, $R_1 = 0.4$ μm, $n_1 = 1.5$ and $n_2 = 1.33$.

	R_2/R_1				
κ_2	1.2	1.5	1.7	2.0	3.0
0.01	0.48	0.26	0.16	0.08	0.02
0.1	0.07	0.019	0.010	0.004	0.0002

The small absorption in the particle core ($\kappa_1 \approx 0.05$) most noticeably manifests itself in the initial period of water-covering, when the particle shell, possessing higher absorption coefficient, is not yet the deciding factor in the common pattern of absorption due to its small width. The contribution of the core absorption at this stage can be estimated by the ratio:

$$\eta = \left(\sigma_{\text{abs}}^{\text{lay}} \Big|_{\kappa_1=0.005} - \sigma_{\text{abs}}^{\text{lay}} \Big|_{\kappa_1=0} \right) \left(\sigma_{\text{abs}}^{\text{lay}} \Big|_{\kappa_1=0} \right)^{-1} \tag{3.2.5}$$

Some typical values of the ratio are presented in Table 3.4. A similar pattern takes place with other core sizes as well. The change of the absorption section with an increase of the core refractive index is most essential at small ratios R_2/R_1.

At the end of this section we shall return once again to hydrometeors with alien fluid films possessing intrinsic absorption. From Figure 3.8 it is evident, that with growth of

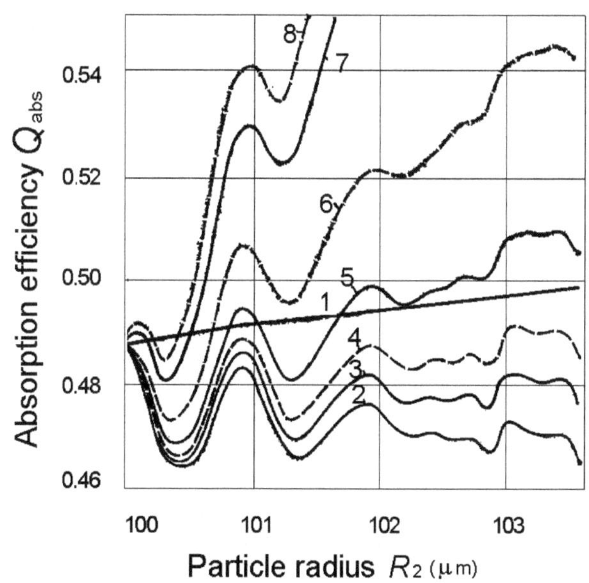

Figure 3.8. Dependencies of absorption efficiencies for homogeneous water droplets (*1*) and droplets with absorbing shell (*2–8*) versus the external radius at $\lambda = 2.36$ μm, $R_1 = 100$ μm, $n_1 = 1.274$, $\kappa_1 = 7.6 \cdot 10^{-4}$, $n_2 = 1.5$, and $\kappa_2 = 0(2)$, $5 \cdot 10^{-4}(3)$, $10^{-3}(4)$, $2 \cdot 10^{-3}(5)$, $4 \cdot 10^{-3}(6)$, $8 \cdot 10^{-3}(7)$, and $10^{-2}(8)$.

κ_2 the absorption efficiency of such TLPs increases, however dependencies $Q_{\text{abs}}^{\text{lay}}(R_2)$ maintain oscillating character in the range of extrema satisfying the condition $j \leq 3n_2$. In this connection we can note, that with growth of κ_2 the minima of dependence $Q_{\text{abs}}^{\text{lay}}(R_2)$ are shifting in the direction of reduced thickness, and maxima, on the contrary, are shifting in the direction of increased thickness. With a decrease in the film refractive index and approximating n_2 to n_1 the oscillations become less obvious.

3.3 EFFECT OF CORE AND SHELL RADII AND OPTICAL CONSTANTS ON SCATTERING AND EXTINCTION BY TWO-LAYERED PARTICLES

At first, let us find out the relation of absorbed and scattered energy at the interaction of light with TLPs. This relation is determined the quantum survival probability $\Lambda = \sigma_{\text{sc}}^{\text{lay}}/\sigma_{\text{ext}}^{\text{lay}}$ (also known as the single-scattering albedo). From Figure 3.9 it is seen, that if the core is small ($x_1 < 1.5$), then at the initial stage of water-covering there is an area of the particle sizes, where at $\kappa_1 = 0.05$ and 0.1 absorption noticeably exceeds scattering. The lowering of albedo to $\Lambda < 0.5$ is also detected for particles with larger cores ($x_1 > 6$) at $\kappa_1 = 0.1$. At $\kappa_1 = 0.05$ in this area of the particle sizes the share of scattered energy exceeds the share of absorbed energy by 4–6% only. In the remaining cases investigated, the scattering predominates above absorption (Prishivalko *et al.*, 1975).

At the first stage of particle water-covering, the albedo Λ rather strongly depends on a specific combination of R_1, n_1, and κ_1 values. At $R_2/R_1 > 2$ the influence of these parameters becomes less noticeable. The further increase of the aqueous shell thickness is accompanied by increasing albedo, which for large particles approaches the asymptotic value: $\Lambda = 1 - \sigma_{\text{abs}}^{\text{asymp}}/2\pi R_2^2$. This implies, that for great particles the single-scattering albedo tends to unity. If the overall TLP radius is approximately 7–10 times more than the core radius, then the influence of the latter on the particle scattering properties becomes negligible. The share of absorbed energy is negligibly small. Increasing the aqueous shell thickness is accompanied by an increase of the scattering section. What is inherent is the greater the core absorption index, the faster the growth of $\sigma_{\text{sc}}^{\text{lay}}$ (Figure 3.10).

The special practical interest represents comparison of scattering properties of the atmospheric aerosol TLPs with the relevant properties of homogeneous water droplets of the same size $\sigma_{\text{sc}}^{\text{lay}} / \sigma_{\text{sc}}^{\text{wat}}$ (Figure 3.11). For TLPs with small cores ($x_1 < 3.5$) the ratio $\sigma_{\text{sc}}^{\text{lay}} / \sigma_{\text{sc}}^{\text{wat}}$ tends to unity at $R_2/R_1 = 2$–4. For larger cores the analyzed dependence has an oscillating nature with gradually decreasing amplitudes of oscillations. The scattering section of a TLP with the moderate aqueous shell thickness can be 1.5–2 times greater, and in some cases 2–2.5 times lower, than for homogeneous water particles of the same size. The ratio $\sigma_{\text{sc}}^{\text{lay}} / \sigma_{\text{sc}}^{\text{wat}}$ becomes practically equal to unity when the external particle size is 7–10 times as large as the core size. For the extinction section of TLPs the results are rather close to those obtained for the scattering section. The basic difference is that the extinction section is influenced less noticeably by the core absorption. This is clearly seen from Figure 3.12.

For TLPs with small cores ($R_1 < 0.25\ \mu m$ at $\lambda = 0.5\ \mu m$) the extinction comparison was carried out both with water droplets, and with HPs from the core substance of the same size (Astafyeva and Prishivalko, 1978). The results are given in Table 3.5, from which it is seen, that the difference in the extinction sections of water-covered particles and water droplets does not exceed 10%, if the aqueous shell thickness is 1.7–3.8 times more than core radius. In the case of non-absorbing cores the difference $\sigma_{\text{ext}}^{\text{lay}}$ from $\sigma_{\text{ext}}^{\text{hom}}$ does not exceed 10% at $R_2/R_1 \lesssim 1.08$. Light extinction by TLPs is comparable with extinction by HPs comprising from a core substance at small water-covering only, and with extinction of water droplets – at considerable water-covering. If we compare the extinction sections of a TLP and a model particle with the same radius and refractive index equal to a volume-averaged refractive index of the relevant TLP, it turns out that in the range $1 \leq R_2/R_1 \leq 10$ the difference does not exceed 15%.

It is shown in Figure 3.9, that with an increase of the non-absorbing aqueous shell the influence of the absorbing core gradually weakens and the properties of TLPs come nearer to the properties of a non-absorbing homogeneous water droplet. In the case of a non-absorbing core, but absorbing shell (Figure 3.13), a somewhat alternative pattern is observed. The increase of the absorbing shell neutralizes, rather rapidly, the non-absorbing cores influence, and the amount of absorbed and scattered energy becomes practically identical ($\Lambda = 0.5$). Naturally, for higher κ_2 this process goes faster. For larger cores at $R_2/R_1 \gtrsim 1.7$–1.9 and $\kappa_2 > 0.05$ the run of curves depends only slightly on quantity κ_2, and at $R_2/R_1 \gtrsim 2.5$ this dependence practically disappears (Prishivalko *et al.*, 1975).

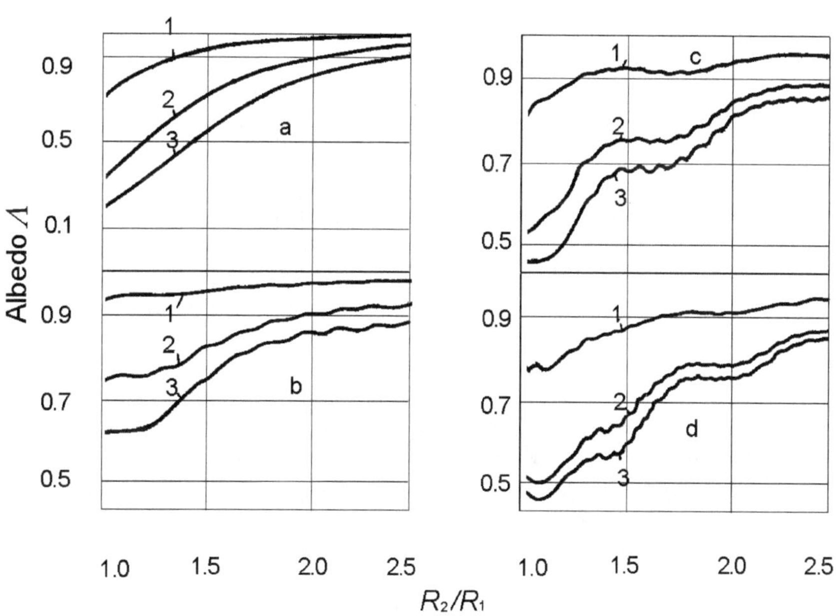

Figure 3.9. Single-scattering albedo versus R_2/R_1 for core radius $R_1 = 0.05(a), 0.4(b), 0.7(c)$, and $1\ (d)\ \mu m$ at $\lambda = 0.5\ \mu m$, $n_1 = 1.5$, $n_2 = 1.33$, $\kappa_2 = 0$, and $\kappa_1 = 0.01(1), 0.05(2)$, and $0.1(3)$.

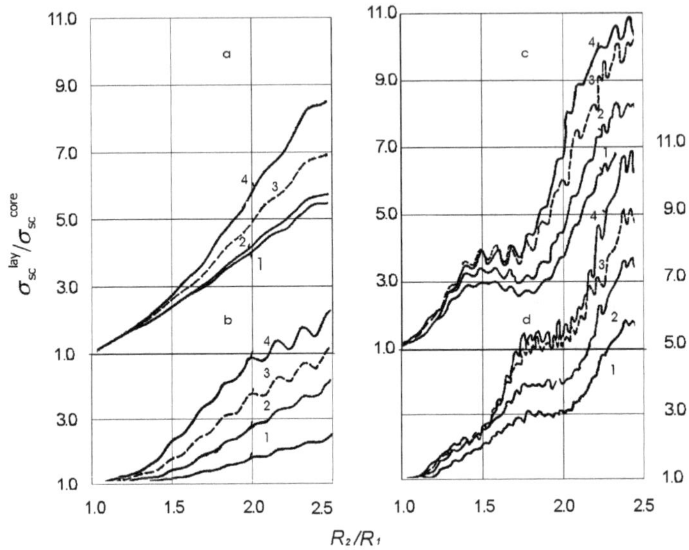

Figure 3.10. Ratio $\sigma_{sc}^{lay}/\sigma_{sc}^{core}$ versus R_2/R_1 for $R_1 = 0.25(a), 0.4(b), 0.7(c)$, and $1.0(d)$ μm at $\lambda = 0.5$ μm, $n_1 = 1.5$, $n_2 = 1.33$, $\kappa_2 = 0$, and $\kappa_1 = 0.1(1), 0.01(2), 0.05(3)$, and $0.1(4)$.

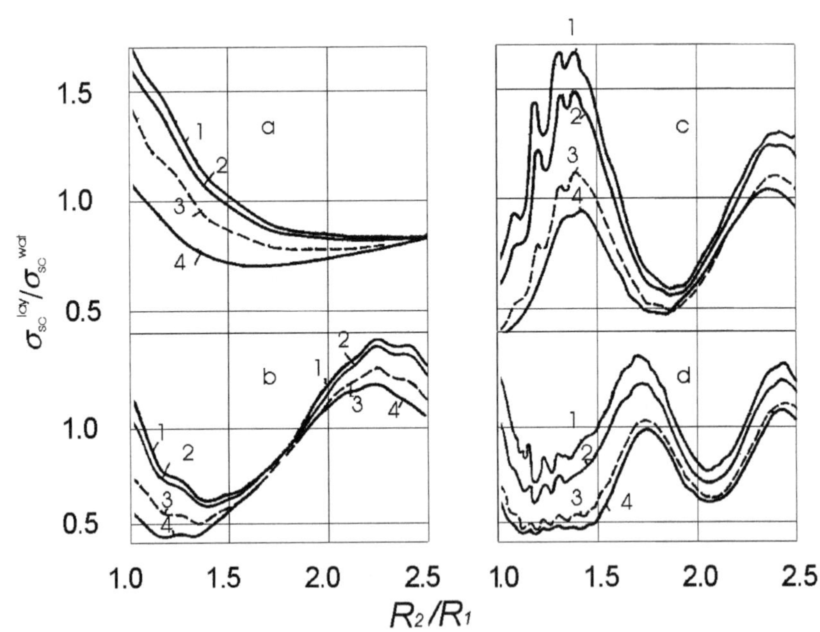

Figure 3.11. Ratio $\sigma_{sc}^{lay}/\sigma_{sc}^{wat}$ versus R_2/R_1 with the same notations, as in Figure 3.10.

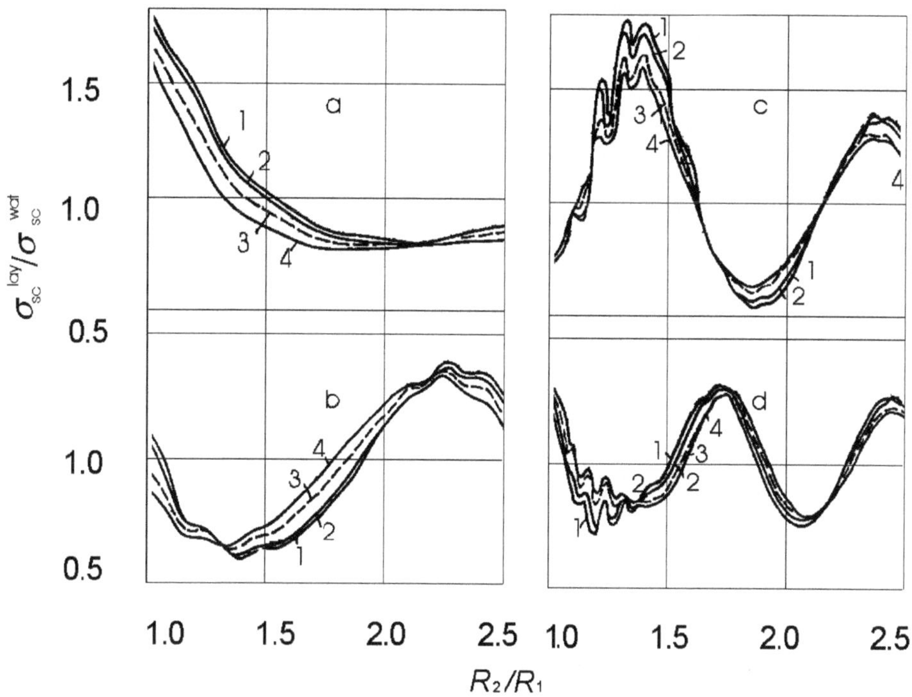

Figure 3.12. Ratio $\sigma_{\text{ext}}^{\text{lay}}/\sigma_{\text{ext}}^{\text{wat}}$ versus R_2/R_1 with the same notation, as in Figure 3.10.

Note, that for small cores the dependence $\Lambda\,(R_2/R_1)$ in a rather wide interval of the ratio R_2/R_1 is practically linear. For example, for $x_1 \approx$ 2–2.5 this linearity takes place at $R_2/R_1 \geq 1.25$ for $\kappa_2 = 0.01$, at $R_2/R_1 \geq 1.5$ for $\kappa_2 = 0.05$, and at $R_2/R_1 \geq 1.9$ for $\kappa_2 = 0.1$. In the case of the core with $x_1 \approx 3.5$ the almost linear nature of the $\Lambda\,(R_2/R_1)$ dependence is observed for $\kappa_2 = 0.01$. For larger cores ($x_1 > 4.5$), there is a drastic decrease of scattering albedo at the first stage of shell formation. At $\kappa_2 = 0.1$ the value $\Lambda = 0.5$ is reached at the ratio $R_2/R_1 \approx$ 1.3–1.6 that is at a shell thickness of about 0.3–0.6 μm.

Figure 3.14 allows us to trace the dynamics of the influence of the shell rise on the particle scattering properties. Its comparison with Figure 3.10 clearly shows the flattening activity of shell absorption on the form of dependencies $\sigma_{\text{sc}}^{\text{lay}}/\sigma_{\text{sc}}^{\text{coRe}}$ versus (R_2/R_1), especially for large cores. An inverse mutual arrangement of curves corresponding to different κ_2 values also obtains our attention. With the growth of the non-absorbing shell the scattering section for a TLP having a high core absorption index increases faster; but if absorbtion in a core is absent then a more rapid increase of scattering section is observed for a TLP with a low-absorbing shell. At core radius $R_1 \geq 0.7$ μm the curves corresponding to $\kappa_2 = 0.05$ and 0.1 almost coincide, that is at $\kappa_2 \geq 0.05$ the difference in the shell absorption index is of no concern.

Table 3.5. Left part: the lower border of the range of the R_2/R_1 values where the ratio $\sigma_{\text{ext}}^{\text{lay}}/\sigma_{\text{ext}}^{\text{wat}}$ is below a specified value F. Right part: the upper border of the range of the R_2/R_1 values where the ratio $\sigma_{\text{ext}}^{\text{lay}}/\sigma_{\text{ext}}^{\text{hom}}$ is below a specified value F ($n_1 = 1.5$, $n_2 = 1.33$, $\kappa_2 = 0$, $\lambda = 0.5$ μm). The first lines correspond to $\kappa_1 = 0$ and the second lines to $\kappa_1 = 0.1$.

	F									
	1.5	1.3	1.2	1.1	1.05	1.5	1.3	1.2	1.1	1.05
R_1 (μm)	minimal R_2/R_1					maximal R_2/R_1				
0.05	1.34	1.66	1.96	2.52	3.08	2.20	1.28	1.16	1.08	1.04
	1.94	2.20	2.44	2.92	3.40	1.38	1.16	1.10	1.06	1.04
0.10	1.48	1.73	1.95	2.40	2.70	4.90	1.29	1.18	1.09	1.04
	1.92	1.80	1.95	2.30	2.50	5.00	1.22	1.16	1.09	1.05
0.15	1.37	1.60	1.73	2.00	2.23	3.28	1.37	1.25	1.09	1.04
	1.36	1.53	1.63	1.80	2.80	3.33	1.43	1.22	1.09	1.04
0.20	1.27	1.40	1.55	1.70	4.80	2.49	2.41	1.32	1.08	1.04
	1.20	1.29	1.36	2.20	4.50	2.90	2.35	1.38	1.12	1.05
0.25	1.16	1.28	1.34	3.80	4.08	2.02	1.95	1.91	1.09	1.05
	1.04	1.15	1.20	3.60	3.92	–	2.00	1.85	1.30	1.06

From a practical point of view, it is interesting to compare the scattering by TLPs and HPs consisting of a shell substance and having the same external size. From Figure 3.15 it is seen that the curves are essentially distinct for $x_1 < 4$ and for $x_1 > 5$. Non-absorbing cores without shells at $x_1 < 4$ have the scattering section 2–3 times higher than that for absorbing HPs with the same size. Occurrence and growth of the absorbing shell results in sharp diminution of this ratio. Though with shell growth and external size increase the scattering section of the TLP increases, but it becomes almost the same, as for HPs of the same size. At $R_2/R_1 > 1.5$ their scattering sections coincide within $\pm 10\%$. Further shell thickness increase is accompanied by small oscillations of the ratio $\sigma_{\text{sc}}^{\text{lay}} / \sigma_{\text{sc}}^{\text{hom}}$ near unity. In the case of cores with $x_1 > 5.5$ the calculated dependence $\sigma_{\text{sc}}^{\text{lay}} / \sigma_{\text{sc}}^{\text{hom}}$ versus (R_2/R_1) oscillates. The least oscillation amplitude, as a result of flattening activity of shell absorption, is observed for higher κ_2 and larger core.

Comparison of the light extinction sections for TLPs and HPs of the shell substance at identical external sizes yields the same pattern, as comparison of scattering sections. The ratios $\sigma_{\text{ext}}^{\text{lay}} / \sigma_{\text{ext}}^{\text{hom}}$ and $\sigma_{\text{sc}}^{\text{lay}} / \sigma_{\text{sc}}^{\text{hom}}$ differ a little at identical values of parameters. In the cases studied by us, these ratios differed no more than by 15–20%. Notice, that the development of small core absorption ($\kappa_1 \lesssim 0.005$) is not reflected on the scattering and extinction sections of this type of TLP. In this case the TLP scattering properties are almost completely determined by the shell optical constants. The small change in core absorbing properties is not exhibited.

Consider further the quite typical example of TLPs with non-absorbing cores and strongly absorbing shells (core – aluminium oxide, shell – soot). From the previous text it follows, that at a particular shell thickness the scattering properties of TLPs practically do not differ from the properties of particles with the same sizes consisting of the shell substance. To distinguish such particles is impossible. This raises the question of at what

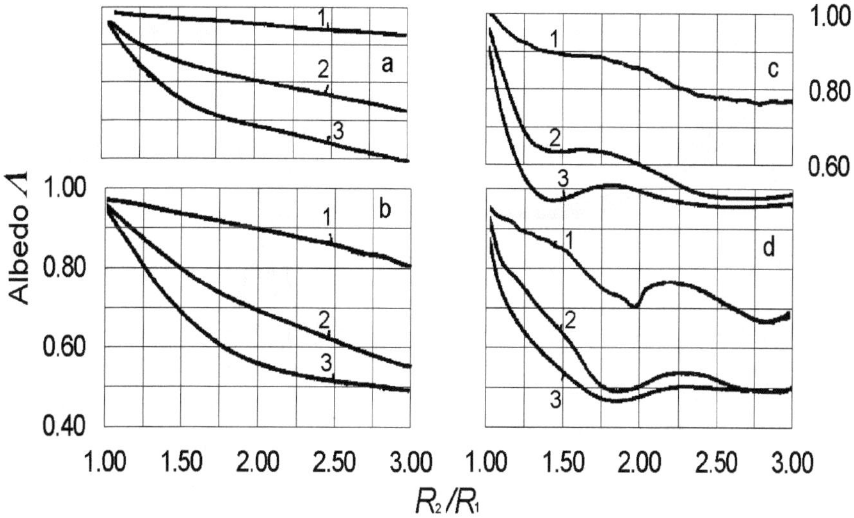

Figure 3.13. Single-scattering albedo versus R_2/R_1 for $R_1 = 0.25(a)$, $0.4(b)$, $0.7(c)$, and $1.0(d)$ μm at $\lambda = 0.6943$ μm, $n_1 = 1.5$, $\kappa_1 = 0$, $n_2 = 1.33$, and $\kappa_2 = 0.01(1), 0.05(2)$, and $0.1(3)$.

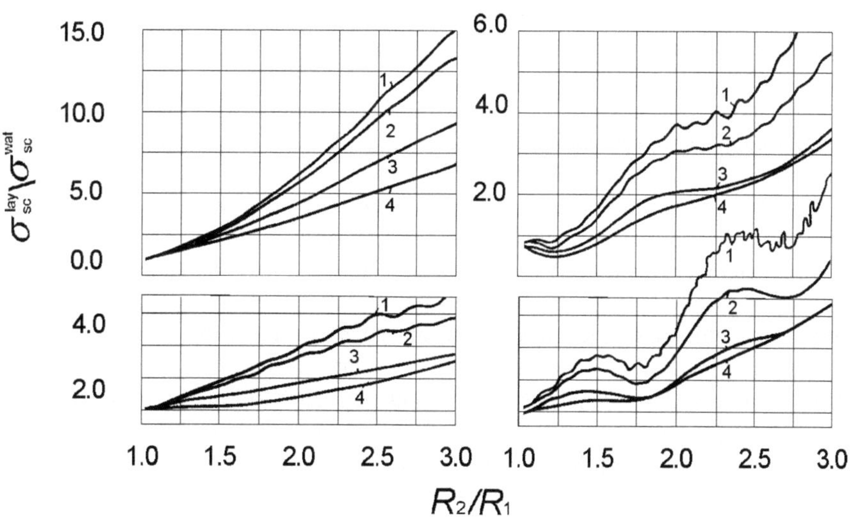

Figure 3.14. The ratio $\sigma_{sc}^{lay}/\sigma_{sc}^{core}$ versus R_2/R_1 for $R_1 = 0.25(a)$, $0.4(b)$, $0.7(c)$, and $1.0(d)$ μm at $\lambda = 0.6943$ μm, $n_1 = 1.5$, $\kappa_1 = 0$, $n_2 = 1.33$, $\kappa_2 = 0(1)$, $0.01(2)$, $0.05(3)$, and $0.1(4)$.

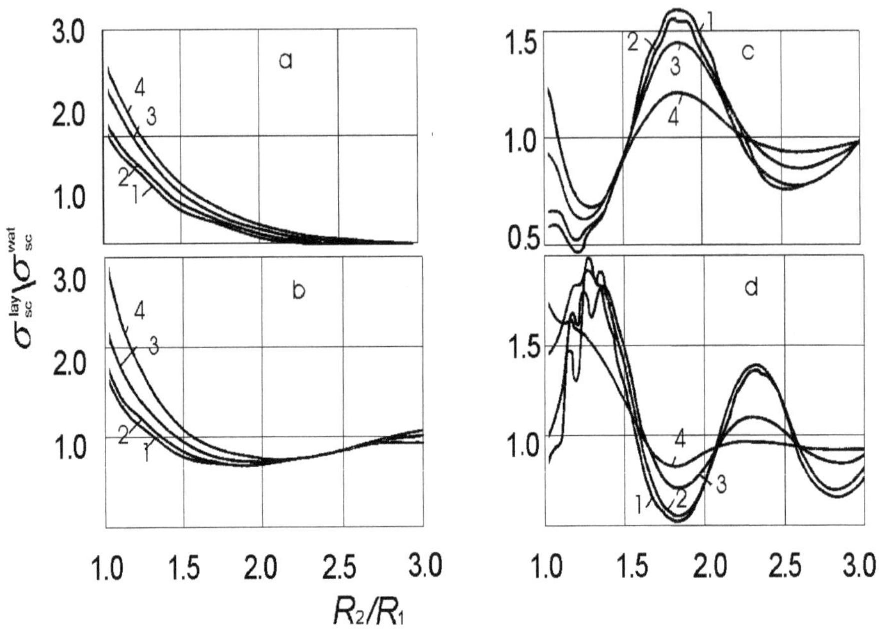

Figure 3.15. The ratio $\sigma_{sc}^{lay}/\sigma_{sc}^{hom}$ versus R_2/R_1 with the same notation, as in Figure 3.14.

ratio of core and shell sizes this distinction disappears. In Table 3.6 the shell thickness and ratio R_2/R_1 are given, starting with which the extinction efficiency of TLPs differs from the extinction efficiency of soot particles by no more than $\Delta Q_{ext} = \pm 0.03$. In percentage terms this difference is no more than 1.4% for great particles and about 1% for small ones. From Table 3.6 it is seen, that if for small particles the shell thickness neutralizing the core influence should exceed its radius by not less than 2.5 times, then at $x_1 = 20$ the shell should constitute 5–7% of the core radius.

Light absorption and scattering by TLPs were compared above with absorption and scattering by HPs, both comprising the core of TLPs, and consisting of the core substance, but having external sizes as TLPs. In contrast to this, some authors discovered the scattering properties of TLPs alone. For example, Ryym (1974) calculated the absorption, scattering, and extinction efficiencies versus the external particle radius for TLPs with iron cores and water shells at $\lambda = 0.589\ \mu m$ and a constant ratio $q = R_1/R_2$. Pointing out the oscillating nature of dependencies $Q_{sc}^{lay}(R_2)$ and $Q_{ext}^{lay}(R_2)$, Ryym explained it by interference of beams transmitting through the shell with the incident primary field. By analogy with reasonings of van de Hulst (1981) for HPs, he derived for TLPs with absorbing cores and non-absorbing shells the expression for external particle radii, at which the maxima and minima of $Q_{sc}(R_2)$ and $Q_{ext}(R_2)$ are observed:

$$R_2 = \frac{\lambda\,(j + 0.5)}{4\,(N_2 - 1)\,\sqrt{1 - q^2}} \tag{3.3.1}$$

Table 3.6. The values of the absorbing shell thickness and ratio R_2/R_1 at which extinction efficiencies for TLPs (core – alumina, shell – soot) and soot HPs coincide. At $\lambda = 1.0\mu m$, $N_1 = 1.755 + 0i$ (Plass, 1964), $N_2 = 1.95 + 1.02i$ (Blokh, 1967) and at $\lambda = 0.5\ \mu m$, $N_1 = 1.75 + 10^{-6}i$ (Plass, 1964), $N_2 = 1.72 + 0.93i$ (Blokh, 1967).

$R_1(\mu m)$	$\Delta R(\mu m)$	R_2/R_1	$R_1(\mu m)$	$\Delta R(\mu m)$	R_2/R_1
$\lambda = 1.0\ \mu m$			$\lambda = 0.5\ \mu m$		
0.1	0.18	2.80	0.15	0.12	1.80
0.3	0.21	1.70	0.25	0.16	1.64
0.5	0.26	1.52	0.40	0.14	1.35
0.8	0.24	1.30	0.50	0.11	1.22
1.0	0.22	1.22	0.70	0.11	1.16
2.0	0.18	1.09	1.00	0.11	1.11
3.0	0.18	1.06	1.50	0.11	1.07

where $q = R_2/R_1$ is treated as a parameter. At fixed core radius, from the theory of light interference in thin films (Rozenberg, 1958) the analogous condition for extrema of $Q_{sc}(R_2)$ and $Q_{ext}(R_2)$ follows:

$$R_2 - R_1 = j\lambda/4N^2 \tag{3.3.2}$$

but in Eq.(3.3.1) the odd j values give the maxima of these dependences, and in Eq.(3.3.2) they correspond to minima. From these relations it is possible to find an approximate distance between the two next extrema that is useful for an estimation of an integration step at calculations of the light scattering characteristics for polydisperse particle systems. From Eq.(3.3.1) we have, that at constant q, the distance between two adjacent maxima (or minima) is approximated by:

$$\Delta R_2 = \frac{\lambda}{2\left(N_2 - 1\right)\sqrt{1 - q^2}} \tag{3.3.3}$$

At constant R_2 we have accordingly:

$$|\Delta R_1| = \frac{\lambda}{2N_2} \tag{3.3.4}$$

Interestingly, the core refractive index does not enter into Eqs.(3.3.3) and (3.3.4). This implies, that the dependencies of the extrema locations and their amplitudes on N_1 are weak. This fact can be attributed to high absorption of cores that, in this case, does not differ in essence from the absolutely absorbing one.

3.4 SPECIFIC FEATURES OF LIGHT SCATTERING AND ABSORPTION BY HOLLOW PARTICLES

One of the simplest types of non-uniform particles are hollow particles (HOPs). Such particles arise at volcanic eruptions as the result of fuming from the not yet solidified particles (volcanic dust). The HOPs are frequently found among hailstones (Rozenberg, 1972). In laboratory conditions the HOPs are formed at the pulverization of solutions of solid substances in volatile fluids (Green and Lane, 1972; Aragon and Elwenspoek, 1982; Marston *et al.*, 1988; Strawbridge and Hallett, 1992).

One of the first detailed investigations of scattering and absorption of light by HOPs was carried out by Dombrowskiy (1974*a*). At first, properties of the hollow Rayleigh particles were studied. Is was shown, that such particles scatter and absorb the radiation more weekly than HPs from the same substance with the same external radius. The ratios $z_{\rm abs} = Q^{\rm lay}_{\rm abs}/Q^{\rm hom}_{\rm abs}$ and $z_{\rm sc} = Q^{\rm lay}_{\rm sc}/Q^{\rm hom}_{\rm sc}$ at all values of the ratio R_2/R_1 are less than unity (notation $Q^{\rm lay}$ refers to HOP and $Q^{\rm hom}$ to the homogeneous one). Distinction between the scattering and absorbing properties of HOPs and HPs reduces, the greater the refractive index.

To compare optical properties of small HOPs and HPs of the same mass, whose radii are connected by the relation $R_{\rm hom} = R_{\rm lay}\sqrt[3]{1-q^3}$, where $q = R_1/R_2$, it is enough to consider the ratios $z'_{\rm abs} = z_{\rm abs}\left(1-q^3\right)^{-1/3}$ and $z'_{\rm sc} = z_{\rm sc}\left(1-q^3\right)^{-1/3}$. As the denominators in these ratios vary from 1 to 0 while increasing q from 0 to 1, then $z'_{\rm abs} > z_{\rm abs}$, $z'_{\rm sc} > z_{\rm sc}$. Under certain conditions (e.g., at $n \approx 2.0$) the absorption efficiency of a HOP can be more, than a HP of the same mass.

In case of great HOPs the cavity can both decrease and increase the light absorption and scattering by HOPs, and also change the phase function. Considering the dependencies of the extinction, scattering, and absorption efficiencies on size parameter x_2 for HOPs and comparing it with the corresponding dependencies for HPs, Dombrovskiy (1974*a*) has come to the conclusion that the cavity causes an increase of the main oscillation period of dependencies $Q_{\rm ext}(x_2)$ and $Q_{\rm sc}(x_2)$ and corresponding shift of the main maximum. As the first approximation, the deformation of the corresponding curves at an increasing of ratio q can be described as a stretching along the abscissa with the coefficient approximately equal to $(1-q^2)$ and weakly dependent on the refractive index. At not too great values of q the extremes of secondary oscillations ("ripple structure") on curves $Q_{sc}(x_2)$ maintain their locations, which testifies their identical origin both in HPs and in HOPs as a result of interference of surface waves and diffracted light (Irvine, 1965; van de Hulst, 1981).

At the relative cavity radius $q \gtrsim 0.5$ the dependencies in hand maintain a periodic structure, but the amplitude of main oscillations diminishes non-monotonically as x_2 increases, and double main maxima evolve. As q increases further the curves regularity is disturbed and the amplitude of the first main maximum noticeably diminishes. Though it is possible to describe qualitatively this stretching of curves by the introduction of an effective refractive index, the calculations indicate that use of the effective refractive index will yield satisfactory results for small particles only. Further we shall return to the question of the effective refractive index for TLPs and shall consider it in greater detail at the analysis of scattering properties of quasi-HPs (Sections 10.2 and 10.3).

The regularities noted above, are exhibited in Figure 3.16 and for absorbing shells. The presence of absorption results in decreasing of oscillation amplitude. It is seen from Figure 3.16*a*, that the HOP absorbs more weakly, than the HP of the same radius. For the values of the complex refractive index studied by Dombrovskiy (1974*a*), the difference of the absorption efficiency of HOPs from the corresponding characteristic of HPs of the same radius becomes noticeable at relative cavity radius $q \gtrsim 0.5$.

The deformation of curves $Q_{\text{abs}}(x_2)$ while q increases can be interpreted also by some stretching along the abscissa with approximately the same coefficient as in the case of $Q_{\text{ext}}(x_2)$ (Figure 3.16*b*). On this basis it is recommended to utilize optical properties of HPs with the effective radius R^* for the estimation of optical properties of HOPs. This effective radius does not depend on the refractive index and can be determined from the relation:

$$R^* = \frac{R_2}{2}\left(1 - q^2 + \sqrt{1 - q^2}\right) \tag{3.4.1}$$

It was shown by Domdrovskiy (1974*b*), that, for optically thick layers, the use of a model of HPs with effective radius R^* allows us to describe radiation in the plane-parallel layer of HOPs with an error of no more than 5%.

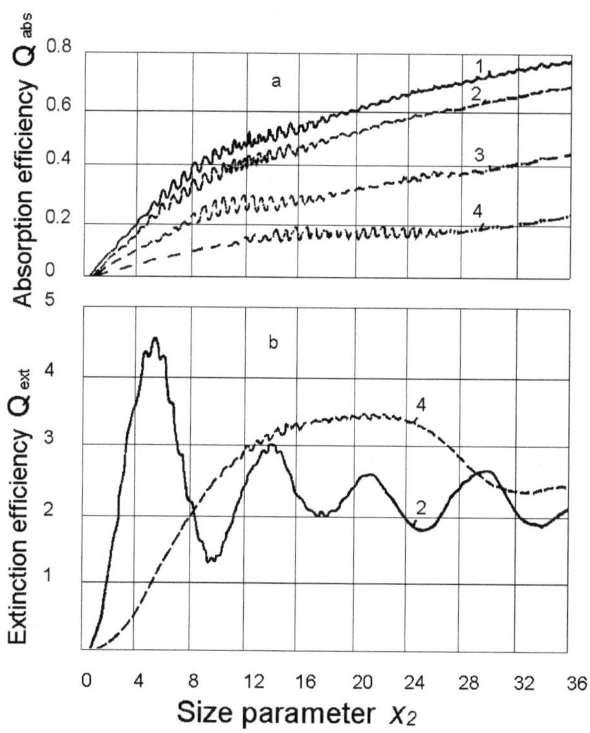

Figure 3.16. Efficiencies of absorption (a) and extinction (b) versus x_2 at $n_1 = 1.0$, $\kappa_1 = 0$, $n_2 = 1.5$, $\kappa_2 = 0.01$, and $q = R_1/R_2 = 0(1), 0.5(2), 0.75(3)$, and $0.9(4)$.

One of the actual problems, bound to radiation scattering and extinction by HOPs, is at of interference effects in non-absorbing or low-absorbing shells, which in a variety of cases can essentially influence aerosol properties (Chew and Kerker, 1976; Kattawar and Hood, 1976; Naumenko and Prishivalko, 1981).

The extinction coefficient of radiation with wavelength λ for monodisperse aerosols from TLPs is equal to (see Eq.(1.4.3)):

$$K_{12}^{\text{ext}} = C_N \pi R_2^2 Q_{12}^{\text{ext}} \left(R_1, R_2, n_1, n_2, \kappa_1, \kappa_2, \lambda\right) \tag{3.4.2}$$

where $C_N(\text{cm}^{-1})$ is the number of particles in the unit volume and Q_{12} is the extinction efficiency for individual TLPs. For HOPs Eq.(3.4.2) may be written as:

$$K_{02}^{\text{ext}} = C_N \pi R_2^2 Q_{02}^{\text{ext}} \left(R_1, R_2, n_2, \kappa_2, \lambda\right) \tag{3.4.2a}$$

At known mass concentration $C_M(\text{g}\cdot\text{cm}^{-3})$ the extinction coefficient of unit volume of such aerosol is equal to:

$$K_{02}^{\text{ext}} = \frac{3}{4} \frac{C_M R_2^2 Q_{02}^{\text{ext}}}{\left(R_2^3 - R_1^3\right) \rho_2} \tag{3.4.3}$$

where $\rho_2(\text{g}\cdot\text{cm}^{-3})$ is the density of the shell substance. For HPs consisting of the shell substance and having the size equal to radius of HOPs, the extinction coefficient looks like:

$$K_2^{\text{ext}} = \frac{3}{4} \frac{C_M Q_2^{\text{ext}} \left(R_2, n_2, \kappa_2, \lambda\right)}{\rho_2 R_2} \tag{3.4.4}$$

To reveal the specifics of radiation scattered by HOPs in comparison with scattering by HPs consisting of the same substance, it is convenient to operate with the ratio of the unit volume extinction coefficients for both aerosol types. From Eq.(3.4.3) and Eq.(3.4.4) we have:

$$\alpha_{02} = \frac{K_{02}^{\text{ext}}}{K_2^{\text{ext}}} = \frac{Q_{02}^{\text{ext}}}{Q_2^{\text{ext}}(1 - q^3)} \tag{3.4.5}$$

The ratio α_{02} characterizes distinctions between the optical properties of aerosols from HPs and HOPs at identical external size and identical mass concentration. The difference of properties of these aerosols at identical numerical concentration is defined primarily by the ratio of the corresponding extinction efficiencies.

If the radiation wavelength is more than the particle size, the HOP can be considered as an inhomogeneity with some effective characteristics. In a first approximation the optical properties are defined by the volume-averaged complex refractive index of the shell substance (Prishivalko and Babenko, 1975; Prishivalko, 1977). The averaging results in decreasing the values n and κ, that is, the particle becomes "more soft". Hence, the extinction coefficient of HOP aerosols with $x \leq 3$–4 is always less than the extinction coefficient of aerosols from solid particles of the corresponding size at identical numerical concentrations.

For small HPs with $\kappa \neq 0$ at $x < 1$ absorption is proportional to particle volume. By analogy it is possible to expect, that the absorption by small HOPs will also be proportional to the absorbing substance volume. Taking into account, that at $x \ll 1$ absorption is much more than scattering, it is possible to consider that $Q_{\text{ext}} \approx Q_{\text{abs}}$. Then the light extinction efficiency for HOP can be associated with Q_2^{ext} for solid particles by the following approximate relation:

$$Q_{02}^{\text{ext}} \approx Q_2^{\text{ext}}(1-q^3) \tag{3.4.6}$$

It follows from the calculated data of Naumenko and Prishivalko (1981), that in the region $1.2 \leq n_2 \leq 1.8$, $0.1 \leq \kappa_2 \leq 0.5$ for particles with $x < 0.5$–1, Eq.(3.4.6) is valid with an error no more than 5%. From here it follows, that $\alpha_{02} \approx 1$.

If the particle size is comparable with the radiation wavelength, or exceeds it, the extinction efficiency of HOPs under certain conditions can be more than the corresponding values for continuous particles owing to interference of beams reflected from outside, and internal, shell boundaries (Chew and Kerker, 1976).

At irradiating by the directional light beam two, almost axial, beams are formed due to reflection on exterior and internal boundaries of the TLP, whose intensities are defined by the reflection coefficients on the opposite shell surfaces. Utilizing the model of the plane layer with thickness equal to shell thickness, we have, that the reflection coefficient from the TLP surface has extreme values under conditions approximately described by the relation:

$$2\,(R_2 - R_1)\,n_2 = j\lambda/2 \tag{3.4.7}$$

Calculations show that maximal influence of interference on extinction efficiency occurs at:

$$d \equiv R_2 - R_1 \approx \frac{\lambda}{2n_2} \tag{3.4.8}$$

when the reflection coefficient from the HOP surface is minimal. It is necessary to note, that the interference effects are observed, if radiation intensities reflected from exterior and internal shell boundaries differ not too greatly.

With the purpose of studying the influence of the interference effects on the extinction coefficients of aerosols from HOPs, the extinction efficiencies for hollow and continuous particles with the sizes $R_2 = 1(0.5)5\ \mu m$ at $\lambda = 0.5\ \mu m$ in the wide interval of aerosol substance optical constants were calculated (Naumenko and Prishivalko, 1981). In Figure 3.17 the dependencies $\Delta Q_{\text{ext}} = \left(Q_2^{\text{ext}} - Q_{02}^{\text{ext}}\right)/Q_2^{\text{ext}}$ versus q value are plotted. In particles, whose shells consist of an absorbing substance with n_2 slightly distinguished from the surrounding medium refractive index (Figure 3.17*a*), the interference effects do not contribute practically to the extinction coefficient increasing ($\Delta Q_{\text{ext}} > 0$), although the condition (3.4.8) is fulfilled. With an increase of the real part of the shell refractive index (Figure 3.17*b*) the coefficients of reflection on TLP boundaries increase. The interference effects become more prominent with the particular relationship between the particle and internal cavity sizes ($\Delta Q_{\text{ext}} < 0$). As κ_2 decreases absorption losses inside the shell lower, thus radiation intensity reflected from internal boundaries increases and the interference contribution to extinction grows (Figure 3.17*c*).

As it is seen from Figure 3.17(*b,c*), the location of the ΔQ_{ext} minimum depends essentially on the particle size, and, consequently, it is practically impossible from the dependencies $\Delta Q_{\text{ext}}(q)$ to reveal conditions at which occur the maximal extinction coef-

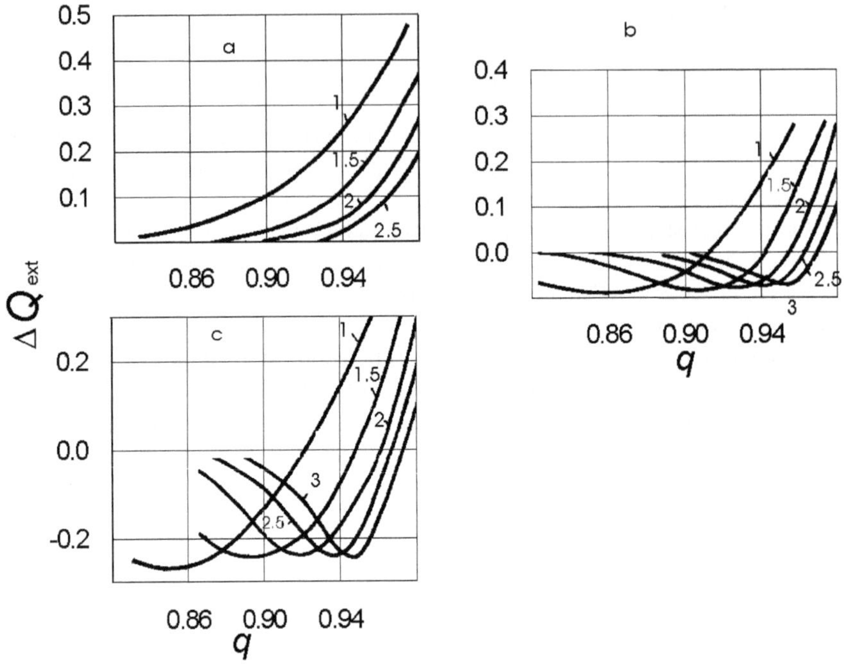

Figure 3.17. ΔQ_{ext} versus q for $n = 1.2(a)$ and $1.56(b, c)$ and $\kappa = 0.45(a, b)$ and $0.2(c)$ at $\lambda = 0.5\ \mu m$. Numerals at curves are values of $R_2\ (\mu m)$.

ficient for HOPs. At the same time on the diagrams of ΔQ_{ext} versus the shell thickness $d = R_2 - R_1$ (at n_2 constant) the ΔQ_{ext} minima for different particle sizes are localized in the narrow range of d values. In so doing, there is a small shift of the minimum towards the greater d values with increasing particle size. As follows from relation (3.4.8), the ΔQ_{ext} minima should lay close to the values $n_2 d = \lambda/2 = 0.25\ \mu m$. The dependencies $\Delta Q_{\text{ext}} = f(n_2 d)$ (Figure 3.18) correlate reasonably with the interference requirement (3.4.8), obtained for a plane layer model, though the ΔQ_{ext} minima take place at d values, a little smaller, than expected from Eq.(3.4.8). For larger particles $(R_2 \gg \lambda)$ the agreement is improved.

In a paper of Naumenko and Prishivalko (1981) on the example of high absorptive substances, such ranges of values of n_2 and κ_2 and shell thickness d/λ were found, where the extinction by HOPs is approximately the same as that by HPs; and also a requirement was obtained, when the extinction coefficient of aerosol from HOPs is 20–40% higher that for HPs of the same size. It is shown, that the interference effects influence on the extinction coefficient decreases almost exponentially as the shell absorption increases. The maximal values of the ratio $Q_{02}^{\text{ext}}/Q_2^{\text{ext}}$ can be approximately described by the expression:

$$\left(\frac{Q_{02}^{\text{ext}}}{Q_2^{\text{ext}}}\right)_{\max} \approx (n_2 - 1)\exp\left(-\frac{2\pi\kappa_2}{n_2}\right) \tag{3.4.9}$$

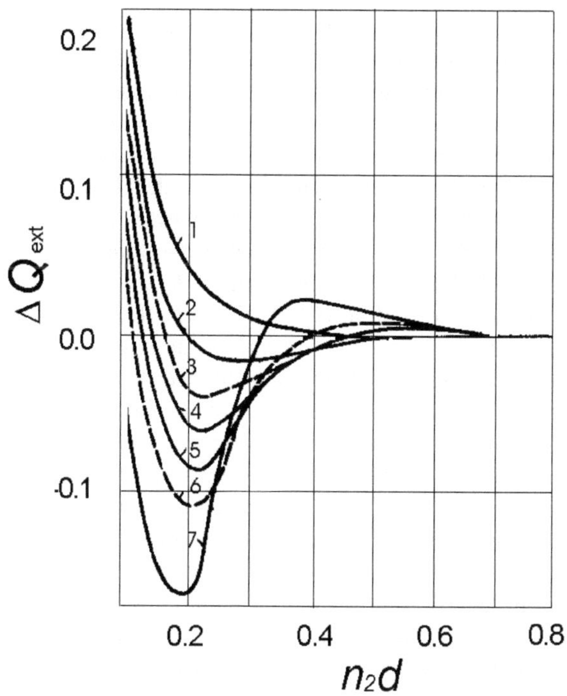

Figure 3.18. ΔQ_{ext} versus $n_2 d$ at $\lambda = 0.5 \mu m$, $R_2 = 3.0 \mu m$, $\kappa = 0.45$, and $n = 1.1(1)$, $1.3(2), 1.4(3), 1.5(4), 1.6(5), 1.7(6)$, and $2.0(7)$.

From the results of this section it follows, that the radiation interaction with HOPs has distinctive features, which under particular conditions can produce essential changes in the pattern of radiation scattering and absorption by aerosol medium and should be taken into account during experimental data interpretation.

3.5 RECURSIVE ALGORITHMS TO CALCULATE THE CHARACTERISTICS OF SCATTERED AND INTERNAL FIELDS FOR MULTILAYERED SPHERES

From the formal point of view the extension of Mie theory to the case of a multilayered sphere is very straightforward (see, e.g., the classical book of Kerker (1969)). Electromagnetic fields in each layer are written as the expansion over the vector spherical wave functions with undetermined amplitude coefficients (if the sphere comprises a core and $(\alpha - 1)$ layers, then there are 2α coefficients for each layer). The boundary conditions at the surfaces between the layers and at the outer sphere surface result in a set of 2α equations that can be solved for the 2α coefficients. This set of linear equations can be written in the matrix form and solved by standard computer codes. Although this formulation of the problem is conceptually simple it suffers from several defects: a) numerical inaccuracies can very quickly develop because of round-off errors; b) some of the calculated quantities can become so large as to cause computer overflow problems; and c) calculation rapidly becomes impractical as α becomes large because of computer memory requirements to store the matrix and because the time required to carry out the matrix calculations, which is approximately proportional to α^3, can become excessive.

These problems have been largely overcome by the development of recursive methods (Wait, 1963; Bhandari, 1985; Sitarski, 1987*a*; Mackowski *et al.*, 1990; Wu and Wang, 1991; Kai and Massoli, 1994; Sinzig and Quinten, 1994; Johnson, 1996, 1999; Wu *et al.*, 1997; Voshchinnikov and Mathis, 1999; Gurwich *et al.*, 2001), where the coefficients for an i-th layer are related to those of an $(i-1)$-th layer. The calculation begins with the core and then develops by recurrence to the surface, adding one layer at a time until the solution for the entire multilayered sphere is constructed. In essence, a single large matrix calculation has been reduced to a series of much smaller calculations, the need for storing a large matrix is eliminated and computational time increased linearly with α.

Nevertheless, the practice of work with existing recursive algorithms and codes have shown that some computational difficulties emerge when the number of layers α is large. There are no problems practically for several tens and even for several hundreds of layers but for $\alpha > 500$ the various errors develop which can distort a result substantially. However, a large number of layers (1000 or more) is required for correct modelling of smooth radial inhomogeneity of the refractive index within a particle (e.g., for optical modelling of large fractal clusters (see Section 8.5)). Hence, an algorithm and computer code needs to be developed, avoiding, if possible, the drawbacks mentioned above.

3.5.1 Formal solution, set of linear equations

1. Let us formulate the problem. The overall number of layers is α. Numeration of layers begins from the lowermost one ($j = 1$ – core, $j = 2$ – the first shell,..., $j = \alpha$ – the uppermost shell, and $j = \alpha + 1$– the surrounding medium). The relative radii (the size parameters) of the layers designate as $x_1, x_2, ..., x_\alpha$ (x_α is the external size parameter), and the complex relative refractive indices as $N_1, N_2, ..., N_\alpha$. The refractive index of the surrounding medium equals $N_{\alpha+1} = 1$. The plane monochromatic electromagnetic wave $(\mathbf{E}_i, \mathbf{H}_i)$ incidents upon the multilayered sphere. Its expansion over the vector spherical functions $\mathbf{M}^{(j)}_{\sigma ml}, \mathbf{N}^{(j)}_{\sigma ml}$ is described by Eqs.(1.3.24)–(1.3.24a). Similar expansion for the scattered field $(\mathbf{E}_{\rm sc}, \mathbf{H}_{\rm sc})$ has the form:

$$\mathbf{E}_{\rm sc} = -\sum_{l=1}^{\infty} E_l \left[a_l^{(\alpha+1)}\mathbf{M}^{(3)}_{o1l}(k_0\mathbf{r}) - ib_l^{(\alpha+1)}\mathbf{N}^{(3)}_{e1l}(k_0\mathbf{r})\right] \tag{3.5.1}$$

$$\mathbf{H}_{\rm sc} = \sqrt{\frac{\varepsilon_0}{\mu_0}}\sum_{l=1}^{\infty} E_l \left[b_l^{(\alpha+1)}\mathbf{M}^{(3)}_{e1l}(k_0\mathbf{r}) + ia_l^{(\alpha+1)}\mathbf{N}^{(3)}_{o1l}(k_0\mathbf{r})\right]$$

where $a_l^{(\alpha+1)}$ and $b_l^{(\alpha+1)}$ are the amplitude coefficients of the scattered field and $E_l = E_0 i^l (2l+1)/\left[l(l+1)\right]$. The expansion of the electric field inside the layer with number

j $(j = 1, 2, ..., \alpha)$ can be written as:

$$\mathbf{E}^{(j)} = \sum_{l=1}^{\infty} E_l(-1)^{\alpha-j} N_j \left[c_l^{(j)} \mathbf{M}_{o1l}^{(1)}(k_j\mathbf{r}) - id_l^{(j)} \mathbf{N}_{e1l}^{(1)}(k_j\mathbf{r}) \right. \\ \left. + a_l^{(j)} \mathbf{M}_{o1l}^{(3)}(k_j\mathbf{r}) - ib_l^{(j)} \mathbf{N}_{e1l}^{(1)}(k_j\mathbf{r}) \right] \tag{3.5.2}$$

where $a_l^{(j)}, b_l^{(j)}, c_l^{(j)}$, and $d_l^{(j)}$ are the amplitude coefficients of the internal field and $k_j = k_0 N_j$. The expansion of the magnetic field inside this layer has the form:

$$\mathbf{H}^{(j)} = -\sqrt{\frac{\varepsilon_0}{\mu_0}} \sum_{l=1}^{\infty} E_l(-1)^{\alpha-j} N_j \left[d_l^{(j)} \mathbf{M}_{e1l}^{(1)}(k_j\mathbf{r}) + ic_l^{(j)} \mathbf{N}_{o1l}^{(1)}(k_j\mathbf{r}) \right. \\ \left. + b_l^{(j)} \mathbf{M}_{e1l}^{(3)}(k_j\mathbf{r}) + ia_l^{(j)} \mathbf{N}_{o1l}^{(1)}(k_j\mathbf{r}) \right]$$

The functions of third kind are absent in expansions for the core, so we can write:

$$a_l^{(1)} = b_l^{(1)} = 0 \tag{3.5.3}$$

The expressions (3.5.1–2) may be formalized in such a way that they give the field outside the particle $(j = \alpha + 1)$ also:

$$\mathbf{E}^{(j)} = \sum_{l=1}^{\infty} E_l(-1)^{\alpha-j} N_j \left[\left(c_l^{(j)} - \delta_{j,\alpha+1} \right) \mathbf{M}_{o1l}^{(1)}(k_j\mathbf{r}) \right. \\ \left. -i \left(d_l^{(j)} - \delta_{j,\alpha+1} \right) \mathbf{N}_{e1l}^{(1)}(k_j\mathbf{r}) + b_l^{(j)} \mathbf{M}_{o1l}^{(3)}(k_j\mathbf{r}) - ia_l^{(j)} \mathbf{N}_{e1l}^{(1)}(k_j\mathbf{r}) \right] \tag{3.5.4}$$

$$\mathbf{H}^{(j)} = -\sqrt{\frac{\varepsilon_0}{\mu_0}} \sum_{l=1}^{\infty} E_l(-1)^{\alpha-j} N_j \left[\left(d_l^{(j)} - \delta_{j,\alpha+1} \right) \mathbf{M}_{e1l}^{(1)}(k_j\mathbf{r}) \right. \\ \left. +i \left(c_l^{(j)} - \delta_{j,\alpha+1} \right) \mathbf{N}_{o1l}^{(1)}(k_j\mathbf{r}) + a_l^{(j)} \mathbf{M}_{e1l}^{(3)}(k_j\mathbf{r}) + ib_l^{(j)} \mathbf{N}_{o1l}^{(1)}(k_j\mathbf{r}) \right]$$

where $j = 1, 2, ..., \alpha + 1$. At $j = \alpha + 1$ Eq.(3.5.4) yields:

$$\mathbf{E}^{(\alpha+1)} = \mathbf{E}_{\text{sc}} + \mathbf{E}_{\text{i}} \qquad \mathbf{H}^{(\alpha+1)} = \mathbf{H}_{\text{sc}} + \mathbf{H}_{\text{i}} \tag{3.5.5}$$

if we put:

$$c_l^{(\alpha+1)} = d_l^{(\alpha+1)} = 0 \tag{3.5.6}$$

Such formulation of fields is not traditional: the factor $(-1)^{\alpha-j} N_j$ is separated out of the amplitude coefficients. It is done for simplification and formalization of sets of linear equations – all coefficients of this set will be positive and the ratios of refractive indices will be absent.

Using the boundary conditions for the tangential components $E_\theta, E_\varphi, H_\theta$, and H_φ at boundaries between different layers gives two sets of linear equations (2α equations in each) for every index $l = 1, 2, \ldots$. In the first set the column of unknowns comprises the coefficients $a_l^{(\alpha+1)}, d_l^{(\alpha)}, a_l^{(\alpha)}, d_l^{(\alpha-1)}, a_l^{(\alpha-1)}, \ldots, a_l^{(2)}$, and $d_l^{(1)}$, and in the second set comprises the coefficients $b_l^{(\alpha+1)}, c_l^{(\alpha)}, b_l^{(\alpha)}, c_l^{(\alpha-1)}, b_l^{(\alpha-1)}, \ldots, b_l^{(2)}$, and $c_l^{(1)}$.

Let us write these sets in a matrix form (here for brevity we omit the lower index l and introduce the simplified form for arguments of the Riccati–Bessel functions $\psi(N_\alpha x_\alpha) \equiv \psi_{\alpha\alpha}, \psi(x_\alpha) = \psi_\alpha$):

$$\begin{bmatrix} \xi_\alpha & N_\alpha\psi_{\alpha\alpha} & N_\alpha\xi_{\alpha\alpha} & 0 & \cdots & 0 & 0 & 0 \\ \xi'_\alpha & \psi'_{\alpha\alpha} & \xi'_{\alpha\alpha} & 0 & \cdots & 0 & 0 & 0 \\ 0 & N_\alpha\psi_{\alpha,\alpha-1} & N_\alpha\xi_{\alpha,\alpha-1} & N_{\alpha-1}\psi_{\alpha-1,\alpha-1} & \cdots & 0 & 0 & 0 \\ 0 & \psi'_{\alpha,\alpha-1} & \xi'_{\alpha,\alpha-1} & \psi'_{\alpha-1,\alpha-1} & \cdots & 0 & 0 & 0 \\ 0 & 0 & 0 & N_{\alpha-1}\psi_{\alpha-1,\alpha-2} & \cdots & 0 & 0 & 0 \\ 0 & 0 & 0 & \psi'_{\alpha-1,\alpha-2} & \cdots & 0 & 0 & 0 \\ \cdots & \cdots & \cdots & \cdots & \cdots & \cdots & \cdots & \cdots \\ 0 & 0 & 0 & 0 & \cdots & N_2\psi_{21} & N_2\xi_{21} & N_1\psi_{11} \\ 0 & 0 & 0 & 0 & \cdots & \psi'_{21} & \xi'_{21} & \psi'_{11} \end{bmatrix} \tag{3.5.7}$$

$$\begin{bmatrix} \xi_\alpha & \psi_{\alpha\alpha} & \xi_{\alpha\alpha} & 0 & \cdots & 0 & 0 & 0 \\ \xi'_\alpha & N_\alpha\psi'_{\alpha\alpha} & N_\alpha\xi'_{\alpha\alpha} & 0 & \cdots & 0 & 0 & 0 \\ 0 & \psi_{\alpha,\alpha-1} & \xi_{\alpha,\alpha-1} & \psi_{\alpha-1,\alpha-1} & \cdots & 0 & 0 & 0 \\ 0 & N_\alpha\psi'_{\alpha,\alpha-1} & N_\alpha\xi'_{\alpha,\alpha-1} & N_{\alpha-1}\psi'_{\alpha-1,\alpha-1} & \cdots & 0 & 0 & 0 \\ 0 & 0 & 0 & \psi_{\alpha-1,\alpha-2} & \cdots & 0 & 0 & 0 \\ 0 & 0 & 0 & N_{\alpha-1}\psi'_{\alpha-1,\alpha-2} & \cdots & 0 & 0 & 0 \\ \cdots & \cdots & \cdots & \cdots & \cdots & \cdots & \cdots & \cdots \\ 0 & 0 & 0 & 0 & \cdots & \psi_{21} & N_2\xi_{21} & N_1\psi_{11} \\ 0 & 0 & 0 & 0 & \cdots & N_2\psi'_{21} & N_2\xi'_{21} & N_1\psi'_{11} \end{bmatrix} \tag{3.5.8}$$

Designate the determinants of the first and second matrixes as K_l and $\bar{K}_l$, respectively. Replace the first columns by the columns of unknowns in these matrixes, that is replace ξ_α and ξ'_α by ψ_α and ψ'_α, respectively, in the two upper left cells. The determinants of these new matrices designate as L_l and $\bar{L}_l$. Then the amplitude coefficients of the scattered field are:

$$a_l^{(\alpha+1)} = \frac{L_l}{K_l} \qquad a_l^{(\alpha+1)} = \frac{\bar{L}_l}{\bar{K}_l} \tag{3.5.9}$$

The determinant L_l is equal:

$$L_l = \begin{vmatrix} \psi_\alpha & N_\alpha\psi_{\alpha\alpha} & N_\alpha\xi_{\alpha\alpha} & 0 & \dots & 0 \\ \psi'_\alpha & \psi'_{\alpha\alpha} & \xi'_{\alpha\alpha} & 0 & \dots & 0 \\ 0 & N_\alpha\psi_{\alpha,\alpha-1} & N_\alpha\xi_{\alpha,\alpha-1} & N_{\alpha-1}\psi_{\alpha-1,\alpha-1} & \dots & 0 \\ 0 & \psi'_{\alpha,\alpha-1} & \xi'_{\alpha,\alpha-1} & \psi'_{\alpha-1,\alpha-1} & \dots & 0 \\ 0 & 0 & 0 & M_{\alpha-1}\psi_{\alpha-1,\alpha-2} & \dots & 0 \\ 0 & 0 & 0 & \psi'_{\alpha-1,\alpha-2} & \dots & 0 \\ \dots & \dots & \dots & \dots & \dots & \dots \\ 0 & 0 & 0 & 0 & \dots & N_1\psi_{11} \\ 0 & 0 & 0 & 0 & \dots & \psi'_{11} \end{vmatrix} \quad (3.5.10)$$

Decompose this determinant by the third column into two determinants and interchange the second and the third columns in the second determinant:

$$L_l = \begin{vmatrix} \psi_\alpha & N_\alpha\psi_{\alpha\alpha} & 0 & 0 & \dots & 0 \\ \psi'_\alpha & \psi'_{\alpha\alpha} & 0 & 0 & \dots & 0 \\ 0 & N_\alpha\psi_{\alpha,\alpha-1} & N_\alpha\xi_{\alpha,\alpha-1} & N_{\alpha-1}\psi_{\alpha-1,\alpha-1} & \dots & 0 \\ 0 & \psi'_{\alpha,\alpha-1} & \xi'_{\alpha,\alpha-1} & \psi'_{\alpha-1,\alpha-1} & \dots & 0 \\ 0 & 0 & 0 & N_{\alpha-1}\psi_{\alpha-1,\alpha-2} & \dots & 0 \\ 0 & 0 & 0 & \psi'_{\alpha-1,\alpha-2} & \dots & 0 \\ \dots & \dots & \dots & \dots & \dots & \dots \\ 0 & 0 & 0 & 0 & \dots & N_1\psi_{11} \\ 0 & 0 & 0 & 0 & \dots & \psi'_{11} \end{vmatrix} \quad (3.5.11)$$

$$- \begin{vmatrix} \psi_\alpha & N_\alpha\xi_{\alpha\alpha} & N_\alpha\psi_{\alpha\alpha} & 0 & \dots & 0 \\ \psi'_\alpha & \xi'_{\alpha\alpha} & \psi'_{\alpha\alpha} & 0 & \dots & 0 \\ 0 & 0 & N_\alpha\psi_{\alpha,\alpha-1} & N_{\alpha-1}\psi_{\alpha-1,\alpha-1} & \dots & 0 \\ 0 & 0 & \psi'_{\alpha,\alpha-1} & \psi'_{\alpha-1,\alpha-1} & \dots & 0 \\ 0 & 0 & 0 & N_{\alpha-1}\psi_{\alpha-1,\alpha-2} & \dots & 0 \\ 0 & 0 & 0 & \psi'_{\alpha-1,\alpha-2} & \dots & 0 \\ \dots & \dots & \dots & \dots & \dots & \dots \\ 0 & 0 & 0 & 0 & \dots & N_1\psi_{11} \\ 0 & 0 & 0 & 0 & \dots & \psi'_{11} \end{vmatrix} \quad (3.5.12)$$

Designate the determinants (2×2) in the left upper angle of L_l and K_l as:

$$\begin{aligned} \psi_l(x_\alpha)\begin{pmatrix}\psi\\ \xi\end{pmatrix}_l(w_a)\left[\begin{pmatrix}D\\ G\end{pmatrix}_l(w_a) - N_a D_l(x_\alpha)\right] &\equiv \psi_l(x_\alpha)\begin{pmatrix}\psi\\ \xi\end{pmatrix}_l(w_a)\begin{pmatrix}f\\ g\end{pmatrix}_l \\ \xi_l(x_\alpha)\begin{pmatrix}\psi\\ \xi\end{pmatrix}_l(w_a)\left[\begin{pmatrix}D\\ G\end{pmatrix}_l(w_a) - N_a G_l(x_\alpha)\right] &\equiv \psi_l(x_\alpha)\begin{pmatrix}\psi\\ \xi\end{pmatrix}_l(w_a)\begin{pmatrix}z\\ u\end{pmatrix}_l \end{aligned} \quad (3.5.13)$$

where $D_l(z)$ and $G_l(z)$ are the log-derivatives of the Riccati–Bessel functions (see Eqs. (1.5.1)) and $w_j = N_j x_j$.

The determinant $(2\alpha \times 2\alpha)$ in the right lower angle of the determinant (3.5.12) designate as $\xi_l^{-1}(w_a)\Phi_l^{(\alpha-1)}$, and the similar determinant in the right lower angle of the determinant (3.5.11) designate as $\psi_l^{-1}(w_a)\Omega_l^{(\alpha-1)}$. The determinants (3.5.11–12) have the zero blocks. But it is known that a determinant with two non-zero square blocks and one zero block (it does not need to be a square one) is equal to the product of the above square blocks. Hence the decomposition is as follows:

$$\begin{pmatrix} L \\ K \end{pmatrix}_l = \begin{pmatrix} \psi \\ \xi \end{pmatrix}_l (w_a) \left[\begin{pmatrix} f \\ z \end{pmatrix}_l \Omega_l^{(\alpha-1)} - \begin{pmatrix} g \\ u \end{pmatrix}_l \Phi_l^{(\alpha-1)} \right] \tag{3.5.14}$$

If we decompose the determinants $\Omega_l^{(j)}$ and $\Phi_l^{(j)}$ by analogy with the decomposition of the determinant (3.5.10) then we have:

$$\begin{pmatrix} \Phi \\ \Omega \end{pmatrix}_l^{(j)} = \begin{pmatrix} \psi \\ \xi \end{pmatrix}_l (y_j) \begin{pmatrix} \xi \\ \psi \end{pmatrix}_l (w_{j+1}) \left[\begin{pmatrix} h \\ p \end{pmatrix}_l^{(j)} \Omega_l^{(j-1)} - \begin{pmatrix} q \\ t \end{pmatrix}_l^{(j)} \Phi_l^{(j-1)} \right] \tag{3.5.15}$$

where:

$$\begin{aligned} \begin{pmatrix} h \\ q \end{pmatrix}_l^{(j)} &= N_{j+1} \begin{pmatrix} D \\ G \end{pmatrix}_l (w_j) - N_j D_l(y_j) \\ \begin{pmatrix} p \\ t \end{pmatrix}_l^{(j)} &= N_{j+1} \begin{pmatrix} D \\ G \end{pmatrix}_l (w_j) - N_j G_l(y_j) \qquad y_j = N_{j+1} x_j \end{aligned} \tag{3.5.16}$$

The expressions (3.5.14–16) are actually the required system of recursions for index j from $j = 2$ to $j = \alpha - 1$. The factors in front of the square brackets in Eq.(3.5.15) may be rewritten in the following way:

$$\begin{aligned} \psi_l(y_j)\xi_l(w_{j+1}) &= [\psi_l(y_j)\xi_l(y_j)] \cdot \left[\frac{\xi_l(w_{j+1})}{\xi_l(y_j)} \right] \\ \xi_l(y_j)\psi_l(w_{j+1}) &= [\xi_l(y_j)\psi_l(y_j)] \cdot \left[\frac{\psi_l(y_j)}{\psi_l(w_{j+1})} \right]^{-1} \end{aligned} \tag{3.5.17}$$

If, as usual, we introduce the ratios:

$$\begin{aligned} R_l(y_j, w_{j+1}) \equiv R_l^{(j)} &= \frac{\psi_l(y_j)}{\psi_l(w_{j+1})} \qquad W_l(w_{j+1}, y_j) \equiv W_l^{(j)} = \frac{\xi_l(w_{j+1})}{\xi_l(y_j)} \\ T_l^{(j)} &= \psi_l(y_j)\xi_l(y_j) \end{aligned} \tag{3.5.18}$$

then we have instead of Eq.(3.5.15):

$$\begin{aligned} \Phi_l^{(j)} &= T_l^{(j)} W_l^{(j)} \left[h_l^{(j)} \Omega_l^{(j-1)} - q_l^{(j)} \Phi_l^{(j-1)} \right] \\ \Omega_l^{(j)} &= \frac{T_l^{(j)}}{R_l^{(j)}} \left[p_l^{(j)} \Omega_l^{(j-1)} - t_l^{(j)} \Phi_l^{(j-1)} \right] \end{aligned} \tag{3.5.19}$$

Consider now the amplitude coefficient for the scattered field $b_l^{(\alpha+1)}$. When the matrix (3.5.7) is compared with matrix (3.5.8) it is apparent that they differ only in that the multipliers N_j stand before functions in the former and before derivatives in the latter. So, the recursion scheme for $\bar{L}_l$ and $\bar{K}_l$ remains the same as for L_l and K_l, when the auxiliary coefficients $f_l, g_l, z_l, u_l, h_l^{(j)}, q_l^{(j)}, p_l^{(j)}$, and $t_l^{(j)}$ are replaced by coefficients $\bar{f}_l, \bar{g}_l, \bar{z}_l, \bar{u}_l, \bar{h}_l^{(j)}, \bar{q}_l^{(j)}, \bar{p}_l^{(j)}$, and $\bar{t}_l^{(j)}$ that can be produced from the corresponding coefficient without accent by interchanging $N_{j+1} \leftrightarrow N_j$ (it implies for f_l, g_l, z_l, and u_l that N_α shifts to the first term). The values Ω_l and Φ_l must be replaced by $\bar{\Omega}_l$ and $\bar{\Phi}_l$ in Eqs.(3.5.15, 16, and 19).

2. Let us go now to the internal amplitude coefficients $d_l^{(j)}$ and $c_l^{(j)} (j = 1, 2, ..., \alpha)$, $a_l^{(j)}$ and $b_l^{(j)}$ $(j = 2, 3, ..., \alpha)$. Consider that the amplitude coefficients enter the expansions of the internal fields in the form of their combinations with the Bessel function of the current coordinate x_c:

$$X_l^{(j)} = \psi_l(N_j x_c) c_l^{(j)} \quad Y_l^{(j)} = \psi_l'(N_j x_c) d_l^{(j)}$$
$$Z_l^{(j)} = \xi_l(N_j x_c) b_l^{(j)} \quad V_l^{(j)} = \xi_l'(N_j x_c) a_l^{(j)} \tag{3.5.20}$$

After introducing the normalized amplitude coefficients:

$$\begin{pmatrix} \tilde{c} \\ \tilde{d} \end{pmatrix}_l^{(j)} = \begin{pmatrix} c \\ d \end{pmatrix}_l^{(j)} \psi_l(w_j) \qquad \begin{pmatrix} \tilde{a} \\ \tilde{b} \end{pmatrix}_l^{(j)} = \begin{pmatrix} a \\ b \end{pmatrix}_l^{(j)} \xi_l(w_j) \tag{3.5.21}$$

Eq.(3.5.20) may be written as:

$$X_l^{(j)} = R_l(N_j x_c, N_j x_j) \tilde{c}_l^{(j)} \qquad Y_l^{(j)} = R_l(N_j x_c, N_j x_j) D_l(N_j x_c) \tilde{d}_l^{(j)}$$
$$Z_l^{(j)} = \frac{\tilde{b}_l^{(j)}}{W_l(N_j x_j, N_j x_c)} \qquad V_l^{(j)} = \frac{G_l(N_j x_c) \tilde{a}_l^{(j)}}{W_l(N_j x_j, N_j x_c)} \tag{3.5.22}$$

The ratios:

$$\tilde{d}_l^{(j)} = \psi_l(w_j) \frac{L_l^{(j)}}{K_l} \qquad \tilde{a}_l^{(j)} = \xi_l(w_j) \frac{\tilde{L}_l^{(j)}}{K_l} \tag{3.5.23}$$

must be calculated to produce the coefficients $\tilde{d}_l^{(j)}$ and $\tilde{a}_l^{(j)}$, where K_l is defined above, and the determinants $L_l^{(j)}$ and $\tilde{L}_l^{(j)}$ are produced by replacing the column with numbers $2(\alpha - j + 1)$ and $2(\alpha - j) + 3$, respectively, in K_l by the column of unknowns. Let us organize the successive interchanging of columns in determinants $L_l^{(j)}$:

$$\begin{aligned} L_l^{(\alpha)} &: \text{ no interchange} \\ L_l^{(\alpha-1)} &: (2 \leftrightarrow 4) \\ L_l^{(\alpha-2)} &: (2 \leftrightarrow 6), (4 \leftrightarrow 6) \\ L_l^{(\alpha-3)} &: (2 \leftrightarrow 8), (4 \leftrightarrow 8), (6 \leftrightarrow 8)...... \\ L_l^{(1)} &: (2 \leftrightarrow 2\alpha), (4 \leftrightarrow 2\alpha), ..., [(2\alpha - 2) \leftrightarrow 2\alpha] \end{aligned}$$

and $\tilde{L}_l^{(j)}$:

$$\begin{aligned}
\tilde{L}_l^{(\alpha)} &: (2 \leftrightarrow 3) \\
\tilde{L}_l^{(\alpha-1)} &: (2 \leftrightarrow 5), (4 \leftrightarrow 5) \\
\tilde{L}_l^{(\alpha-2)} &: (2 \leftrightarrow 7), (4 \leftrightarrow 7), (6 \leftrightarrow 7) \\
\tilde{L}_l^{(\alpha-3)} &: (2 \leftrightarrow 9), (4 \leftrightarrow 9), (6 \leftrightarrow 9), (9 \leftrightarrow 9).... \\
\tilde{L}_l^{(2)} &: [2 \leftrightarrow (2\alpha-1)], ..., [(2\alpha-2) \leftrightarrow (2\alpha-1)]
\end{aligned}$$

As a result the determinant $L_l^{(\alpha)}$ contains the Wronskian in the left upper angle:

$$\xi_l(x_2)\psi_l'(x_2) - \xi_l'(x_2)\psi_l(x_2) = -i \quad \text{that is } L_l^{(\alpha)} = \frac{-i}{\psi_l(w_j)}\Omega_l^{(\alpha-1)} \tag{3.5.24}$$

After the interchange $(2 \leftrightarrow 4)$ in $L_l^{(\alpha-1)}$ the same Wronskian appears in the left upper angle, and to the lower right of this angle the other Wronskian takes place:

$$N_\alpha \xi_l(y_{\alpha-1})\psi_l'(y_{\alpha-1}) - \xi_l'(y_{\alpha-1})\psi_l(y_{\alpha-1}) = -iN_\alpha \tag{3.5.25}$$

When it is considered the change of sign at column interchanging:

$$L_l^{(\alpha-1)} = \frac{N_\alpha}{\psi_l(w_j)}\Omega_l^{(\alpha-2)} \tag{3.5.26}$$

Generally we have:

$$L_l^{(j)} = i^{\alpha-j-1}(N_\alpha N_{\alpha-1}...N_{j+1})\frac{1}{\psi_l(w_j)}\Omega_l^{(j-1)} \tag{3.5.27}$$

(if $j=\alpha$, then $(N_\alpha N_{\alpha-1}...N_{j+1}) = 1$). Finally, for coefficients $\tilde{d}_l^{(j)}$ $(j = 1, 2, ..., \alpha)$:

$$\tilde{d}_l^{(j)} = i^{\alpha-j-1}(N_\alpha N_{\alpha-1}...N_{j+1})\frac{\Omega_l^{(j-1)}}{K_l} \tag{3.5.28}$$

In analogy with the above we obtain:

$$\tilde{L}_l^{(j)} = -i^{\alpha-j-1}(N_\alpha N_{\alpha-1}...N_{j+1})\frac{1}{\xi_l(w_j)}\Phi_l^{(j-1)} \tag{3.5.29}$$

and for coefficients $\tilde{a}_l^{(j)}$ $(j = 2, 3, ..., \alpha)$:

$$\tilde{a}_l^{(j)} = -i^{\alpha-j-1}(N_\alpha N_{\alpha-1}...N_{j+1})\frac{\Phi_l^{(j-1)}}{K_l} \tag{3.5.30}$$

In a similar manner for coefficients $\tilde{b}_l^{(j)}$ $(j = 2, 3, ..., \alpha)$ and $\tilde{c}_l^{(j)}$ $(j = 1, 2, ..., \alpha)$ we obtain:

$$\tilde{c}_l^{(j)} = i^{\alpha-j-1}(N_\alpha N_{\alpha-1}...N_{j+1})\frac{\bar{\Omega}_l^{(j-1)}}{\bar{K}_l} \qquad \tilde{b}_l^{(j)} = -i^{\alpha-j-1}(N_\alpha N_{\alpha-1}...N_{j+1})\frac{\bar{\Phi}_l^{(j-1)}}{\bar{K}_l} \tag{3.5.31}$$

Now we would like to clear up the question of the initial values of recursion over Ω_l and Φ_l. It immediately follows from the definitions of $\Omega_l^{(1)}$ and $\Phi_l^{(1)}$ that:

$$\begin{pmatrix}\Phi\\ \Omega\end{pmatrix}_l^{(1)} = \begin{pmatrix}\psi\\ \xi\end{pmatrix}_l(y_1)\begin{pmatrix}\xi\\ \psi\end{pmatrix}_l(w_2)\begin{pmatrix}h\\ p\end{pmatrix}_l^{(1)}\psi_l(w_1) \tag{3.5.32}$$

On the other hand, from Eq.(3.5.15) it is apparent that:

$$\begin{pmatrix}\Phi\\ \Omega\end{pmatrix}_l^{(1)} = \begin{pmatrix}\psi\\ \xi\end{pmatrix}_l(y_1)\begin{pmatrix}\xi\\ \psi\end{pmatrix}_l(w_2)\left[\begin{pmatrix}h\\ p\end{pmatrix}_l^{(1)}\Omega_l^{(0)} - \begin{pmatrix}q\\ t\end{pmatrix}_l^{(1)}\Phi_l^{(0)}\right] \tag{3.5.33}$$

Comparison of Eqs.(3.5.32–33) gives:

$$\Phi_l^{(0)} = 0 \qquad \Omega_l^{(0)} = \psi_l(w_1) \tag{3.5.34}$$

However, it is easy to see that only ratios Ω_l/Φ_l enter into all final expressions for the amplitude coefficients, that is the value $\Omega_l^{(0)}$ is in essence a normalization constant and may be set as $\Omega_l^{(0)} = 1$.

Summing up, the formal expressions for recursive calculation of the amplitude coefficients of scattered and internal fields in the case of a multilayered sphere have been obtained. The computer realization of this approach will be discussed in the next section.

3.5.2 Algorithm and computer code structure

The experience of many investigators of development of recursive algorithms aimed at calculation of the characteristics of scattering by multilayered spheres suggests that its main difficulties are:

- When the number of layers increases up to $10^3 \div 10^4$ the accumulation of round-off error takes place due to repeated recursion over layers and very little difference in arguments of the cylindrical functions for adjacent layers. There have been the attempts to eliminate this problem by introducing the asymptotic expressions within the limits of very thin layers (Johnson, 1999).
- It is rather difficult to avoid the large two-dimensional complex arrays for calculated functions and coefficients.

However, we managed to overcome these difficulties and developed a fairly fast and stable algorithm. In this section we shall discuss the distinctive features of our computational approach using the example of amplitude coefficients of scattered fields. This algorithm is implemented as the subroutine CB2 (Fortran 77).

The input parameters of this subroutine are:

- α is the number of concentric layers (up to 20 000), there is nothing to prevent this value from increasing by enlarging of corresponding arrays;
- the arrays of real and imaginary parts of complex refractive indices of particle layers $n_1, n_2, ..., n_\alpha$ and $\kappa_1, \kappa_2, ..., \kappa_\alpha$;
- and the array of size parameters for layers $x_1, x_2, ..., x_\alpha$.

The output parameters are:

- the arrays of the amplitude coefficients for scattered fields $a_l^{(\alpha+1)}$ and $b_l^{(\alpha+1)}$ $(l = 1, 2, ..., l_{\alpha+1})$;
- and the maximal value of index l for the scattered field $(l_{\alpha+1})$.

As the first step, the maximum indices for each layer $l_j (j = 1, 2, ..., \alpha)$ and for the surrounding medium $l_{\alpha+1}$ are calculated using the modified estimate of Wiscombe (1980):

$$l_{\alpha+1} = \max[3; 1.1(x_\alpha + 4.05x_\alpha^{1/3} + 2)] \qquad l_j = 1.1(q_j + 4.05q_j{}^{1/3} + 2)] \quad (3.5.35)$$

where $q_j = |(n_j + i\kappa_j)x_j|$. Some "levelling" of l-values occurs: if $l_j > l_{\alpha+1}$, then $l_j = l_{\alpha+1}$; and if $l_j < l_{j-1}$, then $l_j = l_{j-1}$. Such evaluation of the number of terms enables us to circumvent a difficulty arising due to the excess terms over index l for the internal particle layers, particularly with the internal strongly absorbing layers.

Further, the initial values of functions:

$$G_0(x_a) = i \qquad V_0(x_a) = \sin(x_a) / [\sin(x_a) - i\cos(x_a)]$$

are calculated outside the main loops and the subroutine DLD (see Section 1.5.2.1) is used for calculation of arrays of the log-derivative $D_l(x_a) \equiv \psi_l'(x_a)/\psi_l(x_a)$, $(l = 1, 2, ..., l_{\alpha+1})$ (the initial value with maximal index $l_{\alpha+1}$ is computed by the Lentz method (see Section 1.5.2.1) and the remaining terms, via the downward recursion (1.5.3)).

The subroutine external loop is organized over the layer number $(j = 1, 2, ..., \alpha)$, that is, the recursion goes from the particle center to its surface. The following steps are sequentially carried out:

1. Calculation of the initial values of functions $G_0(w_{j+1}) = G_0(w_j) = G_0(y_j) = i$, where $w_j = N_j x_j$ and $y_j = N_{j+1} x_j$.

2. Calculation of the initial values of functions:

$$\tilde{R}_0^{(j)} = \frac{(A_{8j}+1)A_{1j} - iA_{2j}(A_{8j}-1)}{(A_{9j}+1)A_{4j} - iA_{5j}(A_{9j}-1)} \qquad \tilde{W}_0^{(j)} = A_{7j} + iA_{6j} \qquad (3.5.36)$$

where:

$$\begin{aligned}
&A_{1j} = \sin(n_{j+1}x_j) && A_{2j} = \cos(n_{j+1}x_j) && A_{3j} = \sin(2n_{j+1}x_j)\\
&A_{4j} = \sin(n_{j+1}x_{j+1}) && A_{5j} = \cos(n_{j+1}x_{j+1}) && A_{6j} = \sin[n_{j+1}(x_{j+1} - x_j)]\\
&A_{7j} = \cos[n_{j+1}(x_{j+1} - x_j)] && A_{8j} = \exp(-2\kappa_{j+1}x_j) &&\\
&A_{9j} = \exp(-2\kappa_{j+1}x_{j+1}) && A_{10,j} = \exp[\kappa_{j+1}(x_{j+1} - x_j)] &&
\end{aligned}$$

Note that the normalized functions $\tilde{R}_0^{(j)}$ and $\tilde{W}_0^{(j)}$ differ from functions $R_0^{(j)}$ and $W_0^{(j)}$ described by Eq.(3.5.18) with constant factor: $\tilde{R}_0^{(j)} = R_0^{(j)} A_{10,j}$ and $\tilde{W}_0^{(j)} = W_0^{(j)} A_{10,j}$. In the absence of normalization, the possibility to underflow arises at sufficiently great $\kappa_{j+1}(x_{j+1} - x_j)$. This underflow gives rise to erroneous nullifying of all other $R_l^{(j)}$ and $W_l^{(j)}$ due to the multiplicative nature of recursions for $R_0^{(j)}$ and $W_0^{(j)}$.

3. At $j = 1$, (the particle core) call of the subroutine DLD for calculation of arrays of the log-derivatives $D_l(w_1)$, $(l = 1, 2, ..., l_2)$. If $j > 1$, then the array $D_l(w_{j+1})$, calculated at the previous j is used.

4. Call of the subroutine DLD for calculation of arrays $D_l(y_j)$, $(l = 1, 2, ..., l_{j+1})$.

5. If $j = \alpha - 1$ (last but one layer) then call of DLD for calculation of arrays $D_l(w_{j+1})$, $(l = 1, 2, ..., l_{\alpha+1})$. For other layers call of DLD for calculation of arrays $D_l(w_{j+1})$, $(l = 1, 2, ..., l_{j+2})$.

6. Beginning of internal loop over index l ($l = 1, 2, ..., l_s$, where $l_s = l_{j+1}$, but if $j = \alpha$, then $l_s = l_{\alpha+1}$).

7. If $j = 1$ or $l > l_j$, then calculation of the initial values of Ω- and Φ-functions: $\Omega_l^{(j-1)} = \bar{\Omega}_l^{(j-1)} = 1$ and $\Phi_l^{(j-1)} = \bar{\Phi}_l^{(j-1)} = 0$. Hence, the recursion over layers towards the particle surface generally begins not at the particle center, but at a layer immediately preceding some layer where the particular value of index l first emerges. It is very essential point in providing the algorithm stabilization.

If $j = \alpha$, then go to the block for the last layer (point 11).

8. One recursion step for functions:

$$G_l(w_j) = -\frac{l}{w_j} + \frac{w_j}{l - w_j G_{l-1}(w_j)} \qquad G_l(y_j) = -\frac{l}{y_j} + \frac{y_j}{l - y_j G_{l-1}(y_j)}$$

$$G_l(w_{j+1}) = -\frac{l}{w_{j+1}} + \frac{w_{j+1}}{l - w_{j+1} G_{l-1}(w_{j+1})} \tag{3.5.37}$$

$$\tilde{R}_l^{(j)} = \tilde{R}_{l-1}^{(j)} y_j \frac{w_{j+1} D_l(w_{j+1}) + l}{w_{j+1}\left[y_j D_l(y_j) + l\right]} \qquad \tilde{W}_l^{(j)} = \tilde{W}_{l-1}^{(j)} w_{j+1} \frac{y_j G_l(y_j) + l}{y_j\left[w_{j+1} G_l(w_{j+1}) + l\right]}$$

The new values replace the preceding ones.

9. Calculation of auxiliary coefficients:

$$\begin{aligned} p_l^{(j)} &= N_{j+1} D_l(w_j) - N_j G_l(y_j) & h_l^{(j)} &= N_{j+1} D_l(w_j) - N_j D_l(y_j) \\ \bar{p}_l^{(j)} &= N_j D_l(w_j) - N_{j+1} G_l(y_j) & \bar{h}_l^{(j)} &= N_j D_l(w_j) - N_{j+1} D_l(y_j) \end{aligned} \tag{3.5.38}$$

If ($j > 2$ and $l \leq l_j$) then:

$$\begin{aligned} t_l^{(j)} &= N_{j+1} G_l(w_j) - N_j G_l(y_j) & q_l^{(j)} &= N_{j+1} G_l(w_j) - N_j D_l(y_j) \\ \bar{t}_l^{(j)} &= N_j G_l(w_j) - N_{j+1} G_l(y_j) & \bar{q}_l^{(j)} &= N_j G_l(w_j) - N_{j+1} D_l(y_j) \end{aligned} \tag{3.5.39}$$

else $t_l^{(j)} = \bar{t}_l^{(j)} = q_l^{(j)} = \bar{q}_l^{(j)} = 0$.

10. One step of the recursion over layers for Ω- and Φ-functions:

$$\Omega_l^{(j)} = \frac{A_{10,j}}{\tilde{R}_l^{(j)}} \left(p_l^{(j)} - t_l^{(j)} \Phi_l^{(j-1)}\right) \qquad \Phi_l^{(j)} = \frac{\tilde{W}_l^{(j)}}{A_{10,j} \Omega_l^{(j)}} \left(h_l^{(j)} - q_l^{(j)} \Phi_l^{(j-1)}\right) \tag{3.5.40}$$

replacing the previous values. Normalization: $\Omega_l^{(j)} = 1$.

One step of the recursion over layers for Ω- and Φ-functions:

$$\bar{\Omega}_l^{(j)} = \frac{A_{10,j}}{\tilde{R}_l^{(j)}} \left(\bar{p}_l^{(j)} - \bar{t}_l^{(j)} \bar{\Phi}_l^{(j-1)}\right) \qquad \bar{\Phi}_l^{(j)} = \frac{\tilde{W}_l^{(j)}}{A_{10,j} \bar{\Omega}_l^{(j)}} \left(\bar{h}_l^{(j)} - \bar{q}_l^{(j)} \bar{\Phi}_l^{(j-1)}\right) \tag{3.5.41}$$

replacing the previous values. Normalization: $\bar{\Omega}_l^{(j)} = 1$. The intermediate normalization of functions Ω and $\bar{\Omega}$ is very important for the reliability of code because the non-normalized functions grow very rapidly and cause computer overflow problems at $\alpha \sim$ 3000.

Go to point 12 (bypass of the block for the last layer).

11. Block for the last layer. One step of the recursion for functions from x_α:

$$G_l(x_\alpha) = -\frac{l}{x_\alpha} + \frac{x_\alpha}{l - x_\alpha G_{l-1}(x_\alpha)} \qquad V_l(x_\alpha) = V_{l-1}(x_\alpha)\frac{x_\alpha G_l(x_\alpha) + l}{x_\alpha D_l(x_\alpha) + l} \tag{3.5.42}$$

If $l > l_\alpha$, then:

$$G_l(w_{j+1}) = -\frac{l}{w_{j+1}} + \frac{w_{j+1}}{l - w_{j+1}G_{l-1}(w_{j+1})} \tag{3.5.43}$$

Calculation of auxiliary coefficients:

$$\begin{aligned} f_l &= D_l(w_\alpha) - N_\alpha D_l(x_\alpha) \quad & z_l &= D_l(w_\alpha) - N_\alpha G_l(x_\alpha) \\ \bar{f}_l &= N_\alpha D_l(w_\alpha) - D_l(x_\alpha) \quad & \bar{z}_l &= N_\alpha D_l(w_\alpha) - G_l(x_\alpha) \end{aligned} \tag{3.5.44}$$

If $l \le l_\alpha$, then:

$$\begin{aligned} g_l &= G_l(w_\alpha) - N_\alpha D_l(x_\alpha) \quad & u_l &= G_l(w_\alpha) - N_\alpha G_l(x_\alpha) \\ \bar{g}_l &= N_\alpha G_l(w_\alpha) - D_l(x_\alpha) \quad & \bar{u}_l &= N_\alpha G_l(w_\alpha) - G_l(x_\alpha) \end{aligned} \tag{3.5.45}$$

else $g_l = \bar{g}_l = u_l = \bar{u}_l = 0$.

The last step of the recursion over layers:

$$\begin{aligned} K_l &= z_l\Omega_l^{(\alpha-1)} - u_l\Phi_l^{(\alpha-1)} \quad & L_l &= f_l\Omega_l^{(\alpha-1)} - g_l\Phi_l^{(\alpha-1)} \\ \bar{K}_l &= \bar{z}_l\bar{\Omega}_l^{(\alpha-1)} - \bar{u}_l\bar{\Phi}_l^{(\alpha-1)} \quad & \bar{L}_l &= \bar{f}_l\bar{\Omega}_l^{(\alpha-1)} - \bar{g}_l\bar{\Phi}_l^{(\alpha-1)} \end{aligned} \tag{3.5.46}$$

Calculation of amplitude coefficients of the scattered field:

$$a_l^{(\alpha+1)} = V_l(x_\alpha)\frac{L_l}{K_l} \qquad b_l^{(\alpha+1)} = V_l(x_\alpha)\frac{\bar{L}_l}{\bar{K}_l} \tag{3.5.47}$$

12. End of loop over index l ($l = 1, 2, ..., l_s$).
13. End of loop over layers ($j = 1, 2, ..., \alpha$).

Checking the code workability and productivity was done in several ways. This check was as a rule conducted for the most unstable scattering characteristic – the cross section of backscattering σ_{bsc}:

(1) Comparison with Mie theory. Calculations for a homogeneous sphere were made using Mie code and multilayered code dividing the sphere into a very large number of concentric layers with identical refractive indices. Relatively fast convergence of our values to Mie theory was detected as α increased (three significant digits at $\alpha \sim 300$, 4–5 significant digits at $\alpha \sim 1000$).

(2) Investigation of convergence of σ_{bsc} as α increased, modeling an arbitrary profile of radial inhomogeneity. Convergence similar to that in the Mie theory was detected.
(3) Comparison of results obtained for the same inhomogeneity profile using different codes, including the code based on the numerical solution of the radial equations (see Section 8.5) and the code based on the analytical solution of these equations for the degenerate cases.

These tests are sufficient to state that the algorithm and code described in this section allow us to carry out, with great accuracy and reliability, the calculations in the framework of the asymptotic model of a fractal cluster (see Section 8.5), that is questionable regarding some other recursive algorithms for a multilayered sphere.

4

Light scattering by water-containing atmospheric aerosol particles

The main goal of this chapter is to consider and compare, under identical conditions, the different light scattering characteristics for aerosol media consisting of homogeneous and inhomogeneous particles similar in composition and structure. It is suggested, in so doing, that the particles are irregularly located in space, the distances between them are great compared with wavelength, and also much greater than the particles sizes (see, e.g., the paper of Mishchenko *et al.* (1995), which discusses the conditions of independent scattering).

The methodological issues (the choice of the distribution function parameters, the integration limit, and the values of the complex refractive index) associated with the calculation of the light scattering characteristics for the polydisperse systems of homogeneous and inhomogeneous atmospheric particles are considered in Section 4.1. The calculation results for the integral characteristics of scattering by these systems are discussed in Section 4.2. Stimulated regulation of the extinction or absorption by atmospheric aerosol as a result of formation of the very thin layers on the surface of the particles is considered in Section 4.3. The influence of particle humidifying on the angular scattering characteristics are discussed in Sections 4.4 and 4.5. In Section 4.6 the comparison of calculated data with the results of the in-the-field experiment are carried out.

4.1 CALCULATION OF LIGHT SCATTERING AND ABSORPTION BY POLYDISPERSE ENSEMBLES OF HOMOGENEOUS AND INHOMOGENEOUS PARTICLES

For a unit volume of polydisperse medium composed of homogeneous particles (HPs) characterized by a particle size distribution $f(R)$, we have:

$$F(N,\theta,\lambda) = \int_{R_{\mathrm{beg}}}^{R_{\mathrm{end}}} F_1(R,N,\theta,\lambda)\, f\,(R)\, dR \qquad (4.1.1)$$

where $F_1(R, N, \theta, \lambda)$ is the light scattering characteristic for a single particle and $F(N, \theta, \lambda)$ is the corresponding characteristic for a polydisperse medium.

For a polydisperse medium composed of two-layered particles (TLPs) we have in the general case:

$$F(N_1, N_2, \theta, \lambda) = \int\limits_{R_{1.\,\mathrm{beg}}}^{R_{1.\,\mathrm{end}}} \int\limits_{R_{2.\,\mathrm{beg}}}^{R_{2.\,\mathrm{end}}} F_1(R_1, R_2, N_1, N_2, \theta, \lambda)\, f\,(R_1, R_2)\, dR_1\, dR_2 \quad (4.1.2)$$

where $f\,(R_1, R_2)$ is the two-dimensional distribution function of the particles core and shell sizes, the integration is to be taken over the radii of internal and external spheres.

According to the data of Chapter 2, it is possible for real aerosols (at a given relative humidity f and the relative content G of soluble substances in the condensation core) to assume, for simplicity, that the ratio of equilibrium radius of water-containing particles to radius of the initial core is identical for all radii of dry cores R_d. As a result, the integration over two variables in Eq.(4.1.2) is replaced by integration over only one variable – the radius of the internal sphere. In so doing, the decrease of core size owing to its partial dissolution (according to a given G value) and associated increase of solution shell refractive and absorption indices are taken into account. It is suggested, for simplicity, that the densities of soluble and insoluble parts of the condensation core are approximately equal and G determines a core volume share occupied by soluble matter. For systems of TLPs the integration is taken not over the initial core radius, but over the radius of the remaining insoluble part, that is from $R_t^{\mathrm{beg}} = R_d^{\mathrm{beg}}\,(1-G)^{1/3}$ to $R_t^{\mathrm{end}} = R_d^{\mathrm{end}}\,(1-G)^{1/3}$, with the factor $A = R/R_d$ being replaced by $A' = R/R_t = A\,(1-G)^{-1/3}$. The integration step ΔR_t is selected so that the corresponding value Δx on the external particle size does not exceed 0.10–0.15 (Naumenko *et al.*, 1975). The real and imaginary parts of the refractive index of the solution shell in TLPs and corresponding mean-effective characteristic of HPs with the same external size were determined according to Section 2.5. With the purpose of result comparability, the integration limits at the calculation of the light scattering polydisperse characteristics for systems of HPs were selected as equal to the external sizes of the corresponding TLPs. For example, for a system of water-containing particles with $R_d^{\mathrm{beg}} = 0.04\ \mu m$, $R_d^{\mathrm{end}} = 10\ \mu m$, and $R/R_d = 1.216$ the integration limits for the corresponding system of HPs were selected as $R^{\mathrm{beg}} = (0.04 \cdot 1.216)\ \mu m$ and $R^{\mathrm{end}} = 12.46\ \mu m$.

As it follows from above, the comparison of scattering properties was carried out for the same initial ensemble of condensation cores with the given size distribution and high and low limits. The air humidity change resulted in both particle size increasing (at the corresponding shift of high and low limits of the distribution), and in a change of their optical constants. The increasing number of optically active particles owing to Aitken cores growth that was possible in the actual conditions was not taken into account in the first stage of the investigation.

The calculations were carried out for two visual range wavelengths: $\lambda = 0.6328\ \mu m$, which is very often used in laboratory measurements, and $\lambda = 0.55\ \mu m$ which has found wide use in the in-the-field devices (see, e.g., Gorchakov (1966)). The values of R/R_d used, the values of relative humidity, approximately corresponding to them at $G = 0.2$ (see

Section 2.5), the particle volume share of water in the considered range of ratios R/R_d, and the volume-averaged values of the refractive index corresponding to the chosen levels of particle water-content are presented in Table 4.1. As an initial value of the complex refractive index for the absolutely dry aerosol substance the value $N_1 = 1.65+0.005i$ was chosen, which was obtained for the so-called synthetic model of aerosol particles (Ivlev and Popova, 1973; Kondratiev *et al.*, 1973). For comparison in the right part of Table 4.1 the values are given for the volume-averaged $\bar{n}$ and $\bar{\kappa}$, obtained at the same water-content level, but at the initial value of the complex refractive index of the absolutely dry aerosol substance $N_1 = 1.55 + 0.015i$.

As seen in Table 4.1, the difference between the volume-averaged refractive indexes $\bar{n}$ of water-containing particles is insignificant, and as relative humidity increasing from 0.60 to 0.94 decreases from 4.5 to 1.5%. The volume-averaged values of $\bar{\kappa}$ decrease several times as water-content level increases. For water in the visible spectral range the value $N_2 = 1.33$ was chosen (Irvine and Pollack, 1968). The imaginary part of the water refractive index for visible light is very small, and it may be neglected maintaining accuracy sufficient for our calculations.

The choice of integration limits in Eq.(4.1.2) is worthy of notice. Many authors using the Junge type distribution for description of the atmospheric aerosol physical properties (see Section 2.2), integrate in the size interval from $R_d^{\text{beg}} = 0.04\ \mu m$ up to $R_d^{\text{end}} = 10\ \mu m$. In so doing, the formal justification of such a low limit R_d^{beg} is concerned with the fact that more small particles tend to rapid coagulation (Bullrich, 1964; Rozenberg, 1968). The choice of $R_d^{\text{end}} = 10\ \mu m$ as the upper limit is dictated by rapid deposition of larger particles. In this case, the fractional aerosol composition is not usually taken into account. In this connection let us consider the question on the rational and adequate choice of integration limits more fully.

Different authors attempted to estimate the influence of integration limits. Bullrich (1964) has shown, that for forward scattering intensity (in an interval of scattering angles $\theta = 0$–$10°$) the contribution of particles from $R < 0.08\ \mu m$ is not noticeable, but particles with $R > 3\ \mu m$ contribute markedly. For great scattering angles, especially in backward directions ($\theta > 90°$) it is impossible to neglect the influence of small particles.

Table 4.1. Values of R/R_d, $f, \bar{n}$, and $\bar{\kappa}$ for calculation of light scattering by polydisperse systems of particles.

		Water	$N_1 = 1.65 + 0.005i$		$N_1 = 1.55 + 0.015i$	
R/R_d	f	share	$\bar{n}$	$\bar{\kappa}$	$\bar{n}$	$\bar{\kappa}$
1.0	0.	0.	1.65	0.005	1.55	0.015
1.125	0.60	0.289	1.555	0.0035	1.485	0.0105
1.177	0.70	0.387	1.527	0.0031	1.465	0.0092
1.216	0.75	0.444	1.509	0.0028	1.452	0.0083
1.270	0.80	0.512	1.487	0.0024	1.437	0.0073
1.35	0.85	0.594	1.461	0.0020	1.419	0.0061
1.426	0.88	0.655	1.442	0.0017	1.406	0.0052
1.55	0.92	0.731	1.417	0.0013	1.389	0.0040
1.72	0.94	0.803	1.394	0.0010	1.373	0.0029

At the same time for $\theta > 20°$ the scattering intensity weakly depends on particles with $R > 3\ \mu m$ (in 10% limits). Eiden (1966) has considered the problem of the influence of integration upper limit on the vibration ellipse stretching. It was found, that for the three models of aerosol particle size distribution studied (Laktionov and Bogomolov, 1972) at $\lambda = 0.40$–$0.85\ \mu m$, $n = 1.33$, 1.40, 1.44, and 1.50, and $\kappa = 0$, 0.01, and 0.1 the vibration ellipse stretching is determined by particles with $R < 3\mu m$ for all scattering angles. Deirmenjian (1969) with the example of the haze M model has shown, that the values of scattering coefficient and of phase function in different directions (except for that close to $\theta = 0$) do not depend on the upper integration limit at $x > 25$–30. For $\theta = 0°$ the influence of great particles becomes unessential at $x > 50$.

Because, in the indicated cases the systems of HPs were considered, we shall deal with the question as to whether the upper limit can affect the scattering characteristics of TLPs. The results are given in Tables 4.2 and 4.3. Table 4.2 contains the values of the ratios of the extinction (scattering) coefficients calculated with Eq.(4.1.2) for the Junge type particle size distributions at the identical minimum radius of the initial condensation cores $R_d^{\rm beg} = 0.04\ \mu m$ and several values of maximal radius, to the extinction (scattering) coefficient obtained at $R_d^{\rm beg} = 0.04\ \mu m$ and $R_d^{\rm end} = 10\ \mu m$. The value $R/R_d = 1.216$ is accepted, that corresponds to $\bar{n} = 1.509$. From the Table 4.2 it is seen, that at $\nu = 2.5$ and $R_d^{\rm end} = 3\ \mu m$ the values of the extinction coefficient differ from the case when $R_d^{\rm end} = 10\ \mu m$ by no more than 8%, and at $\nu = 3$ the difference does not exceed 2.5%.

It is seen from Table 4.3, that the upper limit $R_d^{\rm end} = 3\ \mu m$ at $\nu = 2.5$ and $\theta = 30$–$170°$ allow for obtaining the values of the angular scattering characteristics differing from the values corresponding to the case $R_d^{\rm end} = 10\ \mu m$ by no more than on 5–6%. At higher ν the difference is insignificant. It is necessary to mention, that the upper limit of the particle size very strongly influences the scattering intensity in the forward direction ($\theta \approx 0$) and more weakly in backward directions ($\theta = 180°$). For these directions the $I(\theta)$ value is tangibly underestimated at $R_d^{\rm end} = 3\ \mu m$. As the calculations have shown, at $\nu = 2$ the upper limit $R_d^{\rm end} = 3\ \mu m$ gives low accuracy for the extinction and scattering coefficients, and also of the scattered light polarization degree for some directions. In several cases it also takes place for the element f_{43}.

Numerous experimental data (Bullrich, 1964; Sedunov, 1972) testify that in atmospheric aerosols the distributions with $\nu = 2.5$–4 are usually implemented and distribution with $\nu = 2$ is rarely implemented. Besides, as in our investigations,the scattering directi-

Table 4.2. The ratios of extinction and scattering coefficients at different upper integration limit $R_d^{\rm end}$.

	$R_d^{\rm end}(\mu m)$							
	3.0	4.0	5.0	6.0	3.0	4.0	5.0	6.0
ν	$K_{\rm ext}^{\rm lay}(R_d^{\rm end})/K_{\rm ext}^{\rm lay}(10)$				$K_{\rm sc}^{\rm lay}(R_d^{\rm end})/K_{\rm sc}^{\rm lay}(10)$			
2.0	0.76	0.82	0.86	0.90	0.81	0.86	0.89	0.92
2.5	0.92	0.94	0.96	0.97	0.94	0.96	0.97	0.98
3.0	0.98	0.99	0.99	1.00	0.98	0.99	1.00	1.00
3.5	1.00	1.00	1.00	1.00	1.00	1.00	1.00	1.00
4.0	1.00	1.00	1.00	1.00	1.00	1.00	1.00	1.00

Table 4.3. Ratios $I(R_d^{\text{end}})/I(10)$ and $f_{ik}(R_d^{\text{end}})/f_{ik}(10)$ at $\lambda = 0.6328\ \mu m$ and $R/R_d = 1.216$.

		$\nu = 2.5$				$\nu = 3.0$			
		$R_d^{\text{end}}\,(\mu m)$							
	θ(deg)	3.0	4.0	5.0	6.0	3.0	4.0	5.0	6.0
$\frac{I(R_d^{\text{end}})}{I(10)}$	0	0.20	0.29	0.39	0.49	0.36	0.46	0.55	0.65
	30	0.98	0.99	0.99	0.99	0.99	1.00	1.00	1.00
	60	0.96	0.98	0.99	0.99	0.99	1.00	1.00	1.00
	90	0.98	0.98	0.99	0.99	0.99	1.00	1.00	1.00
	120	0.98	0.99	0.99	1.00	1.00	1.00	1.00	1.00
	150	0.99	0.99	0.99	1.00	1.00	1.00	1.00	1.00
	170	0.97	0.99	0.99	0.99	0.99	1.00	1.00	1.00
	180	0.75	0.86	0.92	0.97	0.88	0.94	0.97	0.99
$\frac{P(R_d^{\text{end}})}{P(10)}$	30	0.95	0.97	0.97	0.98	0.99	1.00	1.00	1.00
	60	1.03	1.01	1.01	1.00	1.01	1.00	1.00	1.00
	90	1.03	1.03	1.02	1.01	1.01	1.00	1.00	1.00
	120	1.06	1.02	1.01	1.00	1.01	1.00	1.00	1.00
	150	1.02	1.02	1.01	1.01	1.01	1.00	1.00	1.00
	170	1.05	1.07	1.06	1.03	1.01	1.03	1.03	1.03
$\frac{f_{33}(R_d^{\text{end}})}{f_{33}(10)}$	30	1.00	1.00	1.00	1.00	1.00	1.00	1.00	1.00
	60	1.00	1.00	1.00	1.00	1.00	1.00	1.00	1.00
	90	1.00	1.00	1.00	1.00	1.00	1.00	1.00	1.00
	120	1.01	1.01	1.01	1.01	1.00	1.00	1.00	1.00
	150	1.02	1.01	1.01	1.01	1.01	1.01	1.00	1.00
	170	1.10	1.06	1.01	1.01	1.02	1.01	1.00	1.00
$\frac{f_{43}(R_d^{\text{end}})}{f_{43}(10)}$	30	1.02	1.01	1.01	1.01	1.01	1.00	1.00	1.00
	60	1.04	1.02	1.01	1.01	1.01	1.01	1.01	1.00
	90	1.28	1.10	1.06	1.03	1.02	1.01	1.00	1.00
	120	1.01	1.01	1.01	1.01	1.00	1.00	1.00	1.00
	150	1.00	1.00	1.00	1.00	1.00	1.00	1.00	1.00
	170	1.01	0.99	1.00	1.00	1.00	0.99	1.00	1.00

ons are distinct from θ=0 or 180° in most cases, the calculations were carried out for $R_d^{\text{end}} = 3\ \mu m$ (with due regard for the growth coefficient, that is for $R_d^{\text{end}} A$). Tables 4.2 and 4.3 give a rather fair interval of validity of these calculations. The control calculations at other values of $R/R_d = A$ and $\bar{n}$ give results, close to those presented in Tables 4.2 and 4.3. Where the forward or backward scattering intensities were involved, the calculations were executed for $R_d^{\text{end}} = 10\ \mu m$. It is pertinent to note, that the upper integration limit $R_d^{\text{end}} = 3\ \mu m$ approximately corresponds to the maximal particle size in the so-called accumulative fraction of the atmospheric aerosol.

In Table 4.3 attention must be given to the abnormaly high values of the ratio $f_{43}(R_d^{\text{end}}) / f_{43}(10)$ at $\nu = 2.5$ and θ=90°. The reason for this lies in the fact that at $\nu = 2.5$ the dependence $f_{43}(\theta)$ at $R_d^{\rm end} = 10\ \mu m$ passes through zero close to θ=90°. But as the difference of absolute values of $f_{43}(R_d^{\rm end})$ and $f_{43}(R_d^{\rm end} = 10\ \mu m)$ is insignificant, it is possible to consider that the core size upper limit $R_d^{\rm end} = 3\ \mu m$ can be sufficient and in this case.

The recommendations and the conclusions presented here hold true for calculations of the radiation scattering and absorption characteristics for polydisperse systems of atmospheric aerosol particles in the visible spectral range. Naturally, for other spectral ranges and other objects they can be somewhat different.

4.2 EXTINCTION, SCATTERING, AND ABSORPTION COEFFICIENTS FOR POLYDISPERSE SYSTEMS OF TLPs AND HPs

The study of the extinction coefficient variations is one of the basic sources of information about atmospheric aerosol transformations caused by changes of meteorological circumstances, air masses, and so on. The measurements of the extinction coefficient allow us to concretize and forecast atmosphere visibility, to keep track of aerosol contamination. The information about the scattering and absorption coefficients are no less important. For polydisperse medium characterized by the particle size distribution of the Junge type (Section 2.2), we have:

$$K_i^{(2)} = C_N \pi A^2 c \int\limits_{R_{1,\ \rm beg}}^{R_{1,\ \rm end}} Q_i(R_1, R_2, N_1, N_2, \lambda)\, R_1^{1-\nu}\, dR_1 \tag{4.2.1}$$

where $A = R_2/R_1$, $Q_i = Q_{\rm ext}, Q_{\rm sc}$, and $Q_{\rm abs}$ are the extinction, scattering, and absorption efficiencies respectively for a single TLP, and c is the normalizing factor (see Eq.(2.2.2)).

It is shown by Prishivalko *et al.* (1975) that for single radially inhomogeneous particles and TLPs with $x \leq 4$ the values of extinction (scattering) coefficients coincide within 5–6% with the corresponding values for HPs with refractive indices equal to the volume-averaged values of the corresponding non-uniform particles. For particles with higher size parameters it fails to select such an effective refractive index, at which the extinction coefficient of TLPs and HPs with a chosen n would differ no more than 10–15% in a wide enough range of x. However, the difference $\left(Q_{\rm ext}^{(2)} - Q_{\rm ext}^{\rm hom}\right)$ appears to be a sign-alternating quantity with increasing x. Apparently, for this reason, as will be shown below, the values of the extinction and scattering coefficients for polydisperse systems of TLPs and systems of HPs, identical to them in size distribution and having the refractive index equal to the volume-averaged one of TLPs, practically coincide.

In Table 4.4 the values of the ratios of the extinction coefficients for systems of water-containing particles, with given R/R_d and corresponding systems of HPs with averaged refractive indices, are given. It follows from Table 4.4, that at $\nu = 3.5$ the extinction coefficients of homogeneous and TLPs systems coincide within 2% and only at $\nu = 4$

Table 4.4. Ratio $K_{\text{ext}}^{(2)}/K_{\text{ext}}^{\text{hom}}$ at $G = 0.2$, $\lambda = 0.6328\ \mu m$, $R_d^{\text{beg}} = 0.04\ \mu m$, and $R_d^{\text{end}} = 3\ \mu m$. (The asterisks mark the values obtained at $R_d^{\text{end}} = 10\ \mu m$).

		ν				
R/R_d	$\bar{n}$	2.0	2.5	3.0	3.5	4.0
1.0	1.650	1.000	1.000	1.000	1.000	1.000
1.125	1.555	0.982	0.987	0.998	1.011	1.025
1.216	1.509	0.978	0.982	0.994	1.012	1.030
1.216	1.509	0.984*	0.983*	0.994*	1.012*	1.030*
1.27	1.487	0.979	0.982	0.995	1.015	1.036
1.35	1.464	0.979	0.981	0.994	1.015	1.037
1.426	1.442	0.979	0.981	0.995	1.015	1.037
1.55	1.417	0.981	0.983	0.995	1.015	1.037
1.72	1.394	0.989	0.989	0.996	1.013	1.033

may the difference be as much as 4%. The ratio $K_{\text{ext}}^{(2)}/K_{\text{ext}}^{\text{hom}}$ at $R_d^{\text{end}} = 3\ \mu m$ is independent of the integration upper limit.

It is also interesting to clear up the question of how the soluble matter content and core refractive index influence the extinction and scattering coefficients provided the volume-averaged refractive index remains unchanged. Some of computed results, obtained by us, are given in Tables 4.5 and 4.6. Both here and in later tables and figures the absolute values of K_{ext} and K_{sc} are given at $C_N = 1\ \mu m^{-3}$. The data of Tables 4.4–4.6 allow us to make the deduction that the polydisperse scattering and extinction coefficients are very poorly sensitive to particle internal structure. At identical size distributions these characteristics within 5% are determined by the volume-averaged refractive index.

A similar result was obtained by Tuomi (1980), where the influence of relative humidity on the scattering properties of atmospheric particles was studied. It was suggested, that the dry particle consists of a homogeneous spherical insoluble core covered with a soluble spherical shell. With relative humidity increasing, the third layer of the saturated solution is formed on the particle surface. Such three-layered structure exists as long as some relative humidity value will be reached, at which all soluble matter will be dissolved. The particle then becomes two-layered. As the relative humidity rises further the solution concentration within its shell decreases.

Having set the insoluble core refractive index at $N_i = 1.62 + 0.02i$, the soluble one at $N_s = 1.521 + 0i$, and for water $N_w = 1.333 + 0i$, Tuomi compared, in relation to humidity, the scattering properties of model particles of three types with the same external radius:

- a three-layered particle consisting of a core, a not completely dissolved shell, and saturated solution with $n = 1.3867$ ("onion");
- one with a refractive index equal to the volume-averaged one for the three-layered particle ("an average sphere"); and
- an HP with the refractive index equal to the refractive index of the exterior layer of the three-layered particle ("solution sphere").

Table 4.5. Extinction and scattering coefficients at $\bar{n} = 1.51$, $n_1 = 1.65$, $\lambda = 0.55\ \mu m$, $R_d^{\mathrm{beg}} = 0.04\ \mu m$, and $R_d^{\mathrm{end}} = 3\ \mu m$.

	ν							
G	2.5	3.0	3.5	4.0	2.5	3.0	3.5	4.0
	$K_{\mathrm{ext}}^{(2)}\ (\mu m^{-1})$				$K_{\mathrm{sc}}^{(2)}\ (\mu m^{-1})$			
0.0	0.0489	0.0223	0.0115	0.00657	0.0466	0.0216	0.0112	0.00640
0.3	0.0495	0.0224	0.0115	0.00652	0.0472	0.0217	0.0112	0.00635
0.5	0.0495	0.0224	0.0114	0.00645	0.0474	0.0217	0.0111	0.00630
0.7	0.0501	0.0225	0.0114	0.00643	0.0481	0.0219	0.0111	0.00628
1.0	0.0502	0.0225	0.0114	0.00634	0.0483	0.0219	0.0111	0.00618

The calculations are carried out at $\lambda = 0.4$ and $0.7\ \mu m$ for the log-normal particle size distribution with parameters $R_m = 0.0368\ \mu m$ and $\sigma = 2.028$ corresponding to the experimental data. It is obtained, that the extinction coefficient dependencies $K_{\mathrm{ext}}(f)$ for the three-layered particles ("onion" and "average sphere") at $\lambda = 0.4\ \mu m$ are practically superimposed on each other, and at $\lambda = 0.7\ \mu m$ completely coincide. The corresponding curves for "solution sphere" run significantly below and only at $f > 0.9$ do they approach the first two.

So, from essentially different calculations for homogeneous, two-layered and three-layered particles it is obtained that in respect to the extinction (scattering) coefficient the properties of spherical particle systems with any internal structure can be described with the help of the HP model with the volume-averaged refractive index. This conclusion is significant in the sense that all dependencies for the scattering and extinction coefficients deduced experimentally and theoretically for HPs, can be extended for concentrically layered atmospheric aerosol particles.

In many experimental and theoretical works the authors try to establish an interrelation between the extinction (scattering) coefficient of aerosol systems and air relative humidity. Such an interrelationship in natural conditions can only be traced at times of air mass stability. Detailed analysis of the investigations which have been carried out in this direction, have been performed by Hänel (1971). Let us give some reasonings and results from this work.

The radiation extinction coefficient for atmospheric air per unit volume may be represented as the sum:

$$K_{\mathrm{ext}} = K_{\mathrm{ext,a}} + K_{\mathrm{ext,m}} \tag{4.2.2}$$

where $K_{\mathrm{ext,a}}$ is the extinction coefficient for aerosol particles and $K_{\mathrm{ext,m}}$ is the extinction coefficient for air molecules, which is independent of the air relative humidity. The radiation extinction coefficient for aerosol particles is equal to the sum:

$$K_{\mathrm{ext,a}} = \frac{1}{V_{\mathrm{a}}} \sum_{j=1}^{L_{\mathrm{a}}} \pi R_j^2 Q_{\mathrm{ext},j} \tag{4.2.3}$$

where L_{a} is the number of particles in volume V_{a} and R_j, $Q_{\mathrm{ext},j}$ are the radius and extinction efficiencies of single particles.

Table 4.6. Extinction and scattering coefficients at $\bar{n} = 1.51$, $G = 0.2$, $\lambda = 0.55\ \mu m$, $R_d^{\text{beg}} = 0.04\ \mu m$, and $R_d^{\text{end}} = 3\ \mu m$.

	ν							
	2.5	3.0	3.5	4.0	2.5	3.0	3.5	4.0
n_1	$K_{\text{ext}}^{(2)}\ (\mu m^{-1})$				$K_{\text{sc}}^{(2)}\ (\mu m^{-1})$			
1.65	0.0493	0.0224	0.0115	0.00654	0.0470	0.0216	0.0112	0.00636
1.60	0.0495	0.0224	0.0115	0.00650	0.0470	0.0216	0.0112	0.00630
1.55	0.0498	0.0225	0.0115	0.00645	0.0470	0.0216	0.0112	0.00622

The change of the extinction coefficient at relative humidity variations is due to its influence on both the particle size and structure and the particle average refractive index. This influence can be taken into account, if we rewrite Eq.(4.2.3) as:

$$K_{\text{ext,a}} = \pi R_{\text{a}}^2 S_{\text{a}} / V_{\text{a}} \tag{4.2.4}$$

where $S_{\text{a}} = L_{\text{a}} Q_{\text{ext,a}} = \sum_{j=1}^{L_{\text{a}}} (R_j/R_{\text{a}})^2 Q_{\text{ext},j}$, R_{a} is the mean-effective particle radius in the sample, which it is conditionally possible to consider equal to the radius of particles most frequently encountered in the given sample and $Q_{\text{ext,a}}$ is the ensemble-averaged extinction efficiency introduced by Foitzic (1965).

From Eq.(4.2.4) we obtain the ratio of the extinction coefficients at relative humidities of f_1 and f_2 as:

$$\frac{K_{\text{ext,a}}(f_1)}{K_{\text{ext,a}}(f_2)} = \left[\frac{R_{\text{a}}(f_1)}{R_{\text{a}}(f_2)}\right]^2 \frac{S_{\text{a}}(f_1)}{S_{\text{a}}(f_2)} \tag{4.2.5}$$

The first factor in the right-hand side of Eq.(4.2.5) depends only on the size change with relative humidity for particles with a mean-effective radius. The second factor describes the changes of particle radius, structure, and refractive index with the help of the ratio of quantities of view:

$$S_{\text{a}} = \sum_{j=1}^{L_{\text{a}}} \left(\frac{R_j}{R_{\text{a}}}\right)^2 Q_{\text{ext},j} \tag{4.2.6}$$

As the experiment shows, the ratio $(R_j/R_{\text{a}})^2$ is almost independent of humidity at $f \leq 0.95$ and $R_d \geq 0.1\ \mu m$. As influence of particles with $R_d \leq 0.1\ \mu m$ on the extinction coefficients is negligible (Bullrich, 1964), it is possible to consider that the ratio $(R_j/R_{\text{a}})^2$ does not depend on relative humidity.

Investigations of van de Hulst (1981), Penndorf (1962), and others have shown, that $Q_{\text{ext},\,j} \approx Q_{\text{ext},\,j}(\alpha_j)$, where the generalized parameter α_j is expressed as:

$$\alpha_j = \frac{2\pi R_j}{\lambda} \left[(n_j - 1)^2 + \kappa_j^2\right]^{1/2} \tag{4.2.7}$$

Thus, the extinction efficiency $Q_{\text{ext},j}$ undergoes change only if the relative humidity variation results in a change of the generalized parameter α_j.

In Figure 4.1 the average dependencies of the ratio of the generalized parameter α_j for the humidified and dry particles are depicted based on work (Hänel, 1971) for the case of increasing humidity. It is seen, that for continental aerosols the ratio $\alpha_j(f)/\alpha_j(0)$ is almost equal to unity in the interval of relative humidities $f \leq 0.90$ and increases up to 1.2 at $f = 0.95$. The point is that, as it was mentioned above, the particle radius increasing as humidity increases is always accompanied by average refractive index decreasing. The negligibly small change of generalized parameter α_j at humidities smaller than 0.9–0.95 is a consequence of it.

Hence it follows, that for $f \lesssim 0.9 - 0.95$ the extinction efficiency $Q_{\text{ext},j}$ of single particles depends only weakly on relative humidity. Then, the sum S_{a} in Eq.(4.2.6) also depends weakly on relative humidity. It gives rise to the simple approximating formula for the ratio of the extinction coefficients of the given aerosol sample at two different humidity values $f \lesssim 0.90$–0.95:

$$\frac{K_{\text{ext},a}(f_1)}{K_{\text{ext},a}(f_2)} = \left[\frac{R_a(f_1)}{R_a(f_2)}\right]^2 \tag{4.2.8}$$

It means, that the variation of the extinction coefficient with the relative humidity up to $f \lesssim 0.90$–0.95 depends only on the mean-effective radius R_a variation (i.e., on chemical composition and structure of particles of corresponding size).

In view of the formula (2.5.8) Eq.(4.2.8) becomes:

$$\frac{K_{\text{ext},a}(f_1)}{K_{\text{ext},a}(f_2)} = \left[\frac{1 + \varphi\,(f_1)\, f_1(1 - f_1)^{-1}}{1 + \varphi\,(f_2)\, f_2(1 - f_2)^{-1}}\right]^{2/3} \tag{4.2.9}$$

and using the Kasten's approximating formula (Kasten, 1969) (see Section 2.5) it is possible to write:

$$\frac{K_{\text{ext},a}(f_1)}{K_{\text{ext},a}(f_2)} = \left(\frac{1 - f_2}{1 - f_1}\right)^{2\varepsilon} \tag{4.2.10}$$

where the power ε accepts values indicated in Section 2.5.

From Eqs.(4.2.8)–(4.2.10) it is seen that the extinction coefficient change with relative humidity is identical for different wavelengths and particle distributions. Some difference can arise in the field of the first maximum of dependence $Q_{\text{ext},j}(\alpha_j)$, where the small variations of argument α_j can be accompanied by considerable changes of $Q_{\text{ext},j}$ both in the direction of its increase and decrease. In view of this, it is possible to explain the effect of "abnormal" changes of the scattering coefficient sometimes observed in natural conditions at humidities near $f = 0.8$.

Livshits *et al.* (1974) have analysed the results of in situ measurements of the scattering coefficient for aerosol haze. In their judgement, the change of haze scattering with humidity variation may be described schematically as follows. At low humidity the aerosol particle size weakly depends on f. Monotonic particle growth with humidity takes place, starting with some f value and moving up to maximal himidity, and the scattering coefficient grows accordingly. At $f \approx 80\%$ clearing up of the haze is sometimes observed.

Apparently, in some situations depending on the dominant particle size and their optical properties (especially in the field of the first maximum of dependence $Q_{\text{ext},j}(\alpha_j)$),

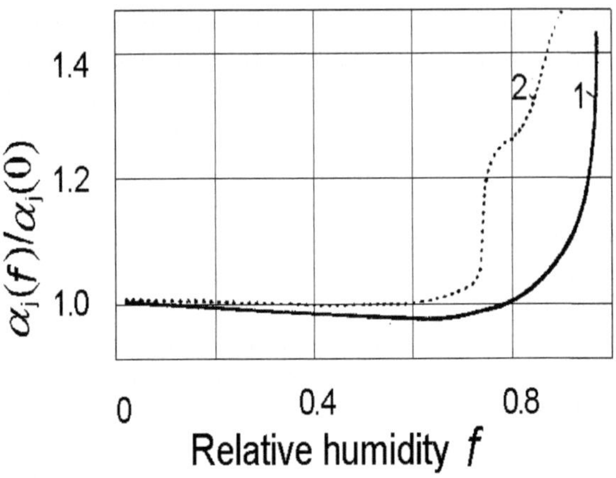

Figure 4.1. Ratio $\alpha_j(f)/\alpha_j\,(0)$ versus relative humidity for continental (*1*) and sea (*2*) aerosols (Hänel, 1971).

the light scattering decrease caused by lowering of the volume-averaged refractive index, not only cancels the scattering increase due to particle size growth, but also prevails above it in some range of relative humidity. As a result, clearing up occurs.

However, as follows from Table 4.1, at considerable particle humidifying the decrease in the volume-averaged refractive index is essentially decelerated as it approaches the water refractive index. But the scattering cross section increases. Thus the scattering coefficient grows with an increased rate.

As is mentioned by Livshits *et al.* (1974), it is a difficult problem to check the dependence $K_{\text{ext}}(f)$ in natural conditions. In many respects the optical properties of the local volume of haze depend on random processes of arrival and departure of different type of aerosols. Therefore it is only possible during a calm period to trace the relative humidity influence on aerosol optical properties "in the pure state". But for field measurements it is difficult to meet such conditions, when in complete calm the relative humidity varies within a wide range.

Having compared small finite differences of f with finite differences of $\log K_{\text{ext}}$, Livshits *et al.* (1974) found the dependence K_{ext} versus f as:

$$K_{\text{ext}}(\,f\,) = K_{\text{ext}}(\,f_0\,)\exp\left[2.303\int_{f_0}^{f}\Phi(\,f\,)df\right] \tag{4.2.11}$$

where $\Phi(\,f\,) = d\ln K_{\text{ext}}/df$. In simplified form this relation can be presented as:

$$K_{\text{ext}}(\,f\,) \approx \gamma\,(1-f)^{\varepsilon} \tag{4.2.12}$$

where γ is the coefficient depending on the total number and nature of the particles. The power ε in the humidity range 0.45–0.9 varies from 0.05 to 0.5, and for the range $0.75 < f < 0.95$ the average ε value is equal 0.33, that is close to the value in the work of Hänel (1971).

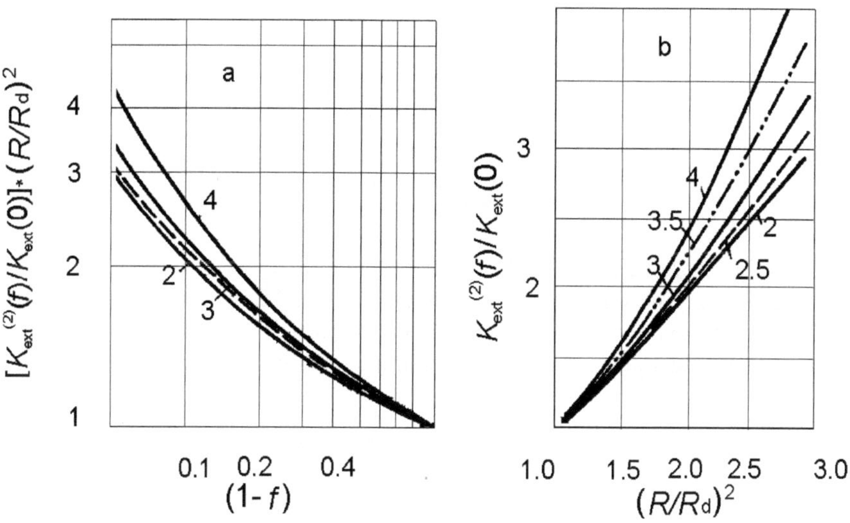

Figure 4.2. Ratio of the extinction coefficients for humidifyed and dry aerosols (solid lines) and $(R/R_d)^2$ value (dashed line) versus $(1-f)$ (a) and versus $(R/R_d)^2(b)$ at $\lambda = 0.6328\ \mu m$. Numbers at curves are ν values.

The empirical formula (4.2.10) is rather simple and obvious. If to take into account the comments of Section 2.5 about dependence of power ε on measurement conditions and Eq.(1.5.10), connecting the ε value with relative humidity, then Eq.(4.2.10) can be used for an estimation of the extinction coefficient variation with change of relative humidity. However, quantities $K_a(f)$ in the left-hand part of Eq.(4.2.10) are to a certain extent the formal ones. The estimates obtained with its help are mere qualitative ones.

In a series of works (Lundgren and Cooper, 1969; Bullrich *et al.*, 1972) the experimental dependence of $\log\,(K_{\text{ext}}(\,f\,)/K_{\text{ext}}(\,0\,))$ versus $\log(1-f)$ is obtained. The comparison of this dependence with the dependence $\log(R/R_d)^2$ versus $\log(1-f)$ have shown, that the corresponding curves are very similar. The observed small differences in curve decline can be explained by the variability of water-absorbing or water-soluble fractions in different air masses.

In Figure 4.2(*a,b*) the calculated dependencies $K^{(2)}_{\text{ext}}(\,f\,)/K_{\text{ext}}(\,0\,)$ versus $(1-f)$ (at different ν) and $(R/R_d)^2$ versus $(1-f)$ are depicted on a logarithmic scale. The most similar points of these two dependencies is observed at $\nu = 2.5$ and 3, that is for not so narrow particle distributions. For greater clarity on Figure 4.2(*b*) the ratios $K^{(2)}_{\text{ext}}(\,f\,)/K_{\text{ext}}(\,0\,)$ versus $(1-f)$ are depicted in the usual scale, from which it is seen, that at $\nu < 3.5$ the discussed dependencies are practically linear. In the interval $1.216 \le R/R_d \le 1.55$ at $G = 0.2$ with an error no more than 2% the approximate relation is valid:

$$\frac{K^{(2)}_{\text{ext}}(\,f\,)}{K_{\text{ext}}(\,0\,)} = S\left(\frac{R}{R_d}\right)^2 - M \qquad (4.2.13)$$

where S and M are the empirical coefficients (Table 4.7). For the ratio $R/R_d = 1.125$

Table 4.7. Coefficients of Eq.(4.2.13).

ν	S	M
2.5	1.029	0.093
3.0	1.141	0.229
3.5	1.310	0.426
4.0	1.528	0.683

$(f \approx 0.60)$ the error of Eq.(4.2.13) is within 4%, and for $R/R_d = 1.125$ $(f \approx 0.60)$ does not exceed 6–8%. The relation (4.2.13), as well as Eq.(4.2.8), confirms the linearity of dependence of the extinction coefficient ratio on the particle cross sections ratio.

Of some practical interest are the approximate relations manifested as:

$$\frac{K_{\text{ext}}^{(2)}(\,f\,)}{K_{\text{ext}}(\,0\,)} = \bar{n}\left[F\left(\nu\right)\bar{n} - L\left(\nu\right)\right]^{-1} \tag{4.2.14}$$

where $F(\nu) = 0.18\nu + 4.65$ and $L(\nu) = 0.336\nu + 5.8$. In the interval $1.125 \leq R/R_d \leq 1.72$ and $1.394 \leq \bar{n} \leq 1.555$, corresponding at $G = 0.2$ to $0.6 \leq f \leq 0.9$, the error of these formulae does not exceed 3% for $\nu = 2.5$–4.

The approximate formulae (4.2.13) and (4.2.14) are obtained for $\lambda = 0.6328\ \mu m$. When passing to $\lambda = 0.55\ \mu m$ the ratio $K_{\text{ext}}^{(2)}(f)/K_{\text{ext}}(0)$ decreases by no more than 1% at $\nu = 3.5$ and by no more than 2% at $\nu = 4$, but the values of $K_{\text{ext}}(0)$ differ rather noticeably (up to 30%). For the $K_{\text{ext}}^{(2)}$ calculation with Eqs.(4.2.13) and (4.2.14) the values of $K_{\text{ext}}(0)$ for these two wavelengths are given in Table 4.8 (Prishivalko, 1978). For $\lambda = 0.55\ \mu m$ the dependence of the extinction coefficient on ν, represented in Table 4.8, with an error no more than ± 1 % is approximated by the relation:

$$K_{\text{ext}}(\,0\,) = 2.21\nu^{-4.53} \tag{4.2.15}$$

As a whole, the extinction coefficient of aerosol medium unit volume can be calculated with the help of the approximate relation:

$$\begin{aligned} K_{\text{ext}}(\bar{n}, \bar{\kappa}, \nu\,) \approx 2.21\nu^{-4.53}\bar{n}\left[1 + \bar{\kappa}\left(0.547\nu^2 - 2.345\nu - 2.394\right)\right] \\ \cdot\left[\bar{n}\left(0.18\nu + 4.65\right) - \left(0.336\nu + 5.8\right)\right]^{-1} \end{aligned} \tag{4.2.16}$$

whose error does not exceed 10–15% at $1.35 \leq \bar{n} \leq 1.65$ and $0 \leq \bar{\kappa} \leq 0.1$. In Eq.(4.2.16) the air relative humidity variations are taken into account through the change of volume-averaged (effective) values of real $\bar{n}$ and imaginary $\bar{\kappa}$ parts of the complex refractive index.

Table 4.8. The values of $K_{\text{ext}}(0)$ at $n = 1.65$, $\kappa_1 = 0.005$, $R_d^{\text{beg}} = 0.04\ \mu m$, $R_d^{\text{end}} = 3\ \mu m$, and $C_N = 1\ \mu m^{-3}$.

	ν			
$\lambda(\mu m)$	2.5	3.0	3.5	4.0
0.55	0.0346	0.0154	0.00764	0.00415
0.63	0.0318	0.0133	0.00622	0.00318

In the end of this section we shall give some results describing the radiation absorption by an ensemble of TLPs at increasing humidity. Inasmuch as the calculations were performed at the values $\kappa = 0.005$ and 0.015 of the imaginary part of the dry aerosol substance refractive index, the absorption efficiency absolute value for single particles is insignificant. The absolute values of the absorption coefficients for polydisperse particle systems are also small. However, as was mentioned in Chapter 3, as humidifying of atmospheric aerosol particles occurs their absorption increases. Similar, but strongly smoothed growth of absorption takes place for polydisperse particle systems.

For the greater clarity, the ratio of the absorption coefficient for systems of TLPs to the absorption index for the initial non-humidified particle system versus the ratio of their cross sections is depicted in Figure 4.3. As seen from this figure, the maximal increase of the absorption index is observed for aerosol systems containing larger particles. It can reach 30–35%, when $\kappa_1 = 0.005$, and 70% at $\kappa_1 = 0.015$. For fine dispersed systems ($\nu > 2.5$) it does not exceed 20% in the former case and 40% in the latter. Apparently, if κ of initial particles is higher, then the increase of the absorption index for an ensemble of water-containing particles can be rather considerable. The comparison of the absorption indices for systems of TLPs and systems of model HPs with the complex refractive index equal to volume-averaged refractive index of TLPs (with identical size distribution), displays that the distinction does not exceed 10–15% at $\kappa_1 = 0.015$ and 8–10% at $\kappa_1 = 0.005$.

Kattawar and Hood (1976) noted the effect of absorption amplification by 8–15% during the formation of aqueous shells of particular thickness on the surfaces of polydisperse coal (sooty) particles. The different characteristics of light scattering for a system of coal particles described by the size distribution $f(R) = AR^{-6}\exp(-6R)$ (R is an external particle radius) were calculated at modal radius $R_0 = 1\ \mu m$ and $\lambda = 0.491\ \mu m$. For water the refractive index was set at $N = 1.33 + 0.0i$ and for coal $N = 1.95 + 0.66i$. The integration limits were chosen equal to 0.1 and 4 μm. Two models were considered. In

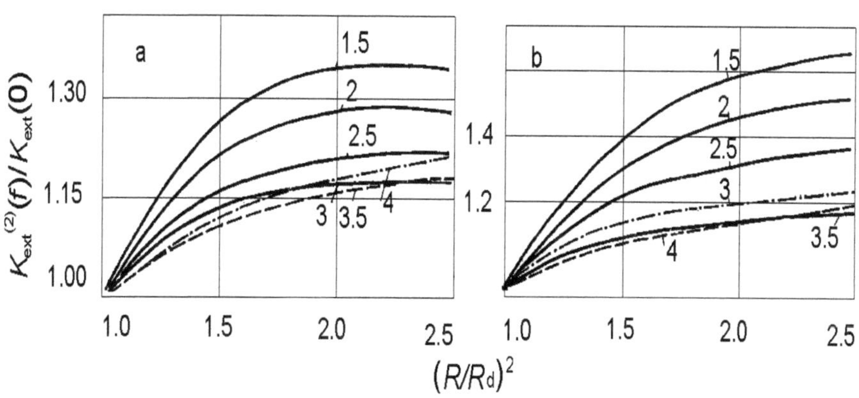

Figure 4.3. Ratio of the absorption coefficients for hymidified and dry aerosols versus $(R/R_d)^2$ value at $\lambda = 0.55\ \mu m$, $N_1 = 1.65 + 0.005i(a)$ and at $\lambda = 0.6328\ \mu m$, $N_1 = 1.55 + 0.015i(b)$. Numbers at curves are ν values.

one case it was suggested that the ratio of the particle core size to its external radius $q = R_1/R_2$ should remain constant, and in the second case all particles in the system have shells of identical thickness $d = R_2 - R_1 = \text{const}$. It is obvious that in the second case there are the favorable conditions for interference effects caused by interaction of fluxes, reflected from shell and core. As a result, the authors have established, that, when $0.79 \leq R_1/R_2 \leq 0.975$, the absorption coefficients for the system of water-containing particles are higher than those for the analogous system of homogeneous coal particles. The scattering coefficient in this range of R_1/R_2 values is minimal.

A similar pattern is observed when the particles in a system have shells of constant thickness. At $d = R_2 - R_1 \leq 0.3\ \mu m$ the absorption coefficient is higher than for an identical system of homogeneous coal particles. As the shell thickness increases further the absorption index for the system of water-containing particles rapidly decreases. As shell thickness exceeds 2 μm the core has no effect on the integral scattering characteristics.

Recently, special attention has been given to the problem of possible regulation of the extinction or absorption of energy incident on the particle. Such problems arise in connection with the problems of simulated action on clouds and fogs, propagation of the radiation of different wavelengths through them, and so on. Two typical variants of such problems are surveyed in the following section.

4.3 THE POSSIBILITY OF PURPOSEFULLY CHANGING THE EFFECTIVENESS OF EXTINCTION AND ABSORPTION BY SYSTEMS OF LAYERED PARTICLES

As was mentioned above, the presence of surface films on atmospheric particles of different kinds can significantly change their optical properties. Therefore, it is of considerable interest to study light scattering and absorption by water droplets or atmospheric haze solid particles upon the formation of thin films from surface-active substances (SAS) (Starojilova, 1969; Bakhanova *et al.*, 1971). Such films covering the particles transform them from homogeneous to two- or three-layered ones, essentially changing results of radiation interaction with such particulate systems.

Consider two types of aerosol: from almost dry or fractionally water-containing condensation cores and water-droplet ones. For the atmospheric haze aerosols we shall consider valid the particle size distribution of the Junge type; for water-droplet aerosols we shall assume, according to numerous literary data, that the particle size spectrum is described by the gamma distribution, already mentioned in Section 2.2:

$$f(R) = aR^{\mu} \exp(-\mu R/R_0) \tag{4.3.1}$$

where R is the droplet radius, μ is the parameter defining the distribution half-width, R_0 is the most probable, or modal, droplet radius, and a is the normalization factor. For simplicity, we shall consider, that the TLPs formed at the SAS deposition have shells of identical thickness: $d = R_2 - R_1 = \text{const}$.

As was mentioned in Chapter 2, in the real atmosphere at air relative humidity $0.6 \leq f \leq 0.95$ many continental aerosol particles consist of a solid core comprising of an insoluble or incompletely dissolved remainder of condensation core and the fluid water-solution shell. If the SAS shells are formed on such water-containing particles, they turn into the three-layered ones, whose properties have been explored (Harrison *et al.*, 1972; Rozenberg, 1972; Lopatin and Sid'ko, 1977, 1988; Terskov *et al.*, 1977; Rakhimov, 1998) insufficiently.

When studying light scattering by the three-layered particle systems for simplification we assumed that at given humidity ($f < 0.95$) the radii of all particles increase by the identical ratio: $A = R_2/R_1 = \text{const}$. The thickness of the outer shell, coating all water-containing particles of the polydisperse system, is constant: $d = R_3 - R_2$ (R_3 is the external particle radius, R_2 is the total radius of core and aqueous shell, and R_1 is the core radius). The radii ratio was chosen equal to $A = 1.216$, which corresponds to an air relative humidity $f \approx 0.75$ (Kasten, 1969).

The calculations are carried out for the visible spectral range: $\lambda = 0.63\ \mu m$ (Astafyeva *et al.*, 1981). The optical constants of the TLP cores are chosen as follows: in case of water $N_1 = 1.33 + 0i$; in case of atmospheric aerosol solid particles $1.4 \leq n_1 \leq 1.7$ and $0 \leq \kappa_1 \leq 0.01$. As the information about optical constants of SAS and thickness of films formed by them is rather limited, the investigations are carried out with the shell refractive index varying over a wide range ($1.05 \leq n_2 \leq 2.0$ and $\kappa_2 = 0$) for shell thicknesses $0 \leq d \leq 0.04\ \mu m$. Selection of wide intervals of n_2 and d is made as for reaching the greater generality of results, and for revealing the tendencies of the light scattering characteristics for TLPs and three-layered particles under a change of optical constants and sizes of cores and shells.

Consider first the scattering properties of TLP polydisperse systems: (1) water droplets covered by SAS shells; and (2) atmospheric aerosol "dry" particles covered with SAS films. Comparison of scattering properties of polydisperse systems of aqueous droplets and polydisperse systems of the same droplets, covered by shells, have shown that the ratio of the extinction coefficients $K_{\text{ext}}^{(2)}/K_{\text{ext}}^{\text{core}}$ ($K_{\text{ext}}^{\text{core}}$ is the extinction coefficient of polydisperse systems of aqueous droplets) depends on the shell thickness and optical constants, and also on the modal radius R_{01} and parameter μ of the particle size distribution.

When increasing the most probable radius R_{01}, the ratio $K_{\text{ext}}^{(2)}/K_{\text{ext}}^{\text{core}}$ sharply decreases, approaching unity at great R_{01} values (Figure 4.4). Only in the field of small values of $R_{01} < 0.2$–$0.3\ \mu m$ is the $K_{\text{ext}}^{(2)}$ value much more than $K_{\text{ext}}^{\text{core}}$. Increase of the shell refractive index and thickness, and of parameter μ also, results in a strong increase of $K_{\text{ext}}^{(2)}$ in comparison with $K_{\text{ext}}^{\text{core}}$ at $R_{01} < 0.6$–$0.7\ \mu m$. For example, at $n_2 = 2.0$, $R_{01} = 0.1\ \mu m$, and $\mu = 20$ the extinction coefficient of the TLP system is 5.8 times ($d = 0.02\ \mu m$) or 11.7 times ($d = 0.03\ \mu m$) larger than that for systems of particles without shells.

The investigations of polydisperse systems of atmospheric haze particles covered with SAS shells, have shown, that the extinction coefficients ratio $K_{\text{ext}}^{(2)}/K_{\text{ext}}^{\text{core}}$ grows with increase of the shell refractive index and thickness, and also with the distribution function parameter ν (Figure 4.5*a*). For example, for values $n_1 = 1.4$ and $\nu = 3$ at $n_2 = 1.7$ the $K_{\text{ext}}^{(2)}$ value is 2.2 times greater than $K_{\text{ext}}^{\text{core}}$, when shell thickness is $d = 0.03\ \mu m$, and 3 times greater at $d = 0.04\ \mu m$. It is necessary to mark also, that an increase of the core re-

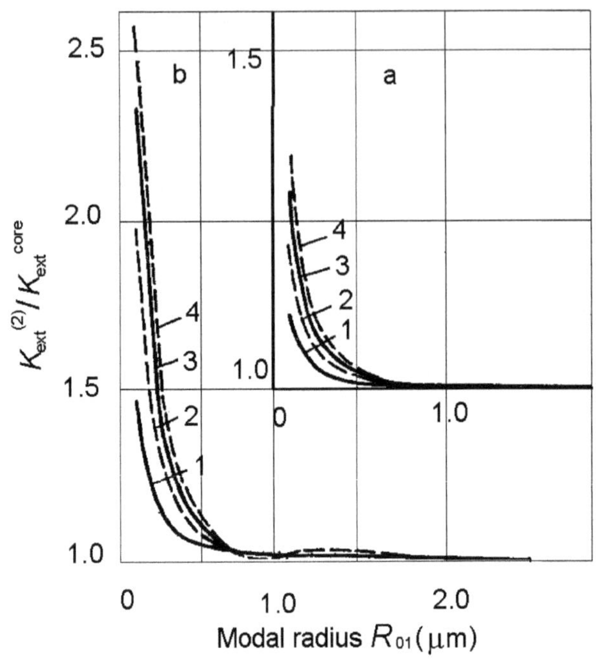

Figure 4.4. Ratio of the extinction efficiencies $K_{\text{ext}}^{(2)}/K_{\text{ext}}^{\text{core}}$ versus modal radius R_{01} for $N_2 = 1.5 + 0i(a)$ and $N_2 = 2.0 + 0i(b)$ at $N_1 = 1.33 + 0i$, with the shell thickness 0.01 μm and $\mu = 2(1), 5(2), 10(3)$, and $20(4)$.

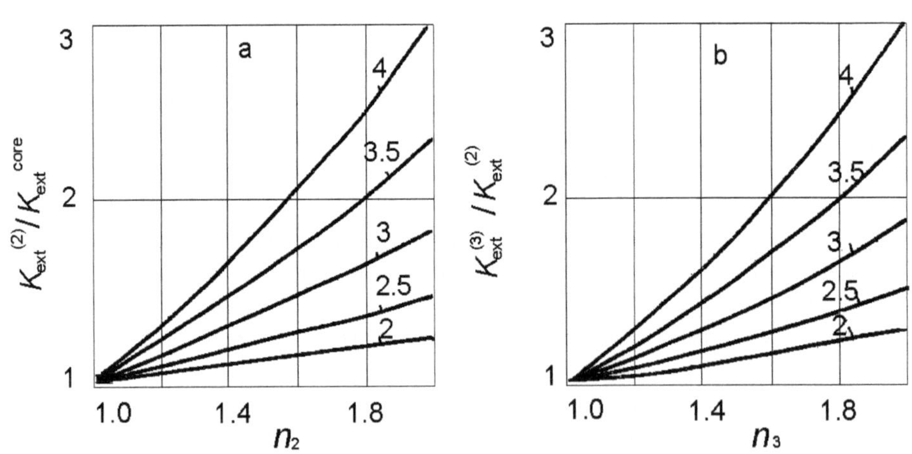

Figure 4.5. Ratio $K_{\text{ext}}^{(2)}/K_{\text{ext}}^{\text{core}}$ versus n_2 at $N_1 = 1.5 + 0.01i(a)$ and ratio $K_{\text{ext}}^{(3)}/K_{\text{ext}}^{(2)}$ versus n_3 at the same N_1 and $N_2 = 1.33 + 0i(b)$, $d = 0.02$ μm. Numbers at curves are the ν values.

fractive index in the range $1.4 \leq n_1 \leq 1.7$ results in a decrease in the extinction coefficient ratio for polydisperse systems of TLPs and their cores.

Thus, the presence of a thin SAS shell on polydisperse water droplets gives an essential difference of light extinction compared with the aqueous particle system only for its finely divided fraction ($R_{01} < 0.2$–$0.3\ \mu m$, $\mu > 10$, $n_2 \approx 2.0$, and $d > 0.02\ \mu m$). The light extinction by atmospheric haze polydisperse systems of particles covered with shells, considerably differs (by the factor of 10–20) from extinction by systems without shells, also only for finely divided atmospheric haze particles ($\nu > 3.5$) at high shell refractive index ($n_2 \approx 2.0$), small core refractive index ($n_1 \approx 1.4$) and great shell thickness ($d > 0.03\ \mu m$).

Let us discuss now the scattering properties of three-layered particle polydisperse systems. For this purpose, consider the extinction coefficient ratio $K_{\rm ext}^{(3)}/K_{\rm ext}^{(2)}$ for a system of water-containing atmospheric aerosol particles covered with shells, and for a system of TLPs without shells. As seen from Figure 4.5b, the extinction by three-layered particle polydisperse systems is greater, than that of the system of water-containing particles without shells, in approximately the same proportion, in which the extinction by the system of dry cores with SAS shells is more than the extinction of the system of cores without shells. The greater the surface layer refractive index, the higher is the extinction coefficient ratio $K_{\rm ext}^{(3)}/K_{\rm ext}^{(2)}$. The same growth of the ratio $K_{\rm ext}^{(3)}/K_{\rm ext}^{(2)}$ is observed at the transition to more finely dispersed aerosol (with increasing of ν).

Thus, the polydisperse system of atmospheric aerosols consisting of three-layered water-containing particles covered by SAS shells has the maximal extinction of the incident radiation if the particles are finely dispersed ($\nu > 3.5$) and the surface layer refractive index is high enough ($n_3 \approx 2.0$).

As was noted in Chapter 3, humidifying of low-absorbing condensation nuclei causes an increase of their absorption cross section. Similar, but strongly smoothed-out absorption growth takes place for polydisperse systems of water-covering aerosol particles.

The opportunity to increase absorption when a wave is incident on a particle frequently acquires the important practical value (e.g., the problem of powerful laser radiation interaction with the water-droplet aerosol). One of the actual aspects of this problem is the search for ways of interaction efficiency enhancement, that is the clearing up of the opportunity to destruct the droplet under condition of maximal lowering of energy consumption and overall power of used radiation sources. Let us find the opportunity to enhance the efficiency of radiation ($\lambda = 10.6\ \mu m$) interaction with clouds and fogs due to artificial creation of the fluid shells on the surface of water droplets. These shells are not soluble in water and have optical constants distinguished from that of water (Prishivalko, 1981). As in the previous case, the particle size spectrum for water-droplet clouds and fogs are described by the gamma distribution (4.3.1).

For the solution of this problem it is necessary to consider the influence of the distribution function parameters, shell thickness and optical constants on the absorption coefficient of the medium unit volume and quantum absorption probability. For polydisperse systems of homogeneous water droplets the absorption coefficient expresses as:

$$K_{\text{abs}}^{\text{hom}} = C_N \pi \frac{\int_{R_{\text{beg}}}^{R_{\text{end}}} Q_{\text{abs}}(R, N, \lambda) R^{\mu+2} \exp(-\mu R/R_0)\, dR}{\int_{R_{\text{beg}}}^{R_{\text{end}}} R^{\mu} \exp(-\mu R/R_0)\, dR} \tag{4.3.2}$$

where R_{beg} and R_{end} are the limiting particle sizes which are taken into account at integration over the ensemble, N and C_N are the complex refractive index and the number density of water droplets, and λ is the incident radiation wavelength. For systems of TLPs we have:

$$K_{\text{abs}}^{(2)} = C_N \pi A^2 \frac{\int_{R_{\text{beg}}}^{R_{\text{end}}} Q_{\text{abs}}^{(2)}(R_1, R_2, N_1, N_2, \lambda) R_1^{\mu+2} \exp(-\mu R_1/R_{01})\, dR_1}{\int_{R_{\text{beg}}}^{R_{\text{end}}} R_1^{\mu} \exp(-\mu R_1/R_{01})\, dR_1} \tag{4.3.3}$$

where R_{beg} and R_{end} are the same limiting sizes of initial water droplets, now the cores of TLPs. In so doing it is supposed, that the radii of all droplets in the ensemble vary by the identical ratio, that is $R_2/R_1 = A = \text{const}$. The calculations were performed for R_{01} = 1 (0.5) 3 (1) 6 (2) 10 μm and μ = 1 (0.5) 3 (1) 6 (2) 10 (i.e., for 100 variants of distributions for the initial ensemble of aqueous droplets).

The water complex refractive index for $\lambda = 10.6\ \mu m$ was taken as $n_1 = 1.173$ and $\kappa_1 = 0.0823$ (Pontier and Dechambenoy, 1966; Irvine and Pollack, 1968). As the calculations have a common character and were not concerned with any concrete fluid, the shell optical constants and thicknesses were varied over the wide ranges: $1.1 \le n_2 \le 1.5$, $0 \le \kappa_2 \le 0.12$, and $0 \le d \le 0.5\ \mu m$, where $d = R_2 - R_1$. The absorption efficiency is determined not only by the absorption index κ. It is important also to know , what part is scattered and spent uselessly in any particular sense. Such information is obtained by the quantum absorption probability $1 - \Lambda$, which is defined as $1 - Q_{\text{sc}}/Q_{\text{ext}} = Q_{\text{abs}}/Q_{\text{ext}}$. It is desirable, certainly, that as much of the radiation energy as possible was absorbed in droplets and as little as possible was scattered.

The consideration of the problem will be carried out under the following assumption: there is a polydisperse system of water droplets characterized by size distribution parameters R_{01} and μ. For this system the values $K_{\text{abs}}^{\text{hom}}$ and $1-\ \Lambda^{\text{hom}}$are known. Then a fluid such as SAS, possessing the ability to form a film of different thickness on the droplet surface, is introduced into this system in a certain way. At first, we consider the non-absorbing shells, and then pass to absorbing ones. As seen from Eqs.(4.3.2) and (4.3.3), the integration and comparison of results will be performed for the same fixed core ensemble, but at varying shell thickness and optical constants. It should be borne in mind, that the formation of a shell on water droplets is exhibited not only by the change of particle optical properties, but also by some increase of their geometrical cross section. In this case the influence of both effects is taken into account.

Figure 4.6 clearly displays the influence of different parameters on the ratio of the absorption coefficient of the TLP system to that of an initial system of homogeneous water droplets $K_{\rm abs}^{(2)}/K_{\rm abs}^{\rm hom}$. As seen from the curves, the absorption coefficient of TLP systems increases with shell thickness for all four studied values of the refractive index. However, for the non-absorbing shell this increase is not so essential. The maximal growth of $K_{\rm abs}^{(2)}$ in comparison with $K_{\rm abs}^{\rm hom}$ makes for $n_2 = 1.1$ no more than 1.5–5%, for $n_2 = 1.3$ no more than 4–9.5%, for $n_2 = 1.4$ no more than 5–11%, and for $n_2 = 1.5$ no more than 5–12%. Thus, the overall growth of $K_{\rm abs}^{(2)}$ is built up with the shell refractive index increasing up to $n_2 = 1.4$. The $K_{\rm abs}^{(2)}$ value was unaffected by further n_2 increase. Practically, in all cases, the noticeably greater growth of $K_{\rm abs}^{(2)}$ is observed for narrower distributions ($\mu = 10$). The smallest increase of $K_{\rm abs}^{(2)}$ takes place more often, when $R_{01} = 10\ \mu m$, that is for larger particles.

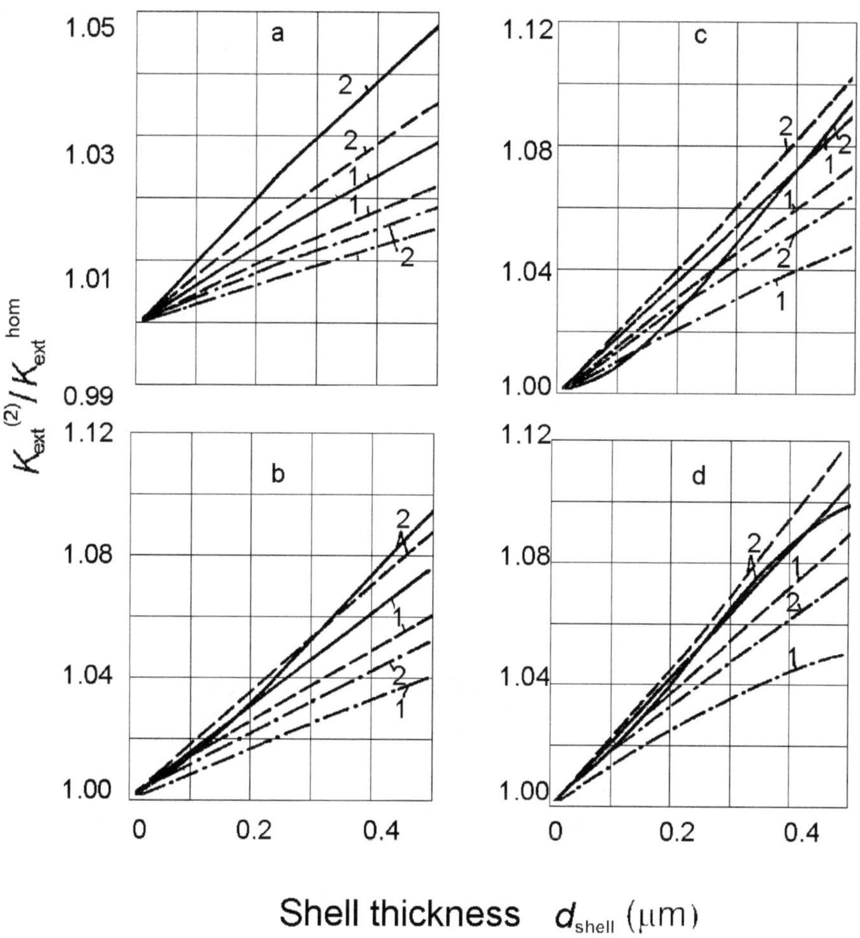

Figure 4.6. Ratio $K_{\rm abs}^{(2)}/K_{\rm abs}^{\rm hom}$ versus the shell thickness for $n_2 = 1.1(a)$, $1.3(b)$, $1.4(c)$, and $1.5(d)$ at $\mu = 1(1)$, $10(2)$, and $R_{01} = 1$ (solid curves), 2 (dashed curves), and 10 μm (dot-dashed curves).

Similarly, in Figure 4.7 no absolute value of $1 - \Lambda = K_{\text{abs}}/K_{\text{ext}}$ is given, but the ratio of these quantities for systems of TLPs and initial homogeneous ones, that is:

$$\delta = \frac{1 - \Lambda^{(2)}}{1 - \Lambda^{\text{hom}}} = \frac{K_{\text{abs}}^{(2)}}{K_{\text{abs}}^{\text{hom}}} \frac{K_{\text{ext}}^{\text{hom}}}{K_{\text{ext}}^{(2)}} \tag{4.3.4}$$

Here the dependencies of the ratio δ versus the shell thickness are given for the same values d, n_2, and R_{01}. Notice, that in all cases this ratio decreases when the shell occurs on water droplets. The point is that the shell formation results in a not so great increase of the absorption coefficient but in a rapid increase of the scattering coefficient, resulting with the ratio $K_{\text{ext}}^{\text{hom}}/K_{\text{ext}}^{(2)} < 1$. The value of δ is determined by the second ratio in Eq.(4.3.4). Despite some increase of the absorption coefficient, the share of absorbed quantums decreases due to an increase of scattered quantums share. The smallest lowering of δ at all n_2 is observed for the particle ensemble with $\mu = 1$ and $R_{01} = 10\ \mu m$, and the greatest one, for the ensemble with $\mu = 10$ and $R_{01} = 1\ \mu m$. The larger particles and wider their distribution, the less the δ decrease. Contrary to this, the smaller the particles and more narrow their distribution, the more strongly δ decreases. In the first case for all n_2 values the decrease of δ does not exceed 3%. In the second case at $n_2 = 1.5$ the δ value impinges up to 0.46, that is decreases by more than half.

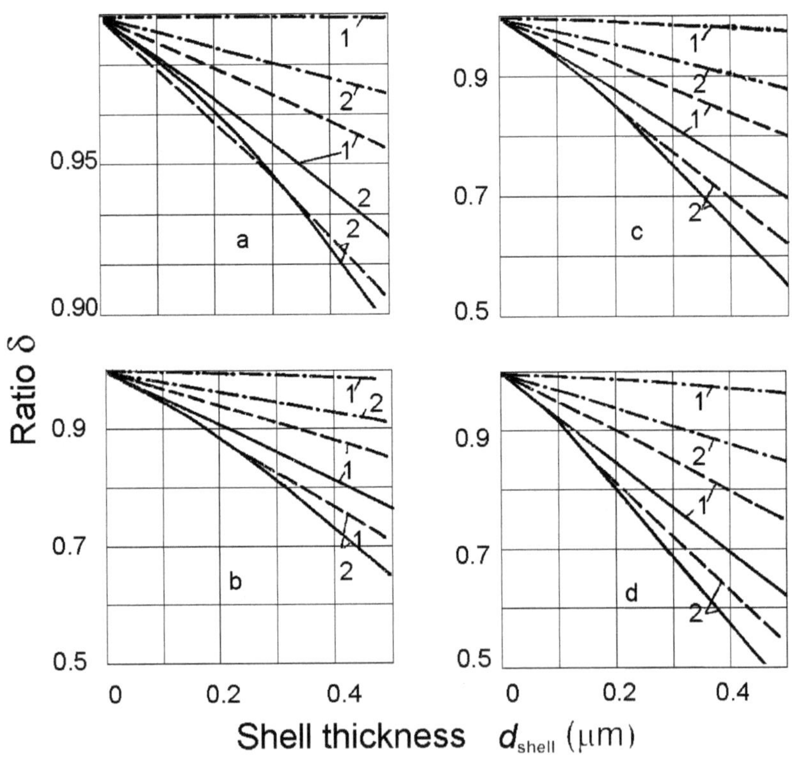

Figure 4.7. Ratio δ versus the shell thickness with the same notation as in Figure 4.6.

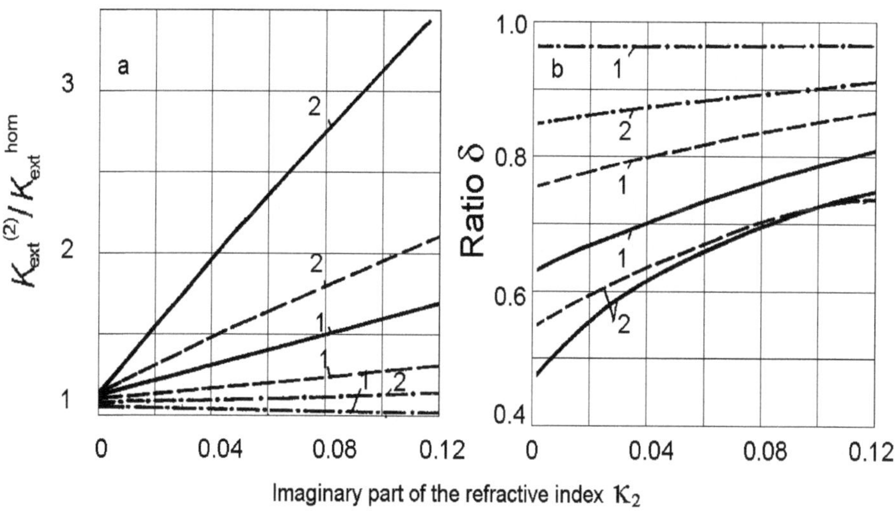

Figure 4.8. Ratios $K^{(2)}_{\text{abs}}/K^{\text{hom}}_{\text{abs}}(a)$ and $\delta(b)$ versus the imaginary part of the shell refractive index at $n_2 = 1.5$ and $d = 0.5\ \mu m$. The values of μ and $R_{01}$are the same as in Figure 4.6.

It follows from Figures 4.6 and 4.7 that formation of the non-absorbing shell on water droplets, with a refractive index both lower and higher than that for water, could give rise to small increases of energy absorption only, however, the efficiency of its use decreases. In so doing, the quantum absorption probability decreases most strongly just when the absorption coefficient increases significantly.

In Figure 4.8 the dependencies of the ratios $K^{(2)}_{\text{abs}}/K^{\text{hom}}_{\text{abs}}$ and δ versus the imaginary part of the shell refractive index κ_2 are plotted for fixed values $n_2 = 1.5$ and $d = 0.5\ \mu m$. These values of n_2 and d are chosen because, for them, the most considerable increase of the absorption coefficient is observed in the case of non-absorbing shells. As the figure shows, if the absorbing shell forms on water droplets, then the absorption coefficient for the ensemble of such TLPs increases rather considerably for systems with distribution parameters $R_{01} = 1\ \mu m$, $\mu = 10$ (i.e., for systems consisting of small particles and described by a narrow size distribution.

For systems formed by large particles ($R_{01} = 10\ \mu m$), the absorbing shell formation does not cause a considerable increase in the absorption coefficient either at narrow, or wide distributions. It has only a weak dependence on κ_2. For large particle systems the δ value also weakly depends on the imaginary part of the shell refractive index. Most strong dependence δ versus κ_2 is observed for systems with $R_{01} = 1\ \mu m$ and $\mu = 10$, that is for systems of small particles with a narrow distribution. In this case there is a strong increase of the absorption coefficient with κ_2 growth, accompanied also by considerable growth of the quantum absorption probability. When κ_2 changes from 0 to 0.12, the ratio $K^{(2)}_{\text{abs}}/K^{\text{hom}}_{\text{abs}}$ for systems of particles with $R_{01} = 1\mu m$ and $\mu = 10$ increases from 1.09 to 3.54, and δ increases from 0.46 to 0.77.

Thus, the formation of absorbing shells on water droplets with the purpose of radiation absorption efficiency enhancement at $\lambda = 10.6\ \mu m$ is suitable in case of finely dispersed clouds and fogs ($R_{01} \approx 1\ \mu m$, $\mu \approx 10$). For coarse-dispersive systems of water droplets with wide size distribution the generation of absorbing shells does not result in essential absorption efficiency enhancement at $\lambda = 10.6\ \mu m$.

4.4 INFLUENCE OF WATER CONTENT ON ANGULAR CHARACTERISTICS OF LIGHT SCATTERING BY AEROSOLS

Unlike the integral characteristics, the angular characteristics of light scattering by polydisperse particle systems are considerably more sensitive to internal particle structure. Let us compare the phase functions, angular dependencies of the linear polarization degree, and the reduced scattering matrix elements f_{33} and f_{43} for TLP systems at different stages of their humidification with the corresponding characteristics for "dry" aerosols, that is for a system of HPs with $N = 1.65 + 0.005i$ (Ivlev and Popova, 1973; Kondratiev *et al.*, 1973).

For elements of reduced scattering matrix we have (see Eq.(1.4.14):

$$f_{\lambda\mu} = \frac{D_{\lambda\mu}}{I_1 + I_2} \qquad \lambda, \mu = 1, 2, 3, 4 \tag{4.4.1}$$

where in the case of the Junge type distribution:

$$I_{1.2} = \frac{C_N}{2} \int_{R_{beg}}^{R_{end}} i_{1.2}(R_1, R_2, N_1, N_2, \omega, \theta) c R_1^{1-\nu}\, dR_1 \tag{4.4.2}$$

$$i_1 = k_0^2 \left|A_{\theta\varphi}\right|^2 \qquad i_2 = k_0^2 \left|A_{\theta\theta}\right|^2 \tag{4.4.3}$$

where c is the distribution constant (2.2.2). Notice, that here the intensities $I_{1.2}$ are normalized in inother way to that of Eq.(1.4.13). It is of no practical concern because the phase function is, as a rule, represented in relative dimensionless units. As well as for HPs, only eight components are distinct from zero in the scattering plane, and only four of them are independent: $f_{11} = f_{22} = 1$, $f_{12} = f_{21}$, $f_{33} = f_{44}$, and $f_{43} = -f_{34}$. In experimental investigations two quantities are used for the most part. It is the linear polarization degree P, which is defined as the ratio of the difference of intensity components of scattered radiation, polarized perpendicular (I_1) and parallel (I_2) to the scattering plane, to their sum and the polarization ellipticity degree $q = S_4/S_1$. Through these quantities it is possible to define components of the reduced scattering matrix $f_{12} = -P$ and $f_{43} = S_4/S_1$, selecting the linearly polarized light beam with the Stokes parameters $\{1, 0, 1, 0\}$ for experimental investigation of the unit volume characteristics. The component f_{33} is defined as the ratio S_3 to S_1.

Based on Eq.(4.4.2), the quantities I_1, I_2, their sum I, and also f_{12}, f_{43}, and f_{33} were evaluated. According to deductions of Section 2.5, the ratio of equilibrium radius of water-

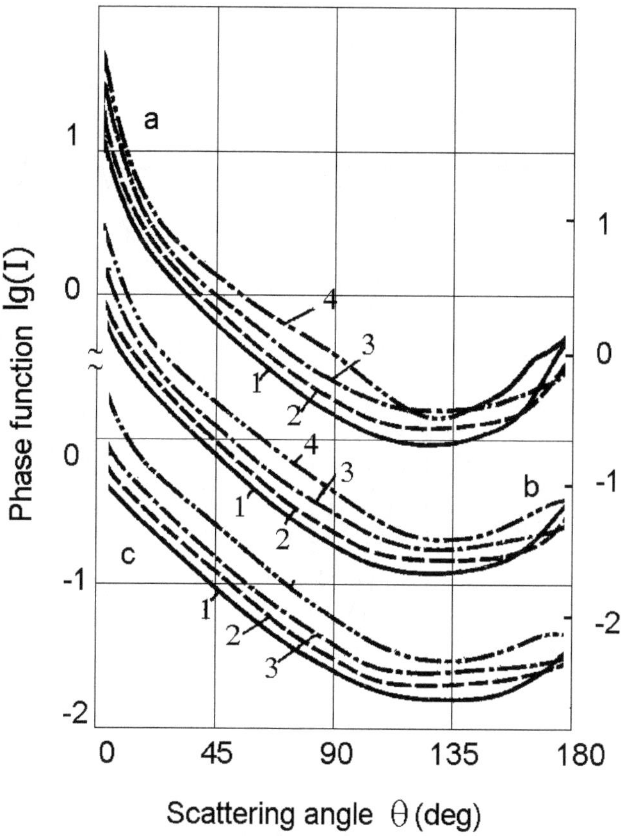

Figure 4.9. Phase functions ($\lambda = 0.6328\ \mu m$) for TLP systems with $\nu = 2.5(a)$, $3.0(b)$, and $3.5(c)$ at $R/R_d = 1(1)$, $1.125(2)$, $1.27(3)$, and $1.55(4)$. The right scale is for (*b*), the upper left one is for (*a*), and lower left one is for (*c*).

containing particles to radius of the initial condensation core was set, as in the previous cases, identical for all R_d values (at given relative humidity f and the relative content G of soluble substances in condensation core).

As seen from Figure 4.9, the particle humidication is accompanied by a rather considerable increase of scattering intensity in all directions. The exception is provided by the scattering angle region $\theta > 160°$, where a scattering decrease can take place, especially for distributions with lower ν values. In the region $\theta > 160°$ the phase function for "dry" aerosol has a clearly expressed increase in steepness. The particle humidification reduces the relative steepness of curves in this region. However, at higher R/R_d the run of curves becomes more composite and non-monotonic (Prishivalko, 1978).

It is shown in Figure 4.10, that humidifying results in narrowing of the scattering angle interval, for which the linear polarization degree is negative; its extreme values decrease in absolute value. The maxima of the positive values of P grow, especially for narrow size

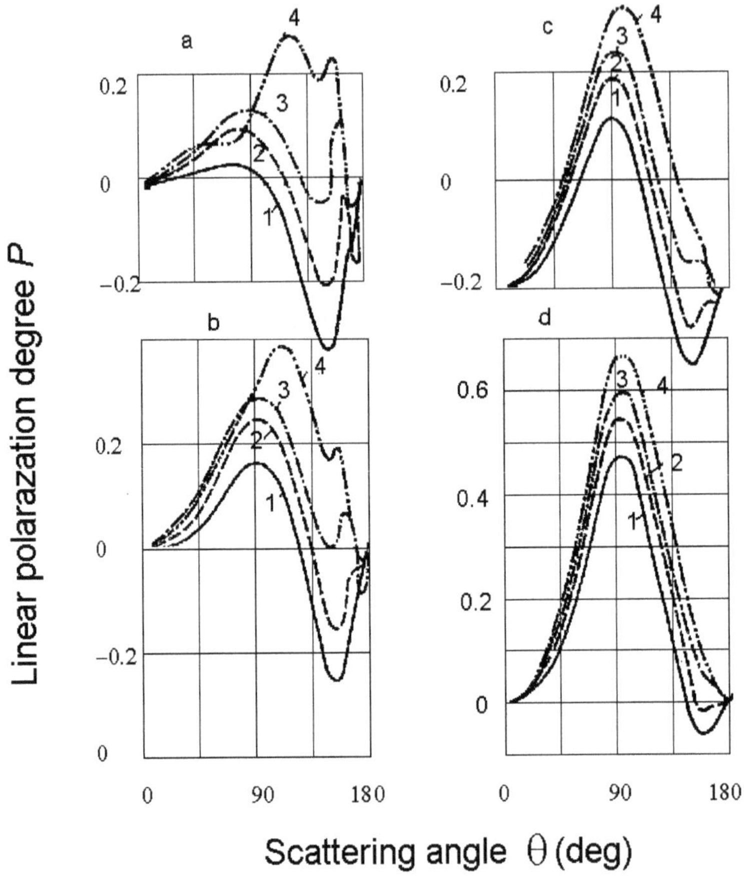

Figure 4.10. Angular dependencies of the linear polarization degree for TLP systems with $\nu = 2.5(a), 3.0(b), 3.5(c)$, and $4.0(d)$ at $R/R_d = 1(1), 1.125(2), 1.27(3)$, and $1.55(4)$

distributions. Polarization behaviour at this stage of humidifying is not determined by particle size increasing, but is due to a rather abrupt lowering of the average refractive index (from 1.65 to 1.42 at $R/R_d = 1.55$).

Quiney and Carswell (1972) measured, in laboratory conditions, the normalized phase function and linear polarization of scattered light for hazes and fogs at the relative humidities $0.77 \leq f \leq 0.99$. It was shown that humidity clearly influences the phase function shape and linear polarization. Unfortunately, for some reason only a qualitative comparison of our calculated data with the results of this work is possible. It displays agreement between the above marked trends in dependencies $I(f)$ and $P(f)$ and the experimental data.

In Figure 4.11 the angular dependencies of the reduced scattering matrix elements f_{33} and f_{43} are depicted. At particle humidifying the point of intersection of the curve $f_{33}(\theta)$ with the abscissa is displaced in the direction of great angles. At transition to narrower

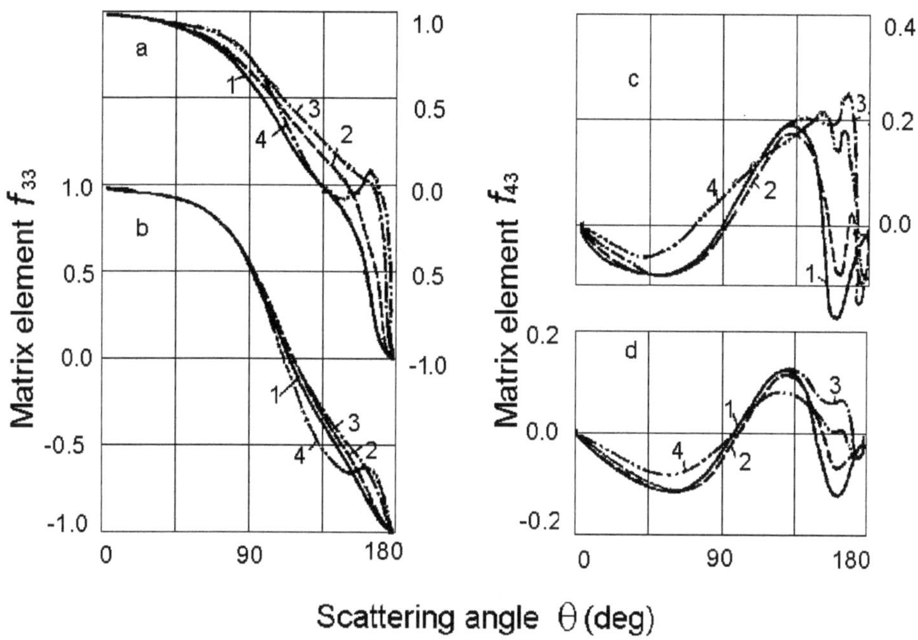

Figure 4.11. Angular dependencies of the reduced scattering matrix elements f_{33} and f_{43} ($\lambda = 0.6328\ \mu m$) for $\nu = 2.5(a, c)$ and $3.0(b, d)$ at $R/R_d = 1(1), 1.125(2), 1.27(3)$, and $1.55(4)$.

distributions the interval of scattering angles, at which the intersection takes place, is shifted to the left and noticeably narrowed. At $\nu = 4$ the curves are rather close to each other, except for the small interval of angles $\theta > 150°$. The angular dependencies of the element f_{43} are weakly sensitive to humidifying at scattering angles up to $\theta \approx 130°$. The greatest sensitivity is observed at $\theta \approx 165°$.

The rather strong changes of the polarization ellipticity degree of scattered light occur at the scattering angle interval $150° < \theta < 180°$ (the same is also valid for the linear polarization degree). As f_{43} essentially depends on the large particles percentage, the maxima of dependence $f_{43}(\theta)$ reduce with an increase of parameter ν . In doing so, the area of the first intersection of curves with the abscissa is shifted in the direction of greater angles. At $\nu \leq 3$ and R/R_d, corresponding to relative humidity $f > 0.75$, the double extremes are observed. From Figure 4.11 it follows, that the linear polarization degree P and the polarization ellipticity degree q are the characteristics most sensitive to change of particle water-content and aerosol microstructure, especially at great scattering angles.

A more complete picture of how these degrees change with yhe variation of the scattering angle and average refractive index (at different humidifying stages) can be inferred from the diagrams of isolines. Examples of such diagrams for $\nu = 2.5$ and 3.5 are given in Figures 4.12–4.15 both for TLPs and for HPs. The characteristic point to note is the more

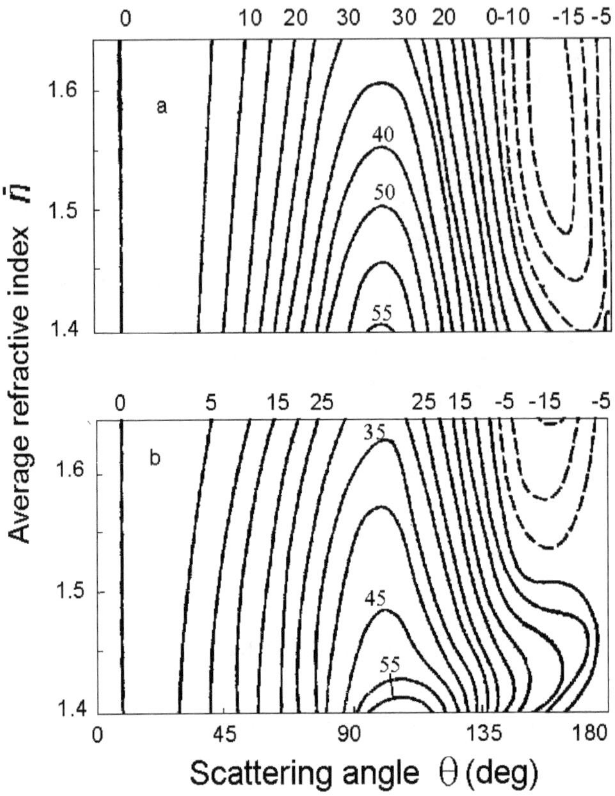

Figure 4.12. Isolines of the linear polarization degree ($\lambda = 0.6328\ \mu m$) for systems of HPs (*a*) and TLPs (*b*) at $\nu = 3.5$. Figures on curves are the P values, (%). The symbol (–) defines the area of small negative P values.

regular run of curves for HPs. For a distribution with $\nu = 3.5$ the P maximum takes place at $\theta = 100°$, but in the case of TLPs with $\bar{n} < 1.45$ some shift of the maximum towards the greater angles is observed. For a distribution with $\nu = 2.5$ the location of polarization degree maximum depends on the average refractive index. With $\bar{n}$ increasing the maximum of P for HPs is displaced in the direction $\theta < 100°$. In the case of the TLPs the maximum location of the positive P varies in a more complicated manner. At $\bar{n} < 1.45$ it moves into a scattering angle area $\theta = 110$–$120°$. For a distribution with $\nu = 2.5$ the negative P can reach –30% in the field of great scattering angles. Its absolute value increases with growth of $\bar{n}$. The size of the region with a negative P depends on the parameter ν.

It is necessary to mark the higher values of the positive P gradients at high ν. At small ν it holds true for negative P also. Increasing the parameter ν, the dependence P versus $\bar{n}$ weakens. When moving off $\theta = 100°$ the curves corresponding to $P =$ const, become steeper. Apparently, such $P(\bar{n})$ dependence is dictated by predominance of small particles.

In Figures 4.14 and 4.15 the isolines, corresponding to constant values of the polarization ellipticity degree, are plotted. As well as in the previous case, it is seen that the isoline pattern is more ordered for HPs, than for TLPs. For systems of HPs the dependence of f_{43} on the average refractive index is much more strong. With decreasing $\bar{n}$ from 1.65 to 1.40 the maximal f_{43} values increase more than three times. The maxima of $f_{43}(\theta)$ correspond to scattering angles $\theta = 130$–$150°$, and are displaced in the direction of greater θ at decreasing $\bar{n}$. For TLPs the extreme values f_{43} are much lower (in 2.5–3 times). The area of negative f_{43} is somewhat wider at small scattering angles and $\theta > 150°$. That is true especially for wide distributions. At $\nu = 2.5$ and $\bar{n} < 1.55$ the "island" character of extrema of f_{43} is typical. At fixed $\bar{n}$ the double maxima of $f_{43}(\theta)$ are observed.

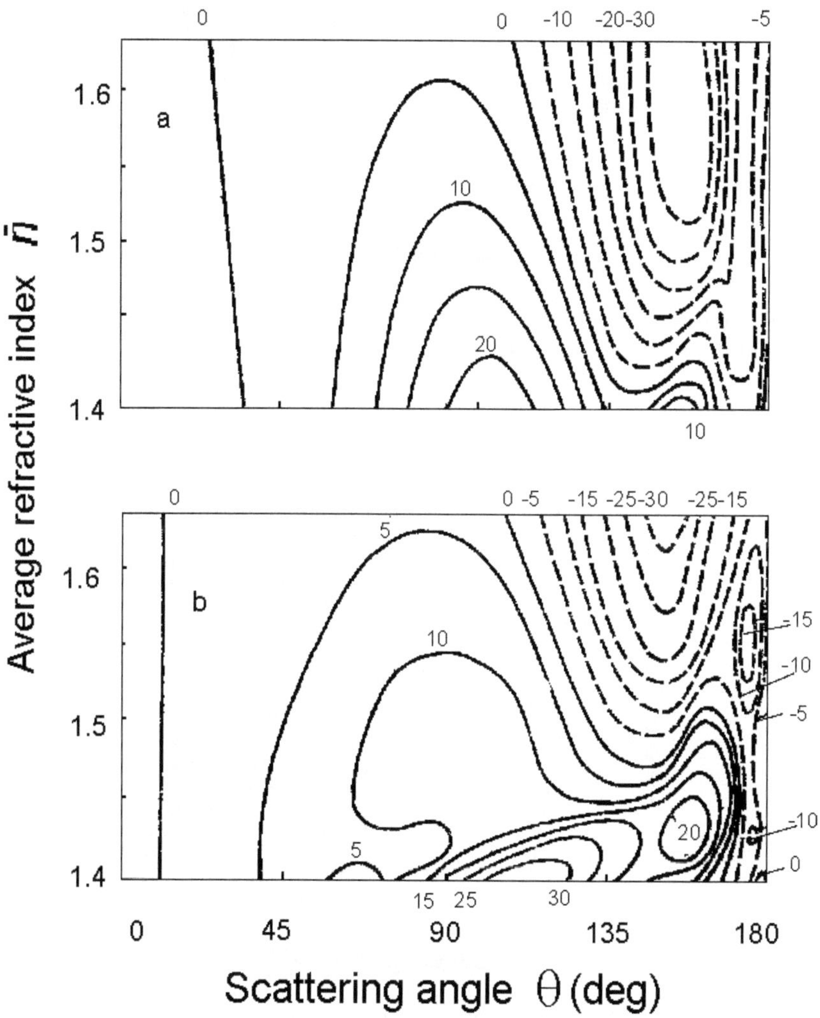

Figure 4.13. Isolines of the linear polarization degree at $\nu = 3.5$. Notation as in Figure 4.12.

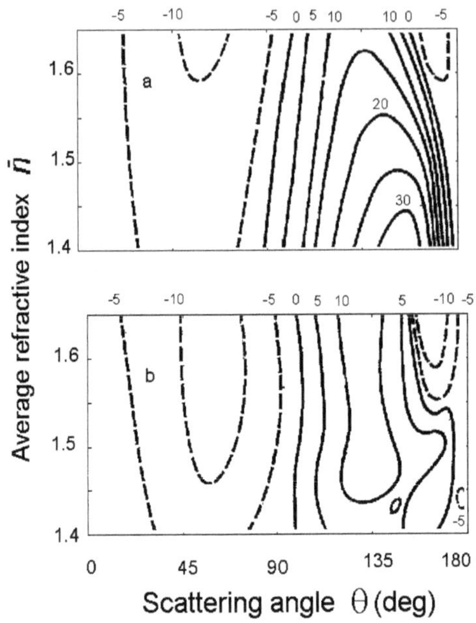

Figure 4.14. Isolines of the polarization ellipticity degree ($\lambda = 0.6328\ \mu m$) for systems of HPs (*a*) and TLPs (*b*) at $\nu = 3.5$. Figure on the curves correspond to the f_{43} value (%).

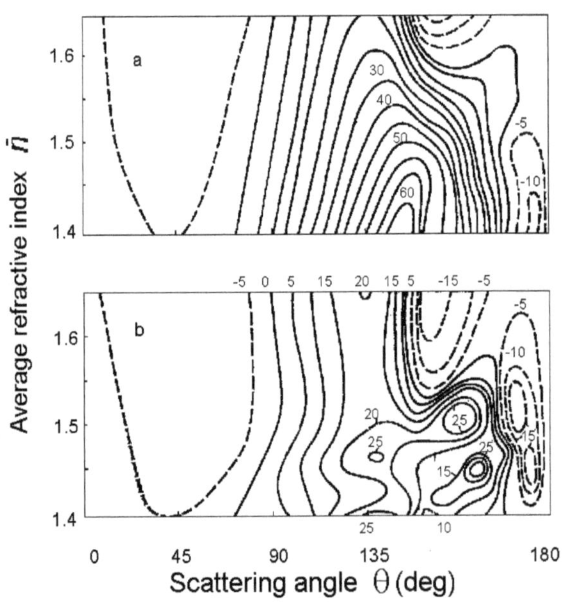

Figure 4.15. Isolines of the polarization ellipticity degree at $\nu = 2.5$. Notation as in Figure 4.14. At $\theta > 160°$ the abscissa axis is on an enlarged scale.

4.5 COMPARISON OF SCATTERING ANGULAR CHARACTERISTICS OF TLP AND HP ENSEMBLES

As was shown above, the influence of particle internal structure on the angular scattering characteristics can be rather essential. It depends on particle water-content, size distribution width and scattering direction, and is different for each of the different characteristics. The results of comparison of light scattering by the TLP systems and HP systems with identical microstructure and the refractive and absorption indices equal to volume-averaged ones for TLPs are given here.

From Table 4.9 it is seen that the distinction in scattering intensities owing to particle inhomogeneity is maximal in the scattering angle interval from 90° to 120°. At wide particle size distributions it is also essential for $\theta \approx 180°$, especially in the presence of large particles. In Table 4.10 the calculated values of the linear polarization degree $P = -f_{12}$ and the reduced scattering matrix elements f_{33} and f_{43} are presented for systems of HPs and TLPs. Since the influence of the particle size upper limit on these quantities is insignificant (see Section 4.1), the results are given only for $R_1^{\text{end}} = 3\ \mu m$. In some cases, the scattering characteristics in question vary essentially at transition from HPs to TLPs. To the greatest degree it is true for the element f_{43}. In Figures 4.16–4.18 the phase functions, angular dependencies of the linear polarization degree P, and the reduced scattering matrix elements f_{33} and f_{43} for TLP systems with $R/R_d = 1.216$ ($f \approx 0.75$) are compared with those for the corresponding HP systems with optical constants equal to the volume-averaged ones for the TLPs: $n = 1.51$ and $\kappa = 0.003$. In the opinion of many authors, the refractive index values, close to $\bar{n} = 1.50$–1.52, are most probable for aerosol particles at average relative humidity.

As it follows from Figure 4.16, at scattering angles $45 < \theta < 150°$ the TLPs scatter more strongly, than HPs. Distinction is maximal close to $\theta = 110$–$120°$ and can reach from 20% at $\nu = 4$ to 55% at $\nu = 2$. In the fields of scattering angles $\theta = 35$–$40°$ and 150–160° the curves $I(\theta)$ for two types of particles are intercrossed and the scattering is identical. The phase functions practically coincide for the forward direction too. At $\theta > 160°$ the phase functions become non-smooth, especially for TLPs.

Table 4.9. Ratio $I^{(2)}/I^{\text{hom}}$ at $G = 0.2$, $\bar{n} = 1.51$, $\lambda = 0.6328\ \mu m$, and $R_1^{\text{beg}} = 0.04\ \mu m$

	R_1^{end} (μm)			
	3.0	10.0	3.0	10.0
θ (deg)	$\nu = 2.5$		$\nu = 3.5$	
0	0.980	0.996	0.928	0.956
30	0.880	0.876	0.940	0.940
60	1.108	1.106	1.118	1.118
90	1.328	1.335	1.230	1.231
120	1.412	1.418	1.237	1.236
150	1.054	1.055	1.081	1.081
170	0.892	0.875	0.952	0.951
180	1.230	1.593	0.919	0.951

Table 4.10. Value of $-f_{12}$, f_{33} and f_{43} at $G = 0.2$, $\bar{n} = 1.509$, $\lambda = 0.6328\ \mu m$, $R_1^{\text{beg}} = 0.04\ \mu m$, and $R_1^{\text{end}} = 3\ \mu m$. First line represents $\nu = 2.5$, second line represents $\nu = 3.5$.

	TLPs			HPs		
θ (deg)	$-f_{12}$	f_{33}	f_{43}	$-f_{12}$	f_{33}	f_{43}
30	0.028	0.982	–0.067	-0.001	0.978	–0.065
	0.047	0.990	–0.064	0.023	0.987	–0.064
60	0.076	0.918	–0.084	0.034	0.899	–0.054
	0.205	0.912	–0.110	0.179	0.904	–0.090
90	0.133	0.738	–0.011	0.117	0.912	0.045
	0.419	0.579	–0.059	0.425	0.518	–0.011
120	0.063	0.429	0.142	0.026	0.148	0.255
	0.317	–0.039	0.082	0.308	0.179	0.168
150	–0.097	0.162	0.176	–0.263	0.124	0.434
	0.001	–0.447	0.077	–0.104	0.534	0.227
170	–0.064	–0.152	0.182	–0.118	0.424	0.072
	–0.013	–0.750	0.037	–0.110	0.829	–0.011

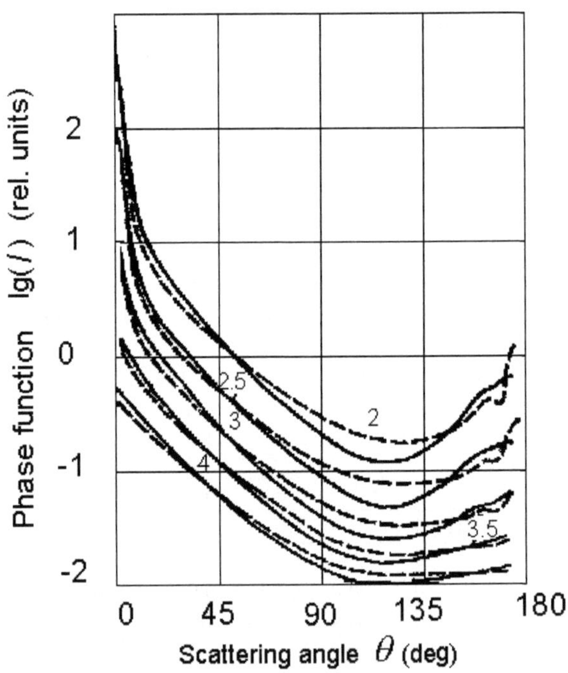

Figure 4.16. Phase functions ($\lambda = 0.6328\ \mu m$) for the TLP system at $R/R_d = 1.216$, $r_{d1} = 0.01\ \mu m$, and $r_{d2} = 10\ \mu m$ (dashed lines) and for the corresponding HP system with $n = 1.51$ (solid lines). The figures on the curves represent the values of ν.

In Section 4.2 the work of Tuomi (1980) was mentioned, in which the author carried out the graphical comparison of scattering properties for particle ensembles described by the identical log-normal size distribution with $R_m = 0.0368\ \mu m$ and $\sigma = 2.028$, but consisting of: (a) three-layered particles – "onions"; (b) "average spheres", that is the HPs with a refractive index equal to the volume-averaged one for three-layered particles; and (c) "solution spheres" that is the HPs with a refractive index coinciding with the refractive index of the exterior layer of three-layered particles – "onions". Comparison of Figure 4.17 with results of Tuomi (1980) (for f=0.7) shows, that the run of dependencies $I(\theta)$ for three-layered particles and "average spheres" is completely the same, as for the TLPs and model HPs with the average refractive index, discussed by us. The intersection of curves occurs at the same angles, and at $\theta > 160°$ the same non-monotones is observed. At transition from $\lambda = 0.4$ to $0.7\ \mu m$ the angle region, where the layered particles scatter more strongly, than homogeneous ones with an average refractive index, is rather extended. Qualitatively, it corresponds to transition to smaller particles or to higher ν values in Figure 4.17. At such transition the angle range, in which the scattering by TLPs is higher, than that by the model HPs, increases also. Note that the behaviour of the corresponding curves for solution spheres strongly differs from that of particles – "onions" and of "average spheres".

In the book of Ivlev and Andreev (1986) the data is given for the calculated light scattering intensity ratios for the two particle ensembles, described by the gamma size distribution with parameters $\mu = 6$ and $R_0 = 0.04\ \mu m$ at $\lambda = 0.5\ \mu m$, at scattering angles from $\theta = 0$ to $180°$ and different relative humidities $f = 0.1(0.1)0.8$.

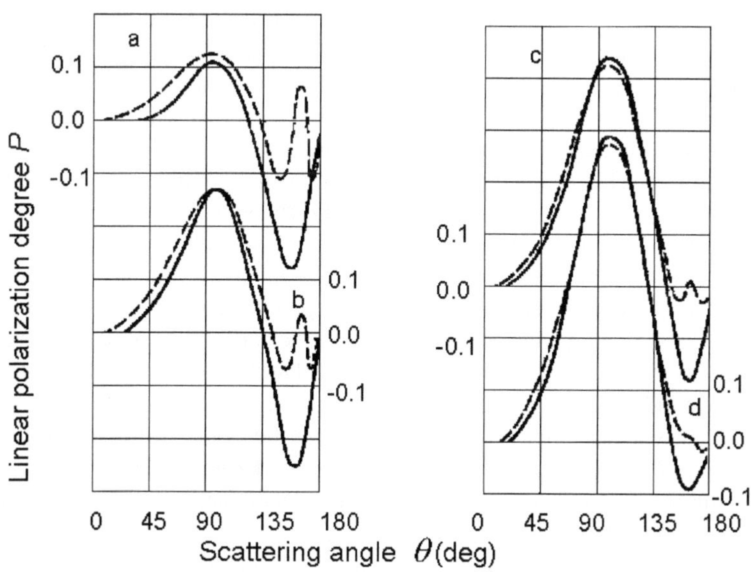

Figure 4.17. Linear polarization degree angular dependencies for light ($\lambda = 0.6328\ \mu m$) scattered by the systems of TLPs with $R/R_d = 1.216$ (dashed lines) and by that of HPs with $n = 1.51$ (solid lines) at $\nu = 2.5(a)$, $3.0(b)$, $3.5(c)$, and $4.0(d)$.

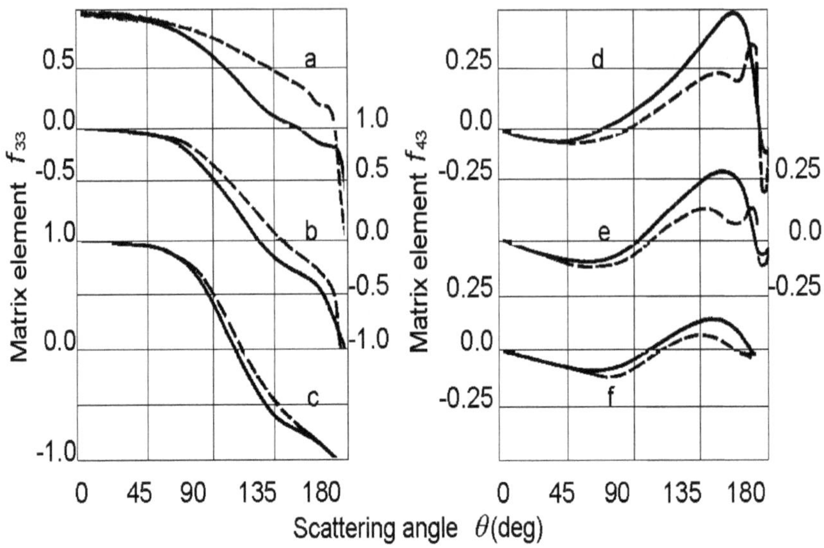

Figure 4.18. Angular dependencies of the reduced scattering matrix elements f_{33} and f_{43} at $\nu = 2.0(a, d)$, $3.0(b, e)$, and $3.5(c, f)$. Notations as in Figure 4.17.

One ensemble involves the TLPs with $N_1 = 1.5 + 0.01i$ and the solution shell width varying depending upon relative humidity. The second ensemble is comprised of particles with the volume-averaged refractive index. The analysis of tabulated data shows, what maximal discrepancies (within 15–17%) between the phase functions of TLPs and model HPs are observed at scattering angles $\theta = 5$–$10°$ and 80–110°. The minimal distinctions take place at θ =30–40° and 150–160°. At $\theta \geq 40°$ the TLPs scatter more strongly, than HPs. Thus, these results are also in good accordance with results of our calculations, depicted in Figure 4.17. A quantitative comparison is not possible due to different values of the particle core complex refractive index and different types of size distributions, but the qualitative agreement may be considered as excellent in all cases.

As in the case of our aerosol particles, the phase functions calculated by Kattawar and Hood (1976) at considerable humidifying of coal particle systems acquire non-monotone character in the scattering angle range $\theta > 150°$. The consideration of angular dependencies of P for TLP systems and HP systems identical to them (in sense of size distribution) with refractive indices equal to the volume-averaged refractive index of TLPs, shows, that at $\theta < 140°$ they are qualitatively identical, though for wider distributions the quantitative difference is noticeable (see Figure 4.17). For scattering angles $\theta > 140°$ the clearly expressed quantitative and qualitative distinctions take place (in the case of TLPs there is an additional maximum in the region $\theta \approx 165°$). It is interesting to note also, that at $\nu = 3.5$–4 the maxima of P for HPs are somewhat higher, than for TLPs, and vice versa at $\nu = 2.5$. At $\nu = 3.5$–4 the curves $P(\theta)$ for TLPs and HPs are intercrossed at $\theta \approx 80°$ and 125°. The area of negative P values for HPs starts at smaller scattering angles, than for TLPs (Prishivalko, 1978). At $\nu \leq 3.5$ in the case of TLPs there is an additional maximum of dependence $P(\theta)$ in the field $\theta \approx 165°$.

Referred again to Tuomi (1980), we see, that at $f = 0.7$ and $\lambda = 0.4\ \mu m$ all features, detected in Figure 4.18 at $\nu = 3.5$–4, are observed. As noted by Ivlev and Andreev (1986), at the scattering angle $\theta = 80°$ the linear polarization degree does not depend on internal particle structure, and most regular differences take place in the angle range $\theta = 10$–$40°$. It is reasoned from this, that with the ratio $P(10\text{–}40°)/P(80°)$ it is experimentally possible to clarify, which from the particle models discussed meets the concrete conditions to a greater extent.

Inasmuch as in the work of Kattawar and Hood (1976) the influence of humidifying on $P(\theta)$ for scattering by a particle system with a very strongly absorbing core (coal particles) was considered, the quantitative comparison with our calculation results is practically impossible. However, there is one qualitative feature, common for both studied cases. The dependencies $P(\theta)$ are complicated and the additional extremes occurred due to the considerable humidifying of particles in the system. Comparison of our calculation results with data of Tuomi (1980) and Ivlev and Andreev (1986) shows, that for systems of fine-dispersed particles with narrow size distributions the marked features of dependencies $I(\theta)$ and $P(\theta)$ have much in common even at considerable relative humidity changes.

The angular dependencies of the scattering matrix element f_{33} (see Figure 4.18) at average relative humidity are least sensitive to internal particle inhomogeneity, however at $\nu = 2$–2.5 in the region $90° < \theta < 170°$ the distinction in f_{33} values is rather essential. As for the angular dependencies of the element f_{43}, they have considerable quantitative and qualitative distinctions. At the TLP distribution with $\nu \leq 3$, the extremum in the scattering angle range $\theta \approx 150$–$170°$ becomes double, that is not the case for HPs. For all four surveyed characteristics in the case of narrow distributions, where the fine particles prevail, the plots corresponding to TLPs and HPs are similar enough.

At $R/R_d = 1.55$, with $\bar{n} = 1.417$ ($f \approx 0.92$), the discrepancy in the curves $P(\theta)$ (Figure 4.19) is more distinct, than in Figure 4.17, especially for wide distributions. Maxima of $P(\theta)$ for TLP systems with $\theta = 3.5$–4 are higher, than at smaller humidities. At $\theta < 3.5$ they are shifted in the direction of greater angles. The area of negative values of P is narrowed significantly down in comparison with HPs. Similar changes with f increasing are mentioned by Tuomi (1980). Qualitative and quantitative discordance of dependencies $f_{43}(\theta)$ for systems of TLPs and HPs with ratio R/R_d growth, increases also (Figure 4.20).

In Figure 4.21 the dependencies of the maximal values of P and polarization ellipticity ($q = f_{43}$) of scattered radiation versus the volume-averaged particle refractive index are plotted over a wide interval of relative humidities. One can see, that at small water-content ($\bar{n} = 1.50$–1.65) the maximal P values for systems of HPs and TLPs are rather close. The noticeable difference is observed at $\bar{n} < 1.46$ ($f > 0.85$), especially for wider distributions. At $\nu \leq 2.5$ the polarization maximum in the region $\theta \approx 100°$ is washed away, but a new one appears, sometimes even higher, in the region $\theta \approx 140°$ (see Figure 4.19*a*). The break of dashed curves in Figure 4.21 is the result of this. The polarization ellipticity extremes are much lower for the TLP systems, than for HP ones. The run of the corresponding curves is more complex, especially at considerable water-content ($\bar{n} < 1.46$).

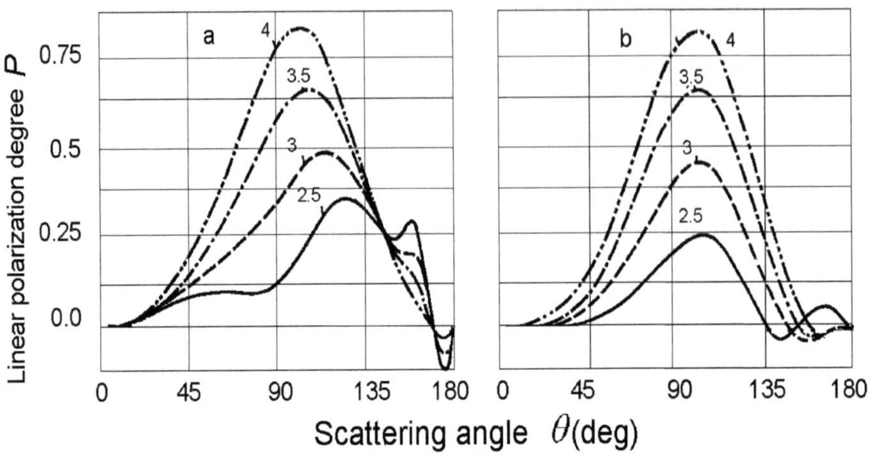

Figure 4.19. The angular dependencies of linear polarization degree ($\lambda = 0.6328\ \mu m$) for light scattered by the systems of TLPs with $R/R_d = 1.55$ (*a*) and by HPs with $\bar{n} = 1.417$ (*b*). The figures on the curves are values of ν.

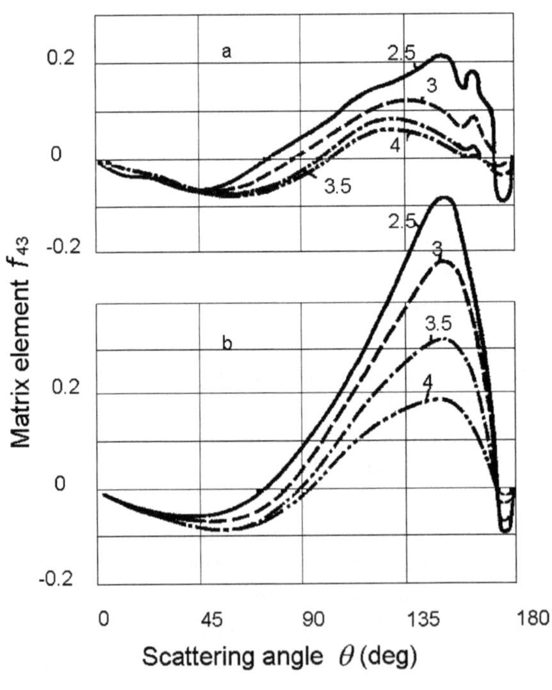

Figure 4.20. The angular dependencies of polarization ellipticity for scattered light. Notation as in Figure 4.19.

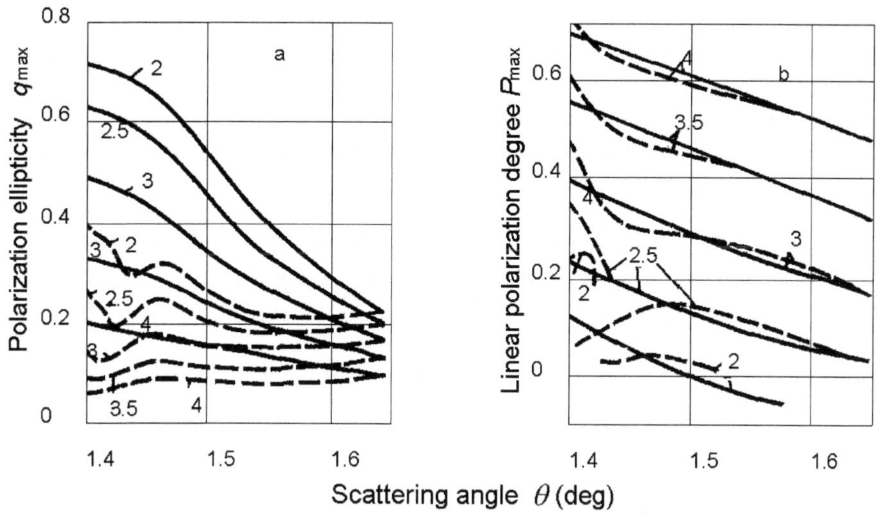

Figure 4.21. The maximal values of polarization ellipticity (a) and linear polarization degree (b) for light ($\lambda = 0.6328\mu m$) scattered by the systems of TLPs (dashed lines) and by HPs (solid lines) with the same volume-averaged refractive index. The figures on the curves are the values of ν.

It follows from the plots presented, that the most typical scattering features of surveyed TLP systems are:

- the considerably lower (compared with HPs) value of the maximal polarization ellipticity degree;
- more complicated dependence of this quantity on the average refractive index;
- the values of the maximal polarization degree different from those for HPs at high humidities;
- considerably different run of dependencies $P(\theta)$ and $q(\theta)$ at $\theta > 150°$ (see Figures 4.18, 4.19); and, what is of particular value,
- splitting of the dependence $q(\theta)$ maximum at $\nu \leq 3$ and $\theta \approx 150$–$170°$ (Prishivalko, 1978).

Let us now dwell on some problems, connected with the backscattering coefficient for polydisperse systems of atmospheric particles. This light scattering characteristic is of particular importance in connection with problems for atmosphere laser sensing, becoming the basis for a new important scientific subject. A great body of publications are devoted to different aspects of this field of science. In particular, the influence of humidity on the atmospheric aerosol scattering characteristics is studied in the book of Krekhov and Rakhimov (1982). Our problem here is restricted to studying only one concrete problem on influence of the particle internal structure on results of remote sensing. This problem is especially important, because at the interpretation of atmosphere laser probing results, one usually describes the aerosols as a plurality of either water droplets or solid particles.

The ratio of backscattering cross sections of TLPs and water droplets $\sigma_{\text{bsc}}^{(2)}/\sigma_{\text{bsc}}^{\text{wat}}$ versus the external radius R_2 oscillates strongly (Astafyeva and Prishivalko, 1978). This ratio slowly comes nearer to unity in accordance with an increasing of the water shell width. The oscillation amplitude increases with increasing TLP core radius. At $R_1 = 0.1\ \mu m$ the distinction, no greater than $\pm 10\%$, between backscattering cross sections of TLPs and water particles is observed only at $R_2/R_1 \gtrsim 9$. At $R_1 = 0.2\ \mu m$ the distinction between backscattering by atmospheric aerosol TLPs and water droplets does not exceed 25%, from $R_2/R_1 \approx 11$. These distinctions do not decrease with the presence of absorption in the core.

The ratio of the backscattering cross sections for the TLPs and HPs, consisting of the core substance, oscillates with increasing R_2 also. Only with enough thin shells the difference between $\sigma_{\rm bsc}^{(2)}$ and $\sigma_{\rm bsc}^{\rm hom}$ is insignificant (does not exceed 20% at $R_2/R_1 < 1.16$ and 1.08 for $R_1 = 0.05$ and $0.2\ \mu m$ accordingly).

The results of light backscattering comparisons for the water-containing and model HPs with optical constants, averaged over volume of the corresponding TLPs, is represented in Figure 4.22. Dependence $\sigma_{\rm bsc}^{(2)}/\sigma_{\rm bsc}^{\rm hom}(R_2)$ strongly oscillates. The oscillation amplitude sharply increases with core radius. For $R_1 = 0.05\ \mu m$, $\sigma_{\rm bsc}^{(2)}$ differs no more than 10% from $\sigma_{\rm bsc}^{\rm hom}$ at $R_2/R_1 \gtrsim 5$, for $R_1 = 0.1$ and $0.15\ \mu m$ at $R_2/R_1 > 8$. For great core radii ($R_1 = 0.20$ and $0.25\ \mu m$) the distinction between light backscattering by TLPs and model HPs increases and reaches 20–40% even at $R_2/R_1 > 11$. The presence of core absorption results in an increase of distinction between the backscattering cross sections for water-containing and model particles in comparison with the case of non-absorbing cores. Notice, that though a difference $\sigma_{\rm bsc}^{(2)} - \sigma_{\rm bsc}^{\rm hom}$ is the quantity of non-fixed sign, the sum of these differences for a particle system can essentially differ from zero even using the HP model with a volume-averaged refractive index.

In Table 4.11 the ratios of backscattering coefficients for the polydisperse TLP systems are given at different degrees of core humidifying with the complex refractive index $N = 1.55 + 0.015i$, and for the systems of model HPs with the same microstructure and with the complex refractive index, averaged over the volume of corresponding TLPs. For comparison, the ratios of their extinction coefficients are given also. From Table 4.11 it follows, that the difference between values $K_{\rm bsc}^{(2)}$ and $K_{\rm bsc}^{\rm hom}$ is least for narrow distributions (at $\nu = 4$ this difference does not exceed 25%). But this difference becomes rather considerable (2–3 times) for systems with wide particle size distributions. It is clear, that if this difference exceeds 30–40%, then it is impossible to receive, on the basis of the backscattering coefficient for the model HPs, reliable information on the TLP system and therefore to interpret the optical remote sensing data with accuracy. The influence of the particle water-content on the backscattering coefficient is especially obvious against the background of practically complete (within 5%) agreement between the extinction coefficients for systems of TLPs and HPs.

As was already mentioned in Section 4.1. the backscattering, as well as forward scattering, essentially depends on presence of large particles in the system. If we restrict the upper integration limit in Eq.(4.4.2) to the value $R_d^{\rm end}$=3 μm, then the difference between $K_{\rm bsc}^{(2)}$ and $K_{\rm bsc}^{\rm hom}$ essentially decreases. In this case it does not exceed 15–20% at $\nu = 4$ to $R/R_d \approx 1.7$, at $\nu = 3.5$ to $R/R_d \approx 1.5$, and at $\nu = 3$ to $R/R_d \approx 1.3$. For wider distributions and higher water-content the $K_{\rm bsc}^{(2)}$ and $K_{\rm bsc}^{\rm hom}$ values differ 1.5–2 times.

Table 4.11. Ratios $K_{\rm bsc}^{(2)}/K_{\rm bsc}^{\rm hom}$ (first line) and $K_{\rm ext}^{(2)}/K_{\rm ext}^{\rm hom}$ (second line) for polydisperse systems of the TLPs and corresponding model HPs at different water-contents. $\lambda = 0.6328\ \mu m$, $n_1 = 1.55$, $\kappa_1 = 0.015$, $R_d^{\rm beg} = 0.04\ \mu m$, $R_d^{\rm end} = 10\ \mu m$, and $G = 0.2$.

			ν				
R/R_d	$\bar{n}$	$\bar{\kappa}$	2.0	2.5	3.0	3.5	4.0
1.125	1.48	0.011	1.19	1.12	1.07	1.06	1.07
			1.00	1.00	1.01	1.03	1.04
1.216	1.45	0.0083	1.58	1.38	1.22	1.14	1.11
			1.00	1.00	1.01	1.03	1.04
1.270	1.44	0.0073	2.15	1.73	1.39	1.20	1.13
			0.99	0.99	1.00	1.01	1.05
1.350	1.42	0.0061	3.13	2.38	1.75	1.36	1.19
			0.99	0.99	0.99	1.01	1.03
1.426	1.41	0.0052	3.05	2.52	1.91	1.46	1.23
			0.99	0.99	1.00	1.01	1.03
1.720	1.37	0.0029	2.24	1.88	1.57	1.33	1.19
			0.99	0.99	1.00	1.01	1.02

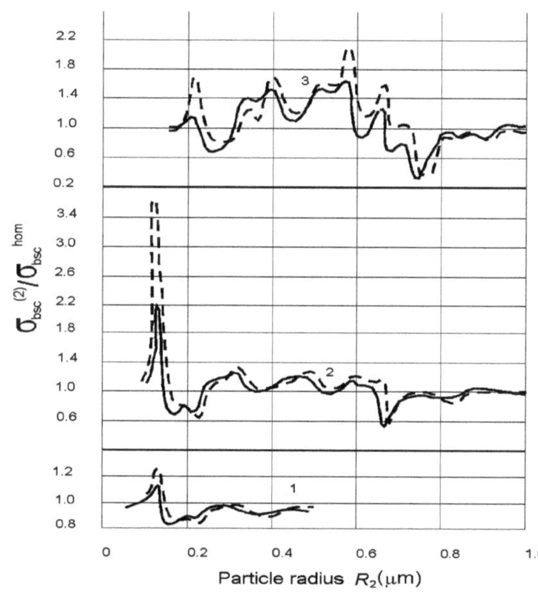

Figure 4.22. Ratio $\sigma_{\rm bsc}^{(2)}/\sigma_{\rm bsc}^{\rm hom}$ versus the particle external radius at the same volume-averaged refractive index for both TLPs and HPs, $\lambda = 0.5\ \mu m$, $n_1 = 1.5$, $n_2 = 1.33$, $\kappa_2 = 0$, $R_1 = 0.05(1)$, $0.1(2)$, and $0.15(3)$, and $\kappa_1 = 0$ (solid lines) and 0.1 (dashed lines). The right scale is for $R_1 = 0.1$, the upper left scale is for $R_1 = 0.15$, and the lower left scale is for $R_1 = 0.05$.

Since at humidity $f \leq 0.95$ the width of a water or water-solution shell on mixed condensation cores, prevailing in continental aerosols, does not usually exceed the doubled core radius (Hänel, 1970; Bullrich *et al.*, 1972), in the majority of actual situations the influence of insoluble inclusions to backscattering appears most significant. The marked distinction of the backscattering coefficients for systems of TLPs and model HPs should be taken into account whilst interpretating data in the optical remote sensing of atmospheric haze, in which the particles, with the sharply expressed internal structure, can predominate. In the case of clouds and radiative fogs, where the external particle sizes are much more than the core sizes, their properties can be identified with the water aerosol properties regards both extinction and backscattering.

Of considerable practical interest is provision of an approximate account of backscattering coefficient dependence on the air relative humidity. Werner (1972) proposed to use, for this purpose, the approach developed by Hänel (1971) in the case of the extinction coefficient (see Section 4.2). He also assumed, that the following relation is valid:

$$\frac{K_{\text{bsc}}(f_1)}{K_{\text{bsc}}(f_2)} = \left[\frac{R_a(f_1)}{R_a(f_2)}\right]^2 \frac{dS_a(f_1)/d\Omega}{dS_a(f_2)/d\Omega} \tag{4.5.1}$$

where Ω is the spatial angle, S_a is defined by relation (4.2.6) and equal to the product $L_a Q_{\text{ext,a}}$ (where L_a is the number of particles in medium unit volume), $Q_{\text{ext,a}}$ is the extinction efficiency averaged over the particle ensemble.

Hänel (1971) has shown, that the extinction coefficient of the single particle is determined by a generalized parameter $\alpha_j = 4\pi R_j \left|N_j - 1\right| / \lambda$, which in the case of continental aerosols does not depend on the air relative humidity at $f \leq 0.9$. By analogy, Werner (1972) assumed, that this proposition may be applied equally well to the backscattering intensity. Then:

$$\frac{K_{\text{bsc}}(f_1)}{K_{\text{bsc}}(f_2)} = \left[\frac{R_a(f_1)}{R_a(f_2)}\right]^2 \tag{4.5.2}$$

Using Kasten's formula (1969), linking the particle radius with the air relative humidity, Werner arrives at the formula:

$$\frac{K_{\text{bsc}}(f_1)}{K_{\text{bsc}}(f_2)} = \left(\frac{1-f_2}{1-f_1}\right)^{2\varepsilon} \tag{4.5.3}$$

which is similar to Eq.(4.2.10).

To test the validity of this assertion, calculations of the ratio $K_{\text{bsc}}^{(2)}(f)/\ K_{\text{bsc}}(0)$ were performed, results are given in Figure 4.23. It is completely clear from this figure, that one would expect the linear relationship between the backscattering coefficient and the particle geometrical cross section (and, consequently, the air relative humidity) for fine-particle systems with $\nu \geq 3.5$ only. Really, in the range $1.125 \leq R/R_d \leq 1.72$, corresponding approximately to the relative humidity interval $0.6 \leq f \leq 0.94$, at $\nu = 4$ with an error of less than 5% the relation $K_{\text{bsc}}^{(2)}(f)/K_{\text{bsc}}(0) = 1.61\,(R/R_d)^2 - 0.83$ is valid. For $\nu = 3.5$ with an error of no more than 7% it is possible to write $K_{\text{bsc}}^{(2)}(f)/K_{\text{bsc}}(0) = 1.44\,(R/R_d)^2 - 0.68$. It is easy to verify, that the coefficients of the two equations defining a decline of approximating curves, with an accuracy of 5–10%, coincide with the corres-

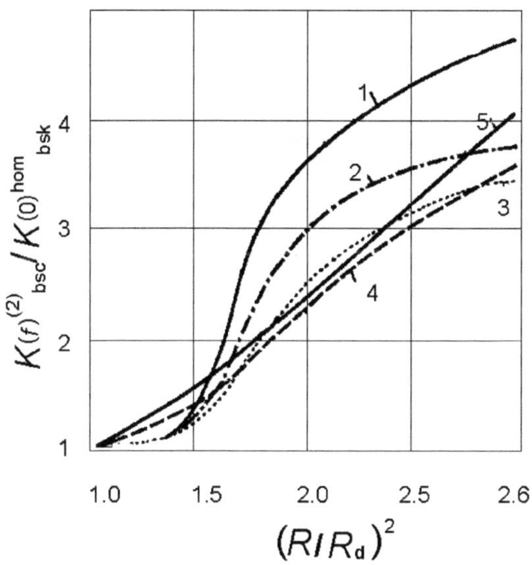

Figure 4.23. Ratio of the backscattering coefficients for the humidified and dry particles versus $(R/R_d)^2$ at $\lambda = 0.6328\ \mu m$, $N = 1.55 + 0.015i$, $R_d^{\text{beg}} = 0.04\ \mu m$, $R_d^{\text{end}} = 10\mu m$, $G = 0.2$, and $\nu = 2(1), 2.5(2), 3(3), 3.5(4)$, and $4(5)$.

ponding coefficients S from Table 4.7, defining the decline of straight lines, fitting the corresponding dependencies for the extinction coefficients. At $\nu < 3.5$ the studied dependence essentially differs from the linear one. This difference became more significant, the greater the large particle share in the polydisperse system. This conclusion remains valid in the case when the integration upper limit in Eq.(4.2) reduces to $R_d^{\text{end}} = 3\ \mu m$. Therefore, Eqs.(4.5.2) and (4.5.3) are applicable only to the fine-dispersed mediums with narrow particle size distributions.

To conclude this section we shall consider the question of the influence of thin fluid films (such as SAS), forming on particles on backscattering. As well as in Section 4.3, consideration will be given to the liquid-droplet aerosol and for hazes at $\lambda = 0.63\ \mu m$, $1.05 \le n_2 \le 2.0$, $\kappa_2 = 0$, and shell width $0 \le d \le 0.04\ \mu m$. The particle size distribution for the liquid-droplet aerosol, as in the previous case, is described by the gamma distribution, and for hazes, by the Junge type distribution. The shell width is supposed identical for all particles in the system. For water-containing haze particles the ratio of the shell water-solution radius to that of the core was selected equal to $A = 1.216$, that approximately corresponds to the relative humidity $f \approx 0.75$.

In the case of liquid-droplet aerosols, the dependence of the backscattering coefficient ratio for the polydisperse systems of droplets with shells and systems of the same droplets without shells, on modal radius R_{01} of the initial particle system, has an oscillating character (Figure 4.24). At different values of the shell refractive index of TLPs the character of oscillations varies. At $n_2 < n_1$ the increase of R_{01} is accompanied at first by decreasing the ratio $K_{\text{bsc}}^{(2)}/K_{\text{bsc}}^{\text{core}}$ down to values smaller than unity; then this ratio somewhat increases.

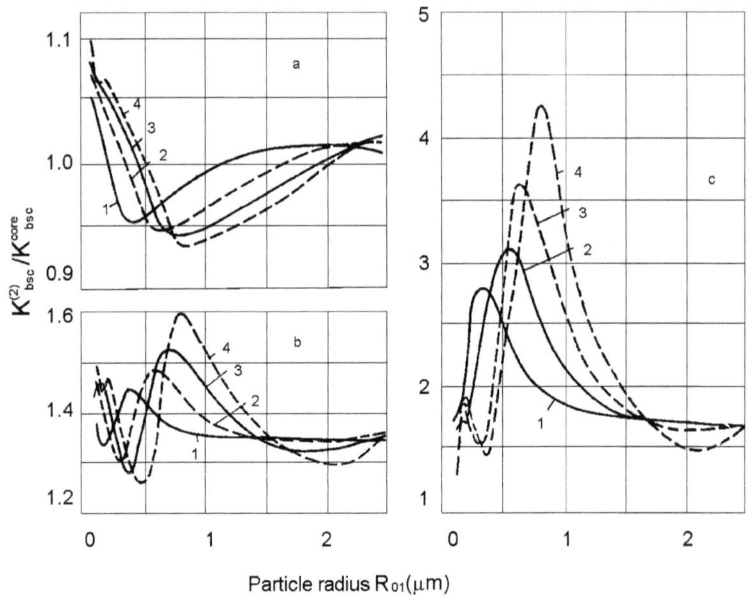

Figure 4.24. Ratio of the backscattering coefficients $K^{(2)}_{\text{bsc}}/K^{\text{core}}_{\text{bsc}}$ versus R_{01} for $n_2 = 1.1(a)$, $1.5(b)$, and $2.0(c)$ at $n_1 = 1.33$, $\kappa_1 = 0$, $\kappa_2 = 0$, $d = 0.01\ \mu m$, and $\mu = 2(1), 5(2), 10(3)$, and $20(4)$.

At $R_{01} < 0.2\ \mu m$ for $\mu = 2$, $R_{01} < 0.4\ \mu m$ for $\mu = 5$, and $R_{01} < 0.6\ \mu m$ for $\mu = 20$ the backscattering coefficient for TLP polydisperse systems exceeds that for their cores by 10–20%. In the case $n_2 > n_1$ the ratio of the backscattering coefficients $K^{(2)}_{\text{bsc}}/K^{\text{core}}_{\text{bsc}}$ oscillates more strongly at increasing R_{01}; there are ranges of values $R_{01}$and μ, where this ratio is maximum. For example, in Figure 4.24, $K^{(2)}_{\text{bsc}}$ exceeds by 3 times $K^{\text{core}}_{\text{bsc}}$ at $\mu = 10$ and $0.5 < R_{01} < 0.9\ \mu m$ and at $\mu = 20$ and $0.6 < R_{01} < 1.1\ \mu m$ in the case $n_2 = 2.0$ and $d = 0.01\ \mu m$. Increasing the shell refractive index the ratio $K^{(2)}_{\text{bsc}}/K^{\text{core}}_{\text{bsc}}$ essentially increases. Increasing the TLP shell width is accompanied by growth of the ratio $K^{(2)}_{\text{bsc}}/K^{\text{core}}_{\text{bsc}}$ at $R_{01} < 0.4\ \mu m$ in all studied range of values d ($0 \le d \le 0.03\ \mu m$), and at $R_{01} \gtrsim 0.4\ \mu m$ only in an interval $0 \le d \le 0.02\ \mu m$. At great radii R_{01} ($R_{01} \approx 2\ \mu m$) the backscattering ratio coefficients do not exceed 2.5–3.0 at any μ and shell thicknesses.

The investigations of properties of the polydisperse atmospheric haze systems comprising of particles covered with shells from SAS, have shown, that the ratio $K^{(2)}_{\text{bsc}}/K^{\text{core}}_{\text{bsc}}$ in most cases grows with an increase in the refractive index, shell width, and parameter ν (Figure 4.25*a*). The backscattering by a system of TLPs is less than that by a system of their cores, 10–12% only, at parameter values $\nu \le 2.5$ and $n_2 < 1.4$.

Note, that the backscattering coefficients ratio $K^{(2)}_{\text{bsc}}/K^{\text{core}}_{\text{bsc}}$ increases much faster, than the ratios $K^{(2)}_{\text{ext}}/K^{\text{core}}_{\text{ext}}$ studied in Section 4.3 and reaches higher values. At $n_1 = 1.4$, $\nu = 4$ and $n_2 = 2.0$ $K^{(2)}_{\text{bsc}}/K^{\text{core}}_{\text{bsc}} \approx 17.5$, but $K^{(2)}_{\text{ext}}/K^{\text{core}}_{\text{ext}} \approx 12.5$. Increasing the core

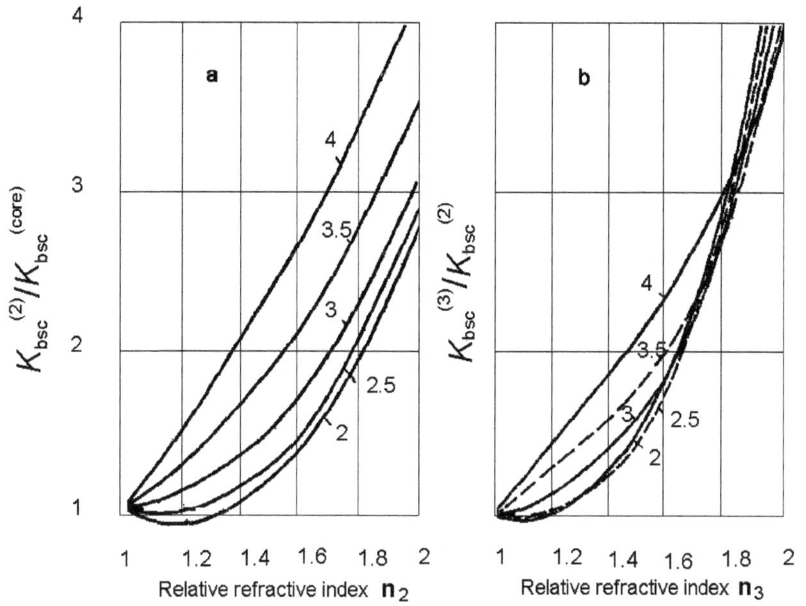

Figure 4.25. Ratio $K_{\text{bsc}}^{(2)}/K_{\text{bsc}}^{\text{core}}$ versus n_2 at $N_1 = 1.5+0.001i$ (a) and ratio $K_{\text{bsc}}^{(3)}/K_{\text{bsc}}^{(2)}$ versus n_3 at the same N_1 with $N_2 = 1.33 + 0i$ (b), and $d = 0.02\mu m$. Figures on curves are ν values.

refractive index of the TLP in the range $1.4 \le n_1 \le 1.7$ results in a decreasing of the backscattering ratio of the TLP polydisperse system and their cores.

Let us consider now the ratios of the backscattering coefficients for systems of three-layered particles, that is for the systems of water-containing particles of atmospheric haze covered with SAS shells, and for the TLP systems without shells $K_{\text{bsc}}^{(3)}/K_{\text{bsc}}^{(2)}$ (Figure 4.25b). Increasing the exterior layer refractive index n_3 is accompanied by growth of the ratio of the backscattering coefficients for systems of three-layered particles and TLPs, but at $\nu \le 3$ this growth is more abrupt than for systems described by values $\nu > 3$. At $n_3 \ge 1.7$ the backscattering by systems with wide size distributions exceeds the backscattering of systems with narrow ones. In the studied area of parameters the maximal value of the ratio $K_{\text{bsc}}^{(3)}/K_{\text{bsc}}^{(2)}$ is reached at $n_3 \approx 2.0$ and $\nu = 2$ and makes approximately 5.3.

It follows from above, that both for the liquid-droplet aerosol, and for the atmospheric haze the formation the thin film with a high refractive index on a particle surface can give considerable backscattering increase. At average n_3 values this effect most strongly manifests itself in the case of hazes (ratios $K_{\text{bsc}}^{(2)}/K_{\text{bsc}}^{\text{core}}$ and $K_{\text{bsc}}^{(3)}/K_{\text{bsc}}^{(2)}$ increase 2–3 times). Obviously, this circumstance should also be taken into account for interpretation of results of optical remote sensing of the atmosphere.

4.6 SCATTERING ANGULAR CHARACTERISTICS OF WATER-CONTAINING PARTICLE ENSEMBLES AND COMPARISONS WITH EXPERIMENTAL RESULTS

The systematic measurements of the light scattering matrix elements for atmospheric aerosols in natural conditions were performed by Gorchakov (1966), Gorchakov and Emilenko (1974), Gorchakov and Sidorov (1980). Unfortunately, these optical measurements in atmospheric hazes were not adequately accompanied by measurements of their microphysical characteristics. Therefore, only qualitative comparisons of experimental dependencies with calculated ones are possible.

Gorchakov and Emilenko (1974) attempted to compare the experimental dependencies $f_{ik}(\theta)$ with calculated ones for the model of HPs. It follows from data of this work, in particular, that the greatest distinctions between calculations and experimental dependencies are detected at scattering angles $\theta > 150°$. In Sections 4.4 and 4.5 it was marked, that the runs of dependencies $P(\theta)$ and $f_{43}(\theta)$ at $\theta > 150°$ for the TLP systems and for HP system are much more distinct. The same, though to a lesser extent, holds true for $I(\theta)$ and $f_{33}(\theta)$. Discrepancy between the experimental data and calculation in this area of scattering angles can be caused by the fact that in experiment TLPs were observed, not HPs (or their contribution to the scattering pattern was prevailing).

A second feature engaging our attention is that on the experimental curves $f_{43}(\theta)$ (see Gorchakov, 1966; Rozenberg, 1968) the double maxima develop with the air relative humidity increasing, closely resembling that obtained in our calculations (see Figures 4.18 and 4.21). It is necessary to underline, that in numerous calculations for systems of HPs such maximum splitting was not observed with any parameter values, that is, the presence of the double maximum of $f_{43}(\theta)$ is, apparently, the typical feature for TLP systems.

Gorchakov and Emilenko (1974) produced the great body of experimental data on the values of $P_{\max}$ and $q_{\max}$ for atmospheric haze observed at the Zvenigorod station of Institute of Atmospheric Physics (Russia) over the course of several years. If we contour all plurality of experimental points presented by them in the form of a closed curve, then some areas depicted in Figure 4.26 will be produced. This area may be named as the field of experimental data (FED). The density of experimental points is not identical in different parts of this field. It is highest in the area corresponding to $0.27 \leq P_{\max} \leq 0.50$ and $0.1 \leq q_{\max} \leq 0.3$. The least point density is observed in ledges from the right part of field. In Figure 4.26a the points are marked, corresponding to maximal calculated values of the linear polarization and polarization ellipticity of light scattered by the polydisperse systems of HPs with different degrees of humidification (without considering the molecular scattering). The same symbols in Figure 4.26 mark the calculated values of $P_{\max}$ and $q_{\max}$ for a light, scattered by polydisperse systems of TLPs. Here the numeral 9 marks the plurality of the same points for systems of particles with cores consisting of insoluble silica and soluble ammonium sulfate.

As was specified above, it was accepted in the calculations for HPs and TLPs, that the initial condensation cores in both cases had $n_1 = 1.65$ and $\kappa_1 = 0.005$. Because our data calculated for $\lambda = 0.63\ \mu m$ enclosed the wider $\bar{n}$ interval than for $\lambda = 0.55\ \mu m$ the calculated points corresponding to the first wavelength are entered on the graphs though the experimental data refer to the second one. The comparison of calculated values of $P_{\max}$ and $q_{\max}$ for $\lambda = 0.63$ and $0.55\ \mu m$, obtained at different values of the meteorological visibility range S_m, has shown, that at transition from $\lambda = 0.63$ to $0.55\ \mu m$ the values of $P_{\max}$ for corresponding $\bar{n}$ and ν decrease on average by 2–3%, and the values $q_{\max}$ increase on average by 2–4%. The account of molecular scattering suggests that at the meteorological visibility range $S_m = 2.5,5$, and 10 km the values $P_{\max}$, calculated for the aerosol component, increase by 2.5–3, 5–6, and 9.5–11.5% respectively, and the values $q_{\max}$ decrease by 4, 7.5–8, and 14–15% respectively. So the values of $P_{\max}$ and $q_{\max}$ plotted in Figure 4.26 may be thought of as practically corresponding to the case of $\lambda = 0.55\ \mu m$ and $S_m = 2.5$ km.

As already mentioned in Section 4.5, the monotonic behaviour of the dependencies $P_{\max}(q_{\max})$ at constant ν and $\bar{n}$ observed in the case of the HP systems allow us to construct a grid of curves connecting the points with identical values of ν and $\bar{n}$. This grid covers almost all FED. In so doing, the grid part, corresponding the greater humidifying of particles (smaller $\bar{n}$), falls within that FED part, where the density of points is small. The line corresponding to $n = 1.33$ (water droplets), would pass outside this field. The reason for this is that the experimental data of Gorchakov and Emilenko (1974) are obtained basically for the weakly humidified hazes.

In Figure 4.26*b* the points corresponding to TLP systems with low water content, are focused in the left-hand part of the FED, where the concentration of experimental points is high. And even the dashed curve, corresponding to $n = 1.39$, still passes through the FED part with the high density of experimental points. Notice, that both in the case of HPs and TLPs, the points corresponding to systems of dry and weakly humidified particles with

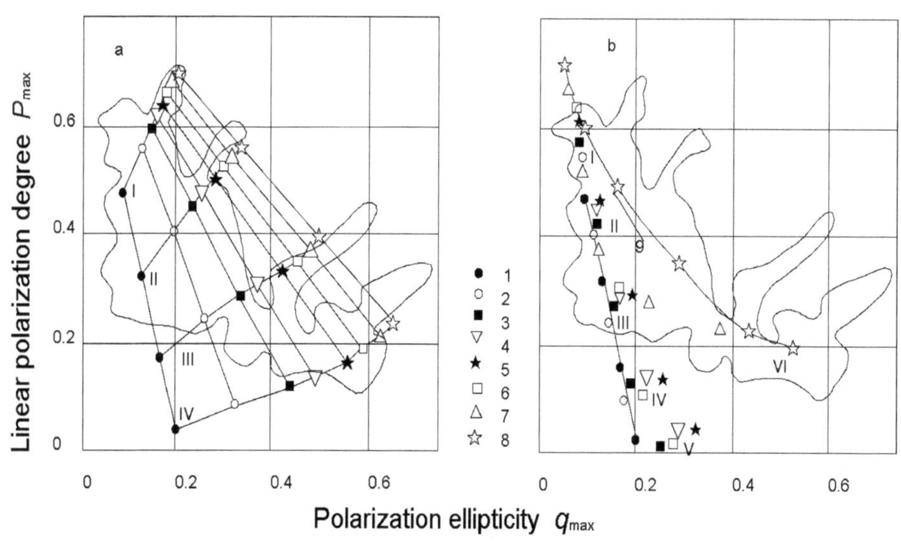

Figure 4.26. FED (Gorchakov and Emilenko, 1974) and calculated values of $P_{\max}$, $q_{\max}$ for systems of HPs (*a*) and TLPs (*b*) with $\bar{n} = 1.65(1), 1.56(2), 1.51(3), 1.49(4), 1.46(5), 1.44(6), 1.42(7)$, and $1.39(8)$ at $\nu = 4(I), 3.5(II), 3(III), 2.5(VI), 2(V)$, and $1.5(VI)$.

$\bar{n} \approx 1.65$ and $\nu < 3$ fall outside the FED. Obviously, the systems of such particles are really characterized by higher parameter ν values, that is the particles of small sizes predominate in them. As the water content increases, there is a reorganization of particle size spectrum (broadening). The points, corresponding to strongly humidified particle systems with $\nu = 4$, also fall outside the FED. But the points corresponding to $\bar{n} = 1.39$ and $\nu = 2$–1.5 (i.e., to systems of particles with high water content and with wide size spectrum) are in this field. Such a transformation of the particle size distribution correlates quite well with conclusions made in the works of Sedunov (1972) and Hänel (1970).

At transition to higher meteorological visibility ranges, the points corresponding to $(P_{\max}, q_{\max})$ should be shifted upwards and to the left on the graphs. Then, points for particle systems with $\bar{n} < 1.51$ and $\nu = 4$ will also be out of the FED. The less the aerosol concentration in the atmosphere, the less probable is the existence of the narrow particle distributions (that is the fine particles) with considerable humidifying degrees. Apparently, at small disperse phase concentration the broadening of the particle size distribution occurs faster, than at high concentrations. This inference is also consistent with the concept about "hierarchy" of condensation cores according to their activity (see Sedunov, 1972) and according to the speed at which they become involved in the condensation process depending on concentration and relative humidity.

It is possible to conclude also, that the existence of TLPs formed from condensation cores, comprising basically of silica mixed with ammonium sulfate, is quite probable. The points corresponding to these particles, lay inside the FED, if $\nu > 2.5$. It signifies, that in actual experimental situations could exist both homogeneous and non-uniform particles of different types. As follows from Figure 4.26, the ranges of values $P_{\max}$ and $q_{\max}$, corresponding to systems of HPs and TLPs, overlap. The measurements of just these two characteristics can be inadequate to clear up the particle type. It was already mentioned, that the distinctions are most appreciable at great scattering angles. It is confirmed by the experimental data of Holland and Gagne (1970), Harrison *et al.* (1972), Waggoner *et al.* (1972), and Ward *et al.* (1973). It is stated, that owing to particle layer structure, as well as because of their irregular shape, the backscattering can vary by an order of magnitude compared with spherical HPs (at the same refractive index). It is clear, that for reliable diagnostics of particles it is impossible to be restricted by measurements of one or two scattering characteristics only. Only complex measurements involving the angular and spectral dependencies can give the most complete information on properties of particles forming aerosols.

Gorchakov and Emilenko (1974) gave the example of satisfactory concurrence of angular dependencies for all four measured non-zero components of the scattering matrix with ones calculated for the model of HPs with a refractive index $n = 1.5$, distributed under the Junge type power law with $\nu = 3.5$. In Figure 4.27, are the results of comparing the experimental data with calculations for identical systems of HPs and TLPs with $\bar{n} = 1.51$. We see from the graph, that the curve $P(\theta)$ for TLPs displays a somewhat better agreement with experimental data, than that for HPs, especially at $\theta < 90°$ and $\theta > 130°$. As for $f_{43}(\theta)$, here the curve run for systems of TLPs noticeably differs from the experimental dependence. The reasons for such disagreement between calculations and experimental data may consist in failure of the values of real and imaginary parts of the

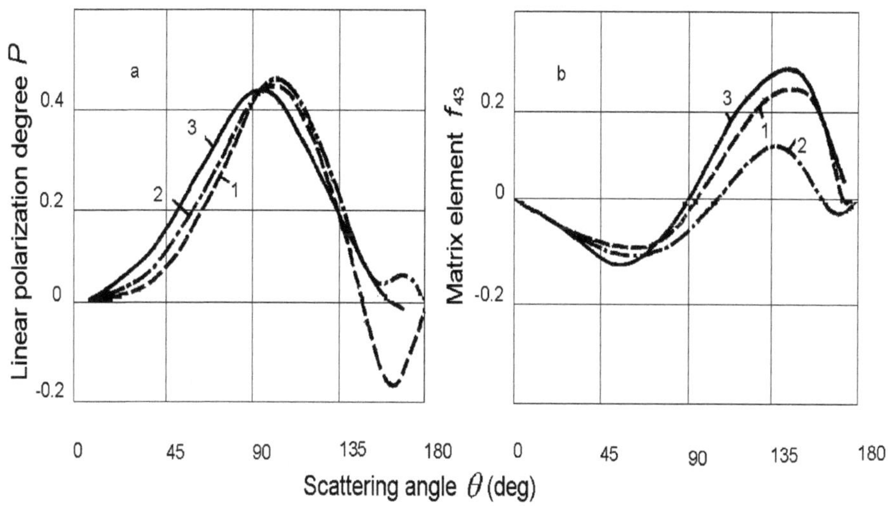

Figure 4.27. Dependencies $P(\theta)$ (*a*) and $f_{43}(\theta)$ (*b*). Calculations at $\lambda = 0.55\ \mu m$, $\bar{n} = 1.51$, and $\nu = 3.5$ for HPs (*1*), TLPs with $G = 0.3$ (*2*), and experimental results (Gorchakov and Emilenko, 1974) (*3*).

condensation cores refractive index or percentage of soluble matter, accepted in calculations, to meet the experimental conditions.

From Figure 4.28 we notice that the increase in the core absorption index (at constant $\bar{n}$ and G) results in a progressively higher distinction in the scattering characteristics. The curves run in Figure 4.29*a,b* would testify that the decrease in the real part of the core refractive index from $n = 1.65$ to 1.55 results in the scattering properties of the HPs and TLPs systems becoming noticeably merged. (At calculations, the volume-averaged refractive index of TLPs was constant: $\bar{n} \approx 1.51$ at the expense of the water shell width change.) Notice, that dependence $P(\theta)$ at $\theta < 150°$ is rather low-sensitive to core refractive index. The noticeable changes are observed only at $150° < \theta < 175°$. The dependence $f_{43}\,(\theta)$ is considerably more sensitive in this respect.

In Figure 4.29*c,d* the dependencies $P(\theta)$ and $f_{43}\,(\theta)$ are shown at constant values of refractive and absorption indices, but at varying soluble matter percentages in condensation cores. In this case at $\theta <$ 145–150° the discrepancy between the $P(\theta)$ curves is not so great: the $P_{\max}$ differs by no more than 0.05. The essential differences of this dependence for different G takes place at $\theta > 145°$. The behaviour of $f_{43}\,(\theta)$, as well as in the previous case, is more sensitive to change in the condensation core properties, than $P(\theta)$. Merging of curves $f_{43}\,(\theta)$ with the corresponding curve for HPs occurs with some oscillations of maximum height upwards and downwards. At $G = 0.9$ the curves distinction in maximum values is approximately 0.09; at $G \gtrsim 0.95$ the curves practically coincide.

Thus, agreement between the calculations and experimental data observed by Gorchakov and Emilenko (1974) is probable, if the soluble matter content in the condensation cores exceeds 20–30%, and the bulk mass comprises the substances with refractive indices,

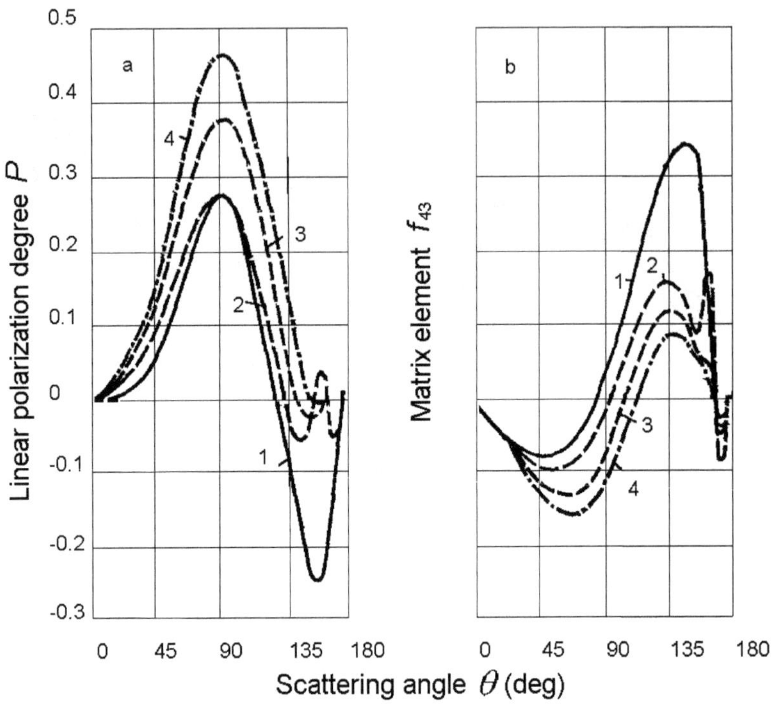

Figure 4.28. Dependencies $P(\theta)$ (a) and $f_{43}(\theta)$ (b) for HPs (1) and TLP $(2–4)$ at $\nu = 3$, $\lambda = 0.55\ \mu m$, $n_1 = 1.65$, $G = 0.2$, $\bar{n} = 1.51$, and $\kappa_1 = 0.005(2), 0.05(3)$, and $0.1(4)$.

close to $n = 1.55$ (e.g., NaCl, $(NH_4)_2SO_4$, and $MgCl_2$, whose refractive index equals accordingly 1.54, 1.52, and 1.54 in the visual spectral range). Laktionov (1972) has shown, that $(NH_4)_2SO_4$ predominates among soluble matter in continental aerosol particles. It is known that the ions of SO_4 and NH_4 make more than 70% of the soluble matter contained in great particles of atmospheric aerosols. As regards the share of soluble matter in condensation cores, it can achieve 60% or more for the European continental aerosol as stated by Winkler and Junge (1972) and Winkler (1973).

The difference of core optical constants and the soluble matter percentage from those accepted in calculations can also be bound with the influence of industrial discharges into atmosphere. It is possible, that such an influence may be the case for hazes observed by Gorchakov and Emilenko (1974).

As was mentioned above, the analysis and the conclusions of the present section carry qualitative character in view of the lack of the necessary information about conditions, in which the measurements were performed. However, the examples studied convincingly show that application of the TLP model and the comparison of results with experimental data open new opportunities for the study of the properties and diagnostics of atmospheric aerosols. These opportunities are much broader, than when using the traditional model of HPs.

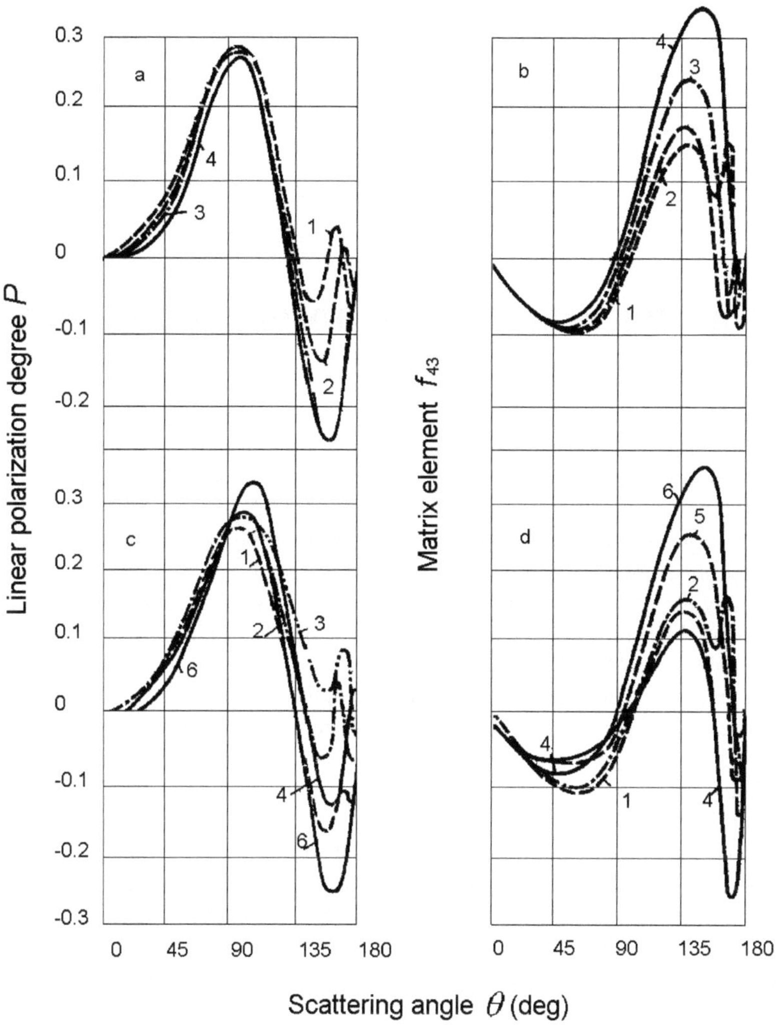

Figure 4.29. Dependencies $P(\theta)$ (*a,b*) and $f_{43}(\theta)$ (*c,d*) for HPs and TLPs at $\nu = 3$, $\lambda = 0.55$ μm: *a,b* – $G = 0.2$, $\kappa_1 = 0.005$, $n_1 = 1.65(1)$, $1.60(2)$, $1.55(3)$, and HPs (4); *c,d* – $n_1 = 1.65$, $\kappa_1 = 0.005$, $G = 0(1)$, $0.2(2)$, $0.5(3)$, $0.7(4)$, $0.9(5)$, and HPs with $G = 1(6)$.

5

Integral characteristics of absorption by spheres

As was stated earlier, the extinction of electromagnetic radiation by a particle is governed by two principally different physical processes: by formation of the scattered field outside the particle and by absorption of radiation inside it. The second process is intimately connected with the so-called internal field (see Eq.(1.3.32)). The studies of internal fields and attendant problems began much later than those of the scattered ones. However, the pace of such investigations continues to increase in recent years, because it turns out that the great body of diversified physical phenomena (both linear and non-linear) is connected with the internal fields. The review of such effects one can find, for example, in the book of Barber and Chang (1990).

In the preceding chapters we largely dealt with the scattered field. Here and in the next two chapters we shall consider the problems concerning the internal field. In particular, Chapter 5 is devoted to the integral characteristics of absorption. The electromagnetic fields inside homogeneous and inhomogeneous particles will be considered in Chapters 6 and 7 accordingly.

The investigation of electromagnetic wave absorption is a classical one in the optics of dispersive media. The Mie theory is used for absorption calculation in the simplest case of spherical scatterers (Bohren and Huffman, 1983). In so doing the particle is considered as a whole and characterized by the absorption cross section $\sigma_{\rm abs}$. The alternative approach consists in the detailed description of the internal field distributions inside the particle (Prishivalko, 1983). The former gives too little information about the absorption process, and in the latter connected with sufficiently great computer time consuming the information is on frequent occasions abundant and hard to interpret. In this regard it is interesting to introduce the sufficiently simple additional characteristics of absorption. Examples are found in the papers of Babenko and Astafyeva (1985), Sitarski (1987*a,b*), Mackowski *et al.* (1990), Babenko (1991, 1999), where the absorption inside the restricted spherical sub-layers of a multilayer sphere is calculated. This section is devoted to derivation of analytical expressions for several additional absorption characteristics including the angle-averaged intensity of the internal field (Section 5.3), the absorption asymmetry (Section 5.4), and the absorption in the restricted region of the spherical particle (Section 5.5). All these quantities are obtained using the analytical or semi-analytical integration of the internal field over a volume or a surface enclosing this volume (Section 5.1). In addition, in Section 5.2 the calculation of absorption using the so-called formula of Kattawar and Eisner is discussed.

5.1 ABSORPTION IN SELECTED REGIONS OF THE PARTICLE: VOLUME AND SURFACE INTEGRAL REPRESENTATIONS

As will be indicated in Chapters 6 and 7, the rather inhomogeneous distribution of internal energy can emerge in some restricted regions of a single particle. The construction of a detailed picture of this distribution requires relatively complex computations of an electromagnetic field induced by the incident wave inside the particle. The other, much more simple but nevertheless informative, approach consists in obtaining some *integral characteristics of absorption*, in particular, power absorbed in some bounded region of the particle. To obtain this value, we must integrate the internal field over the volume of interest. Let us write down the basic formulae needed for correct setting up of the problem.

If the first Maxwell equation multiplied by $\mathbf{H}$ subtract from the second one multiplied by $\mathbf{E}$ then we have:

$$\mathbf{E}\cdot(\nabla\times\mathbf{H})-\mathbf{H}\cdot(\nabla\times\mathbf{E})=\mathbf{E}\cdot\dot{\mathbf{D}}+\mathbf{H}\cdot\dot{\mathbf{B}}+\mathbf{E}\cdot\mathbf{j} \tag{5.1.1}$$

With vector identity $\mathbf{b}\cdot(\nabla\times\mathbf{a})-\mathbf{a}\cdot(\nabla\times\mathbf{b})=\nabla\cdot[\mathbf{a}\times\mathbf{b}]$, Eq.(5.1.1) may be rewritten as:

$$\nabla\cdot\mathbf{S}=\mathbf{E}\cdot\dot{\mathbf{D}}+\mathbf{H}\cdot\dot{\mathbf{B}} \tag{5.1.2}$$

where $\mathbf{S}=\mathbf{E}\times\mathbf{H}$ and $q=\mathbf{E}\cdot\mathbf{j}$. Integrate Eq.(5.1.2) over the arbitrary volume V:

$$\int_V \nabla\cdot\mathbf{S}\,dV+\int_V\left(\mathbf{E}\cdot\dot{\mathbf{D}}+\mathbf{H}\cdot\dot{\mathbf{B}}\right)dV+\int_V q dV=0 \tag{5.1.3}$$

The Gauss theorem $\int_V \nabla\cdot\mathbf{F}\,dV=\int_V \mathbf{F}\cdot\mathbf{n}\,dS$, where S is the surface bounding the volume V and $\mathbf{n}$ is the unit vector of outward normal, enables us to formulate this relation in the form:

$$\int_S \mathbf{S}\cdot\mathbf{n}\,dS+\int_V\left(\mathbf{E}\cdot\dot{\mathbf{D}}+\mathbf{H}\cdot\dot{\mathbf{B}}\right)dV+\int_V q dV=0 \tag{5.1.4}$$

which is valid regardless of the form of the material equations.

Let us introduce the ordinary linear material equations $\mathbf{D}=\varepsilon\varepsilon_0\mathbf{E}$, $\mathbf{B}=\mu\mu_0\mathbf{H}$, and $\mathbf{j}=\sigma\mathbf{E}$. Then:

$$\mathbf{E}\cdot\dot{\mathbf{D}}+\mathbf{H}\cdot\dot{\mathbf{B}}=\frac{1}{2}\frac{\partial}{\partial t}(\mathbf{E}\cdot\mathbf{D}+\mathbf{H}\cdot\mathbf{B})$$

If we denote $(\mathbf{E}\cdot\mathbf{D})/2=w_e$, $(\mathbf{H}\cdot\mathbf{B})/2=w_m$, and $w=w_e+w_m$, then Eq.(5.1.4) becomes:

$$-\frac{\partial}{\partial t}\int_V w dV=\int_S \mathbf{S}\cdot\mathbf{n}\,dS+\int_V q dV \tag{5.1.5}$$

To this point all manipulations were formal. Now consider the physical background of quantities involved. It is shown in the book of Born and Wolf (1999) that w is the volume density of the electromagnetic energy, and w_e and w_m can be considered as the volume densities of electric and magnetic energies. Then the left-hand term in Eq.(5.1.5) is the velocity of the decreasing electromagnetic field energy inside the volume V. This energy decrease is due to: (a) Joule heat release in conductors immersed in the field (it is shown in courses on electrodynamics that this volume integral to the right describes the work done by a field in V per unit time, that is, q is the volume density of the heat power of the current); and (b) the energy transmission through the closed surface S, bounded by the volume V. The vector $\mathbf{S}$, named the Pointing vector, describes the instantaneous density of energy flux, that is the energy, transferred through the surface perpendicular to the direction of energy transport per unit time.

Hence, Eq.(5.1.5) is actually an integral representation of the law of energy conservation in the electromagnetic field. The differential form of this law frequently named the Pointing theorem readily follows from above equations:

$$\nabla \cdot \mathbf{S} + \frac{\partial w}{\partial t} + q = 0 \tag{5.1.6}$$

In the absence of free charges and currents Eq.(5.1.6) transforms to the continuity equation for the volume energy density:

$$\nabla \cdot \mathbf{S} = -\frac{\partial w}{\partial t} \tag{5.1.7}$$

Note that the interpretation of $\mathbf{S}$ as the density of energy flux introduces arbitrariness to some extent. A specific physical meaning can be assigned to the integral of quantity $\mathbf{S} \cdot \mathbf{n}$ over a closed surface rather than $\mathbf{S}$ itself. But it is clear that knowing this integral does not allow us to reach a conclusion on the detailed distribution of $\mathbf{S}$. So other definitions of the density of energy flux are possible. Rotor of an arbitrary vector can be added to $\mathbf{S}$ because such a term does not contribute to the surface integral, that follows from the Gauss theorem and identity $\nabla \cdot (\nabla \times \mathbf{a}) = 0$.

In the stationary case when a particle is illuminated by the electromagnetic wave with constant parameters, the derivative in the left-hand part of Eq.(5.1.5) is equal to zero:

$$\int_V q dV = -\int_S \mathbf{S} \cdot \mathbf{n} \, dS \tag{5.1.8}$$

The meaning of this relation is simple – the difference between the fluxes entering a particle and leaving it is spent on heat release inside the particle, that is on energy absorption. Let us clear up the sense of the minus sign on the right-hand side. If $\mathbf{S}$ is in opposition to the vector of outward normal $\mathbf{n}$ then $\mathbf{S} \cdot \mathbf{n} < \mathbf{0}$ and the minus sign gives positiveness to the right-hand side of Eq.(5.1.8) if the overall transport of electromagnetic energy is directed inside the volume.

Hence, the absorption power inside some volume V may be calculated employing the volume integral for heat sources:

$$W_{\text{abs}} = \int_V q dV = \sigma \int_V \mathbf{E} \cdot \mathbf{E} \, dV \tag{5.1.9}$$

or using the surface integral for the Pointing vector over the closed surface S bounding the volume V:

$$W_{\mathrm{abs}} = -\int_S [\mathbf{E} \times \mathbf{H}] \cdot \mathbf{n}\, dS \tag{5.1.10}$$

Up to this point we took no account of the time dependence of fields. Let us assign its dependence to be:

$$\mathbf{E}_c(\mathbf{r}, t) = \mathbf{E}_{0c}(\mathbf{r}) \exp(-i\omega t) \qquad \mathbf{H}_c(\mathbf{r}, t) = \mathbf{H}_{0c}(\mathbf{r}) \exp(-i\omega t) \tag{5.1.11}$$

Naturally, the real parts of these complex fields $\mathbf{E}(\mathbf{r}, t) = \mathrm{Re}\, \mathbf{E}_c(\mathbf{r}, t)$ and $\mathbf{H}(\mathbf{r}, t) = \mathrm{Re}\, \mathbf{H}_c(\mathbf{r}, t)$ appear in Eqs.(5.1.9–10). It is easily seen that:

$$\mathbf{E}(\mathbf{r}, t) = \left[\mathbf{E}_{0c}(\mathbf{r}) \exp(-i\omega t) + \mathbf{E}_{0c}^*(\mathbf{r}) \exp(i\omega t)\right] /2$$

(and similarly for $\mathbf{H}(\mathbf{r}, t)$). This relation will be used later.

The optical frequencies are very high $\left(\omega \sim 10^{15} c^{-1}\right)$. So the physical reality may be attributed only to quantities averaged over some time interval much more than $\tau' = 2\pi/\lambda$. Let us average the Pointing vector:

$$\langle \mathbf{S} \rangle = \frac{1}{2\tau} \int_{-\tau}^{\tau} \mathbf{S} dt = \frac{1}{8\tau} (\mathbf{E}_{0c} \times \mathbf{H}_{0c}) \int_{-\tau}^{\tau} e^{-2i\omega t} dt + \frac{1}{8\tau} (\mathbf{E}_{0c}^* \times \mathbf{H}_{0c}^*) \int_{-\tau}^{\tau} e^{2i\omega t} dt$$
$$+ \frac{1}{4} (\mathbf{E}_{0c} \times \mathbf{H}_{0c}^*) + + \frac{1}{4} (\mathbf{E}_{0c}^* \times \mathbf{H}_{0c})$$

It is evident that integrals:

$$\frac{1}{\tau} \int_{-\tau}^{\tau} e^{-2i\omega t} dt = \frac{1}{\tau} \int_{-\tau}^{\tau} e^{2i\omega t} dt = \frac{\sin 2\omega t}{\omega t} = \frac{\tau' \sin 2\omega t}{2\pi\tau}$$

may be neglected ($\tau' \ll \tau$) and we have as a result:

$$\langle \mathbf{S} \rangle = \frac{1}{4} (\mathbf{E}_{0c} \times \mathbf{H}_{0c}^*) + \frac{1}{4} (\mathbf{E}_{0c}^* \times \mathbf{H}_{0c}) = \frac{1}{2} \mathrm{Re} \left(\mathbf{E}_{0c} \times \mathbf{H}_{0c}^*\right). \tag{5.1.12}$$

Similar averaging of $\mathbf{E} \cdot \mathbf{E}$ gives:

$$\langle \mathbf{E} \cdot \mathbf{E} \rangle = \frac{1}{2} (\mathbf{E}_{0c} \cdot \mathbf{E}_{0c}^*) \tag{5.1.13}$$

Substituting Eqs.(5.1.12–13) into Eqs.(5.1.9–10) we have for time-averaged absorption power (the subscripts $0c$ are omitted):

$$W_{\mathrm{abs}} = \frac{\sigma}{2} \int_V \mathbf{E}(\mathbf{r}) \cdot \mathbf{E}^*(\mathbf{r}) dV \tag{5.1.14}$$

$$= -\frac{1}{2} \mathrm{Re} \int_S \left[\mathbf{E}(\mathbf{r}) \cdot \mathbf{H}^*(\mathbf{r})\right] \cdot \mathbf{n}\, dS \tag{5.1.15}$$

Equation (5.1.14) may be rewritten in another form when it is considered that $\sigma = \omega\varepsilon_0\varepsilon'' = 2\omega\varepsilon_0 n\kappa = 2k_0 n\kappa\sqrt{\varepsilon_0/\mu_0}$:

$$W_{\text{abs}} = k_0 n\kappa\sqrt{\frac{\varepsilon_0}{\mu_0}}\int\limits_V \mathbf{E}(\mathbf{r})\cdot\mathbf{E}^*(\mathbf{r})dV, \text{ where } k_0 = 2\pi/\lambda. \tag{5.1.16}$$

Thus, we obtain the expressions for the power absorbed in some bounded region of the particle in two different forms (through the surface integral (5.1.15) and through the volume integral (5.1.16)).

5.2 KATTAWAR–EISNER FORMULA

Let us apply Eqs.(5.1.14–15) to a spherical homogeneous particle. The question arises of what integral (volume or surface) is more convenient to use? We prefer the surface one (Eq.(5.1.15)) because the integration of angular functions over the angles θ and φ is much more simple than the integration of Bessel functions over the radial coordinate r. In principle one can integrate in Eq.(5.1.15) over a surface bounding any arbitrary volume inside the particle, but in so doing it is very doubtful to obtain some analytical relations. If the integration region is in some coordination with the spherical coordinate system then the expressions suitable for calculations can be obtained in some cases. The simplest variant is: integration over the surface of sphere of radius $R_1 \geq R$ (R is the particle radius). Designate the energy absorbed inside this surface per unit time as $W_{\text{abs}}^{(R_1)}$. In this case the outward normal has an r-component only $\mathbf{n} = \mathbf{e}_r$ and therefore $[\mathbf{E}\times\mathbf{H}]\cdot\mathbf{n} = E_\theta H_\varphi^* - E_\varphi H_\theta^*$. As a result, Eq.(5.1.15) takes the form:

$$W_{\text{abs}}^{(R_1)} = \frac{R_1^2}{2}\int\limits_0^\pi \sin\theta\, d\theta \int\limits_0^{2\pi} \left(E_\theta H_\varphi^* - E_\varphi H_\theta^*\right)\big|_{R=R_1} d\varphi \tag{5.2.1}$$

Now we can substitute the expressions for the electric and magnetic fields inside the homogeneous sphere (see Eq.(1.3.32)):

$$\mathbf{E}_{\text{int}} = E_I\sum_{l=1}^\infty \mu_l\left[\alpha_l^{31}\mathbf{M}_{o1l}^{(1)}(Nk_0\mathbf{r}) - i\beta_l^{31}\mathbf{N}_{e1l}^{(1)}(Nk_0\mathbf{r})\right] \tag{5.2.2}$$

$$\mathbf{H}_{\text{int}} = -NE_I\sqrt{\frac{\varepsilon_0}{\mu_0}}\sum_{l=1}^\infty \mu_l\left[\beta_l^{31}\mathbf{M}_{e1l}^{(1)}(Nk_0\mathbf{r}) + i\alpha_l^{31}\mathbf{N}_{o1l}^{(1)}(Nk_0\mathbf{r})\right]$$

where $\mu_l = i^l\dfrac{(2l+1)}{l(l+1)}$

into Eq.(5.2.1) and integrate it. The internal field components may be reformulated as:

$$\begin{pmatrix} E_\theta \\ E_\varphi \end{pmatrix} = \frac{E_I}{Na_1x}\begin{pmatrix} \cos\varphi \\ -\sin\varphi \end{pmatrix}\sum_{l=1}^{\infty}\mu_l\left[X_l\begin{pmatrix} \pi_l \\ \tau_l \end{pmatrix} - iV_l\begin{pmatrix} \tau_l \\ \pi_l \end{pmatrix}\right] \tag{5.2.3}$$

$$\begin{pmatrix} H_\theta \\ H_\varphi \end{pmatrix} = \frac{E_I}{a_1x}\begin{pmatrix} \sin\varphi \\ \cos\varphi \end{pmatrix}\sqrt{\frac{\varepsilon_0}{\mu_0}}\sum_{l=1}^{\infty}\mu_l\left[Z_l\begin{pmatrix} \pi_l \\ \tau_l \end{pmatrix} - iY_l\begin{pmatrix} \tau_l \\ \pi_l \end{pmatrix}\right]$$

where:

$$X_l(a_1) = \alpha_l^{31}\psi_l(Na_1x) \qquad Z_l(a_1) = \beta_l^{31}\psi_l(Na_1x) \tag{5.2.4}$$

$$Y_l(a_1) = \alpha_l^{31}\psi_l'(Na_1x) \qquad Z_l(a_1) = \beta_l^{31}\psi_l'(Na_1x)$$

a is the relative radial coordinate $a = r/R$ and $a_1 = R_1/R$. If we replace the variable cos $\theta = \mu$ and to integrate over φ-angle then:

$$W_{\text{abs}}^{(R_1)} = -A\,\text{Re}\,\frac{1}{N}\int\limits_{-1}^{1}\sum_{l=1}^{\infty}\sum_{j=1}^{\infty}\mu_l\mu_j^*\left[(X_l\tau_l - iV_l\pi_l)\,(Z_j\pi_j - iY_j\tau_j)^*\right. \tag{5.2.5}$$

$$\left. +(X_l\pi_l - iV_l\tau_l)(Z_j\tau_j - iY_j\pi_j)^*\right]_{a=a_1} d\mu$$

where:

$$A = \frac{\pi E_I}{2k_0^2}\sqrt{\frac{\varepsilon_0}{\mu_0}}$$

Combining the terms and interchanging the integration and summation we have:

$$W_{\text{abs}}^{(R_1)} = A\,\text{Re}\,\frac{1}{N}\sum_{l=1}^{\infty}\sum_{j=1}^{\infty}\mu_l\mu_j^*\left[i\beta_{jl}\int\limits_{-1}^{1}(\pi_j\pi_l + \tau_j\tau_l)d\mu - \alpha_{jl}\int\limits_{-1}^{1}(\pi_j\tau_l + \tau_j\pi_l)d\mu\right] \tag{5.2.6}$$

where:

$$\alpha_{jl} = \left(X_lZ_j^* + V_lY_j^*\right)_{a=a_1} \qquad \beta_{jl} = \left(V_lZ_j^* + X_lY_j^*\right)_{a=a_1}$$

The integrals over the angular functions in Eq.(5.2.6) are taken in closed form (see Eq.(B.47)):

$$\int\limits_{-1}^{1}(\pi_j\pi_l + \tau_j\tau_l)d\mu = \frac{2j^2(j+1)^2}{2j+1}\delta_{jl} \qquad \int\limits_{-1}^{1}(\pi_j\tau_l + \tau_j\pi_l)d\mu = 0 \tag{5.2.7}$$

and Eq.(5.2.22) is simplified to:

$$W_{\text{abs}}^{(R_1)} = 2A\,\text{Re}\left[iN^{-1}\sum_{l=1}^{\infty}(2l+1)\beta_{ll}\right] \tag{5.2.8}$$

where:

$$\beta_{ll} = \left|\beta_l^{31}\right|^2 \psi_l'(Nxa_1)\,\psi_l^*(Nxa_1) - \left|\alpha_l^{31}\right|^2 \psi_l(Nxa_1)\,\psi_l'^*(Nxa_1)$$

If we substitute the explicit expressions for the amplitude coefficient of the internal field:

$$\alpha_l^{31} = \frac{iN}{\xi_l'(x)\psi_l(Nx) - N\xi_l(x)\psi_l'(Nx)} \qquad \beta_l^{31} = \frac{iN}{N\xi_l'(x)\psi_l(Nx) - \xi_l(x)\psi_l'(Nx)} \tag{5.2.9}$$

into Eq.(5.2.8) and use, in the ordinary way, the log-derivatives of the Riccati–Bessel functions, then this equation transforms to:

$$W_{\text{abs}}^{(R_1)} = -2A\sum_{l=1}^{\infty}(2l+1)\left|\frac{R_l(Nxa_1)}{\xi_l(x)}\right|^2 \left[\frac{\text{Im}\left[N^* D_l(Nx)\right]}{\left|NG_l(x) - D_l(Nx)\right|^2} + \frac{\text{Im}\left[ND_l(Nx)\right]}{\left|G_l(x) - ND_l(Nx)\right|^2}\right] \tag{5.2.10}$$

where (see Section 1.5.2.3):

$$R_l(Nxa_1) = \frac{\psi_l(Nxa_1)}{\psi_l(Nx)}$$

If we need to calculate the absorption power in a spherical concentric layer from R_2 to R_1 then it is obvious that:

$$W_{\text{abs}}^{(R_2,R_1)} = W_{\text{abs}}^{(R_2)} - W_{\text{abs}}^{(R_1)} \tag{5.2.11}$$

But if $R_2 = 0$ and $R_1 = R$ (particle as a whole), then $R_l(Nxa_1) = 1$ and Eq.(5.2.10) transforms into the well known formula due to Kattawar and Eisner (1970), which is usually written as:

$$Q_{\text{abs}} = -\frac{2}{x^2}\sum_{l=1}^{\infty}\frac{(2l+1)}{\left|\xi_l(x)\right|^2}\left[\frac{\text{Im}\left[N^* D_l(Nx)\right]}{\left|NG_l(x) - D_l(Nx)\right|^2} + \frac{\text{Im}\left[ND_l(Nx)\right]}{\left|G_l(x) - ND_l(Nx)\right|^2}\right] \tag{5.2.12}$$

Transformation of Eq.(5.2.10) into Eq.(5.2.12) is as follows. First, we shall obtain the expression for the incident wave intensity. Let the incident wave be:

$$\mathbf{E}_{\text{inc}} = E_I e^{ik_0 z}\mathbf{e}_x \qquad \mathbf{H}_{\text{inc}} = \frac{1}{i\omega\mu_0}\nabla\times\mathbf{E}_{\text{inc}} = E_I e^{ik_0 z}\sqrt{\frac{\varepsilon_0}{\mu_0}}\mathbf{e}_y$$

and the energy flux in the incident wave per unit surface area is:

$$\mathbf{S} = \frac{1}{2}\,\text{Re}\,\left(\mathbf{E}_{\text{inc}}\times\mathbf{H}_{\text{inc}}^*\right) = \frac{1}{2}E_I^2\sqrt{\frac{\varepsilon_0}{\mu_0}}\mathbf{e}_z$$

that is, the incident wave intensity equals:

$$I_{\text{inc}} = \frac{1}{2} E_I^2 \sqrt{\frac{\varepsilon_0}{\mu_0}} = \frac{k_0^2}{\pi} A \tag{5.2.13}$$

Then the absorption cross section is $C_{\text{abs}} = W_{\text{abs}}/I_{\text{inc}}$ and absorption efficiency factor is $Q_{\text{abs}} = C_{\text{abs}}/\pi R^2 = W_{\text{abs}}/\pi R^2 I_{\text{inc}}$, whence Eq.(5.2.12) follows.

From a computational point of view, the advantage of the Kattawar–Eisner formula lies in the fact that all terms under the sum are positive because they are the integrals from sums of positive functions. Therefore, the computations with this formula are free from loss of significant digits. This is of particular value for small absorption indices ($\kappa < 10^{-5}$), when the values of Q_{sc} and Q_{ext} are so close that calculating Q_{abs} as their difference may be computationally incorrect.

The formula (5.2.28) was obtained by Kattawar and Eisner (1970) by a method which bears no resemblance to ours and relates to thermodynamics rather than to optics. This relation was many times re-derived independently by other investigators (e.g., by Karam and Fung (1993) and Perelman (1994)).

Let us mention in passing some useful quantities related to absorption by a sphere. The quantity:

$$q_{\text{abs}} = \frac{W_{\text{abs}}}{V} = \frac{I_{\text{inc}} S_{\text{geom}}}{V} Q_{\text{abs}} \tag{5.2.14}$$

where S_{geom} is the area of particle shadow, is referred to as the specific absorption power (that is power per unit volume). For a sphere:

$$q_{\text{abs}} = \frac{3}{4R} I_{\text{inc}} Q_{\text{abs}} = \frac{3\pi}{2\lambda x} I_{\text{inc}} Q_{\text{abs}} \tag{5.2.15}$$

Normalized intensity of the local electrical field inside the particle is:

$$B(r,\theta,\varphi) = \frac{1}{E_I^2} \mathbf{E}_{\text{int}}(r,\theta,\varphi) \cdot \mathbf{E}_{\text{int}}^*(r,\theta,\varphi) \tag{5.2.16}$$

Then the volume integral in Eq.(5.1.16) can be written as:

$$W_{\text{abs}} = \frac{2\pi n\kappa}{\lambda} E_I^2 \sqrt{\frac{\varepsilon_0}{\mu_0}} \int_V B(r,\theta,\varphi)\, dV \tag{5.2.17}$$

If introduce the volume-averaged normalized intensity $\bar{B}$ then:

$$W_{\text{abs}} = \frac{4\pi n\kappa}{\lambda} I_{\text{inc}} V \bar{B} \qquad q_{\text{abs}} = \frac{4\pi n\kappa}{\lambda} I_{\text{inc}} \bar{B}$$

In view of Eq.(5.2.15) we obtain the very interesting formula:

$$\bar{B} = \frac{3}{8n\kappa} \frac{Q_{\text{abs}}}{x} \tag{5.2.18}$$

interrelating the characteristics of the internal ($\bar{B}$) and scattered (Q_{abs}) fields.

5.3 ANGLE-AVERAGED INTENSITY OF INTERNAL FIELD

5.3.1 Theory

The normalized intensity $B(r, \theta, \varphi)$ of the local electrical field inside the particle, or more simply, the internal intensity (see Eq.(5.2.15)) is a function of three variables: a radial coordinate r and two angular coordinates (θ and φ). Thus, the graphic representation of value B requires the construction of relatively complex 3D pictures (Prishivalko, 1983). However, often such detailed data are not needed, whereas the angle-averaged internal intensity is still required:

$$B(r) = \frac{1}{4\pi} \int_0^{2\pi} d\varphi \int_0^{\pi} B(r, \theta, \varphi) \sin\theta \, d\theta \tag{5.3.1}$$

The quantity $B(r)$ that depends only on the radial coordinate has clear physical meaning in two different situations:

(1) The particles located in a spherically symmetric (or, diffusive) radiation field (e.g., carbon particles under exposure to thermal radiation in furnaces or atmospheric particles in a cloud).
(2) The incident wave is directional but solid particles themselves are chaotically rotating quite quickly due to Brownian motion, photophoresis, and similar reasons. An important point here is that the characteristic time of such rotation should be much less than the characteristic time of those thermal processes which determine the thermodynamics of a particle.

Besides, knowledge of $B(r)$ dependencies appear also to be useful outside these two situations because it gives a generalized internal field characteristic which is free from excess details. Note, that calculation of $B(r)$ is much more simple than construction of detailed 3D pictures of internal fields. Let us obtain the analytical expression for $B(r)$. After substituting the components of the internal field (Eq.(5.2.3)) into Eq.(5.3.1) we have:

$$B(r) = \frac{E_I^2}{4\pi} \sum_{l=1}^{\infty} \sum_{j=1}^{\infty} \mu_l \mu_j^* \Bigg[\alpha_l^{31} \alpha_j^{31*} \int_{(4\pi)} \mathbf{M}_{o1l}^{(1)} \cdot \mathbf{M}_{o1j}^{(1)*} d\Omega + \beta_l^{31} \beta_j^{31*} \int_{(4\pi)} \mathbf{N}_{e1l}^{(1)} \cdot \mathbf{N}_{e1j}^{(1)*} d\Omega$$
$$+ i\alpha_l^{31} \beta_j^{31*} \int_{(4\pi)} \mathbf{M}_{o1l}^{(1)} \cdot \mathbf{N}_{e1j}^{(1)*} d\Omega - i\beta_l^{31} \alpha_j^{31*} \int_{(4\pi)} \mathbf{N}_{e1l}^{(1)} \cdot \mathbf{M}_{o1j}^{(1)*} d\Omega \Bigg] \tag{5.3.2}$$

where we omit the argument $k\mathbf{r}$ of wave functions and (4π) indicates the integration over all solid angle. It is shown in Appendix C that $\int_{(4\pi)} \mathbf{M}_{\sigma ml}^{(1)} \cdot \mathbf{N}_{\sigma' ml}^{(1)} d\Omega = 0$. Complex conjugation changes nothing in this formula and consequently the third and forth terms in Eq.(5.3.2) are canceled. If you reformulate the remaining terms using vector angular functions $\mathbf{m}$, $\mathbf{n}$,and $\mathbf{l}$ (see Section 1.3) you will get:

$$B(r) = \frac{1}{4\pi}\sum_{l=1}^{\infty}\sum_{j=1}^{\infty}\mu_l\mu_j^*\left[\alpha_l^{31}\alpha_j^{31*}\frac{\psi_l(kr)\psi_j^*(kr)}{kk^*r^2}\int\limits_{(4\pi)}\mathbf{m}_{o1l}\cdot\mathbf{m}_{o1j}d\Omega\right. \tag{5.3.3}$$

$$+\beta_l^{31}\beta_j^{31*}\frac{\psi_l(kr)\psi_j^*(kr)}{k^2k^{*2}r^4}\int\limits_{(4\pi)}\mathbf{l}_{e1l}\cdot\mathbf{l}_{e1j}d\Omega+\beta_l^{31}\beta_j^{31*}\frac{\psi_l'(kr)\psi_j'^*(kr)}{kk^*r^2}\int\limits_{(4\pi)}\mathbf{n}_{e1l}\cdot\mathbf{n}_{e1j}d\Omega$$

$$\left.+\beta_l^{31}\beta_j^{31*}\frac{\psi_l(kr)\psi_j'^*(kr)}{k^2k^*r^3}\int\limits_{(4\pi)}\mathbf{l}_{e1l}\cdot\mathbf{n}_{e1j}d\Omega+\beta_l^{31}\beta_j^{31*}\frac{\psi_l'(kr)\psi_j^*(kr)}{kk^{*2}r^3}\int\limits_{(4\pi)}\mathbf{n}_{e1l}\cdot\mathbf{l}_{e1j}d\Omega\right]$$

Two last terms are zero. After using the integrals (see Appendix C):

$$\int\limits_{(4\pi)}\mathbf{m}_{\sigma 1l}\cdot\mathbf{m}_{\sigma 1j}d\Omega=\int\limits_{(4\pi)}\mathbf{n}_{\sigma 1l}\cdot\mathbf{n}_{\sigma 1j}d\Omega=\delta_{jl}\frac{2\pi j^2\left(j+1\right)^2}{2j+1}$$

$$\int\limits_{(4\pi)}\mathbf{l}_{\sigma 1l}\cdot\mathbf{l}_{\sigma 1j}d\Omega=\delta_{jl}\frac{2\pi j^3\left(j+1\right)^3}{2j+1}$$

Eq.(5.3.3) becomes:

$$B(r)=\frac{1}{2\left|kr\right|^2}\sum_{l=1}^{\infty}(2l+1)\left[l(l+1)\frac{\left|\beta_l^{31}\psi_l(kr)\right|^2}{\left|kr\right|^2}+\left|\beta_l^{31}\psi_l'(kr)\right|^2+\left|\alpha_l^{31}\psi_l(kr)\right|^2\right] \tag{5.3.4}$$

or, utilizing the notations (5.2.4), it becomes:

$$B(a)=\frac{1}{2\left|N\right|^2x^2a^2}\sum_{l=1}^{\infty}(2l+1)\left[l(l+1)\frac{\left|Z_l(a)\right|^2}{\left|N\right|^2x^2a^2}+\left|V_l(a)\right|^2+\left|X_l(a)\right|^2\right] \tag{5.3.5}$$

Some notes on computation of this quantity can be found in Section 5.4

The specific absorption power was defined in Section 5.2 as a ratio of the absorption power in the volume of the particle to the volume itself. One should use the following formula to generalize it for the case of a spherically symmetric layer with internal R_1 and external R_2 radii:

$$q_{\text{abs}}^{(R_2,R_1)}=\frac{W_{\text{abs}}^{(R_2,R_1)}}{V^{(R_2,R_1)}} \tag{5.3.6}$$

where $W_{\text{abs}}^{(R_2,R_1)}$ are defined by Eqs.(5.2.8) and (5.2.11) and $V^{(R_2,R_1)}=4\pi\left(R_2^3-R_1^3\right)/3$.

Let $R_2 \to R_1$. Then:

$$q_{\rm abs}^{(R_1)} = \frac{3}{4\pi} \lim_{R_2 \to R_1} \frac{W_{\rm abs}^{(R_2,R_1)}}{R_2^3 - R_1^3}$$

$$= \frac{3}{4\pi} \lim_{R_2 \to R_1} \frac{W_{\rm abs}^{(R_2,R_1)}}{(R_2 - R_1)(R_2^2 + R_1 R_2 + R_1^2)} = \frac{1}{4\pi R_1^2} \frac{dW_{\rm abs}(R)}{dr}\bigg|_{R=R_1}$$

Introducing the arbitrary distance r from the particle center instead of fixed distance R_1 we have:

$$q_{\rm abs}(r) = \frac{1}{4\pi r} \frac{dW_{\rm abs}(r)}{dr} \tag{5.3.7}$$

Similar to Eq.(5.2.14), the function $q_{\rm abs}(r)$ may be related to the angle-averaged internal intensity $B(r)$:

$$q_{\rm abs}(r) = \frac{4\pi n\kappa}{\lambda} I_{\rm inc} B(r) \tag{5.3.8}$$

The relation:

$$q_{\rm abs}(r,\theta,\varphi) = \frac{4\pi n\kappa}{\lambda} I_{\rm inc} B(r,\theta,\varphi) \tag{5.3.9}$$

between these quantities is quite obvious in the general case. In fact, $q_{\rm abs}(r,\theta,\varphi)$ describes the local rate of heat release in point (r,θ,φ) and $q_{\rm abs}(r)$ is a result of angle-averaging of this quantity.

It is not difficult to understand that Eq.(5.3.7) gives us another way to obtain an expression for $B(r)$ (Mackowski *et al.*, 1990). Upon substituting Eq.(5.2.8) into Eq.(5.3.8), performing the differentiation, and using the well known relation for Riccati–Bessel functions:

$$\psi_l''(z) = \left[\frac{l(l+1)}{z^2} - 1\right] \psi_l(z)$$

after some cumbersome transformations we can obtain Eq.(5.3.4).

5.3.2 Some results

Let us move on to the analysis of some typical results. We present below the results for four matters with widely different absorption indexes: (1) very high absorption – Al, $\lambda = 2.36$ μm and $N = 2.83 + i24.4$ (Palik, 1985); (2) high absorption – ice, $\lambda = 10.6$ μm and $N = 1.097 + i0.134$ (Schaaf and Williams, 1973); (3) moderate absorption – ice, $\lambda = 2.36$ μm and $N = 1.246 + i5.8 \cdot 10^{-4}$ (Warren, 1984); and (4) low absorption – water, $\lambda = 0.69$ μm and $N = 1.33 + i3.0 \cdot 10^{-8}$ (Irvine and Pollack, 1968). These variants are in the order of decreasing absorption index κ and cover the entire practical range of this quantity variation. We shall consider them in this order.

The dependencies of the angle-averaged relative intensity of internal field $B(a)$ on the relative radial coordinate a, measured from the particle center, for the spherical Al particles are shown in the Figure 5.1 for some size parameters. The almost homogeneous field (at very low $B \sim 3 \cdot 10^{-5}$) are realized at $x = 0.01$. The field becomes markedly inhomogeneous as x grows – the intensity increases near the surface and abruptly decreases in the internal region. The surface-like field is observed at $x = 0.2$ already. This effect becomes more pronounced with further increase of particle size. The intensity of surface field B_{surf} depends on size very slightly and is equal to $2{\cdot}10^{-3}$ at $x > 10$. The dependencies $B(a)$ for particles with sizes $x \sim 10 \div 500$ are very similar in appearance: the averaged intensity decrease linearly on a logarithmic scale, that corresponds to the exponent $B(a) = B_{\text{surf}} \exp\left[-\beta(1-a)\right]$. If the absolute thickness Δr of a skin-layer is arbitrarily defined as the distance from the sphere surface to a depth where $B = 10^{-5}$, then it is seen that Δr is approximately 40 Å in this range ($x = 10 \div 500$).

For ice particles at $\lambda = 10.6\mu m$ (Figure 5.2) the field homogeneity (at the level $B \sim 0.9$) is still retained up to $x \sim 1.0$. The field intensity drops (e.g., $B \sim 0.3 \div 0.6$ at $x = 5.0$), and its inhomogeneity increases with increasing x. As in the two cases above, the progressive displacement of the field into the surface takes place at $x > 10$ but this displacement is less pronounced. The exponential character of field decreasing is very approximate in this case – even at $x = 500$ the dependencies $B(a)$ are not straight lines on log-scale. The depth of skin layer and B_{surf} itself are much more than in case (1).

The value of absorption index for variant (3) (ice at $\lambda = 2.36\mu m$, Figure 5.3) is in the transition region between the cases of strong and low absorption. The field homogeneity (at level $B \sim 0.8$) is observed at $x = 0.5$, but even at $x = 1.0$ it is degraded due to some field growing in the particle center. The field gradually grows (particularly in the center) with x (up to $x \sim 3$). The sharp maximum appears at the center at $x \sim 5$ which persists up to $x \sim 30$. The oscillating structure pronounced at $x \sim 0.1 \div 1.0$ is gradually smoothed out as x increases. As distinct from cases with strong absorption (variants (1)–(2)), the field minimum instead of maximum is realized at the particle surface . At sufficiently great x ($\sim 50 \div 300$) the internal intensity is practically constant in the region $a = 0 \div 0.8$; this value is slowly falls at greater x ($B \sim 1.14$, 1.10 and 1.00 at $x = 50$, 100, and 200, accordingly). The field decreases both inside the internal region and at the surface with further increasing of x.

Now consider the case of very low ($\kappa \sim 10^{-6} \div 10^{-8}$) absorption. Variant (4) (water at $\lambda = 0.69\mu m$, Figure 5.4) corresponds to this situation. Because, as seen from the figures, the curves $B(a)$ for these variants are very similar, we shall consider them together. The field homogeneity ($B \sim 0.7$) begins to break down after $x = 0.5$ and in so doing the field intensity grows. On the whole, the picture is similar to that for the moderate absorption. This analogy remains intact at moderate sizes ($x \sim 5 \div 30$) also; however the pronounced minimum in the particle center typical for moderate absorption is replaced now by a maximum (Figure 5.4*c*). The field levelling in the internal part of the particle is in evidence at $x \sim 15 \div 20$ already; this effect becomes well defined at $x > 50$. Except the small oscillating region near the center (these oscillations decay with increasing x), the intensity value $B(a)$ is practically constant ($B(a) \sim 1.3 \div 1.35$). There is a sharp peak on the edge of this homogeneity region (at $a = 0.8 \div 0.85$). This peak totally disappears at

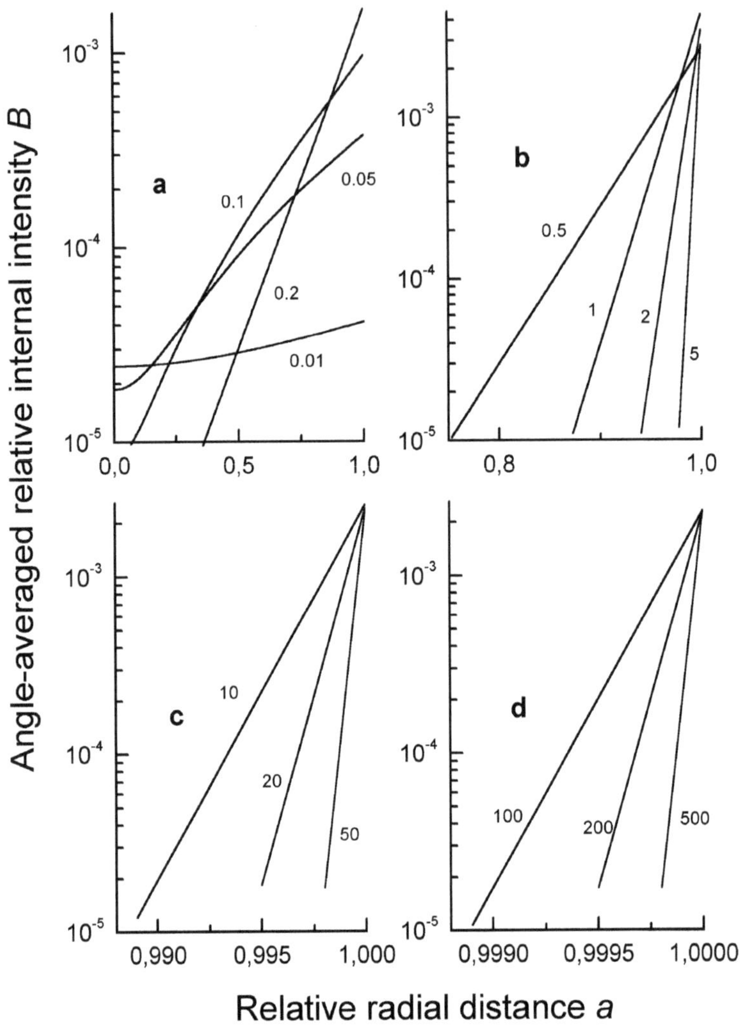

Figure 5.1. Angle-averaged relative intensity of internal field B versus relative distance a from the center of the spherical aluminium particle at $\lambda = 2.36\ \mu m$. Figure on curve is size parameter x.

large x $(x > 700)$. This narrow peak region appears to be indirectly connected with the morphological resonances of internal fields situated on the main axis of particles approximately at these values of a both in the shadowed and in the illuminated hemispheres. The direct manifestation of such resonances is not observed in our calculations for small κ, because it is necessary to use the extremely fine step over x ($\Delta x \sim 10^{-6}$ and low) to reveal the resonances clearly. Nevertheless, the large value $B \sim 5$ was detected at $x = 100.0$ and $a = 0.8$ even with a relatively large step Δx that may be treated as the consequence of a resonance located very near to $x = 100$. We replace the value $x = 100$ on $x = 99$ in Figure 5.4d to eliminate the influence of the resonance. The steep descending of $B(a)$ to the surface level $B_{\rm surf} \sim 0.5$ is detected in the near-surface region ($x \sim 0.8 - 1.0$); in so doing, $B_{\rm surf}$ is very slightly dependent on x.

To illustrate the influence of the morphological resonance the dependencies $B(a)$ are presented in Figure 5.5*a* for the model spherical particle with the real part of the refractive index $n = 1.40$ and the size parameter $x = 31.789230$, that corresponds to the first resonance of the internal amplitude coefficient β_{39}^{31} (Barber and Hill, 1990). The calculations were conducted for the sequence of κ-values (from 10^{-3} to 10^{-8}) and also for the particle out of resonance ($N = 1.4 + i10^{-8}$, $x = 31.80$). At relatively strong absorption ($\kappa \sim 10^{-3}$) the behaviour of the curve $B(a)$ and the absolute values of B are similar, by and large, to the dependencies described above both in resonance and out of resonance. But in resonance the quite broad and high maximum appears in the near-surface region at $a = 0.95$. The maximal value of B is very high ($\sim 9, 90$, and 170 at $\kappa = 10^{-4}$, 10^{-5}, and 10^{-8}, accordingly). The dependency of the resonant internal intensity $\bar{B}$ averaged over the complete particle volume on the lg κ is shown in Figure 5.5*b* for comparison. It is seen

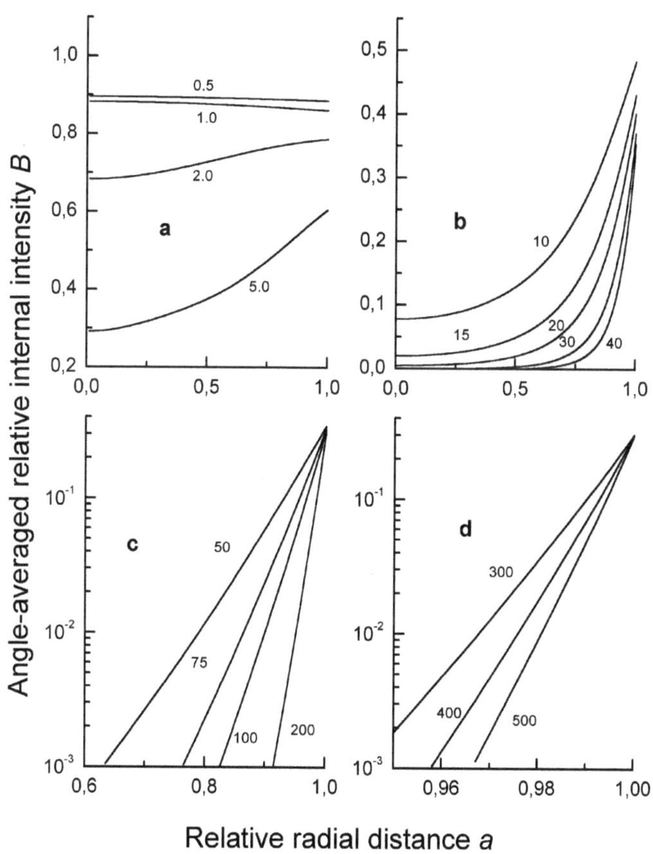

Figure 5.2. Angle-averaged relative intensity of internal field B versus relative distance a from the center of sphere for particles of ice at $\lambda = 10.6\ \mu m$. Figure on curve is size parameter x.

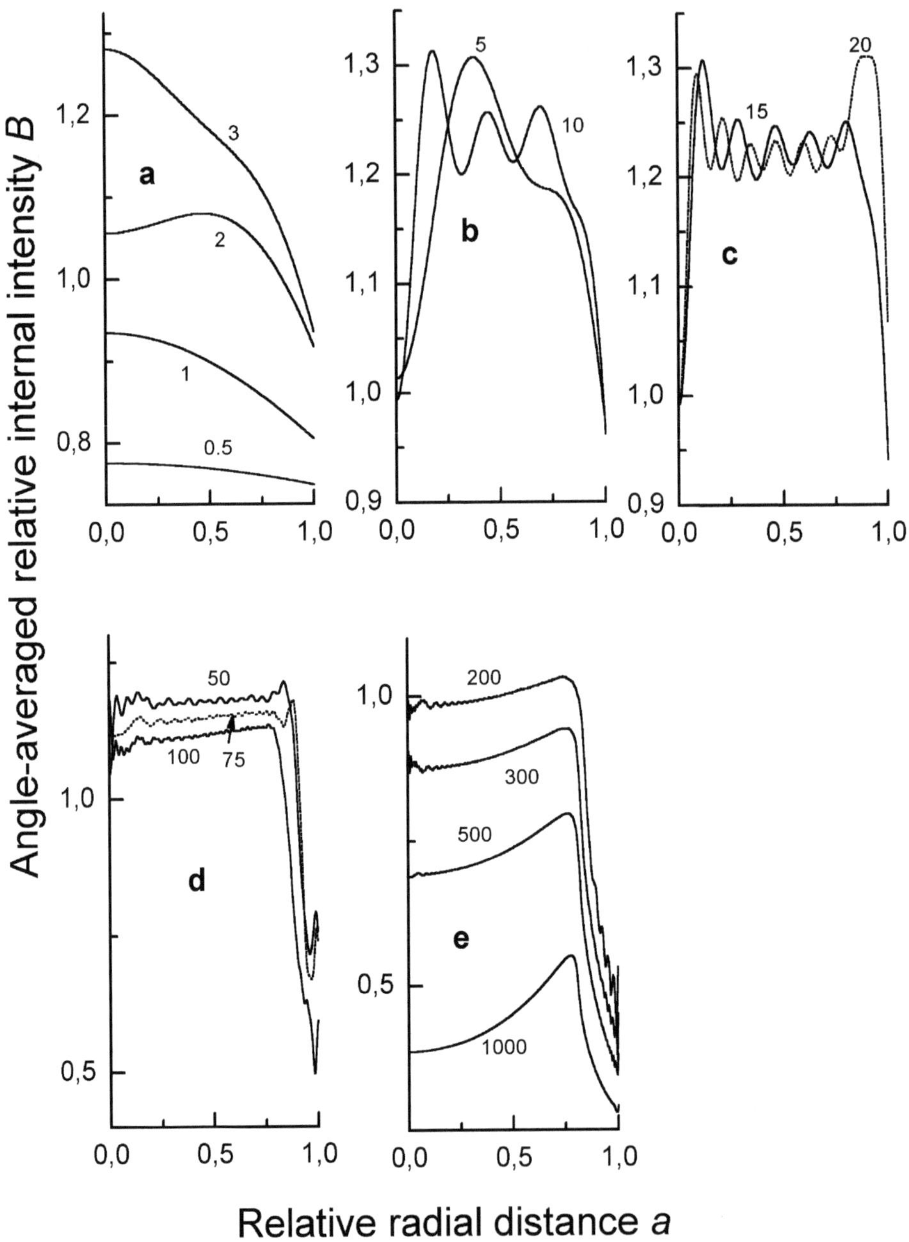

Figure 5.3. Angle-averaged relative intensity of internal field B versus relative distance a from the center of sphere for particles of ice at $\lambda = 2.36\ \mu m$. Figure on curve is size parameter x.

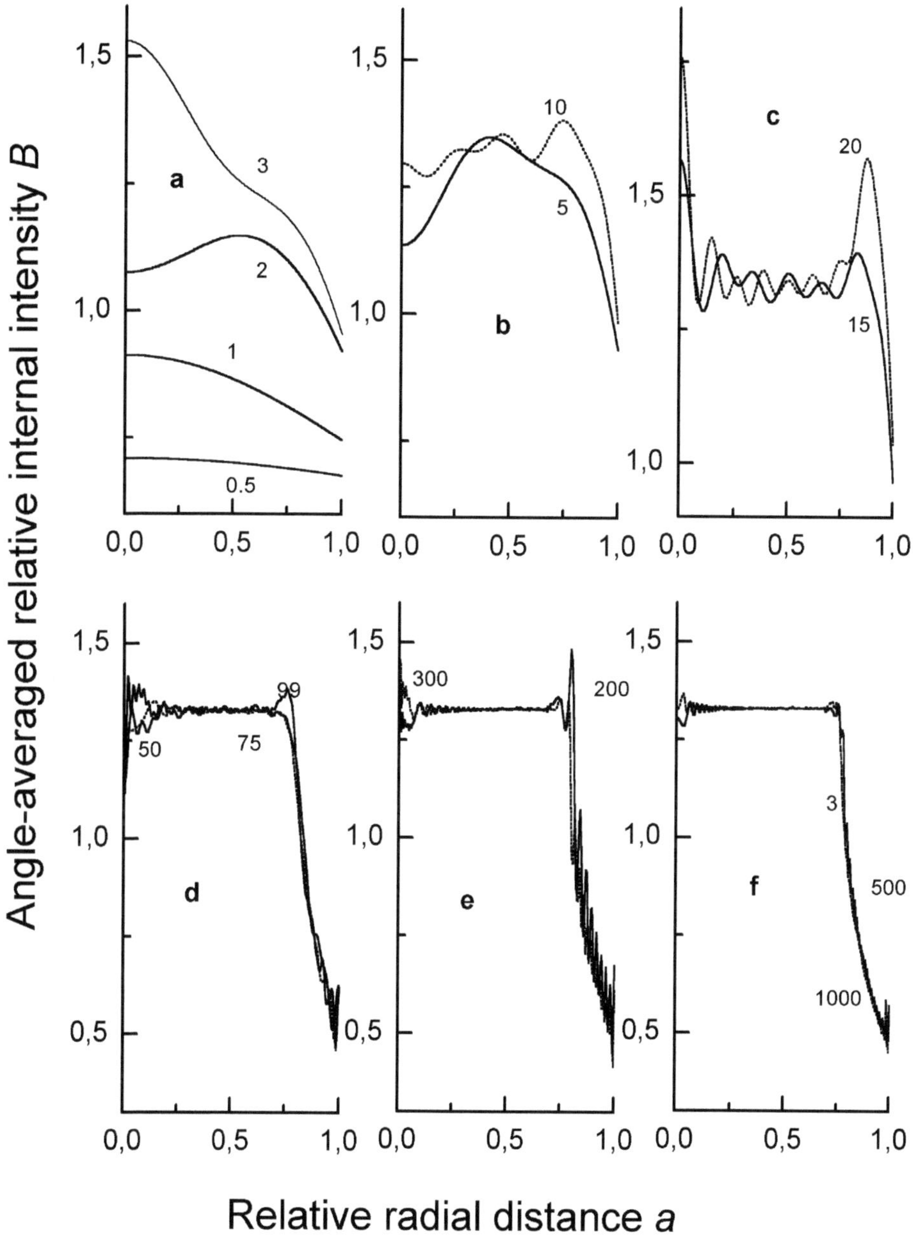

Figure 5.4. Angle-averaged relative intensity of internal field B versus relative distance a from the center of sphere for particles of water at $\lambda = 0.69\ \mu m$. Figure at curve is size parameter x.

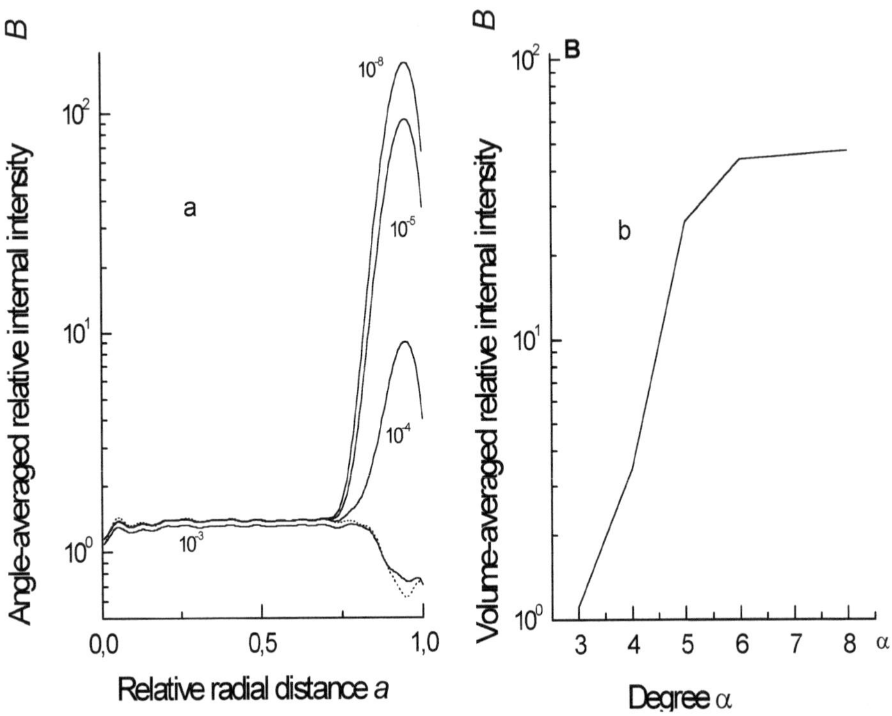

Figure 5.5. Model spherical particle with the real part of the refractive index $n = 1.40$ and the size parameter $x = 31.789230$ (resonance of β_{39}^{31}): (a) Angle-averaged relative intensity of internal field B versus relative distance a from the center of sphere. Figures at curves are the values of absorption index κ. Dashed line is the dependence $B(a)$ for a non-resonant sphere with $N = 1.4 + i10^{-8}$ and $x = 31.80$; (b) Average internal intensity $\bar{B}$ versus the degree α of absorption index κ ($\kappa = 10^{-\alpha}$).

that $\bar{B}$ grows drastically at first and then (at $-\lg\kappa > 6$) reaches the asymptotic value ~ 44.

From these results it follows that the behaviour of $B(a)$ is largely governed by the value of the absorption index κ and is quite different for strongly and low absorbing particles. Our calculations of $B(a)$ for very large x and very low κ have revealed a very interesting regularity. If $n > 1$, then function $B(a) = n$ with great accuracy in the range of the relative internal distance $a = 0 \div 1/n$ with an exponential decrease of B towards the particle surface. This tendency is illustrated in Figure 5.6 ($x = 7500$, $\kappa = 10^{-8}$). If, in contrast, $n < 1$ then such a constant value of $B = n$ is obtained in the range $a = 0 \div 0.98$ with a sharp increase in near-surface region. The small ripple structure added to such behaviour is attenuated while x increases up to several thousands. To our knowledge, although such regularity was described to some extent by others (e.g., Liu, Tan and Tong (2002) in the framework of the ray tracing method) its explanation on the basis of exact theory is absent up to now.

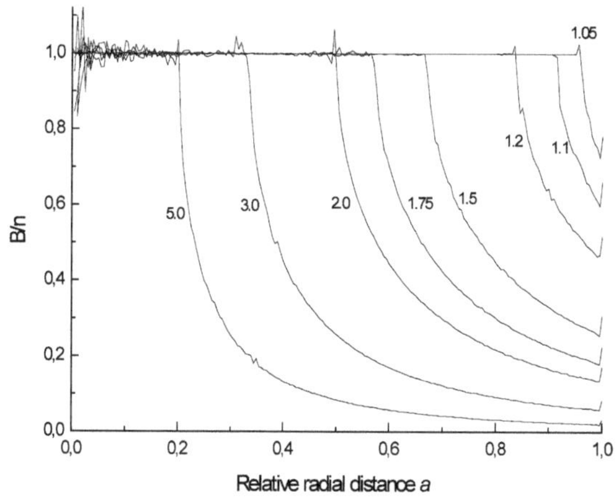

Figure 5.6. Ratio $B(a)/n$ for spherical particles with size parameter $x = 7500$ and $\kappa = 10^{-8}$ versus relative distance a from the center of the sphere. The figures at the curves are the values of real part n of the refractive index.

5.4 THE ABSORPTION ASYMMETRY

5.4.1 Theory

This section is devoted to derivation of expressions describing the absorption power in forward and back hemispheres of spherical particles (Babenko, 1991). Denote the absorption power in the forward (shadow) hemisphere $(0 \le \theta \le \pi/2)$ as W_{forw}, and that in the back (illuminated) hemisphere $(\pi/2 \le \theta \le \pi)$ as W_{back}. We use the surface integral (5.1.17) to calculate these quantities. In doing so the integration surface for W_{forw} is $S_{\text{forw}} = S_1 + S_3$, and for W_{back} is $S_{\text{back}} = S_2 + S_3$, where S_1 is the surface of forward hemisphere with radius R $(a = 1,\ 0 \le \theta \le \pi/2,\ 0 \le \varphi \le 2\pi)$, S_2 is the surface of back hemisphere with radius R $(a = 1,\ \pi/2 \le \theta \le \pi,\ 0 \le \varphi \le 2\pi)$, and S_3 is the plane interface of both hemispheres $(0 \le a \le 1,\ \theta = \pi/2,\ 0 \le \varphi \le 2\pi)$. Clearly the outward normal to S_1 and S_2 is $\mathbf{e}_r$, and that to S_3 is $\pm\mathbf{e}_\theta$ (+ sign for W_{forw} and $-$ sign for W_{back}). Designate the respective integrals as W_1, W_2, and W_3, that is:

$$W_{\text{forw}} = W_1 + W_3 \qquad W_{\text{back}} = W_2 + W_3 \tag{5.4.1}$$

Let us begin with W_1. Applying the mathematics similar to that in Section 5.2 (Eqs. (5.2.1–6)) we have:

$$W_1 = A\,\text{Re}\,\frac{1}{N}\sum_{l=1}^{\infty}\sum_{j=1}^{\infty}\mu_l\mu_j^*\left[i\beta_{jl}\int_0^1(\pi_l\pi_j + \tau_l\tau_j)\,d\mu - \alpha_{jl}\int_0^1(\pi_l\tau_j + \tau_l\pi_j)\,d\mu\right] \tag{5.4.2}$$

where:

$$\alpha_{jl} = (X_l Z_j^* + V_l Y_j^*)_{a=1} \qquad \beta_{jl} = (V_l Z_j^* + X_l Y_j^*)_{a=1}$$

Use the integrals from Appendix B:

$$\int_0^1 (\pi_l \tau_j + \tau_l \pi_j)\, d\mu = \pi_j(0)\pi_l(0) \tag{5.4.3}$$

$$\int_0^1 (\pi_l \tau_j + \tau_l \pi_j)\, d\mu \tag{5.4.4}$$

$$= \begin{cases} [l(l+1)\pi_l(0)\tau_j(0) - j(j+1)\pi_j(0)\tau_l(0)] \,/\, [(l-j)\,(l+j+1)] & j \neq l \\ l^2(l+1)^2/\,(2l+1) & j = l \end{cases}$$

On the other hand (see Eqs.(B.14–15) in Appendix B):

$$\pi_l(0) = \begin{cases} (-1)^{(l-1)/2}\,\nu_l \\ 0 \end{cases} \qquad \tau_l(0) = \begin{cases} 0 & l \text{ odd} \\ (-1)^{l/2}\, l\nu_{l+1} & l \text{ even} \end{cases} \tag{5.4.5}$$

Then after some transformations we obtain:

$$W_1 = A\,\mathrm{Re}\,\frac{1}{N}\left[\sum_{il=1}^{\infty} \frac{i\beta_{ll}}{2l+1} - {\sum_{l=1}^{\infty}}'{\sum_{j=1}^{\infty}}' \frac{\alpha_{jl}}{jl\nu_{j+1}\nu_{l+1}} + {\sum_{l=2}^{\infty}}''{\sum_{j=1}^{\infty}}' \frac{\nu_j\beta_{jl} - \beta_{lj}}{\nu_l(l-j)(j+l+1)}\right] \tag{5.4.6}$$

instead of Eq.(5.4.2), where summations over odd and even indexes are symbolized by $\sum'$ and $\sum''$ respectively. It is easy to see that the first double sum in Eq.(5.4.3) decomposes into a product of two single sums:

$$\mathrm{Re}\,\frac{1}{N}{\sum_{l=1}^{\infty}}'{\sum_{j=1}^{\infty}}' \frac{\alpha_{jl}}{jl\nu_{j+1}\nu_{l+1}} \tag{5.4.7}$$

$$= \mathrm{Re}\,\frac{1}{N}\left[{\sum_{l=1}^{\infty}}' \frac{X_l\nu_l}{l(l+1)} \cdot {\sum_{l=1}^{\infty}}' \frac{Z_l^*\nu_l}{l(l+1)} + {\sum_{l=1}^{\infty}}' \frac{V_l\nu_l}{l(l+1)} \cdot {\sum_{l=1}^{\infty}}' \frac{Y_l^{*}\nu_l}{l(l+1)}_{a=1}\right]$$

Consider W_2 now. This integral differs from W_1 in the integration limits only. However, with $(-1 \div 0)$ in place of $(0 \div 1)$ the integrals (5.4.3) and (5.4.4, $j \neq l$) change sign with integral (5.4.4, $j = l$) remaining unchanged. Then we have:

$$W_2 = A\,\mathrm{Re}\,\frac{1}{N}\left[\sum_{l=1}^{\infty} \frac{i\beta_{ll}}{2l+1} - {\sum_{l=1}^{\infty}}'{\sum_{j=1}^{\infty}}' \frac{\alpha_{jl}}{jl\nu_{j+1}\nu_{l+1}} - {\sum_{l=2}^{\infty}}''{\sum_{j=1}^{\infty}}' \frac{\nu_j\beta_{jl} - \beta_{lj}}{\nu_l(l-j)(j+l+1)}\right] \tag{5.4.8}$$

as a result, that means, W_2 differs from W_1 by sign before the double sums.

Let us go to W_3:

$$W_3 = -\frac{1}{2}\,\mathrm{Re}\int\limits_{S_3} (\mathbf{E}\times\mathbf{H}^*)_{\mu=0}\cdot \mathbf{e}_\theta dS = \frac{R^2}{2}\,\mathrm{Re}\int\limits_0^1 a da \int\limits_0^{2\pi} \left(E_r H_\varphi^* - E_\varphi H_r^*\right)_{\mu=0} d\varphi. \tag{5.4.9}$$

Substituting the field components at $\mu = 0$ (with account of Eq.(5.4.5)):

$$E_r|_{\mu=0} = \frac{E_0\cos\varphi}{(Nxa)^2}\sum_{l=1}^{\infty}{}' \nu_l Z_l \qquad H_\varphi|_{\mu=0} = \frac{E_0\cos\varphi}{Na}\sqrt{\frac{\varepsilon_0}{\mu_0}}\left[\sum_{l=1}^{\infty}{}' \frac{Y_l}{l\nu_{l+1}} + \sum_{l=2}^{\infty}{}'' \frac{Z_l}{\nu_l}\right]$$

$$H_r|_{\mu=0} = \frac{E_0 N\sin\varphi}{(Nxa)^2}\sqrt{\frac{\varepsilon_0}{\mu_0}}\sum_{l=1}^{\infty}{}' \nu_l X_l \quad E_\varphi|_{\mu=0} = -\frac{E_0\sin\varphi}{Nxa}\sum_{l=1}^{\infty}{}' \frac{V_l}{l\nu_{l+1}} + \sum_{l=2}^{\infty}{}'' \frac{X_l}{\nu_l} \tag{5.4.10}$$

into Eq.(5.4.9), after integration over angle φ we have:

$$W_3 = \frac{A}{x}\,\mathrm{Re}\int\limits_0^1 \frac{da}{a^2}\left\{\frac{1}{N^2}\sum_{l=1}^{\infty}\nu_l Z_l(a)\left[\sum_{j=1}^{\infty}{}' \frac{Y_j(a)\nu_j}{j(j+1)} + \sum_{j=2}^{\infty}{}'' \frac{Z_j(a)}{\nu_j}\right]^* \right. \tag{5.4.11}$$

$$\left. + \frac{1}{NN^*}\sum_{l=1}^{\infty}\nu_l X_l(a)\left[\sum_{j=1}^{\infty}{}' \frac{V_j(a)\nu_j}{j(j+1)} + \sum_{j=2}^{\infty}{}'' \frac{X_j(a)}{\nu_j}\right]^*\right\}$$

We can, in principle, interchange integral and sum here (as in W_1), but the integrals $\int\limits_0^1 \psi_l(Nxa)\psi_j^*(Nxa)a^{-2}da$ which will appear as a result have neither close analytical representation nor some recursive relations. So W_3 must be calculated through numerical integration. Note that this integration occurs in a "quiet" region, where the interference structure of the internal field appears only slightly.

From a formal point of view, the factor $1/a^2$ in Eq.(5.4.11) gives rise to singularity as $a \to 0$. But from the asymptotes of the Riccati–Bessel functions:

$$\lim_{z\to 0}\psi_l(z) = \frac{z^{l+1}}{(2l+1)!!} \qquad \lim_{z\to 0}\psi_l'(z) = \frac{(l+1)\,z^l}{(2l+1)!!} \tag{5.4.12}$$

it follows that:

$$\frac{1}{a^2}\sum_{l=1}^{\infty}\nu_l\begin{pmatrix}Z_l\\X_l\end{pmatrix} \to N^2x^2\begin{pmatrix}d_1\\c_1\end{pmatrix}$$

$$\left[\sum_{j=1}^{\infty}{}' \frac{\nu_j}{j(j+1)}\begin{pmatrix}Y_l\\V_l\end{pmatrix} + \sum_{j=2}^{\infty}{}'' \frac{1}{\nu_j}\begin{pmatrix}Z_l\\X_l\end{pmatrix}\right]^* \to 0$$

and, consequently, the integrand of W_3 and a go to zero simultaneously.

Therefore, the relations (5.4.6, 8, 11) provide a means of analytical calculation of W_{abs} for illuminated and shadow hemispheres and of the absorption asymmetry coefficient $\eta = W_{\text{forw}}/W_{\text{back}}$. The calculation of functions X_l, Y_l, Z_l, and V_l in integrand points a_i permits the parallel calculation of the absorption power in a spherical region of radius aR (see Eq.(5.2.10)):

$$W_{\text{abs}} = 2A\,\text{Re}\left\{\frac{i}{N}\sum_{j=1}^{\infty}(2l+1)\left[V_l(a)Z_l^*(a) - X_l(a)Y_l^*(a)\right]\right\} \tag{5.4.13}$$

and of the angle-averaged internal intensity (see Eq.(5.3.5)).

Let us obtain the relations for the intensity of the internal field averaged over back $(\bar{B}_{\text{back}})$ and forward $(\bar{B}_{\text{forw}})$ hemispheres. We can write expressions for the absorption power in the hemisphere by analogy with Eq.(5.3.18):

$$\begin{pmatrix} W_{\text{back}} \\ W_{\text{forw}} \end{pmatrix} = \frac{2\pi n\kappa}{\lambda} I_0 V \begin{pmatrix} \bar{B}_{\text{back}} \\ \bar{B}_{\text{forw}} \end{pmatrix} \tag{5.4.14}$$

Considering that $\bar{B}_{\text{back}} = \eta \bar{B}_{\text{forw}}$ and $W_{\text{abs}} = W_{\text{back}} + W_{\text{forw}}$, we obtain:

$$\bar{B}_{\text{forw}} = \frac{2\bar{B}}{1+\eta} \qquad \bar{B}_{\text{back}} = \frac{2\eta\bar{B}}{1+\eta} \tag{5.4.15}$$

The algorithm and code KPOG for computing the values η, W_{back}, W_{forw}, $W(a)$, and $B(a)$ are described in Appendix D. Here we shall mention some computational features only. After introducing the log-derivatives of the Riccati–Bessel functions and ratio R_l (see Section 1.5.2.3) and taking into account the explicit relations for the amplitude coefficients of the internal field (Eq.(5.2.9)) we have:

$$\begin{aligned} X_l(a) &= \frac{iNR_l(Nxa)}{\xi_l(x)\left[G_l(x) - ND_l(Nx)\right]} & Y_l(a) &= X_l(a)D_l(Nxa) \\ Z_l(a) &= \frac{iNR_l(Nxa)}{\xi_l(x)\left[NG_l(x) - D_l(Nx)\right]} & V_l(a) &= Z_l(a)D_l(Nxa) \end{aligned} \tag{5.4.16}$$

Hence, the arrays of $D_l(Nx)$, $G_l(x)$, $\xi_l(x)$ and ν_l are necessary for computing $W_{1,2}$. The values of functions $D_l(Nxa_i)$ and $R_l(Nxa_i)$ at each division point a_i of the integration interval $a = 0 \div 1$ are needed for numerical evaluation of the integral W_3. The process of generating of arrays $D_l(Nx)$, $G_l(x)$, $\xi_l(x)$, $D_l(Nxa_i)$, and $R_l(Nxa_i)$ is conventional and it is described above (Section 1.5). The coefficients ν_l were computed using up-recurrence:

$$\nu_{l+1} = \frac{l+1}{\nu_l} \tag{5.4.17}$$

5.4.2 Some results

We conducted the extensive calculations of the absorption asymmetry for the spherical homogeneous particles of ice, water, soot, aluminium, and aluminium oxide at laser wavelengths λ =1.06, 2.36, and 10.6 μm over a wide range of size parameters. As the optical constants for the five matters mentioned above we used the data from reviews of Irvine and Pollack (1968) for water and of Warren (1984) for ice (with exception of ice for $\lambda = 10.6$ μm, where the data of Schaaf and Williams (1973) were used in analogy with previous papers on radiation absorption by ice particles). Data concerning n and κ of Al_2O_3 are from papers of Belov (1989) and Lingart *et al.* (1982), accordingly. Data from the reference book of Palik (1985) were utilized for aluminium. Our estimates based on the analysis of a great deal of experimental results of many authors were used for soot that is characterized by great variability of optical constants depending on conditions of the sample preparation. As seen from the Table 5.1, the very broad range of optical constants are covered by these matters and wavelengths. Because our concern here is only with absorption, the overall range of constants may be divided into four parts in accordance with the value of κ: (1) very high absorption (Al and soot at all wavelengths); (2) high absorption (water, ice, and Al_2O_3 at λ = 10.6 μm); (3) moderate absorption ($\kappa \sim 10^{-4}$, water and ice at $\lambda = 2.36$ μm); and (4) low absorption ($\kappa \sim 10^{-6} \div 10^{-8}$, water and ice at λ =1.06 μm and Al_2O_3 at λ = 1.06 μm and 2.36 μm). As will be shown below, the behaviour of the calculated dependencies is largely governed by the value of κ.

The dependencies of $\bar{B}_{\rm forw}$, $\bar{B}_{\rm back}$, and η versus size parameter x are depicted in Figures 5.7–5.9. The range of x was very broad (from 10^{-2} and even less for strongly absorbing particles) to 10^3, that is from the Rayleigh region to the geometro-optical one. Step over x was very small and sufficient to reveal the ripple structure (except the variants with $\kappa \sim 10^{-6} \div 10^{-8}$, where this structure was not fully revealed). Go to the analysis of the results.

Begin with $\lambda = 10.6$ μm (Figure 5.7). Aluminium and soot have very great values of κ at this wavelength; κ for water, ice, and Al_2O_3 is in the range 0.05-0.13; κ-values for ice

Table 5.1. Optical constants used in calculations (the first line holds a value of n, the second line holds a value of k).

	Wavelength (μm)			
	0.69	1.06	2.36	10.6
Water	1.33	1.319	1.274	1.173
	$3{\cdot}10^{-8}$	$1.46{\cdot}10^{-8}$	$7.6{\cdot}10^{-4}$	0.0823
Ice	1.3071	1.3005	1.246	1.097
	$2.4{\cdot}10^{-8}$	$1.96{\cdot}10^{-6}$	$5.8{\cdot}10^{-4}$	0.134
Aluminium oxide	1.762	1.755	1.728	0.78
	$2.0{\cdot}10^{-8}$	$2.25{\cdot}10^{-8}$	$2.0{\cdot}10^{-8}$	0.05
Aluminium	1.74	1.21	2.83	30.63
	8.21	11.8	24.4	64.84
Soot	1.69	1.80	2.0	2.8
	0.65	0.65	0.8	2.0

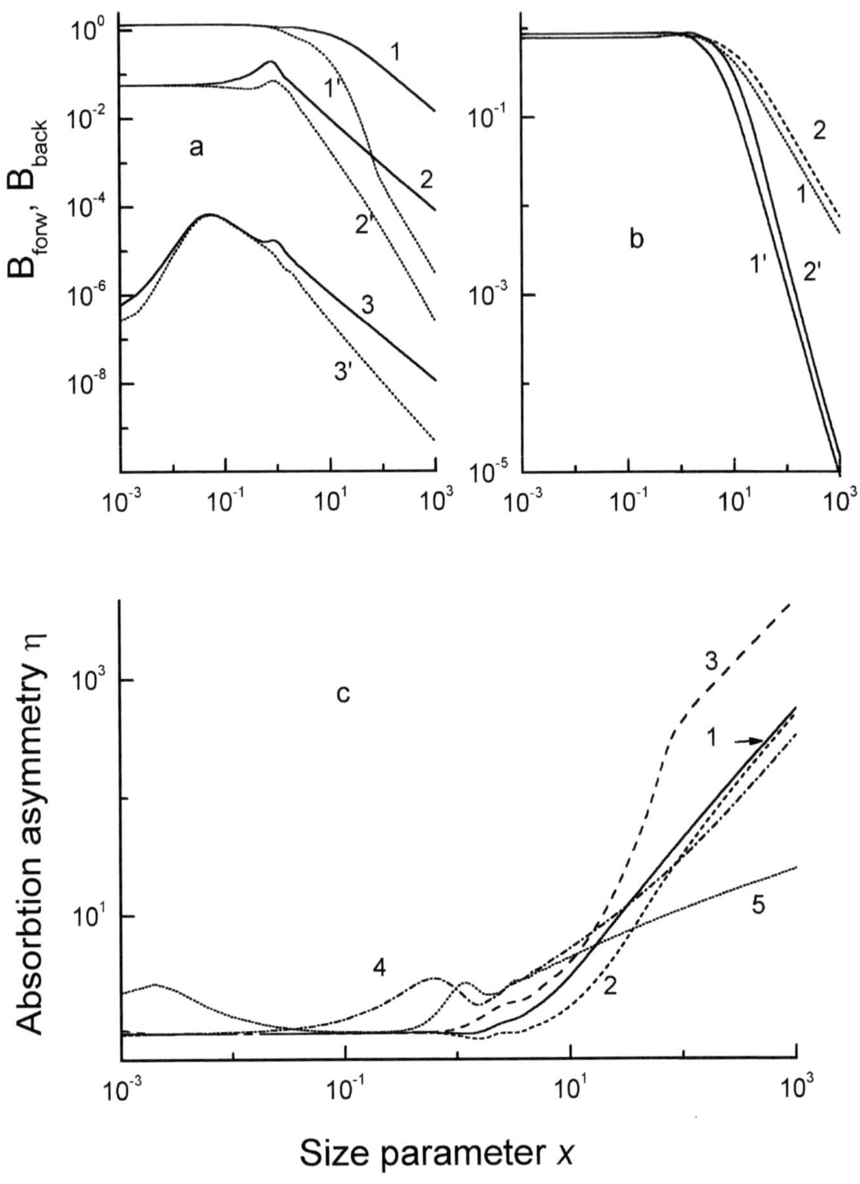

Figure 5.7. Internal intensity averaged over forward ($\bar{B}_{\text{forw}}$) and back ($\bar{B}_{\text{back}}$) hemispheres for: (*a*) ice (*1*, *1'*), water (*2*, *2'*); (*b*) Al_2O_3 (*1*, *1'*), soot (*2*, *2'*), Al (*3*, *3'*); and (*c*) absorption asymmetry η for spherical particles of ice (*1*), water (*2*), Al_2O_3 (*3*), soot (*4*), and Al (*5*) versus size parameter x at wavelength $\lambda = 10.6\ \mu m$.

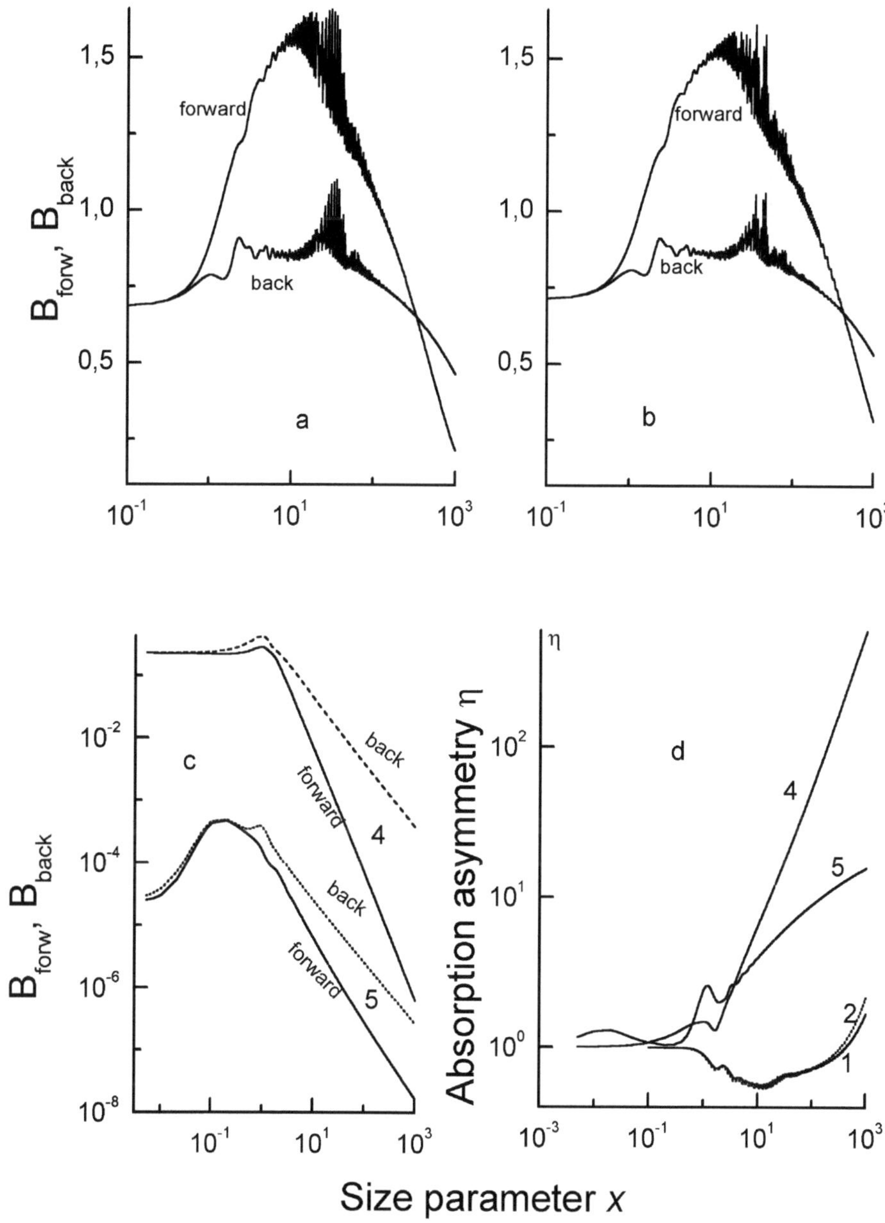

Figure 5.8. Internal intensity averaged over forward ($\bar{B}_{\text{forw}}$) and back ($\bar{B}_{\text{back}}$) hemispheres for water (*a*), ice (*b*), soot (*c*, *4*), and Al (c, *5*) particles and (*d*) absorption asymmetry η for spherical particles of ice (*1*), water (*2*), soot (*4*), and Al (*5*) versus size parameter x at wavelength $\lambda = 2.36\ \mu m$.

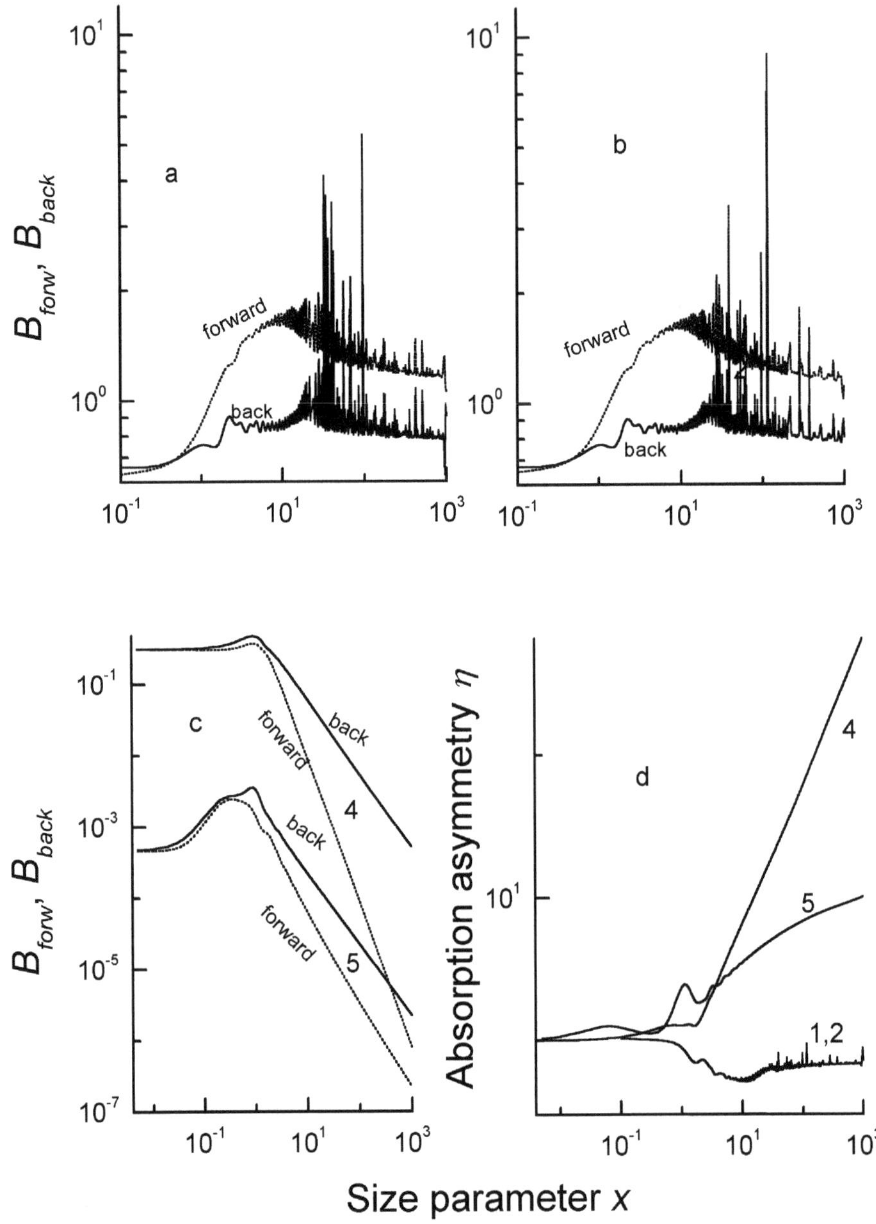

Figure 5.9. Internal intensity averaged over forward ($\bar{B}_{\text{forw}}$) and back ($\bar{B}_{\text{back}}$) hemispheres for water (*a*), ice (*b*), soot (*c*,*4*), Al (*c*,*5*), and (*d*) absorption asymmetry η for spherical particles of ice (*1*), water (*2*), soot (*4*), and Al (*5*) versus size parameter x at wavelength $\lambda = 1.06\ \mu m$.

and water are very close. As expected, for ice and water particles $\bar{B}_{\text{forw}} = \bar{B}_{\text{back}}$ (that is, asymmetry $\eta = 1$) in the Rayleigh region (up to $x \sim 1$). At larger x the dependencies $\bar{B}_{\text{forw}}(x)$ and $\bar{B}_{\text{back}}(x)$ behave quite differently. The intensity averaged over the illuminated hemisphere $\bar{B}_{\text{back}}$ decreases as $1/x$, but $\bar{B}_{\text{forw}}$ decreases much faster (as $\sim 1/x^2$). As a consequence, the asymmetry grows proportionally to x and reaches great values (~ 1000) at $x \sim 1000$. The behaviour of $\bar{B}_{\text{forw}}(x)$ and $\bar{B}_{\text{back}}(x)$ for other substances is approximately the same. Note also, that $\bar{B}_{\text{forw}}(x)$ and $\bar{B}_{\text{back}}(x)$ have the small maximum near $x \sim 0.8$ for soot, and the divergence of these quantities (that is, the non-homogeneity of absorption) begins at small $x \sim 7 \cdot 10^{-2}$. The asymmetry grows as $1/x$ with a further increase of x. A different situation arises with Al particles – the small asymmetry ($\eta \sim 2 - 3$) is detected for very small $x \sim 10^{-3}$, asymmetry is practically absent in the range $x \sim 10^{-2} \div 5 \cdot 10^{-1}$, and the increase of η as $x^{0.4}$ takes place at great x, which is much slower than for other substances. The result is, the absorption asymmetry for Al spheres at the end of x-range is not high ($\eta \sim 20$).

Go now to $\lambda = 2.36\mu m$ (Figure 5.8). Water and ice absorb moderately ($\kappa \sim 5 \cdot 10^{-4}$) at this wavelength; absorption in soot and Al is very high and that in Al_2O_3 is very low ($\sim 10^{-8}$). Our numerical experiments showed that calculations of absorption asymmetry at such small κ may be unstable at some x-values. The treatment of results at small x is severely complicated by the presence of the morphologically dependent resonances (Hill and Benner, 1988). It would be reasonable in this situation to use the volume Pointing integral over heat sources (see Section 5.1) instead of the surface one.

Figures 5.8*a,b*, where the dependencies $\bar{B}_{\text{forw}}(x)$ and $\bar{B}_{\text{back}}(x)$ are plotted for water and ice particles, is very demonstrative. The curve $\bar{B}_{\text{forw}}(x)$ has the distinct main maximum at $x \sim 10$ and decrease with z. But absorption $\bar{B}_{\text{back}}(x)$ in the illuminated hemisphere has no main maximum and remained at a level $\sim 0.7 \div 0.8$ at the broad range of x ($2 \div 100$). In connection with this, the absorption asymmetry (Figure 5.8.*c*) is less than 1 up to $x \sim 400$, that is the release of energy in the shadow hemisphere is markedly greater than in the illuminated one up to this great size parameter. The dependencies $\bar{B}_{\text{forw}}(x)$, $\bar{B}_{\text{back}}(x)$, and $\eta(x)$ for strongly absorbing particles of soot and Al do not contribute something tangibly new in comparison with the results for $\lambda = 10.6\ \mu m$.

The results of calculations for ice, water, soot, and Al particles at $\lambda = 1.06\ \mu m$ are presented in Figure 5.9. Soot and Al remain strongly absorbing, as at the two previous wavelengths. The absorption coefficient κ for water and ice is 2.5 orders of magnitude lower than for $\lambda = 2.36\ \mu m$. This results in growth of oscillations, related to the morphological dependent resonances. Note that we were not making efforts to resolve the resonant structure totally though the calculation was conducted with rather small steps over x. Hence, the high and very narrows peaks in Figures 5.9*a,b* are random to some extent and represent, at most, the presence and overall properties of a resonant structure. It is seen, in particular, that the region of maximal resonant manifestation is shifted to the greater x compared with results for $\lambda = 2.36\ \mu m$, and the resonances are not suppressed at large x though they become lower. The absorption asymmetry η is lower than 1 for the water and ice particles for all x-values (excluding the Rayleigh region), that is, the shadow hemisphere absorb radiation more than illuminated one up to $x \sim 1000$. In the very broad size range ($x \sim 20 \div 1000$) the absorption asymmetry depends only weakly on x and is equal approximately to 0.7. It is notable that the oscillating structure of curve $\eta(x)$ is much less pronounced than that of the curves for internal fields in both hemispheres.

Thus, it may be concluded that the behaviour of dependencies of absorption asymmetry on the size parameter is largely governed by the value of the imaginary part κ of the complex refractive index. The particles with strong absorption have pronounced absorption asymmetry (η may be as mush as $10^2 \div 10^3$), but the relative intensity of the internal field is very low in this case. On the contrary, the asymmetry is quite small for particles with low and moderate absorption, the intensity of the average internal field being relatively high.

5.5 ABSORPTION IN THE RESTRICTED REGION OF THE SPHERICAL PARTICLE

The absorption of light in illuminated and shadow hemispheres of the spherical particle was investigated in Section 5.4. However, on frequent occasions it is of interest to calculate the absorption in the restricted region not coinciding with the hemisphere. This is of particular importance for the morphologically dependent resonances. As stated in Section 5.2, the problem of calculation of the absorption power in the arbitrary region of the spherical particle has no analytical solution in the general case and in practice reduces to the numerical integration of the internal field intensity over some volume. At small absorption indices, that is just when the morphological resonances are most conspicuous, such integration is difficult due to sharply oscillating structures of the internal field.

Nevertheless, this problem may be substantially simplified considering that the resonances are situated on the sphere axis coinciding with the direction of propagation of the incident wave. In this case we can take the conical body of revolution V with the revolution axis made coincident with the z-axis.

To be precise, the geometry of the problem is as follows. The monochromatic (exp $(-i\omega t)$) linearly polarized (the vector $\mathbf{E}$ oscillates along x-axis) electromagnetic wave with the amplitude E_I incidents upon the spherical homogeneous particle with radius R and complex refractive index $N = n + i\kappa$. The particle center is in the origin of Cartesian (x, y, z) and spherical (r, θ, φ) coordinate systems. The volume V is restricted by the conical surface $\theta = \theta_0$ and by two spherical surfaces $r = R_1$ and $r = R_2$. Clearly the variant discussed in Section 5.4 is a special case of this problem at $\theta = \pi/2$, $R_1 = 0$, and $R_2 = R$. Because our geometry is totally the same as that in the book of Bohren and Huffman (1983), we can use their expression for the expansions of the internal electric $\mathbf{E}_{\text{int}}$ and magnetic $\mathbf{H}_{\text{int}}$ fields over the basis of vector spherical functions (see Eq.(5.2.2) for designations):

$$\begin{pmatrix} \mathbf{E}_{\text{int}} \\ \sqrt{\mu_0/\varepsilon_0}\mathbf{H}_{\text{int}} \end{pmatrix} = E_I \begin{pmatrix} 1 \\ N \end{pmatrix} \sum_{l=1}^{\infty} \mu_l \left[\begin{pmatrix} \alpha_l^{31} \\ -\beta_l^{31} \end{pmatrix} \mathbf{M}_{o1l}^{(1)}(Nk_0\mathbf{r}) - i \begin{pmatrix} \beta_l^{31} \\ \alpha_l^{31} \end{pmatrix} \mathbf{N}_{e1l}^{(1)}(Nk_0\mathbf{r}) \right] \tag{5.5.1}$$

Radial components of fields $\mathbf{E}_{\text{int}}$ and $\mathbf{H}_{\text{int}}$ in the spherical coordinate system are:

$$\begin{pmatrix} E_r \\ \sqrt{\mu_0/\varepsilon_0}H_r \end{pmatrix} = \frac{E_I \sin\theta}{(Nax)^2}\begin{pmatrix} \cos\varphi \\ N\sin\varphi \end{pmatrix}\sum_{l=1}^{\infty} i^{l-1}\begin{pmatrix} Z_l(a) \\ X_l(a) \end{pmatrix}\pi_l \tag{5.5.2}$$

where $Z_l(a)$ and $X_l(a)$ are defined by Eq.(5.2.4) and a is the relative radial coordinate. The other components are described by Eqs.(5.2.3).

In line with Eq.(5.1.15) the absorption power inside a volume surrounded by a closed surface S is expressed in terms of the Pointing surface integral:

$$W_{\text{abs}} = -\frac{1}{2}\,\text{Re}\int_S (\mathbf{E}\times\mathbf{H}^*)\cdot\mathbf{n}dS \tag{5.5.3}$$

Clearly in our case $S = S_1 + S_2 + S_3$, where S_1 is some part of the internal spherical surface $(a = a_1 = R_1/R,\ 0 \le \theta \le \theta_0,\ 0 \le \varphi \le 2\pi)$, S_2 is some part of the external spherical surface $(a = a_2 = R_2/R,\ 0 \le \theta \le \theta_0,\ 0 \le \varphi \le 2\pi)$, and S_3 is the conical surface $(a_1 \le a \le a_2,\ \theta = \theta_0,\ 0 \le \varphi \le 2\pi)$. The outward normal $\mathbf{n}$ coincides with the unit vectors $-\mathbf{e}_r, \mathbf{e}_r, \mathbf{e}_\theta$ for S_1, S_2, S_3, respectively. If the appropriate integrals designate as W_1, W_2, and W_3, then $W_{\text{abs}} = W_1 + W_2 + W_3$.

If we substitute the expansions (5.2.3) into the expression for the Pointing integral W_1:

$$W_1 = \frac{1}{2}\,\text{Re}\int_0^{\theta_0}\int_0^{2\pi}\left(E_\theta H_\varphi^* - E_\varphi H_\theta^*\right)_{a=a_1} R^2\sin\theta\,d\theta\,d\varphi \tag{5.5.4}$$

and integrate over the azimuthal angle φ, then:

$$\begin{aligned} W_1 = A\,\text{Re}\,\frac{1}{N}\int_{\nu_0}^{1}\sum_{l=1}^{\infty}\sum_{j=1}^{\infty}\mu_l\mu_j^*\,&\left[(X_l\tau_l - iV_l\pi_l)\,(Z_j\pi_j - iY_j\tau_j)^*\right. \\ &\left.+i\,(X_l\pi_l - iV_l\tau_l)\,(Z_j\tau_j - iY_j\pi_j)^*\right]_{a=a_1} d\mu \end{aligned} \tag{5.5.5}$$

where $\nu_0 = \cos\theta_0$ and $A = \pi E_I^2\sqrt{\varepsilon_0/\mu_0}/2k_0^2$. Rearranging the terms and interchanging integration and summation, we have as a result:

$$W_1 = A\,\text{Re}\,\frac{1}{N}\sum_{l=1}^{\infty}\sum_{j=1}^{\infty}\mu_l\mu_j^*\left[\alpha_{jl}(a_1)f_{jl}(\nu_0) - i\beta_{jl}(a_1)h_{jl}(\nu_0)\right] \tag{5.5.6}$$

where:

$$\alpha_{jl}(a_1) = X_l(a_1)Z_j^*(a_1) + V_l(a_1)Y_j^*(a_1) \qquad \beta_{jl}(a_1) = V_l(a_1)Z_j^*(a_1) - X_l(a_1)Y_j^*(a_1) \tag{5.5.7}$$

$$h_{jl}(\nu_0) = \int_{\nu_0}^{1}(\pi_l\pi_j + \tau_l\tau_j)\,d\mu \qquad f_{jl}(\nu_0) = \int_{\nu_0}^{1}(\pi_l\tau_j + \tau_l\pi_j)\,d\mu \tag{5.5.8}$$

In perfect analogy we have:

$$W_2 = A\,\mathrm{Re}\,\frac{1}{N}\sum_{l=1}^{\infty}\sum_{j=1}^{\infty}\mu_l\mu_j^*\left[-\alpha_{jl}(a_2)f_{jl}(\nu_0)+i\beta_{jl}(a_2)h_{jl}(\nu_0)\right] \tag{5.5.9}$$

Calculation of the definite integral $f_{jl}(\nu_0)$ should present no problems, because it is known (Appendix B, Eq.(B.38)) that $\int(\pi_l\tau_j+\tau_l\pi_j)\,d\mu = -\left(1-\mu^2\right)\pi_j\pi_l$. Consequently, $f_{jl}(\nu_0) = \left(1-\nu_0^2\right)\tau_j(\nu_0)\pi_l(\nu_0)$. The calculation of $h_{jl}(\nu_0)$ is much more complicated. The indefinite integral (Appendix B, Eq.(B.39)):

$$\int(\pi_l\pi_j+\tau_l\tau_j)\,d\mu = \frac{1-\mu^2}{(l-j)\,(j+l+1)}\left[j(j+1)\tau_l\pi_j - l(l+1)\pi_l\tau_j\right] \tag{5.5.10}$$

holds true for $l \neq j$ only. So we can put at $l \neq j$:

$$h_{jl}(\nu_0) = \frac{1-\nu_0^2}{(l-j)\,(j+l+1)}\left[l(l+1)\pi_l(\nu_0)\tau_j(\nu_0) - j(j+1)\pi_j(\nu_0)\tau_l(\nu_0)\right] \tag{5.5.11}$$

and use the approach described by Wiscombe and Chylek (1977) for $l = j$. Designate the indefinite integral (5.5.10) at $l = j$ as $J_l(\mu)$ and the auxiliary indefinite integral $\int\left(1-\mu^2\right)\pi_l^2 d\mu$ as $T_l(\mu)\,/\,(2l+1)$. It is evident that these functions are connected by:

$$J_l(\mu) = \frac{l(l+1)}{2l+1}H_l(\mu) - \left(1-\mu^2\right)\pi_l(\mu)\,\tau_l(\mu) \tag{5.5.12}$$

The function $H_l(\mu)$ obeys the following upward recurrence (see Appendix B, Eq.(B.42)):

$$(l-1)\,H_l(\mu) = (l+1)H_{l-1}(\mu) + \left(1-\mu^2\right)\left\{\mu\left[(l+1)\pi_{l-1}^2 + (l-1)\pi_l^2\right] - 2l\pi_{l-1}\pi_l\right\}$$

$$H_1(\mu) = 3\int\left(1-\mu^2\right)\pi_1^2 d\mu = \mu\left(3-\mu^2\right) \tag{5.5.13}$$

Hence, using the initial values $H_1(1) = 2$ and $H_1(\nu_0) = \nu_0\left(3-\nu_0^2\right)$ one can find $H_l(1)$ and $H_l(\nu_0)$ via the recursion (5.5.13), then $J_l(1)$ and $J_l(\nu_0)$ via Eq.(5.5.12) and, finally, $h_{ll}(\nu_0) = J_l(1) - J_l(\nu_0)$.

Thus, the calculation of integrals $W_{1,2}$ reduces to the computing of the double series over the combinations of Bessel and Legendre functions. The Pointing integral for the conical surface S_3 takes the form:

$$W_3 = \frac{R^2}{2}\sqrt{1-\nu_0^2}\,\mathrm{Re}\int_{a_1}^{a_2}\int_0^{2\pi}\left(E_r H_\varphi^* - E_\varphi H_r^*\right)_{\theta=\theta_0} a\,da\,d\varphi \tag{5.5.14}$$

Substitution of the field components (Eqs.(5.2.3 and 5.5.2) at $\theta = \theta_0$ and integration over angle φ yields:

$$W_3 = \frac{A}{x}\left(1-\nu_0^2\right)\mathrm{Re}\int_{a_1}^{a_2}\left[\frac{1}{N^2}S_1(a)\,S_2^*(a) + \frac{1}{|N|^2}S_3^*(a)\,S_4(a)\right]\frac{da}{a^2} \tag{5.5.15}$$

where:

$$S_1(a) = \sum_{l=1}^{\infty} i^{l-1} Z_l(a)\pi_l(\nu_0) \qquad S_2(a) = \sum_{l=1}^{\infty} \gamma_l \left[Z_l(a)\tau_l(\nu_0) - iY_l(a)\pi_l(\nu_0)\right]$$

$$S_3(a) = \sum_{l=1}^{\infty} i^{l-1} X_l(a)\pi_l(\nu_0) \qquad S_4(a) = \sum_{l=1}^{\infty} \gamma_l \left[X_l(a)\tau_l(\nu_0) - iV_l(a)\pi_l(\nu_0)\right] \tag{5.5.16}$$

Unlike the integrals $W_{1,2}$ there is no sense in interchanging the summation and integration here because the definite integrals:

$$\int_{a_1}^{a_2} \psi_j\left(Nax\right)\psi_l^*\left(Nax\right)\frac{da}{a^2} \tag{5.5.17}$$

resulting from this change have, as far as we know, neither closed analytical representation nor recursive relations over indices. So, the formula (5.5.15) must be taken as ultimate.

Let us dwell on some computational aspects of the problem. Introducing the log-derivatives $D_l(Nx)$, $D_l(Nax)$, $G_l(x)$, and the function ratio $R_l(Nax)$ we conclude that arrays $D_l(Nx)$, $G_l(x)$, $\xi_l(x)$, $D_l(Na_1x)$, $D_l(Na_2x)$, $R_l(Na_1x)$, $R_l(Na_2x)$, $\pi_l(\nu_0)$, $\tau_l(\nu_0)$, and $h_{ll}(\nu_0)$ are necessary to obtain $W_{1,2}$, and the values of functions $D_l(Na_ix)$ and $R_l(Na_ix)$ at each dividing point a_i of the integration interval $a_1 \div a_2$. The upmost values of l-index, that is the number of terms in sums over l large enough for convergence of these sums, is defined by the relation $L = \min\left[fL_W(x), fL_W(|N|\,xa)\right]$, where $L_W(x) = x + 4.05x^{0.333} + 2$ is the Wiscombe estimate for the number of terms in the Mie external problem (see Eq.(1.5.5)), and f is the empirical coefficient larger than 1. The appearance of this coefficient is due to slower convergence of the series over the amplitude coefficients of the internal field by comparison with the similar series over the coefficients of the scattered field for which the estimate $L_W(x)$ was in introduced initially. Our experience of the internal field calculations demonstrates that $f \sim 1.2$.

The functions $\xi_l(x)$ and $G_l(x)$ are calculated via the ordinary upward recursions (1.5.28) and (1.5.38) accordingly, and the log-derivatives $D_l(Nx)$, $D_l(Na_1x)$, $D_l(Na_2x)$, and $D_l(Na_ix)$ via the downward recursion (1.5.17). In so doing the initial values D_L are generated using Lentz's method of continued fractions (see Section 1.5.2.1). The ratios $R_l(Nax)$, $R_l(Na_1x)$, and $R_l(Na_2x)$ are calculated via the upward recursion (1.5.41). Equation (5.5.15) was numerically integrated with the Gauss quadrature (Press *et al.*, 1986). The number of quadrature nodes was first evaluated on the grounds of our experience and adjusted on the basis of convergence of Eq.(5.5.15) at increasing the number.

When the absorption index κ is very small ($\kappa \sim 10^7 \div 10^8$) the calculation of absorption in the restricted region (and calculation of absorption asymmetry (see Section 5.4)) using the surface Pointing integral (5.1.15) becomes computationally non-stable at some values of x. This is not surprising because in this approach the absorption is evaluated as a difference between the radiation flux entering the volume and that leaving it. At $\kappa \sim 10^7 \div 10^8$ this difference becomes very small ($\approx 10^{-12} \div 10^{-14}$) in comparison with these fluxes that cause the computational non=stability. In this situation, the approach based on the volume integral:

$$W_{\text{abs}} = \frac{2\pi n\kappa}{\lambda}\sqrt{\frac{\varepsilon_0}{\mu_0}}\int\limits_V (\mathbf{E}\cdot\mathbf{E}^*)\, d^3\mathbf{r} \tag{5.5.18}$$

offers an alternative. Substituting the field components into Eq.(5.5.18) and integrating over φ-angle we have after little manipulation:

$$W_{\text{abs}} = \frac{2n\kappa x A}{|N|^2}\int\limits_{a_1}^{a_2}\int\limits_{\nu_0}^{1}\left[\frac{1-\mu^2}{|N|^2 x^2 a^2}\sum_{l=1}^{\infty}\sum_{j=1}^{\infty} i^{l-1} Z_l Z_j^* \pi_l \pi_j \right. \tag{5.5.19}$$
$$\left. + \sum_{l=1}^{\infty}\sum_{j=1}^{\infty}\mu_l\mu_j^*\left(\omega_{jl}H_{jl} + i\nu_{jl}F_{jl}\right)\right] da\, d\mu$$

where:

$$\omega_{jl}(a) = X_l(Nxa)X_j^*(Nxa) + V_l(Nxa)V_j^*(Nxa)$$
$$\nu_{jl}(a) = X_l(Nxa)V_j^*(Nxa) + V_l(Nxa)X_j^*(Nxa)$$
$$H_{jl}(\mu) = \pi_l\pi_j + \tau_l\tau_j \qquad F_{jl}(\mu) = \pi_l\tau_j + \tau_l\pi_j$$

Let us interchange the integration over μ and the double summation:

$$W_{\text{abs}} = \frac{2n\kappa X A}{|N|^2}\int\limits_{a_1}^{a_2}\sum_{l=1}^{\infty}\sum_{j=1}^{\infty} i^{l-1} \tag{5.5.20}$$
$$\times\left\{\frac{Z_l Z_j^*}{|N|^2 x^2 a^2} t_{jl}(\nu_0) + \mu_l\mu_j^*\left[\omega_{jl}(a)h_{jl}(\nu_0) + i\nu_{jl}(a)f_{jl}(\nu_0)\right]\right\} da$$

where:

$$f_{jl}(\nu_0) = \int\limits_{\nu_0}^{1} F_{jl}(\mu)d\mu \qquad h_{jl}(\nu_0) = \int\limits_{\nu_0}^{1} H_{jl}(\mu)d\mu \qquad t_{jl}(\nu_0) = \int\limits_{\nu_0}^{1}\left(1-\mu^2\right)\pi_l\pi_j d\mu$$

As stated above, the function $f_{jl}(\nu_0)$ has the closed analytical representation, and the function $h_{jl}(\nu_0)$ has the analytical formula (5.5.11) at $j \neq l$ and the recursion (5.5.12–13) at $j = l$. Similarly, the function $t_{jl}(\nu_0)$ has analytical representation at $j \neq l$ because:

$$\int\left(1-\mu^2\right)\pi_l\pi_j d\mu = \frac{1-\mu^2}{(l-j)(j+l+1)}\left(\tau_l\pi j + \pi_l\tau_j\right) \tag{5.5.21}$$

At $j = l$ we can use for $t_{jl}(\nu_0)$ the same recursion as for $h_{jl}(\nu_0)$.

Interchanging the summation and integration over a in Eq.(5.5.20) yields the integrals:

$$\int_{a_1}^{a_2} \psi_l(Nxa)\,\psi_l^*(Nxa)\,\frac{da}{a^2} \qquad \int_{a_1}^{a_2} \psi_l(Nxa)\,\psi_l^*(Nxa)\,da$$

$$\int_{a_1}^{a_2} \psi_l'(Nxa)\,\psi_l^*(Nxa)\,da \qquad \int_{a_1}^{a_2} \psi_l'(Nxa)\,\psi_l'^*(Nxa)\,da$$

which appear to have no closed analytical representation. So the linear integral over a in Eq.(5.5.20) must be taken numerically through the Gauss quadrature. This is a drawback of the approach, nevertheless the calculation of the linear integral is much simpler than the investigation of the initial volume integral. A more sophisticated treatment of relation (5.5.20) shows that the imaginary parts of the double sums reduce to zero, as expected from the reality of W_{abs}. So we have as the net result:

$$W_{\text{abs}} = \frac{2n\kappa xA}{|N|^2}\int_{a_1}^{a_2}\sum_{l=1}^{\infty}\sum_{j=1}^{\infty} \tag{5.5.22}$$

$$\times\left[\frac{t_{jl}(\mu_0)}{|N|^2 x^2 a^2}\operatorname{Re}\left(i^{l-1}Z_l Z_j^*\right) + h_{jl}(\mu_0)\operatorname{Re}\left(\mu_l\mu_j^*\omega_{jl}\right) + f_{jl}(\mu_0)\operatorname{Re}\left(i\mu_l\mu_j^*\nu_{jl}\right)\right]da$$

This relation is similar to Eq.(5.5.15) from a computational point of view, so no additional difficulties emerge here.

As discussed in many works on the morphology dependent resonances (MDR) (see, e.g., Hill and Benner, 1988), the internal field at condition of MDR is entirely described by the resonant mode (at least, at points near the particle main axis coinciding with the direction of the incident wave). In this connection, it is of specific interest to simplify Eq.(5.5.22) with regard to one resonant mode only.

Let us begin with the magnetic resonance. Assume that only one amplitude coefficient α_L^{31} is distinct from zero. Then the double summation and the first and the third terms in Eq.(5.5.22) disappear. As a result we have:

$$W_{\text{abs}}^{(L,\text{mag})} = \frac{2n\kappa xA}{|N|^2}\cdot\frac{h_{LL}(\nu_0)}{L^2(L+1)^2}\int_{a_1}^{a_2}|X_L|^2\,da \tag{5.5.23}$$

or considering $X_L = \alpha_L^{31}\psi_L(Nxa)$:

$$W_{\text{abs}}^{(L,\text{mag})} = \frac{2n\kappa xA}{|N|^2}\cdot\frac{h_{LL}(\nu_0)}{L^2(L+1)^2}\left|\alpha_L^{31}\right|^2\int_{a_1}^{a_2}|\psi_L(Nxa)|^2\,da \tag{5.5.24}$$

Taking the integral (see Appendix A, Eq.(A.29)):

$$\int \psi_l(\alpha r)\psi_l(\beta r)dr = \frac{1}{\alpha^2-\beta^2}\left[\beta\psi_l(\alpha r)\psi_l'(\beta r) - \alpha\psi_l'(\alpha r)\psi_l(\beta r)\right]$$

and using the relations $\psi_l^*(Nxa) = \psi_l(N^*xa)$, $\psi_l'^*(Nxa) = \psi_l'(N^*xa)$ we finally arrive at:

$$W_{\text{abs}}^{(L,\text{mag})} = \frac{Ah_{LL}(\nu_0)}{|N|^2 L^2 (L+1)^2} \operatorname{Im}\left[N^* X_L(a) Y_L^*(a)\right]\Big|_{a_1}^{a_2} \tag{5.5.25}$$

Let us go to the electric MDR (only one amplitude coefficient β_L^{31} is distinct from zero). Then:

$$W_{\text{abs}}^{(L,\text{elec})} = \frac{2n\kappa x A}{|N|^2} \int\limits_{a_1}^{a_2} \left[\frac{|Z_L|^2}{|N|^2 x^2 a^2} t_{LL}(\nu_0) + \frac{|V_l|^2}{L^2(L+1)^2} h_{LL}(\nu_0)\right] da \tag{5.5.26}$$

It follows from above that $t_{ll}(\nu_0) = \left[h_{ll}(\mu_0) - (1-\nu_0^2)\pi_l(\nu_0)\tau_l(\nu_0)\right] / \left[l(l+1)\right]$. Then

$$W_{\text{abs}}^{(L,\text{elec})} = \frac{2n\kappa x A}{|N|^2 L(L+1)} |\beta_L^{31}|^2 \left\{ \frac{h_{LL}(\nu_0)}{l(l+1)} \int\limits_{a_1}^{a_2} \left[\frac{L(L+1)}{|N|^2 x^2 a^2} |\psi_L(Nxa)|^2 \right.\right.$$
$$\left.\left. + |\psi_L'(Nxa)|^2\right] da - \frac{(1-\nu_0^2)\pi_L(\nu_0)\tau_L(\nu_0)}{|N|^2 x^2} \int\limits_{a_1}^{a_2} |\psi_L(Nxa)|^2 \frac{da}{a^2} \right\} \tag{5.5.27}$$

The first integral in Eq.(5.5.27) may be evaluated analytically (see Appendix A, Eq.(A.40)) but the analytics for the second one is unknown. As a result we have:

$$W_{\text{abs}}^{(L,\text{ elec})} = \frac{Ah_{LL}(\nu_0)}{|N|^2 L^2(L+1)^2} \operatorname{Im}\left[N Z_L(a) V_L^*(a)\right]_{a_1}^{a_2}$$
$$- \frac{2n\kappa x A(1-\nu_0^2)\pi_L(\nu_0)\tau_L(\nu_0)}{|N|^4 x L(L+1)} |\beta_L^{31}|^2 \int\limits_{a_1}^{a_2} |\psi_L(Nxa)|^2 \frac{da}{a^2} \tag{5.5.28}$$

where the numerical quadrature can be applied for the second term that is not a particular problem here.

In the previous sections of this chapter we were dealing with the homogeneous spherical particles solely. Consider as a generalization the absorption inside the restricted volume of the shell of the two-layered sphere. Let the internal radius of a sphere be R, the core radius be R_1, and the relative complex refractive indices of core and shell be N_1 and N_2 respectively. As noted in Section 3.1, the expansion of field inside the shell of the two-layered sphere in terms of the spherical vector functions differs from the similar expansion for the homogeneous sphere in that it contains not only the functions $\mathbf{M}^{(1)}$and $\mathbf{N}^{(1)}$ regular in the origin, but the non-regular ones $\mathbf{M}^{(3)}$and $\mathbf{N}^{(3)}$ as well. Here we shall use the volume integral for calculation of the absorption power. The electric field inside the shell is described by Eq.(3.5.1):

$$\mathbf{E}_{\text{sh}}^{\text{lay}}(\mathbf{r}) = E_I \sum_{l=1}^{\infty} \mu_l \left[\alpha_{2l}\mathbf{M}_{o1l}^{(1)}(k_2\mathbf{r}) + \alpha_{3l}\mathbf{M}_{o1l}^{(3)}(k_2\mathbf{r}) - i\beta_{2l}\mathbf{N}_{e1l}^{(1)}(k_2\mathbf{r}) - i\beta_{3l}\mathbf{N}_{e1l}^{(3)}(k_2\mathbf{r})\right] \tag{5.5.29}$$

where $\alpha_{2l}, \alpha_{3l}, \beta_{2l}$, and β_{3l} are the amplitude coefficients of the shell internal field (see Eqs.(3.1.7–3.1.9)). The spherical components of this field have the form:

$$E_r^{(2)} = \frac{E_I \sin\theta \cos\varphi}{(N_2 xa)^2} \sum_{l=1}^{\infty} i^{l-1} \bar{Z}_l \pi_l \tag{5.5.30}$$

$$\begin{pmatrix} E_\theta^{(2)} \\ E_\varphi^{(2)} \end{pmatrix} = \frac{E_I}{N_2 xa} \begin{pmatrix} \cos\varphi \\ -\sin\varphi \end{pmatrix} \sum_{l=1}^{\infty} \gamma_l \left[\bar{X}_l \begin{pmatrix} \pi_l \\ \tau_l \end{pmatrix} - i\bar{V}_l \begin{pmatrix} \tau_l \\ \pi_l \end{pmatrix} \right]$$

where:

$$\begin{aligned} \bar{Z}_l &= \alpha_{3l}\psi_l(N_2xa) + \beta_{3l}\xi_l(N_2xa) \qquad \bar{X}_l = \alpha_{2l}\psi_l(N_2xa) + \beta_{2l}\xi_l(N_2xa) \\ \bar{V}_l &= \alpha_{3l}\psi_l'(N_2xa) + \beta_{3l}\xi_l'(N_2xa) \qquad \bar{Y}_l = \alpha_{2l}\psi_l'(N_2xa) + \beta_{2l}\xi_l'(N_2xa) \end{aligned}$$

Comparison of the expansion (5.5.29) with the similar expressions for the homogeneous sphere shows that the detailed mathematics are not necessary here because the structures of these expansions for homogeneous and two-layered spheres is identical. Hence, substitution of the values $\bar{X}_l, \bar{Y}_l, \bar{Z}_l$, and $\bar{V}_l$ in Eqs.(5.5.22) instead of X_l, Y_l, Z_l, and V_l gives us the expression for the absorption power in the restricted region of the shell. The only real additional difficulty here is the calculation of the internal amplitude coefficients (see Section 7.1).

The practical realization of the algorithms discussed in this section have shown that at not very small values of the absorption index ($\kappa > 10^{-6}$) the code based on the surface Pointing integral is preferable because it is faster than the code based on the volume integral. This difference in speed is not so significant at relatively small size parameters but increases steeply with x and reaches one order of magnitude at $x \sim 100$. The different speed of these similar (at first glance) codes is dictated by a necessity to calculate the single and double integrals, respectively, at every point of the integration interval. On the other hand, at very small absorption indices ($\sim 10^{-7}$ and under) the surface integral approach becomes, as noted above, numerically unstable. Hence, at such small absorption the calculation may be realized with the volume integral only. However, the relative slowness of this algorithm hinders, to some extent, the calculation at $x > 100$.

Some difficulties arise with the treatment of the results obtained. It is not a priori evident what quantity must be chosen as an absorption measure in the restricted volume; this choice may be governed by the situation. In our opinion, the rather informative and simple characteristic is the ratio γ of the mean relative intensity of field $\bar{B}_V$ in the given volume V to the analogous quantity $\bar{B}$ averaged over the entire volume of the spherical particle: $\gamma = \bar{B}_V/\bar{B}$. Actually, this value shows how many times the field intensity in the volume is greater than the mean intensity. The relations of γ with other absorption characteristics are rather evident.

The dependencies of γ on the spherical angle θ_0 counted from the forward scattering direction are shown in the Figures 5.10–11 for water particles of different sizes at wavelengths $\lambda = 10.6$ and $2.36\ \mu m$. The volume of interest in the shadow hemisphere was restricted by the conical surface ($\theta = \theta_0$, $\varphi = 0 \div 2\pi$, $a = a_1 \div a_2$) and by two spherical surfaces ($a_1 = 0.5$, $\theta = 0 \div \theta_0$, $\varphi = 0 \div 2\pi$ and $a_2 = 0.5$, $\theta = 0 \div \theta_0$, $\varphi = 0 \div 2\pi$). The angle $\theta_0$varied from 0 to 45°.

In the region of strong absorption ($\lambda = 10.6\ \mu m$, Figure 5.10*a,b*) the values of γ are relatively low. At $x \sim 1$ the value of γ is equal 1.05 and does not depend on the angle θ_0 that is indicative of the almost homogeneous field distribution. However, the distinct maximum ($\theta = 0$) already takes place at $x \sim 2 \div 5$ corresponding to the development of maxima at the main axis in the shadow hemisphere. This maximum is absent at $x = 10$ already, and the γ-value itself is less than 1 because most of the energy begins to be released in the illuminated hemisphere (see Figure 5.7). The ratio γ continues to decrease

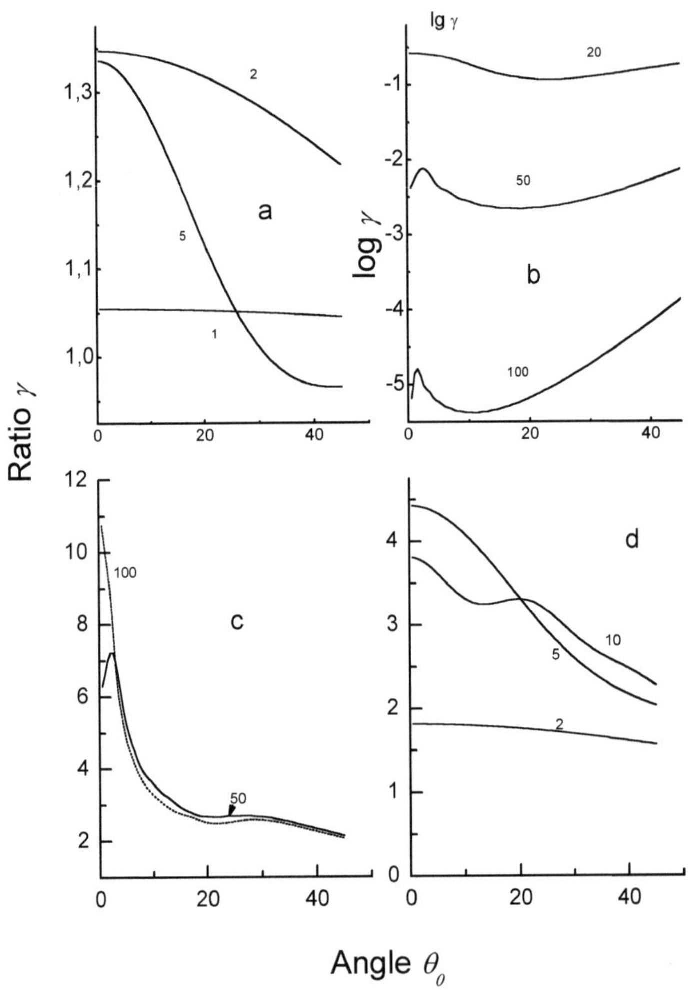

Figure 5.10. Water particles at λ=10.6 μm (*a,b*) and λ=2.36 μm (*c,d*). Ratio γ of the mean relative intensity of field $\bar{B}_V$ in the conical volume with angle θ_0 to the analogous quantity $\bar{B}$ averaged over the entire volume of the spherical particle versus angle θ_0. The figure at the curve is the size parameter x.

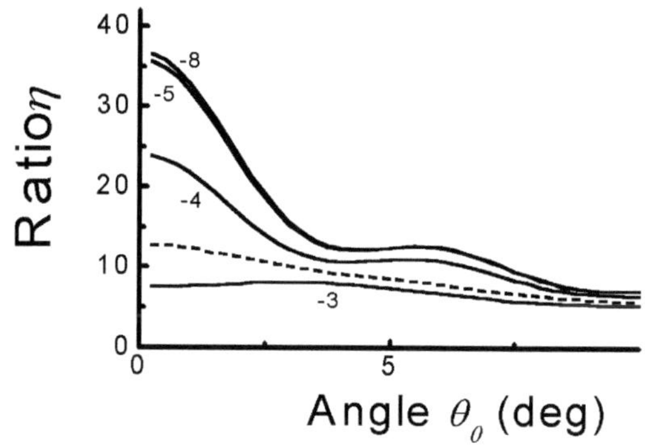

Figure 5.11. Dependence of ratio γ on angle θ_0 for model spherical particle with the real part of the refractive index $n = 1.40$ and size parameter $x = 31.789230$. The figures near the curves are the values of $\log \kappa$. The dashed curve corresponds to the spherical particle with $m = 1.4 + i10^{-8}$ and $x = 31.80$.

with increasing x; a little maximum occurs at $x \sim 50$ near $\theta_0 \sim 2°$, that is associated with the shift of maxima from the particle main axis.

At the wavelength $\lambda = 2.36$ μm, where the absorption of water is two orders of magnitude lower, a qualitatively different picture is observed (Figure 5.10*c,d*). The ratio γ rather slightly depends on θ_0 at small $x \sim 1 \div 2$, however the maximum γ at angle $\theta_0 = 0$ appears for larger particles ($x \sim 5$). This maximal value is much greater than at $\lambda = 10.6$ μm ($\gamma \sim 4$, $6 \div 7$, and 11 for $x = 5 \div 20$, 50, and 100, respectively). Other maximum ($\theta_0 = 2 \div 3$, $x \sim 20 \div 30$) may be attributed to the shift of the field maxima from the main axis.

The above dependencies $\gamma\,(\theta_0)$ relate to the non-resonant case. The resonant case (the particle is the same as in Figure 5.5 ($n = 1.40$, $x = 31.789230$)), is illustrated in Figure 5.11. The calculations are carried out for several values of the absorption index (from 10^{-3} to 10^{-8}) and also for a particle out of resonance ($m = 1.40 + i10^{-8}$, $x = 31.80$). At relatively large absorption ($\kappa \sim 10^{-3}$) the ratio γ depends on θ_0 sufficiently weakly both in out of resonance. But the sharp maximum of γ is built up while κ is decreasing in resonance conditions. The maximal values of γ is very high here (~ 25 at $\kappa = 10^{-4}$ and ~ 37 at $\kappa = 10^{-6}$ and less). If it is remembered that the internal intensity $\bar{B}$ averaged over the entire volume of this resonating particle is 44 at $\kappa = 10^{-6}$ and less (see Figure 5.5*b*), then it turns out that the mean relative intensity of the internal field is very great (~ 1600) in the narrow cone with $\theta_0 \sim 0.5°$.

6

Internal intensity distributions for homogeneous spherical particles

In contrast to the previous chapter where the simplified averaged characteristics of light absorption by the spherical particles were discussed, in Chapter 6 and 7 we shall consider some results of calculations of the detailed distribution of electromagnetic fields inside the diversified spherical particles on exposure to the incident wave. As will be shown below, the quite non-uniform energy distribution can appear inside the volume of the individual particle. At some combinations of values of size parameter and refractive index the energy concentrates in some regions ("hot points") with local intensity of the internal field tens and hundreds of times greater than the incident intensity. Therefore, the interaction of quite powerful waves with particles become non-linear. The calculation of the internal field (other name is *source function*) is the first (electrodynamic) step in the theoretical modeling of many processes which may be pooled together and named "the non-linear optics of scattering media" (Zuev *et al.*, 1980; Prishivalko, 1983; Barber and Chang, 1990). The list of these physical phenomena is vast. The points that should be mentioned first are the optical effects associated with the radiative heating, evaporation, dissociation, fragmentation, and ionization of atmospheric aerosols (Andreev *et al.*, 1977; Belov, 1986; Zardecki and Pendelton, 1989; Gashevskaya *et al.*, 1990; Lange and Schweiger, 1994). Of great practical importance are the issues concerning the kinetics of heating, evaporation, and explosive destruction of water droplets and ice crystals – the typical elements of the aerosol atmosphere (Volkovitskiy *et al.*, 1982; Prishivalko, 1983; Loskutov and Strelkov, 1984; Prishivalko *et al.*, 1988). The internal fields inside water spheres and particles with $n < 1$ are considered in Sections 6.1 and 6.3, accordingly. Section 6.5 is devoted to consideration of the similarity criterion for optical fields in weakly absorbing aerosol particles.

The other interesting objects are the metal particles (Astafyeva and Prishivalko, 1991; Astafyeva *et al.*, 1991; Prishivalko and Astafyeva, 1991) which will be discussed in Section 6.2. Immediate relation of internal "hot points" with the morphologically dependent resonances (Prishivalko *et al.*, 1984*a*, 1984*b*, *1984c*; Chýlek *et al.*, 1985; Prishivalko, 1986; Drobnik *et al.*, 1989; Astafyeva and Lednyeva, 1999) are treated in Section 6.4.

As was already stated above, the sharp inhomogeneity of the internal field is a peculiar kind of "seed" for initiation of non-linear effects, such as photophoresis (Kerker and Cooke, 1973; Greene *et al.*, 1985), laser breakdown (Prishivalko *et al.*, 1988; Pinnick *et al.*, 1992; Kopytin *et al.*, 1990), luminescence and Raman-effect (McNulty *et al.*, 1980; Arnold and Folan, 1989; Schweiger, 1990), and laser effect (Ching *et al.*, 1987; Kotomtseva *et al.*, 1998).

Noteworthy also are the following papers on exact (Dusel *et al.*, 1979; Folan *et al.*, 1985; Benincasa *et al.*, 1987; Bott and Zdunkowski, 1987; Dobson and Lewis, 1989; Lai *et al.*, 1991; Hoekstra *et al.*, 1998) and geometro-optical (Jarzembski and Srivastava, 1989; Chowdhury *et al.*, 1992; Velesco *et al.*, 1997; Roll *et al.*, 1998*b*; Lock, 2001; Liu *et al.*, 2002) calculations of internal fields of homogeneous spheres.

Regrettably, in this and the next chapters we have restricted ourselves to be quite brief. A voluminous book is necessary for sufficiently detailed consideration of the issues concerned with internal fields.

6.1 INTENSITY DISTRIBUTION INSIDE WATER DROPLETS

Recall that the internal electric field for a sphere with radius R and complex refractive index N has the form:

$$\mathbf{E}_{\text{int}} = E_I \sum_{l=1}^{\infty} \mu_l \left[\alpha_l^{31} \mathbf{M}_{o1l}^{(1)}(k\mathbf{r}) - i\beta_l^{31} \mathbf{N}_{e1l}^{(1)}(k\mathbf{r})\right] \tag{6.1.1}$$

where $\mu_l = i^l\,(2l+1)\,/n(l+1)$, and the amplitude coefficients for the internal field α_l^{31} and β_l^{31} are:

$$\begin{aligned}\alpha_l^{31} &= iN\left[\xi_l'(x)\psi_l(Nx) - N\xi_l(x)\psi_l'(Nx)\right]^{-1} \\ \beta_l^{31} &= iN\left[N\xi_l'(x)\psi_l(Nx) - \xi_l(x)\psi_l'(Nx)\right]^{-1}\end{aligned} \tag{6.1.2}$$

where size parameter $x = k_0 R$. If we write Eq.(6.1.1) in the spherical coordinate system, then:

$$\begin{aligned}E_r^{\text{int}} &= E_I \frac{\sin\theta\cos\varphi}{N^2x^2a^2} \sum_{l=1}^{\infty} i^{l-1}(2l+1)Z_l(a)\pi_l(\cos\theta) \\ E_\theta^{\text{int}} &= E_I \frac{\cos\varphi}{Nxa} \sum_{l=1}^{\infty} \mu_l\left[X_l(a)\pi_l(\cos\theta) - iV_l(a)\tau_l(\cos\theta)\right] \\ E_\varphi^{\text{int}} &= -E_I \frac{\sin\varphi}{Nxa} \sum_{l=1}^{\infty} \mu_l\left[X_l(a)\tau_l(\cos\theta) - iV_l(a)\pi_l(\cos\theta)\right]\end{aligned} \tag{6.1.3}$$

where a is the relative radial coordinate $a = r/R$ and the notations used in Chapter 5 are introduced:

$$X_l = \alpha_l^{31}\psi_l(Nax) \quad Z_l = \beta_l^{31}\psi_l(Nax) \quad V_l = \beta_l^{31}\psi_l'(Nax) \tag{6.1.4}$$

If the sums in Eq.(6.1.3) are symbolized by S_r, S_θ, and S_φ respectively then the relative intensity B (see Eq.(5.2.16)) of the electric field in any point (a, θ, φ) inside the sphere is:

$$\begin{aligned} B(a,\theta,\varphi) &= \frac{1}{E_I^2}(\mathbf{E}_{\text{int}} \cdot \mathbf{E}_{\text{int}}^*) \\ &= \frac{1}{|N|^2 x^2 a^2}\left[|S_r|^2 \frac{\sin^2\theta\cos^2\varphi}{|N|^2 x^2 a^2} + |S_\theta|^2 \cos^2\varphi + |S_\varphi|^2 \sin^2\varphi\right] \end{aligned} \tag{6.1.5}$$

Simplify this formula for some special cases:

(1) Let the incident wave be non-polarized. Then the averaging over angle φ gives:

$$B(a,\theta) = \frac{1}{2|N|^2 x^2 a^2}\left[|S_r|^2 \frac{\sin^2\theta}{|N|^2 x^2 a^2} + |S_\theta|^2 + |S_\varphi|^2\right] \tag{6.1.6}$$

(2) Let $\theta = 0°$. Such a value corresponds to the main axis in the shadow half-sphere. In this case the radial component E_r^{int} is equal zero and the sums S_θ and S_φ are:

$$S_\theta(a,1) = S_\varphi(a,1) = \frac{1}{2}\sum_{l=1}^{\infty} i^l(2l+1)(X_l - iV_l) \tag{6.1.7}$$

because $\pi_l(1) = \tau_l(1) = l(l+1)/2$. Then the formula for B has the form:

$$B(a,1) = \frac{1}{|N|^2 x^2 a^2}|S_\theta(a,1)|^2 \tag{6.1.8}$$

(3) Let $\theta = 180°$ (the main axis in the illuminated half-sphere). In this case the radial component $E_r^{\text{int}} = 0$ and $\pi_l(-1) = (-1)^{l+1} l(l+1)/2$, $\tau_l(-1) = (-1)^l l(l+1)/2$. Consequently:

$$-S_\theta(a,-1) = S_\varphi(a,-1) = \frac{1}{2}\sum_{l=1}^{\infty} (-i)^l (2l+1)(X_l + iV_l) \tag{6.1.9}$$

and

$$B(a,-1) = \frac{1}{|N|^2 x^2 a^2}|S_\varphi(a,-1)|^2 \tag{6.1.10}$$

Evidently, that above formulae do not include any new functions. Thus, no new computational difficulties can emerge compared with Chapter 5. However, it must be borne in mind that the convergence of series S_r, S_θ, and S_φ is somewhat slower than that in Mie theory for scattered fields. Thus, the Wiscombe estimate (1.5.5) must be enlarged by some factor $f \sim 1.2$. When calculating the internal field at resonant conditions, f may grow still further (Roll *et al.*, 1998*a*). In addition, it is appropriate to replace the size parameter x in Eq.(1.5.5) with $y = |N|\,xa$ to circumvent some numerical difficulty arising due to the excess terms over index l for the points close to the particle center, particularly for strongly absorbing particles (Barber and Hill, 1990):

$$L = f\left(y + 4.05y^{1/3} + 2\right)$$

Let consider the intensity distributions inside water droplets illuminated by a continuous beam of the *non-polarized light*. All illustrations in Chapters 6 and 7 will be for non-polarized incident fields, unless otherwise stated. In this case the energy distribution inside the spherical particle is symmetrical about its diameter coinciding with the propagation direction of the incident plane wave. It is explained by the symmetry of the object. Calculations of intensity distributions inside water droplets were performed for the spheres with $0.5 \le R \le 30\ \mu m$ in a radius illuminated by two laser wavelengths: $\lambda = 0.63\ \mu m$ and $\lambda = 10.6\ \mu m$. Complex refractive indices: $N = 1.332 + i1.43 \times 10^{-8}$ for $\lambda = 0.63$ μm and $N = 1.178 + i0.0705$ for $\lambda = 10.6\ \mu m$ were taken from Hale and Querry (1973), Kou *et al.* (1993), and Huibers (1997).

The results of these calculations give the source function for the internal fields (Prishivalko and Astafyeva, 1972*a*, 1972*b*; Astafyeva, 1973).

Figures 6.1–6.2 show the internal intensity distribution in spherical droplets for 0.63 μm. The incident wave propagates in the $+z$ direction. The intensity within the particle depends only on two coordinates r and θ, because the radiation is not polarized and the third coordinate φ is absent. The angles θ from 90° to 180° correspond to the illuminated hemisphere of the particle. The opposite hemisphere is called shadow hemisphere.

Figures 6.1–6.2 make it possible to see how the energy distribution inside a spherical particle changes with a size increase. The water droplets are weakly absorbing at wavelength 0.63 μm. In this case even for the small particles with $R = 1\ \mu m$ the intensity distribution is substantially inhomogeneous. The main maximum of internal intensity is located on the shadow surface of the droplet. Then the intensity decreases in going to the illuminated hemisphere and the second maximum begins to form near the illuminated surface. With increasing the droplet radius both maxima in the illuminated and in shadow hemispheres grow. For large particles ($R > 20\ \mu m$) the maxima in both hemispheres becomes practically the same and the intensity distribution is essentially inhomogeneous. A picture of the intensity distribution contains two regions of energy concentration near the illuminated and shadow surfaces of the particle where the maximum value of the intensity is hundreds of times more than that in the surroundings inside the droplet.

Research in internal intensity distributions show that the maximum inhomogeneity inside water droplets takes place in the direction of the propagation of incident radiation ($\theta = 180° - 0°$) (Figure 6.3). Notice, that in this and the following figures the relative coordinate A takes the negative values $(-1 \div 0)$ in the illuminated hemisphere and positive ones $(0 \div 1)$ in the shadow hemisphere. At the main axis the difference between maximum and minimum values of intensity is the greatest. For the other directions this difference is much smaller. The most homogeneous distribution is observed in the direction normal to the propagation direction.

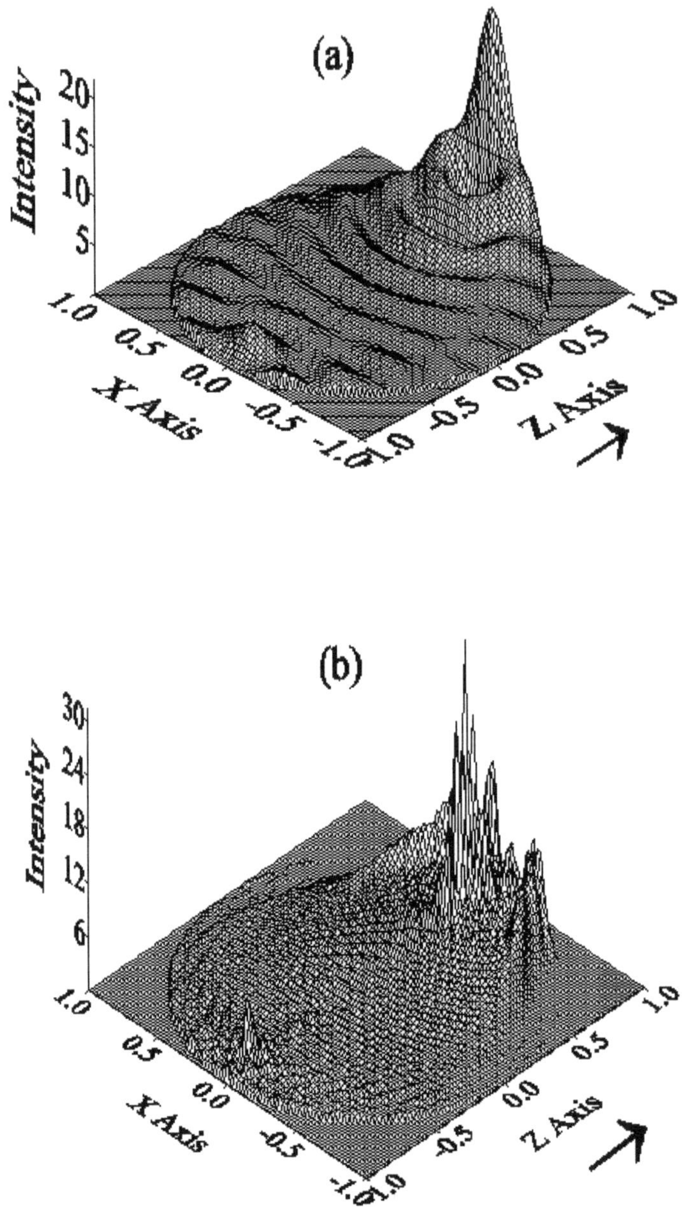

Figure 6.1. Relative intensity distribution inside the water droplets with the radii: $a = 1\ \mu m$ (a) and $a = 5\ \mu m$ (b). The incident wavelength is $\lambda = 0.63\ \mu m$. The arrow shows the direction of the propagation of incident radiation.

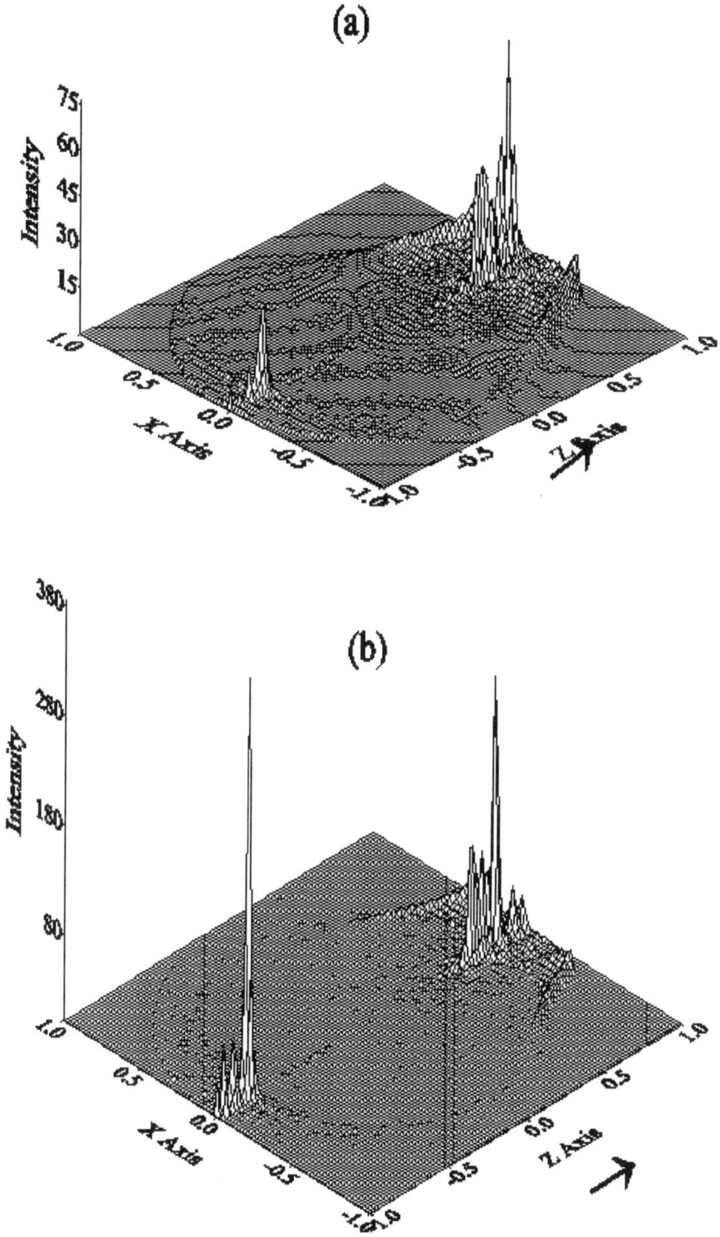

Figure 6.2. Relative intensity distribution inside the water droplets with the radii: $a = 15\ \mu m$ (a) and $a = 30\ \mu m$ (b). The incident wavelength $\lambda = 0.63\ \mu m$. The arrow shows the direction of the propagation of incident radiation.

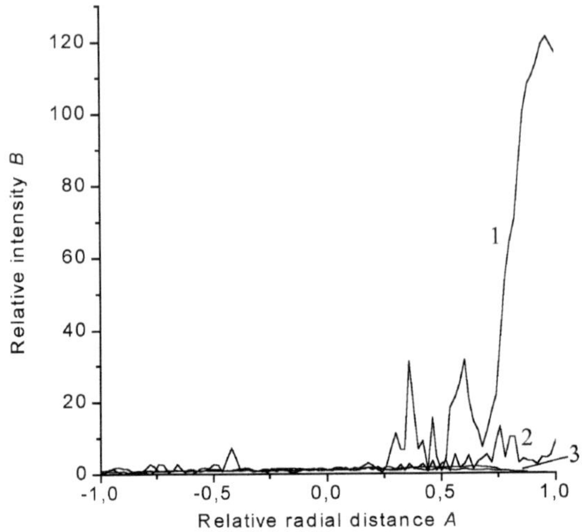

Figure 6.3. Relative intensity B versus the relative radial coordinate A for different directions inside the water droplet: $1 - \theta = 0$–$180°$; $2 - \theta = 5$–$185°$; and 3 – all the rest angles. The incident wavelength is $\lambda = 0.63\ \mu m$.

In Figure 6.4 the dependence of the maximum value of the internal intensity (on the particle radius) is shown at wavelength 0.63 μm. We see that the dependence $B_{\max}(R)$ has a greatly oscillating character with an increase of the amplitude of the oscillation when the particle size grows.

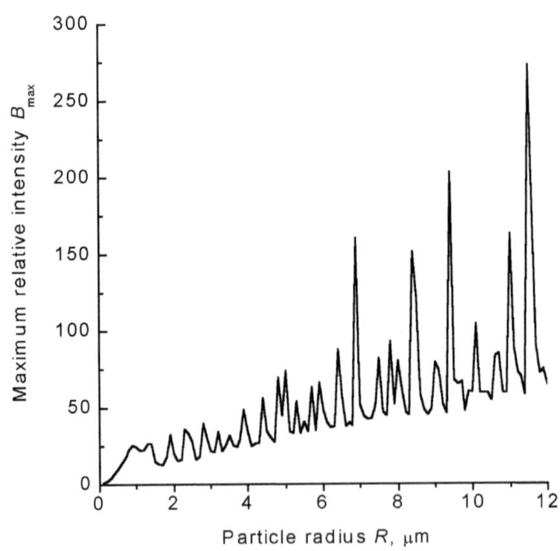

Figure 6.4. Maximum relative intensity $B_{\max}$ versus the particle radius R. The incident wavelength is $\lambda = 0.63\ \mu m$.

Somewhat different trends are observed when water droplets are under the action of laser radiation with the wavelength 10.6 μm. Figures 6.5 and 6.6 allow us to imagine the change of the intensity distribution inside the spherical water particle with increasing particle size. In the case of the particle with $R = 1\ \mu m$ the distribution is practically homogeneous with the maximum value of the intensity at its center. The gradient of the intensity is very small (within the limits of 2 per cent). By increasing the particle size the internal distributions become complicated, the gradient of the intensity value increases. At first the maximum of intensity appears in the shadow hemisphere near its surface. Then the intensity values near the illuminated surface begin to grow. As before, the maximal inhomogeneity in the intensity distribution is in the propagation direction (Figure 6.7). For other directions the dependences of $B\,(A)$ are more smooth and differences of the maximum and minimum values of B are essentially smaller. The most homogeneous intensity distributions are observed in the direction normal to the direction of propagation.

In Figure 6.8 the intensity distribution along the particle diameter coinciding with the incident beam direction (we will call it the main diameter) is shown. Various lines in Figure 6.8 correspond to the particles of different sizes. From this figure it follows that for the particles with $R < 2\ \mu m$ the intensity distributions along the main diameter are practically homogeneous. Beginning from $R = 3\ \mu m$ to $R = 14\ \mu m$ it is observed that there is an appreciable increase of the energy release in the shadow hemisphere near the particle surface. The further rise of R leads to the decrease of energy release in the shadow hemisphere. In the case of $R > 14\ \mu m$ the greatest energy release already takes place in the illuminated hemisphere. As the particle radius is increased, the dependences of $B\,(A)$ are smoothed out. With increasing the particle radii, the energy derived by the shadow particle surface reduces progressively. The maximum energy is now localized on the illuminated surface of the droplet.

Detected peculiarities are confirmed by the experimental data obtained by Kafalas and Ferdinand (1973). In this article, water droplets in the interval of sizes 7 to 50 μm were under the action of laser radiation with wavelength $\lambda = 10.6\ \mu m$. The results obtained show that in the case of droplets with $R \leq 12\ \mu m$ volumetric heating is observed. In the interval of sizes 15–17 μm the front surface of the droplet is largely heated. For droplet radii more than 20 μm only the front particle surface is heated.

The dependence of the maximum intensity inside the particles on their radii for $\lambda = 10.6\ \mu m$ is shown in Figure 6.9. At first, $B_{\max}$ increases with increasing droplet radius, and attains the maximum value $B_{\max} \approx 1.72$ at $R = 8\ \mu m$, then decreases to the minimum value $B_{\max} \approx 0.84$ at $R = 15\ \mu m$ and practically remains unchanged with further rise of the particle size.

The results obtained show that the intensity distribution inside the water droplets are inhomogeneous. It essentially depends on particle size. It is necessary to take this into account when performed the experiments and explaining their results.

As the water droplets often contain micro-admixtures which can change their optical properties it is of interest to elucidate the influence of such admixtures on the intensity distributions inside the water particles. Therefore in Figure 6.10 is shown the intensity distributions inside particles for two values of the imaginary parts of the refractive index. It should be noted that with a decrease in the value of κ (from $\kappa = 0.07$ to $\kappa = 0.2$) the

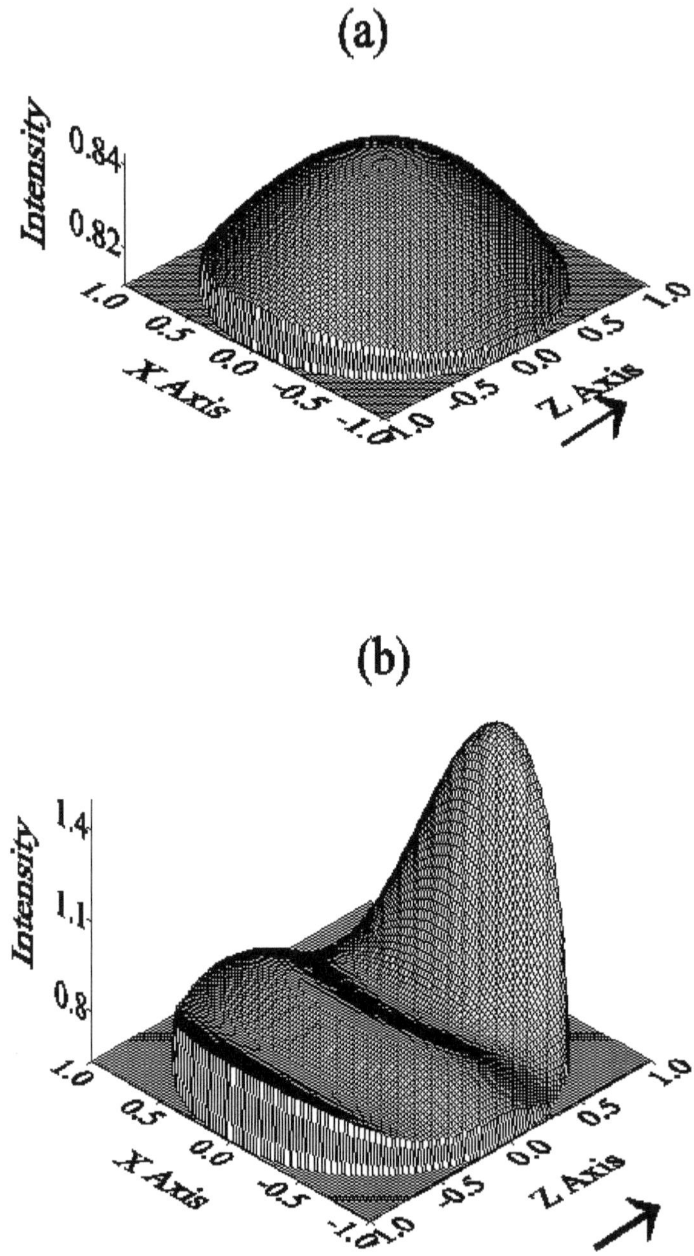

Figure 6.5. Relative intensity distribution inside the water droplets with the radii: $a = 1\ \mu m$ (a) and $a = 5\ \mu m$ (b). The incident wavelength is $\lambda = 10.6\ \mu m$. The arrow shows the direction of the propagation of incident radiation.

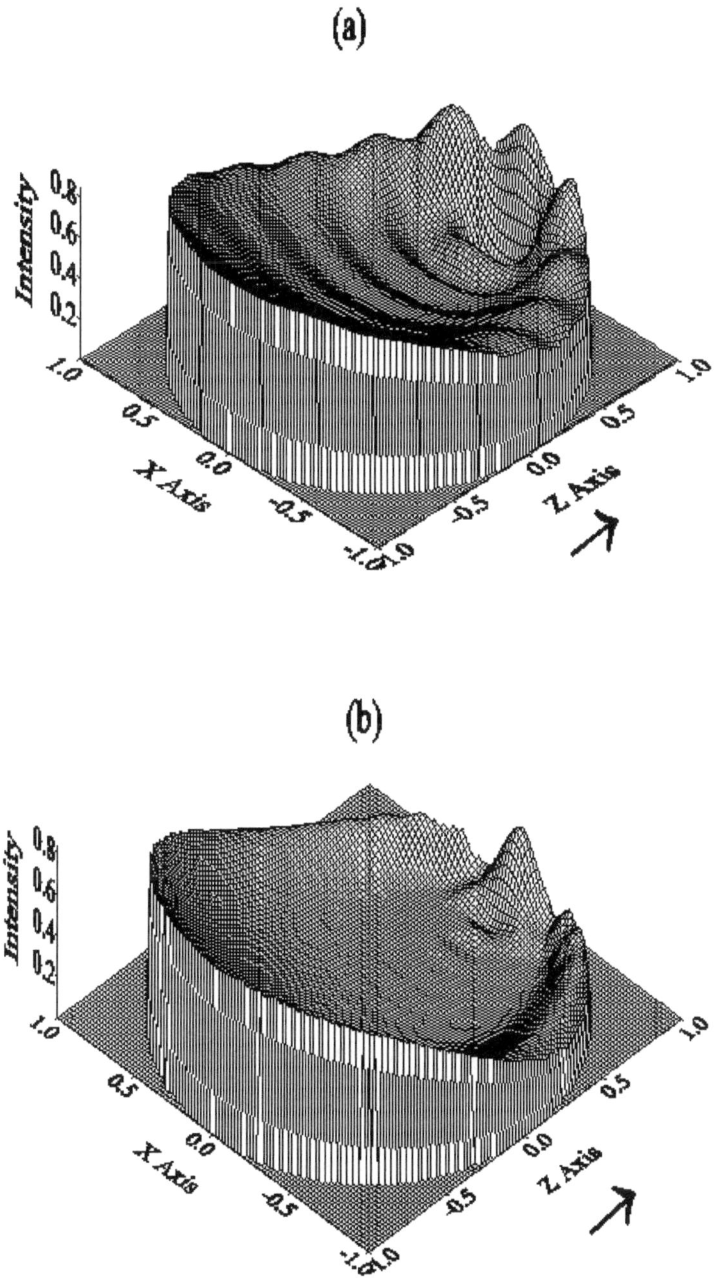

Figure 6.6. Relative intensity distribution inside the water droplets with the radii: $a = 15\ \mu m$ (*a*) and $a = 30\ \mu m$ (*b*). The incident wavelength is $\lambda = 10.6\ \mu m$. The arrow shows the direction of the propagation of incident radiation.

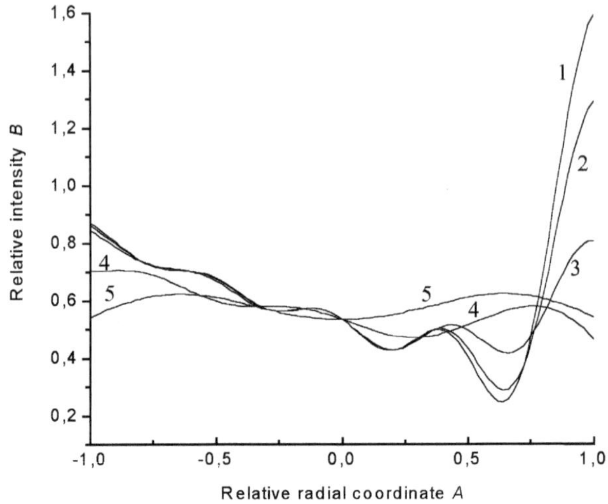

Figure 6.7. Relative intensity B versus the relative radial coordinate A for different directions inside the water droplet: *1* – $\theta = 0$–$180°$; *2* – $\theta = 20$–$200°$; *3* – $\theta = 40$–$220°$; *4* – $\theta = 60$–$240°$; and *5* – $\theta = 90$–$270°$. The incident wavelength is $\lambda = 10.6\ \mu m$.

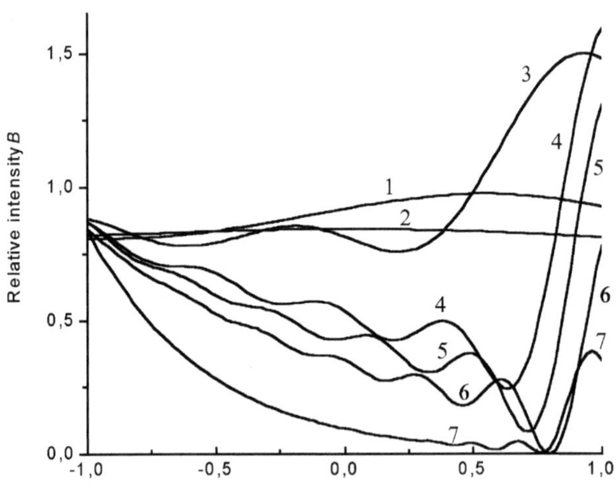

Figure 6.8. Relative intensity B versus the relative radial coordinate A for the direction ($\theta = 0° - 180°$), coinciding with the incident radiation, inside water droplets with different radii: 1 – $a = 0.5$ μm; 2 – $a = 2\ \mu m$; 3 – $a = 5\ \mu m$; 4 – $a = 10\ \mu m$; 5 – $a = 12\ \mu m$; 6 – $a = 15\ \mu m$; and 7 – $a = 30\ \mu m$. The incident wavelength is $\lambda = 10.6\ \mu m$.

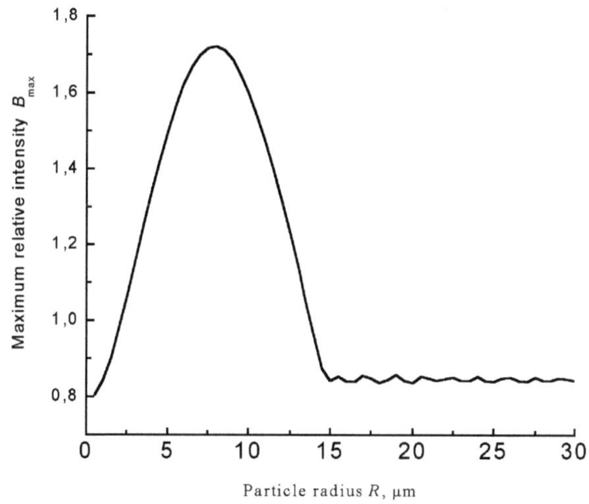

Figure 6.9. Maximum relative intensity $B_{\max}$ versus the particle radius R. The incident wavelength is $\lambda = 10.6\ \mu m$.

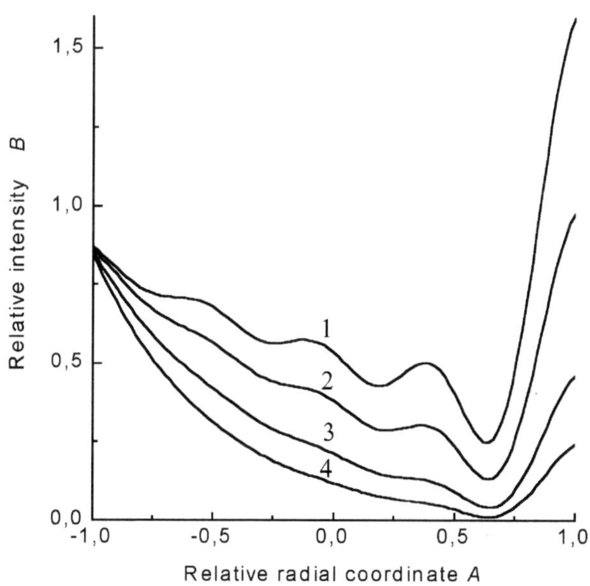

Figure 6.10. Relative intensity B versus the relative radial coordinate A for the direction ($\theta = 0° - 180°$), coinciding with the incident radiation, inside water droplets (radius $R = 10\ \mu m$) with different values of the imaginary parts of the refractive index: *1* – $\kappa = 0.0705$; *2* – $\kappa = 0.1$; *3* – $\kappa = 0.15$; and *4* – $\kappa = 0.2$. The incident wavelength is $\lambda = 10.6\ \mu m$.

maximum value of the energy distribution is shifted from the shadow hemisphere to the illuminated and decreases in magnitude from $B \approx 1.6$ to ≈ 0.84. The energy distribution becomes more smooth. From Figure 6.10 it is seen the decrease of the intensity values with growing the absorption of the particle matter. For $\kappa \approx 0.11$ the maximum values of intensity in the illuminated and shadow hemispheres are approximately equal.

6.2 INTERNAL FIELDS OF METALLIC PARTICLES

Peculiarities of the intensity distribution inside metallic particles are considered as an example of the particles of aluminium and titanium. The calculations (Astafyeva *et al.*, 1989, 1991) have shown that the intensity distributions inside metallic particles were sufficiently inhomogeneous. Practically all the absorbed energy is released in a thin spherical layer near the particle surface. A maximum the released energy is localized in the illuminated hemisphere on the particle surface in the direction coinciding with the propagation direction of incident radiation. As we recede from the particle surface to its center the quantity of the released energy decreases exponentially.

Figure 6.11 shows the intensity distribution inside titanium particles along the different angular directions for two wavelengths of incident radiation ($\lambda = 0.69\ \mu m$ and 10.6 μm). The central part of the metallic particle is excluded from this figure, because the density of energy is very small there. The intensity distribution inside titanium particles depends on the angular coordinate. The maximum inhomogeneity of the energy distribution is in the direction of $\theta = 180$–$0°$ the and minimum inhomogeneity is in the direction of $\theta = 90°$. The energy quantity decreases essentially with distance away the particle surface.

As the imaginary part of the complex refractive index κ of titanium for $\lambda = 0.69\ \mu m$ ($N = 2.67 + i3.2$) is essentially smaller than for $\lambda = 10.6\ \mu m$ ($N = 4.85 + i20.1$) (Ordal *et al.*, 1985), considerably more radiation penetrates into the particle in the former case and more rapid drops are observed as the distance from particle surface is increased. For example, the energy density decreases at a distance of 0.04 μm by more than tenfold for $\lambda = 0.69\ \mu m$ and by a factor of 2.6 for $\lambda = 10.6\ \mu m$ for all angular directions.

The maximum density of energy inside the titanium particle is localized at $\theta = 180°$ and the minimum value is at the angular interval $\theta = 12 \div 10°$ (Figure 6.12). As the intensity inside the titanium particle is maximal on the particle surface in the illuminated hemisphere for all angular directions, in Figure 6.12 the intensity distribution is shown on the particle surface for two wavelengths. As the angle θ changes from $\theta = 0°$ to $\theta = 180°$, the intensity is increased smoothly in the interval $\theta < 120°$. For $\lambda = 10.6\ \mu m$ the angular dependences of the intensity becomes oscillatory with increasing angle. In the case of $\lambda = 0.69\ \mu m$ the intensity continues to decrease to the minimum value, then it is slightly increases.

It should be noted that the intensity distributions inside particles of different sizes are qualitatively the same as in Figure 6.12. The values of the intensity inside the particles of various radii are decreased e times when moving to depth 0.017 μm from the particle surface to its center for $\lambda = 0.69\ \mu m$ and 0.042 μm for $\lambda = 10.6\ \mu m$. With an increase

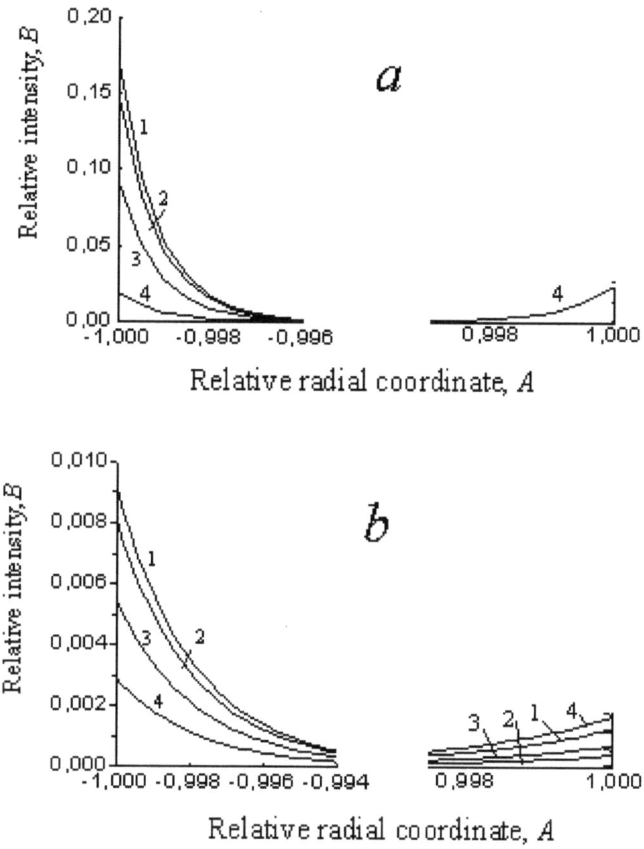

Figure 6.11. Relative intensity B versus the relative radial coordinate A for different directions inside the titanium spherical particle with radius $R = 10\ \mu m$: *1* – $\theta = 0\text{–}180^0$; *2* – $\theta = 30\text{–}210^0$; *3* – $\theta = 60\text{–}240^0$; and *4* – $\theta = 90\text{–}270^0$. The incident wavelength is $\lambda = 0.69\ \mu m$ (*a*) and $\lambda = 10.6\ \mu m$ (*b*).

of the particle radius in the considered interval from $0.5\ \mu m$ to $50\ \mu m$ the value of the intensity on the particle surface in the shadow hemisphere is reduced by 4 times for $\lambda = 10.6\ \mu m$ and by some orders of magnitude in the case of $\lambda = 0.69\ \mu m$. As for the change of the maximum values of the intensity inside the particles of the different sizes, it is shown in Figure 6.13. The dependences of $B_{\max}$ on the particle radius have oscillating character with diminishing amplitude of the oscillations. Beginning from some value of the particle radius R the maximum intensity approximates to the constant value depending on the incident wavelength. For $\lambda = 0.69\ \mu m$ this value of the maximum intensity is $B_{\max} \approx 0.17$, for $\lambda = 10.6\ \mu m$ $B_{\max} \approx 0.0091$ that is less by a factor of 18.5. With a further increase of the particle radius the maximum value of internal intensity remains practically the same.

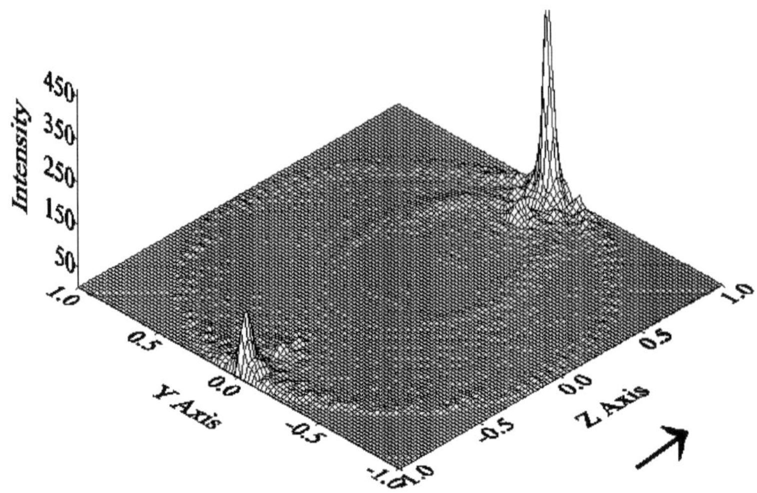

Figure 6.12. Relative intensity distribution on the titanium particle (radius $R = 10\ \mu m$) surface for two incident wavelengths: *1* – $\lambda = 0.69\ \mu m$ and *2* – $\lambda = 10.6\ \mu m$.

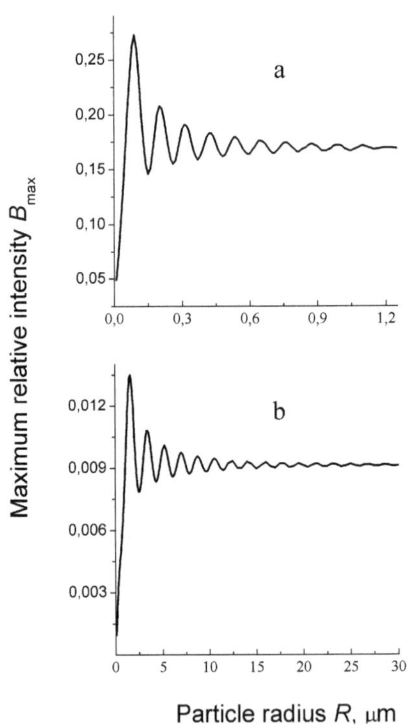

Figure 6.13. Maximum relative intensity $B_{\max}$ inside the titanium particle versus its radius R for two incident wavelengths: *1* – $\lambda = 0.69\ \mu m$ and *2* – $\lambda = 10.6\ \mu m$.

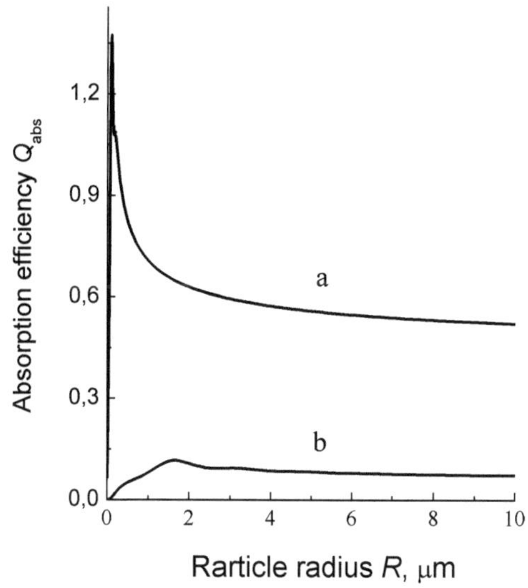

Figure 6.14. Absorption efficiency Q_{abs} of the titanium sphere versus the particle radius R for two incident wavelengths: $\lambda = 0.69\ \mu m$ (*a*) and $\lambda = 10.6\ \mu m$ *(b)*.

Consider now the integral characteristic of absorption of titanium particles – efficiency of absorption Q_{abs}. Note that the dependence $Q_{\text{abs}}(R)$ (Figure 6.14) correlates in general outline with the dependence $B_{\max}(a)$ (Figure 6.13). For small particles efficiency of absorption grows at first to the maximum value then falls and with an increasing of the particle size tends to the asymptotic value. This value is equal to 0.52 for $\lambda = 0.69\ \mu m$ and 0.12 for $\lambda = 10.6\ \mu m$, that is less by a factor of 4.3. The maximum efficiency of absorption is realized for $R = 0.1\ \mu m$ in the case of $\lambda = 0.69\ \mu m$ and for $R = 1.6\ \mu m$ for $\lambda = 10.6\ \mu m$. This bears resemblance to the behaviour of the maximum intensity inside the titanium particles in relation to the particle size.

In conclusion we would like to note that the intensity distribution inside other metallic particles will be different only quantitatively. The same is also true for other characteristics considered. For example, for aluminium particles asymptotic values of the efficiencies of absorption are 4 times less than for titanium, and the asymptotic maximum intensity is less by almost a factor of 12. The reason is that the refractive index of aluminium is essentially larger that for titanium.

A characteristic feature of metallic particles exposed to the radiation is the fact that due to high values of optical constants the intensity is distributed in a very thin surface layer within which the energy decreases exponentially. The main maximum of the internal intensity is located on the illuminated surface of the particle at the point nearest to the source of radiation and the second maximum on the shadow surface. The appearance of the second surface intensity maximum on the shadow surface of highly absorbing particles has not yet found a simple explanation. Evidently, here occurs something that reminds the "slide" of some portion of the energy penetrated inside the skin-layer (due to the so-called surface waves).

6.3 FEATURES OF ELECTROMAGNETIC FIELDS INSIDE PARTICLES WITH *n* < *1*

It should be noted that for the above-cited cases (water and metallic particles) the real part of the complex refractive index of the particle matter n is greater than 1. For these situations the following regularities of energy distributions inside spherical particles are typical:

- the energy distribution is homogeneous for small spheres, becomes substantially inhomogeneous for the particles with the diffraction parameters $x = 2\pi R/\lambda > 3 \div 5$;
- the highest values of the energy density are observed on the main diameter of the particle (the diameter parallel to the propagation of incident laser radiation); and
- on entering the radiation inside the particle in the center of its illuminated surface $(r = R, \theta = 180°)$ and in some region near this point $B < 1$.

For many metals and their oxides in some spectral intervals the real part of the refractive index $n < 1$. As this takes place and the particle material absorption is great ($\kappa > 0.6$), then the usual picture occurs wherein the main part of the penetrated (inside the particle) energy is absorbed in a sufficiently thin layer near the illuminated particle surface. A maximum of the energy is located on the main diameter of the particle. But if $\kappa < 0.6$, the internal intensity distribution has an unusual character with the extreme values located in the illuminated hemisphere, not on the main diameter, but on the surface of a certain circular zone, with its position, described by the angular coordinate θ, being dependent on the size and optical constants of the particle material (Astafyeva and Prishivalko, 1990*a*).

The typical distribution is shown in Figure 6.15. It is the internal intensity distribution in the aluminium oxide particle irradiated by laser radiation with $\lambda = 10.6\ \mu m$ in air. The complex refractive index for aluminium oxide is $N = 0.78 + i0.05$ (Belov, 1989). Figure 6.15 shows that in contradistinction to the spheres with $n > 1$, the relative intensity $B > 1.0$ in the center of the illuminated surface of the particle and in some region near this point. The density of energy rapidly decreases in the direction from the particle surface to its center. The maximum values of the internal intensity are localized not on the main diameter, but in two regions, situated symmetrically about the main diameter and characterized by the angular coordinates $\theta \approx 120 \div 100°$. The maximum value of the internal intensity in this case $B \approx 1.7$ is localized at the angle $\theta \approx 117°$.

Calculations, fulfilled for the aluminium oxide particles of various sizes, show that in the considered region of particle size the highest values of B are observed on the particle surface. At the point of radiation entry into the particle $(r = R,\ \theta = 180°)$ the energy density is more than 1 and weakly depends on the particle radius. For the aluminium oxide spheres $B \approx 1.24 \div 1.27$. Some results of these calculations are illustrated in Table 6.1; it is evident from these that $B_{\max}$ increases with the particle radius, and the location of this maximum, at first, weakly oscillates and then becomes stable at $\theta = 117°$. The values of the internal intensity for so-called "side" maxima are substantially higher than on the main diameter.

Similar calculations were also made for the particles with $R = 10\ \mu m$ in a wide range of the real and imaginary parts of the refractive index (Table 6.2). This table shows that

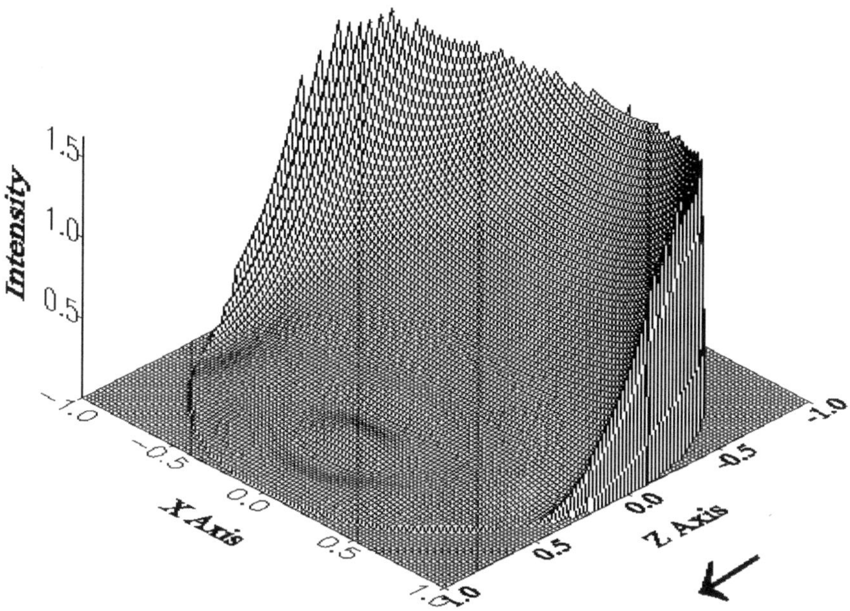

Figure 6.15. Relative intensity distribution inside the aluminium oxide spherical particle with radius $R = 25\ \mu m$. The incident wavelength is $\lambda = 10.6\ \mu m$. The arrow shows the direction of the propagation of incident radiation.

the increase of n and κ lead to the decrease of the side internal energy maxima. As for the angular coordinate of the maximum, it drops with increasing κ, and grows at first with increasing n decreasing further as n approximates to 1.

The mentioned peculiarities of the intensity distribution are explained by the specific character of the diffraction in the case of weakly absorbing particles with the refractive index $n < 1$. Figure 6.15 shows that presence of the absorption strongly simplifies the picture of the intensity distribution. If we take the limiting case, when $\kappa = 0$, this internal distribution is substantially complicated. In support of this, the intensity distribution inside the air bubble of radius $R = 2\ \mu m$ in polystyrene under the action of radiation with

Table 6.1. Maximum value of relative intensity B and its angular coordinate for aluminium oxide particles of different radii.

$R(\mu m)$	B_{max}	θ_{max}(deg)	$R(\mu m)$	B_{max}	θ_{max}(deg)
4	1.37	121	25	1.66	117
6	1.34	114	30	1.72	117
8	1.42	121	35	1.77	117
10	1.43	113	40	1.82	117
16	1.53	119	45	1.86	117
20	1.59	117	50	1.90	117

Table 6.2. Maximum value of relative intensity B and their angular coordinate for $\lambda = 10.6\ \mu m$, $R = 10\ \mu m$ and variations of optical constants of particle material.

$n = 0.78$			$\kappa = 0.05$		
κ	$B_{\max}$	$\theta_{\max}$(deg)	n	$B_{\max}$	$\theta_{\max}$ (deg)
0	1.643	104	0.3	2.922	136
0.005	1.615	105	0.4	2.641	129
0.01	1.589	106	0.5	2.261	126
0.02	1.542	107	0.6	1.991	125
0.05	1.434	113	0.7	1.69	118
0.1	1.339	125	0.8	1.374	112
0.2	1.266	146	0.85	1.243	126
0.3	1.235	167	0.9	1.152	130
0.5	1.188	180	0.95	1.062	132

$\lambda = 0.69\ \mu m$ is shown in Figure 6.16. We assume that the absorption of polystyrene is practically absent at this wavelength. In this case, in parallel with the side maxima of internal intensity ($B \approx 3.7$), there is the region of the maximum energy density on the main particle diameter in its shadow hemisphere ($B \approx 2.1$). For some particle sizes the maxima of the intensity on the main diameter become higher than the side maxima. Even so, the side intensity maxima still occur, although they are not observed inside the spherical particles with $n > 1$.

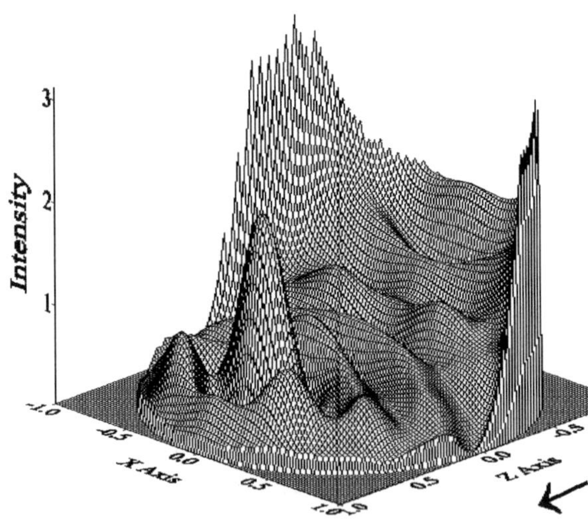

Figure 6.16. Relative intensity distribution inside the air bubble in polystyrene. The radius of the bubble is $R = 2\ \mu m$. The incident wavelength is $\lambda = 0.69\ \mu m$. The relative refractive index in the bubble is $\bar{N} \simeq 0.625 + i0$. The arrow shows the direction of the propagation of incident radiation.

It follows that presence of the side intensity maxima with $B > 1$ is the specific character of weakly absorbed and transparent spherical particles with the refractive index $n < 1$.

6.4 ANALYSES OF THE EFFECTS OF ABSORPTION ON SPATIAL STRUCTURES OF MDRs IN SPHERICAL MICRO-PARTICLES

Resonance peaks near the natural frequencies of oscillation in dielectric spheres have been observed in spectra of the optical radiation pressure (Fahlen and Bryant, 1968; Ashkin and Dziedzic, 1977, 1981; Chylek *et al.*, 1983), elastic scattering and inelastic fluorescence (Benner *et al.*, 1980; Hill *et al.*, 1984, 1985), and Raman scattering (Thurn and Kiefer, 1984, 1985; Lettiery and Preston, 1985). The resonance frequencies depend on the structure (size, shape, and refractive index) of the particles. Hence, the resonances have been termed *morphology-dependent resonances* (MDR's), or structure or dimensional resonances. The resonant modes give rise to the ripple structure that appears in scattering and absorption spectra (Chylek, 1976; Chylek *et al.*, 1978). The resonances are also termed partial wave resonances to emphasize that they are attributable to resonances in particular expansion coefficients of the field.

A geometro-optical interpretation of MDR's is that of rays or beams propagating around the sphere, confined by an almost total internal reflection. After circumnavigating the sphere the beam returns to its starting position in phase. A beam approaches the internal surface of the sphere beyond the critical angle and so is almost totally reflected. Hence, it takes longer for the energy of a mode to leak out of the sphere, and very large energy densities can accumulate in the resonant mode.

Electromagnetic diffraction theory for micron-sized homogeneous spherical particles (Kerker, 1969; van de Hulst, 1981; Bohren and Huffman, 1983; Barber and Hill, 1990) predicts a large number of the partial-wave resonances for the defined particle radius and complex refractive index of its matter at different values of the size parameter $x = 2\pi R/\lambda$. The resonances of a dielectric sphere are modes with transverse magnetic (TM) and transverse electrical (TE) polarization for different combinations of the azimuthal mode number l and radial mode number, or order, s. The resonant modes have different widths, or quality-factor (Q) values. The resonant Q values increase slowly with l and decrease more sharply with s.

In parallel with the spectrum of natural frequencies and their quality factors an important electrodynamic characteristic of a micro-resonator is the spatial structure of the electromagnetic fields inside it. This structure is formed by the summation of the spatial structures of natural resonant modes and non-resonant background. The resonant modes usually occur inside the medium with small values of the absorption index (i.e., in almost transparent medium). Therefore, in studies of the resonances the absorption is rather commonly ignored. However, the investigation (Astafyeva and Lednyeva, 1999) has shown that the absorption of the particle matter may substantially change the resonant modes characteristics.

The spatial structure of the internal fields of the spherical micro-particle (on condition that the resonant modes are realized) is considered here. In so doing we take into account the particle matter absorption. The spatial structure of each mode is characterized by its number l and order s. The former determines the quantity of peaks in the internal intensity distribution as the polar angle θ changes from 0 to 180°, and the latter equals the quantity of peaks occuring under a change of radial coordinate from 0 to R. If the incident wave is non-polarized, then the internal fields can be considered as the superposition of TM and TE waves. In the case of the TM resonant mode, the spatial distribution of the energy density has the form (see Eqs.(6.1.3–5)):

$$\begin{aligned} B_l^{\mathrm{TM}}(a,\theta) = \frac{1}{|N|^2 x^2 a^2} \frac{(2l+1)^2}{l^2 (l+1)^2} \left|\alpha_l^{31} \psi_l (Nax)\right|^2 \\ \times \left[\pi_l^2(\cos\theta)\cos^2\varphi + \tau_l^2(\cos\theta)\sin^2\varphi\right] \end{aligned} \tag{6.4.1}$$

and

$$\begin{aligned} B_l^{\mathrm{TE}}(a,\theta) = \frac{(2l+1)^2}{|N|^2 x^2 a^2} \left|\beta_l^{31}\right|^2 \left\{ \frac{\sin^2\theta\cos^2\varphi}{|N|^2 x^2 a^2} \left|\psi_l (Nax)\right|^2 \pi_l^2(\cos\theta) \right. \\ \left. + \frac{\left|\psi_l' (Nax)\right|^2}{l^2 (l+1)^2} \left[\tau_l^2(\cos\theta)\cos^2\varphi + \pi_l^2(\cos\theta)\sin^2\varphi\right] \right\} \end{aligned} \tag{6.4.2}$$

in the case of the TE mode. One can characterize the internal intensity distribution of the spherical micro-particle by calculating the quantities $\left|\alpha_l^{31}\right|^2$ (TM polarization) or $\left|\beta_l^{31}\right|^2$ (TE polarization). The magnitude of these coefficients depend on the variations of particle size and the complex refractive index of the particle matter and determines the spatial distribution of the internal field in the microsphere.

Calculations of modules of the amplitude coefficients $\left|\alpha_l^{31}\right|^2$ at resonances were performed for the case of the dye droplet with the refractive index $N = 1.626 + i10^{-7}$ (Astafyeva and Lednyeva, 1999). Investigations have shown that with increase in the number of mode l the amplitude coefficient grows at first, then attains the maximum value (e.g., $\left|\alpha_{34}^{31}\right|^2 \approx 899.8$ for $s = 1$) and finally begins to decrease to small values of less than 1. With a further increase in l they slowly fall. Such change of the amplitude coefficients at TM-resonance does not depend on the mode order s, as shown in Figure 6.17, where the envelopes of squared modules of the amplitude coefficients at TM-resonances of three orders ($s = 1, 2, 3$) are illustrated. Figure 6.18 also shows that with increasing mode order their frequencies shift to the right on the frequency scale (i.e., to blue region of the spectrum). Therefore with increasing mode order the squared modules of the amplitude resonant coefficients of the internal field attain the maximum value for progressively higher numbers of modes. So, for the second order mode the maximum value is less by a factor of 2.9 than for the first order mode, attains the value of $\left|\alpha_l^{31}\right|^2 \approx 308.9$, and is realized for the higher number of $l = 47$. In the case of the resonant mode of the third order the peak value is still less by a factor of 2.2 $\left|\alpha_l^{31}\right|^2 \approx 138.2$ and is attained for an even higher mode number $l = 58$. The decrease of the real part of the complex refractive index of the particle matter leads to a shift of the peak values of the amplitude coefficients of resonant

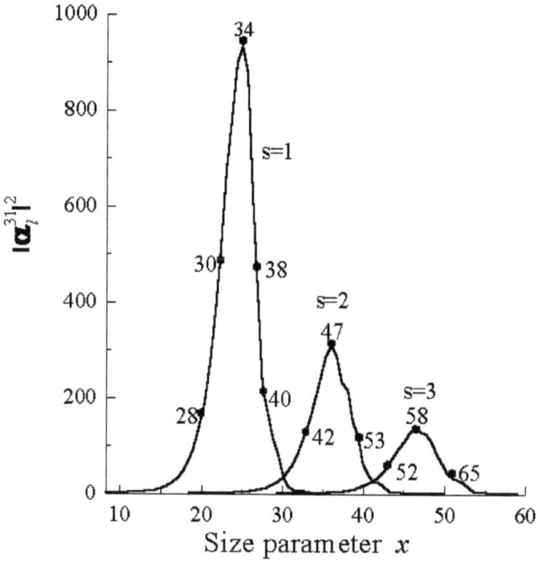

Figure 6.17. Dependences of squares moduli of the amplitude coefficients of the magnetic resonant modes of three orders ($s = 1, 2, 3$) on the size parameter ($x = 2\pi R/\lambda$) for $N = 1.626 + i10^{-7}$. The asterisks indicate positions of the resonant modes; the figures near the asterisks are the numbers of resonances.

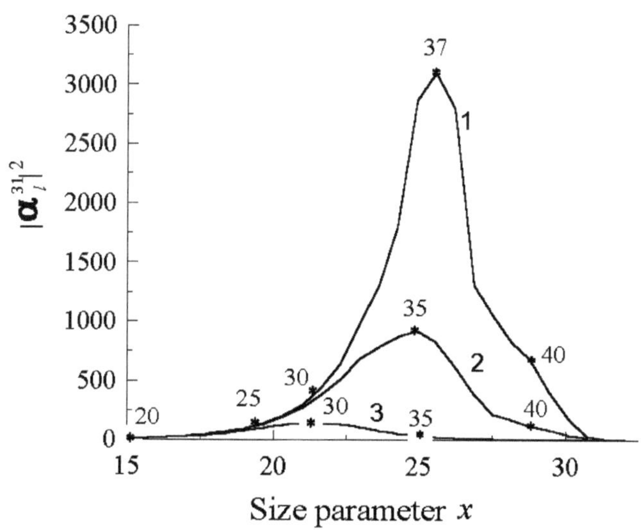

Figure 6.18. Dependences of squares of moduli of the amplitude coefficients of the magnetic resonance modes on the size parameter for order $s = 1$, $N = 1.626 + i\kappa$, and $\kappa = 10^{-7}(1)$, $10^{-6}(2)$, and $10^{-5}(3)$. The asterisks indicate positions of the resonant modes; the figures near the asterisks are the numbers of resonances.

modes in the direction of higher mode numbers and leads at the same time to decreasing magnitudes of the peak values.

Figure 6.18 clearly illustrates the influence of the change of absorption index on the values of the amplitude resonant mode coefficients by the example of magnetic resonances. The coefficients of the resonant modes with peaks and nearby values are more sensitive to the absorption change. First, with an increase of the absorption index κ the coefficient values $\left|\alpha_l^{31}\right|^2$ decreased, second, the resonant mode number, when the maximum coefficient value $\left|\alpha_l^{31}\right|^2$ occurs, decreases also. For example, for the resonant modes of the dye droplet with $N = 1.626 + i10^{-7}$ the tenfold increase of κ leads to the decrease of the maximum value of the coefficient $\left|\alpha_l^{31}\right|^2$ by a factor of 6.3 and the mode number, when the peak value of the coefficient is taken place, shifts from $l = 34$ to $l = 30$.

Now consider the variations of the internal spatial structures of the dye micro-droplet in accordance with changes of the resonant mode amplitude coefficients with increasing mode numbers. For small amounts of resonance ($l < 10$) the coefficients $\left|\alpha_l^{31}\right|^2$ can be less than 1, and the internal intensity distributions in such particles have a non-resonant character (Figure 6.19). In this case the field pattern inside the micro-droplet is to a greater degree formed by non-resonant waves. Really, there are two regions of energy localizations which represent the summation of the spatial structures of resonant and non-resonant modes: less intensive in the illuminated hemisphere and more intensive (1–2 orders of magnitude higher) in the shadow hemisphere.

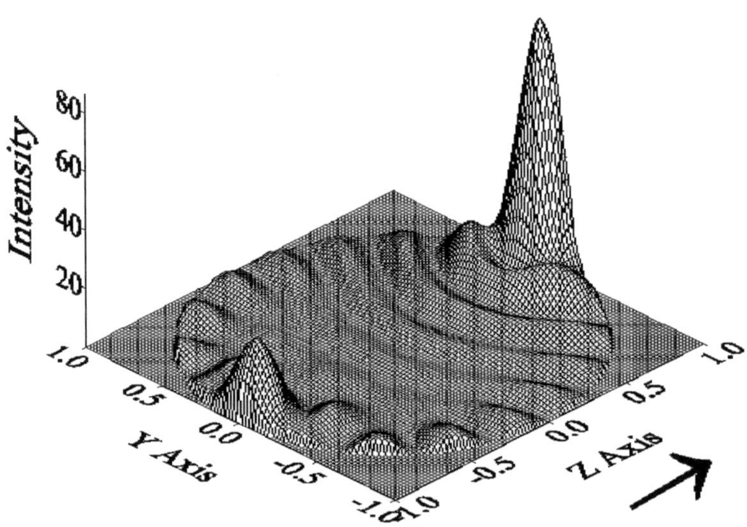

Figure 6.19. Relative intensity distribution inside the spherical particle ($N = 1.626 + i10^{-7}$, size parameter $x = 7.54871911$) for the magnetic resonant mode with number $l = 9$ and order $s = 1$. The arrow shows the direction of the propagation of incident radiation.

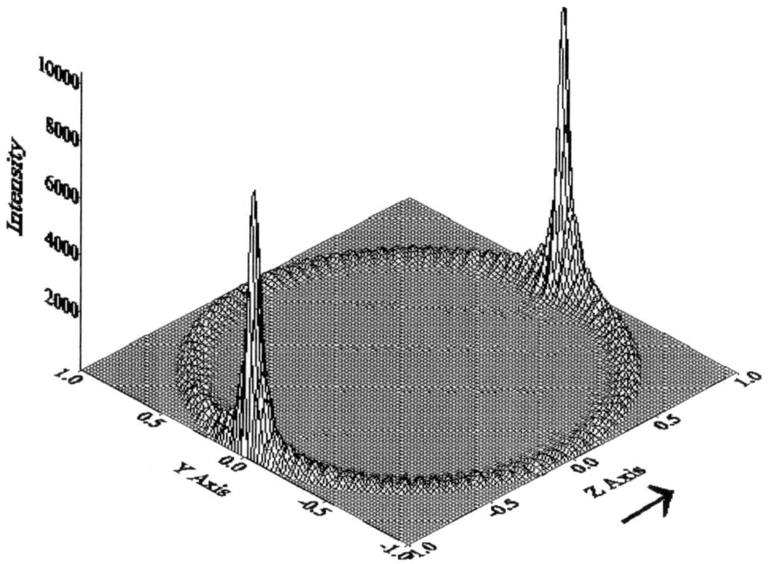

Figure 6.20. Relative intensity distribution inside the spherical particle ($N = 1.626 + i10^{-7}$, size parameter $x = 32.678425$) for the magnetic resonant mode with number $l = 47$ and order $s = 1$. The arrow shows the direction of the propagation of incident radiation.

A further increase in the numbers of resonant modes leads to the domination of the intensities of the resonant modes over the intensities of non-resonant waves. In this case the internal intensity distribution in the micro-droplet is determined for the most part by the spatial structure of the resonant mode: on the main diameter in the illuminated and in the shadow hemispheres two high intensive regions of energy localization near the particle surface are realized (Figure 6.20). The maxima of the internal intensity in these regions take place when the coefficients $\left|\alpha_l^{31}\right|^2$ are maximal. With increasing mode number l the internal intensity of the resonance decreases and the resonance intensity value becomes commensurate with the non-resonant waves intensity, focused in the shadow hemisphere (Figure 6.21).

It is also of interest to consider the change in the internal intensity distributions in the sphere with increasing mode number. As illustrated in Figure 6.17, the internal amplitude coefficient values for the modes with higher orders can be both lower and higher than for the modes with lower orders. Therefore the intensity of the spatial structure of the resonance in the microsphere with the high order value may be both lower and higher than the intensity of the resonance with a low order value in spite of the fact that quality factors decrease with increasing mode order. For example, for the second mode order with mode number $l = 47$ and quality factor $Q \approx 3 \times 10^6$ the maximum value of the internal intensity distribution becomes $B_{\max} \approx 4 \times 10^5$ (i.e., it is higher than $B_{\max} \approx 7.6 \times 10^3$ for the same numbered order of the first mode with a higher quality factor $Q \approx 5.4 \times 10^6$). In this case, as illustrated in Figures 6.20 and 6.22, the spatial structure of the resonant mode of higher order ($s = 2$) is more intensive.

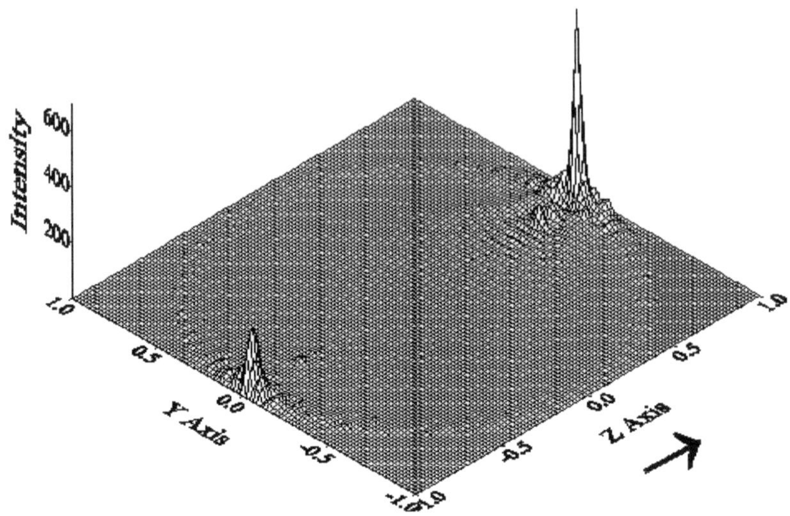

Figure 6.21. Relative intensity distribution inside the spherical particle ($N = 1.626 + i10^{-7}$, size parameter $x = 36.534049$) for the magnetic resonant mode with number $l = 53$ and order $s = 1$. The arrow shows the direction of the propagation of incident radiation.

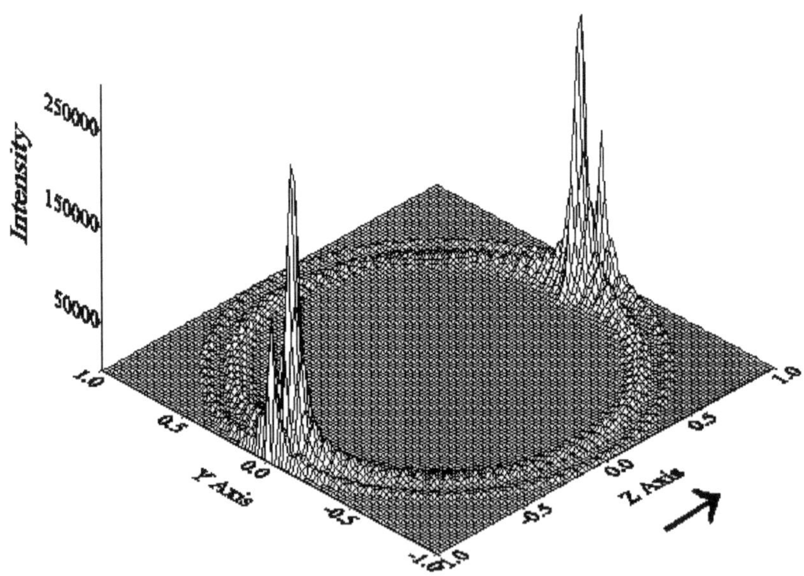

Figure 6.22. Relative intensity distribution inside the spherical particle ($N = 1.626 + i10^{-7}$, size parameter $x = 36.077442$) for the magnetic resonant mode with number $l = 47$ and order $s = 2$. The arrow shows the direction of the propagation of incident radiation.

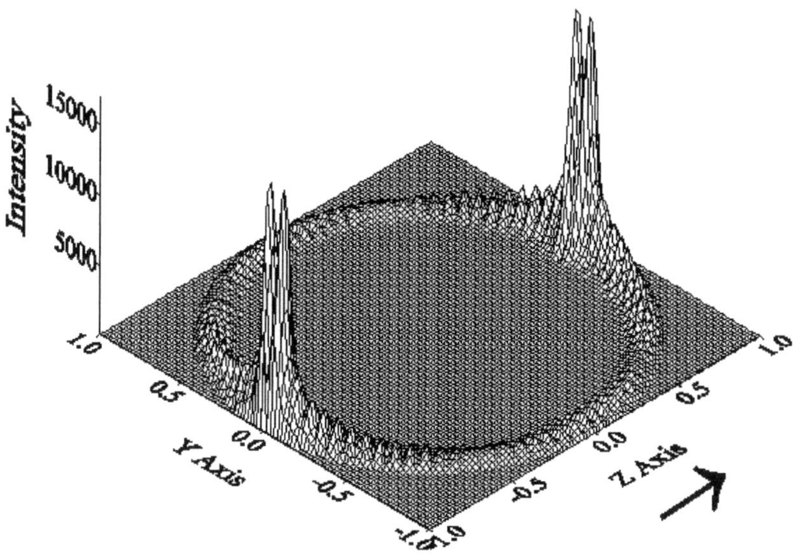

Figure 6.23. Relative intensity distribution inside the spherical particle ($N = 1.626 + i10^{-7}$, size parameter $x = 30.572146$) for the magnetic resonant mode with number $l = 43$ and order $s = 1$. The arrow shows the direction of the propagation of incident radiation.

Similarly, for TE-resonant modes the envelopes of the squared modules of the amplitude coefficients of the internal fields were obtained for a large range of the numbers. They are shifted on the frequency scale to the right with regard to the TM resonances, and the maximum values of the coefficients $|\beta_l^{31}|^2$ are lower than for the TM modes. This causes a lower quality factor for TE modes in comparison with TM modes because of the presence of the radial field component. This component causes some dissimilarity of the spatial structure of the resonant TE mode from the TM mode. In Figure 6.23 the spatial field distribution of the resonant TE mode inside the spherical particle is shown. We see that in each region of high energy density of the TE field two subregions of the maximum energy density, shifted symmetrically through some small angle relative to the main particle diameter, are resolved. Maximum value of the energy density of the TE field, as a rule, is lower than for the TM mode.

Thus these results show that in the presence of some small absorption ($\kappa \lesssim 10^{-7}$), the mode amplitude coefficients and the corresponding spatial field structures inside the spheres can be reduced by two or more orders of magnitudes especially for modes with high quality factors. The influence of the weak absorption of particle material on the spatial field structures inside the spheres differs from the influence on quality factors. Considering the quality factor only does not permit prediction the internal intensity in the particle with greater or lesser accuracy.

6.5 SIMILARITY CRITERION FOR OPTICAL FIELDS IN WEAKLY ABSORBING SPHERICAL AEROSOL PARTICLES

The aim of this section is to describe on the quantitative level the relations between size parameter x and absorption index κ for fixed values of refractive index n which, being fulfilled, provide the similarity of optical fields inside particles in the sense of small variations (within the limits of several percent) of internal intensity B at the internal point with the same coordinates. Such similarity, when present, enable us to evaluate the field inside the low absorbing particle through the use of reference data, which describes the field distribution in non-absorbing ($\kappa = 0$) particles with the same n and x.

For this purpose, we carried out a detailed study of optical fields inside uniform spherical particles whose optical constants varied within the ranges $0.15 \leq n \leq 2.5$ and $10^{-10} \leq \kappa \leq 10^{-1}$; the size parameter varied within the range $0.1 \leq x \leq 1000$ (Babenko and Prishivalko, 1997). It is convenient to use average absorption characteristics for obtaining the most general results and conclusions for such a wide range of n, κ, and x. If optical fields of two particles are similar in the sense mentioned above, then volume averaged internal fields $\bar{B}$ also have close values. It is evident that the closeness of $\bar{B}$ is the necessary, but not sufficient, condition of similarity. However, if a continuous variation of κ from a negligibly small value of about 10^{-10} ($\kappa = 0$ is not used because it has no physical sense) to a certain κ_1 (in what follows, it is called the boundary value), with parameters n and x being fixed, does not remove the value of $\bar{B}$ beyond the limits of a sufficiently narrow interval $\bar{B} = \bar{B}_0 \pm \Delta\bar{B}$, where $\bar{B}_0$ is the volume-averaged relative intensity of the field inside a particle with $\kappa = 10^{-10}$, it is quite reasonable to assume that internal optical fields are similar throughout the range of $\kappa = 10^{-10} \div \kappa_1$.

An additional the advantage of the quantity $\bar{B}$ is that it is directly related to the absorption efficiency Q_{abs} by the expression (5.2.18):

$$\bar{B} = 3Q_{\text{abs}}/8n\kappa x \tag{6.5.1}$$

from which it follows that laborious calculation of internal field intensity averaged over the particle volume may be replaced by simple calculation of Q_{abs}. However, it is good to bear in mind that for $\kappa \leq 10^{-6}$ the conventional calculation of Q_{abs} through the difference of extinction and scattering efficiencies is numerically unstable, and one should use in this case other techniques, such as the Kattawar–Eisner formula (Section 5.2). From Eq.(6.5.1) it follows that the dependence of Q_{abs} versus κ must be linear in the "similarity region", which can be realized only for sufficiently small values of κ.

To make concrete estimates, one should specify the value of relative deviation $\Delta\bar{B}/\bar{B}_0$. In our study, we set:

$$\delta\bar{B} \equiv \Delta\bar{B}/\bar{B}_0 \leq 0.01 \tag{6.5.2}$$

The results of calculations of optical fields inside low absorbing particles (the incident wave is naturally polarized) showed that, in the case where condition (6.5.2) is fulfilled, the absolute value of intensity $B(n, \kappa, x)$ at any internal point (r, θ) of a particle with n, κ, and x lying within the limits of the intervals noted above (with the exception of some special cases) differed from $B_0(n, x)$ by no more than 5–6%. In most cases, the difference is considerably smaller: $\delta B \equiv |B(n, \kappa, x) - B_0(n, x)| / B_0(n, x) < 3\%$.

One such special situation is observed for the large refractive indices $n \geq 2$ and sufficiently large x. In the case of $50 \leq x \leq 200$, values of B in a small neighborhood of separate points on the major particle diameter (as a rule, in the illuminated hemisphere) may be understated by up to 15–20% with respect to B_0. This region of strongly understated values of B is the only one and is rather small in comparison with the particle ($\delta V/V < 1\%$). As x increases further ($x \geq 200$) the regions of understated values of B are localized both directly on the illuminated hemisphere and near it. For $x > 450$, the difference between B and B_0 on the illuminated particle surface (for $\theta = 180°$) may reach $\sim 25\%$, but the relative size of the region where B is understated decreases by 15–25%. However, one more region of understated intensities of B may be formed. In this region, the intensity differs from B_0 by 10–15%, and it is found on the surface of spherical layer of the particle (or near it) with angular coordinates $\theta \sim 90$–$130°$.

One more special situation is realized for particles with a large refractive index but small values of x. For $x \sim 0.5$ and $n \geq 2$, condition (6.5.2) is insufficient to provide the similarity of internal optical fields within the limits of $\pm(5$–$6)\%$ mentioned above. In this case, one must fulfill the relation $\delta\bar{B} \leq 0.005$, and when $n \geq 2.4$, one must satisfy the stronger inequality $\delta\bar{B} \leq 0.003$.

It should be emphasized that the presence of special regions in which the relative deviation δB may reach 15–25% (in some cases, it may be even larger) does not impair the criterion (6.5.2). The point is that, in the case where one evaluates the influence of any non-linear phenomena inside particles, it suffices to know the absolute magnitude and the position of main maxima of the internal intensity. Test calculations show that, for κ varied in the range $10^{-10} \div \kappa_1$, the deviations at principal maxima $\delta B_{\max}$ were smaller, as a rule, than deviations at other points of the particle. If condition (6.5.2) is fulfilled, $\delta B_{\max} \leq (3 \div 4)\%$. As for the aforementioned small volume regions in which relative deviations δB may reach rather large values, they are observed far from main maxima and for B that are considerably smaller than $B_{\max}$. For example, in the specific case of $x = 474.2$ and $n = 2.4$ (the boundary value is $\kappa_1 = 4.365 \cdot 10^{-6}$), the relative deviation for the principal maximum, which is positioned on the major diameter in the shadow hemisphere ($\theta = 0$, $r/R \sim 0.6$) and has the magnitude $B_{\max} = 4410.59$ for $\kappa = 10^{-10}$, is as small as $\delta B_{\max} = -0.4\%$. In this case the largest deviation of the internal intensity is obtained on the illuminated surface ($\theta = 180°$, $r = R$) and equals $\delta B = -23.9\%$. However, the value of B at this point is rather small and $B/B_{\max} = 6.35 \cdot 10^{-4}$. It is obvious that the relative deviation of B of about 24%, the absolute value being small, cannot have a substantial influence on the general picture and those characteristics of the internal field that are associated with the realization of non-linear effects. In other cases of large relative deviations δB, which we mentioned above, the situation is quite the same ($B/B_{\max}$ is always smaller than 0.1).

Thus, it may be concluded that relation (6.5.2) may be used with a high degree of reliability as the similarity condition for internal optical fields in weakly absorbing spherical particles. Note that if condition (6.5.2) is fulfilled, the maximum deviations of B in large weakly absorbing particles are observed not for the side directions but, as a rule, at the points found on the major particle diameter but far removed from the point with the principal maximum B_{max}. Even in the case where the principal maximum of relative intensity is not localized on the major diameter (for $n < 1$) (see Section 6.2), the maximum relative deviations for δB, as a rule, are observed for the direction $\theta = 0$–$180°$.

Note that the transition from condition (6.5.2) to a less rigid condition $\delta \bar{B} = \Delta \bar{B}/\bar{B}_0 \leq 0.02$ does not strongly deteriorate the similarity of the internal fields. Sample estimates made for this case show that the relative deviations for δB_{max} increase by a factor of about two; at other points inside particles, the increase does not exceed a factor of two.

The employment of similarity criterion in practice requires knowledge of the boundary values of absorption index κ_1. In what follows, we present the plots illustrating the calculations of κ_1 as a function of the real part n of the refractive index for a number of fixed values of size parameter x. The calculations show that κ_1 is mainly determined by the value of x, and the dependencies $\kappa_1(n)$ is substantially different in the regions of small and large x.

In the Rayleigh region (Figure 6.24) the dependencies $\kappa_1(n)$ are smooth; κ_1 gradually decreases with increasing x (for $x = 0.1$ and 0.5, κ_1 is found in the range of 0.1–1.0, whereas for $x = 1.0$, typical values of κ_1 are about 0.05). One can see two characteristic features in the plots: a dip that gradually moves in the direction of small n with increasing x, and the presence of points where $\bar{B}$ exceeds $\bar{B}_0$ (the dashed curves correspond to the ratio $\bar{B}/\bar{B}_0 = 1.01$). Note in this regard that, for $x \geq 1.2$, the domain of parameters under study always contains the points where $\bar{B}$ is smaller than $\bar{B}_0$ (in subsequent figures, the curves correspond to $\bar{B}/\bar{B}_0 = 0.99$). The data of Figure 6.24 suggest an obvious (in view of the fact that the internal field is virtually uniform in the Rayleigh region) conclusion

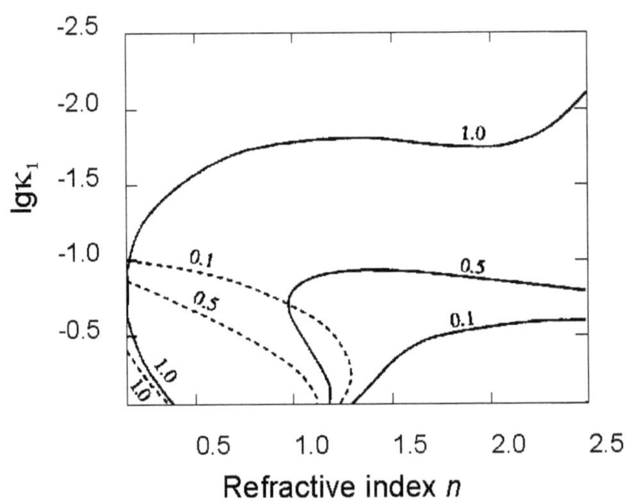

Figure 6.24. Boundary value of absorption coefficient κ_1 versus the refractive index n for small size parameter x. The values of x are specified near the curves. $\bar{B}/\bar{B}_0 = 0.99$ (solid curves) and 1.01 (dashed curves).

that internal fields of weakly absorbing spherical particles are similar in this region, because the aforementioned values of κ_1 most likely belong to the region of strong absorption.

As the size parameter increases further (Figure 6.25) the picture becomes more complex due to the emergence of the κ_1 resonances (MDRs) whose positions on the n-axis coincides with the positions of the so-called morphological absorption resonances (see Section 5.4) of the dependence $Q_{\text{abs}}(n)$ for fixed x The resonant amplitude coefficients α_l^{31} and β_l^{31} of the internal field (see Eq.(6.1.2)) are specified near the corresponding peaks. Note that the resonant modes have the indices $l \geq x$. As x and l increase, resonances increase in height, but their half-width decreases. The similarity of internal fields sharply deteriorates in the presence of MDR. In the particular case of $x = 5$, the background non-resonant value of κ_1 (for $n > 1$) is of the order of 10^{-3}, and, for narrow MDRs the corresponding value decreases to approximately $5{\cdot}10^{-5}$. In this connection, we note that the similarity of optical fields inside the rather large weakly absorbing particles cannot be correctly studied without giving the inclusion of the resonant absorption structure into consideration. One more feature of Figure 6.25 is that κ_1 increases by almost an order of magnitude in the region of small $n < 0.5$.

As x increases, the number of resonant modes in the dependencies $\kappa_1(n)$ becomes greater. Figure 6.26 presents the comparison of the dependencies $\kappa_1(n)$ and $\bar{B}(n)$ in a narrow range of n near $n = 1.55$ for $x = 50$. In this case, $\bar{B}(n)$, the dependence for the internal intensity averaged over the particle, contains two extremely narrow resonances for

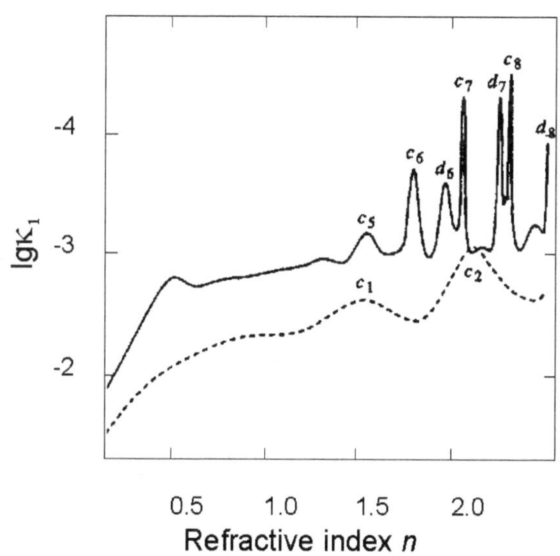

Figure 6.25. The same as in Figure 6.25, but for the size parameters $x = 2$ (dashed curve) and 5 (solid curve). $\bar{B}/\bar{B}_0 = 0.99$. The resonant modes of the internal field are specified near the peaks (the upper index ^{31} is omitted).

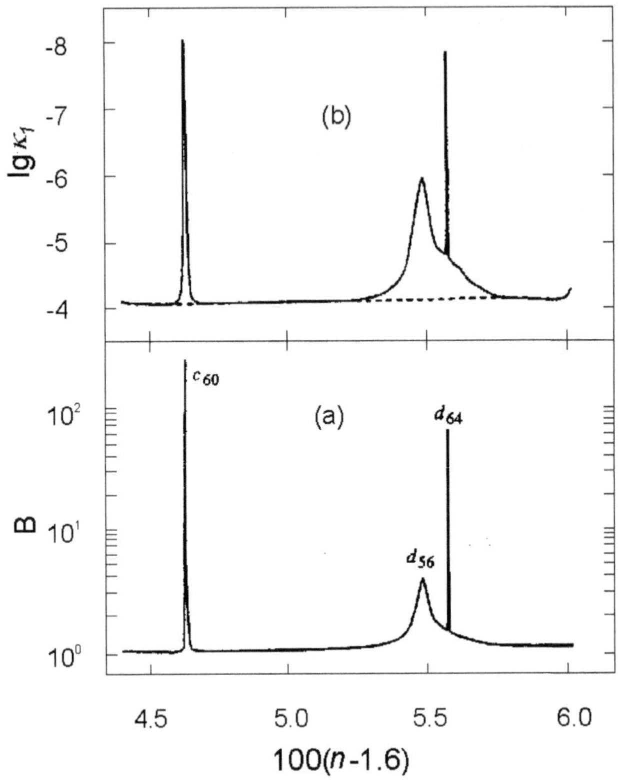

Figure 6.26. The relative intensity $\bar{B}(\kappa = 10^{-10})$ of the internal field averaged over the particle volume (*a*) and the boundary value of absorption coefficient κ_1 (*b*) versus the refractive index n in a narrow region near $n \approx 1.65$ for size parameter $x = 50$. The resonant modes of the internal field are specified near the peaks (the upper index ^{31} is omitted). $\bar{B}/\bar{B}_0 = 0.99$.

amplitude coefficients α_{60}^{31} and β_{64}^{31} and a considerably wider (but less pronounced) MDR for the coefficient β_{56}^{31}. The dependence $\kappa_1(n)$ is nearly identical in shape to the dependence $\bar{B}(n)$. We did not notice the slightest relative shift for the resonances of these two dependencies either in this case or in other cases under consideration.

Thus, the form of the dependencies presented in Figure 6.26 strongly supports the above statement that the similarity of internal fields sharply deteriorated under conditions of MDRs (κ_1 decreases in magnitude by several orders). For extremely narrow MDRs, similarity is out of the question. However, despite the great importance of MDRs to the theory of light scattering and various related non-linear phenomena (Hill and Benner, 1988), one should bear in mind that these resonances can be realized only under well controlled laboratory conditions. Under natural conditions, they are blurred due to small deviations from the spherical shape and small inhomogeneity of particles, and non-monochromaticity of incident radiation. Because of this it is quite reasonable, in our opinion, to make estimates of κ_1 for large particles without considering MDRs (such non-resonant "background" dependence is shown by the dashed lines in Figure 6.26*b*).

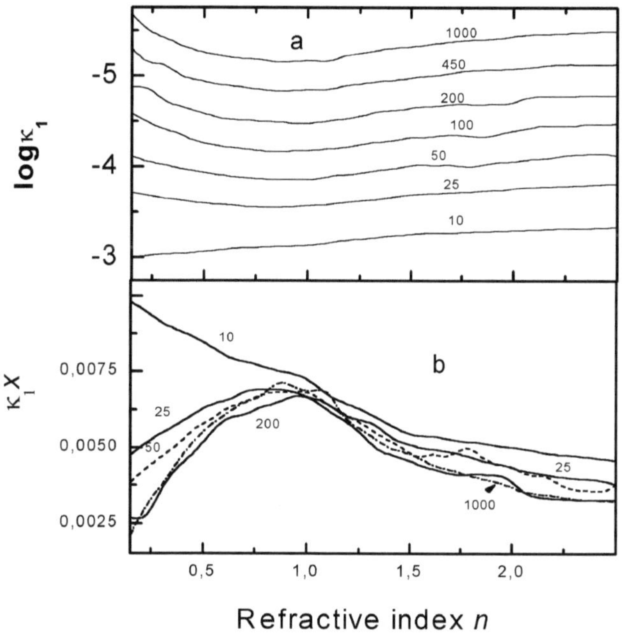

Figure 6.27. Non-resonant dependencies of the boundary value of the absorption coefficient κ_1 (*a*) and the product $\kappa_1 x$ (*b*) on the refractive index n for moderate and large values of the size parameter x. The values of x are specified near the curves. $\bar{B}/\bar{B}_0 = 0.99$.

Figure 6.27*a* illustrates non-resonant dependencies $\kappa_1(n)$ for a number of size parameters x in the range 10 to 1000 (a weak oscillation structure, which is present for small $n < 0.5$, is smoothed out in this figure). One can see from the form of the curves that, for $x \geq 25$, the boundary value κ_1 has a relatively weak dependence on the refractive index n and is determined predominantly by size parameter x (as x increases, κ_1 monotonically decreases (i.e., the similarity becomes worse)).

Baev and Belov (1995) conjectured that the product κx is the characteristic quantity making it possible to estimate the similarity of internal fields. In this regard, we present, in Figure 6.27*b,* the dependencies of $\kappa_1 x$ versus n for rather large x. One can see that the shapes of the dependencies are nearly the same (with a wide peak near $n = 1$) for $x \geq 25$. Consequently, for such values of x, the product κx indeed may be considered as a characteristic quantity (a weak dependence on n is neglected). To make a very rough estimate (especially in the case of refractive indices $n > 1$, which is often the case in practice), one may use the inequality:

$$\kappa x < 0.005 \tag{6.5.3}$$

as the similarity condition for internal fields of weakly absorbing spherical particles. It is considerably stronger and more concrete than the condition $\kappa x \ll 0.5$, which was proposed by Baev and Belov (1995).

7

Internal intensity distributions inside spherical two-layered particles (TLPs)

Some typical results of our calculations of internal electromagnetic fields inside spherical TLPs are presented in this chapter. In particular, the fields within metallic particles with an oxide surface layer are considered in Section 7.1 and fields inside hollow metallic and metallized spheres are considered in Section 7.2. The issue on morphology dependent absorption resonances in hollow spheres, that are closely related with internal fields, is discussed in Section 7.4.

Other results of computation of internal fields inside multilayered spheres can be found, for example, in the papers of Sitarski (1987*a*,*b*); Lock (1990); Mackowski *et al.* (1990); Kaiser *et al.* (1994); Johnson (1996); Skarlopoulos *et al.* (1996); Burlak *et al.* (2001).

The electric fields inside the core and shell of TLPs are described by expressions reported at the beginning of Chapter 3 (Eqs. (3.1.5–3.1.9)).

7.1 ABSORPTION AND ENERGY DISTRIBUTION INSIDE METALLIC PARTICLES WITH OXIDE FILMS

As is known, the majority of metals and, consequently, metallic particles are oxidized in air. Depending on specific physical conditions (temperature, humidity, pressure, *etc.*) and time, the thickness of the oxide layer formed can differ. For example, on aluminum particles at room temperature, the thickness of the oxide film is 0.005–0.01 μm and at melting temperature it is of about 0.2 μm (Pochil *et al.*, 1972). Presence of the oxide film leads to the substantial variation of the absorptivity of such particles, and so, to the change of the energy distribution within them. Since at the present time the powdery metals are finding increasing use in new technology (and often in active and corrosive media), it is of interest to investigate the regularities in absorption and heat evolution inside metallic particles with oxide films under the action of laser radiation of different wavelengths. The computation of energy distribution inside such TLPs was made by Astafyeva and Prishivalko (1989, 1995). The optical properties of the oxidizing metallic particles may be described in the first approximation with the help of the model of TLP consisting of metallic core and oxide cover.

In this section the absorption of such particles are estimated and the internal fields in them are considered depending on their radii and thickness of oxide film. Calculation was fulfilled for aluminum particles at two wavelengths of the incident radiation: $\lambda = 0.53$ and $10.6\ \mu m$. The optical properties of aluminum and aluminum oxide differ greatly ($N_{Al} = 0.7 + i5.2$ (Schulz, 1954; Schulz and Tangherlini, 1954) and $N_{\mathrm{Al_2O_3}} = 1.77 + i0$ (Lingart *et al.*, 1982) for $\lambda = 0.53\ \mu m$; $N_{Al} = 30.6 + i64.8$ (Palik, 1985) and $N_{\mathrm{Al_2O_3}} = 0.78 + i0.049$ (Belov, 1989) for $\lambda = 10.6\ \mu m$).

In Figure 7.1 the dependencies of the absorption efficiency factors of TLPs on their outer radius R_2 and shell thickness h in the case of $\lambda = 0.53\ \mu m$ are shown. The curves are plotted at the constant shell thickness of TLPs $h = R_2 - R_1 = \text{const}$. It follows from Figure 7.1 that with an increase of particle radius the absorption efficiency Q_{abs} of an aluminum homogeneous particle (HP) grows up to a maximum value which is attained at a small value of particle radius ($R_2 \sim 0.08\ \mu m$), and then decreases. The appearance and growth of the oxide layer results in the increase of absorption. For example, for $R_2 = 0.4$

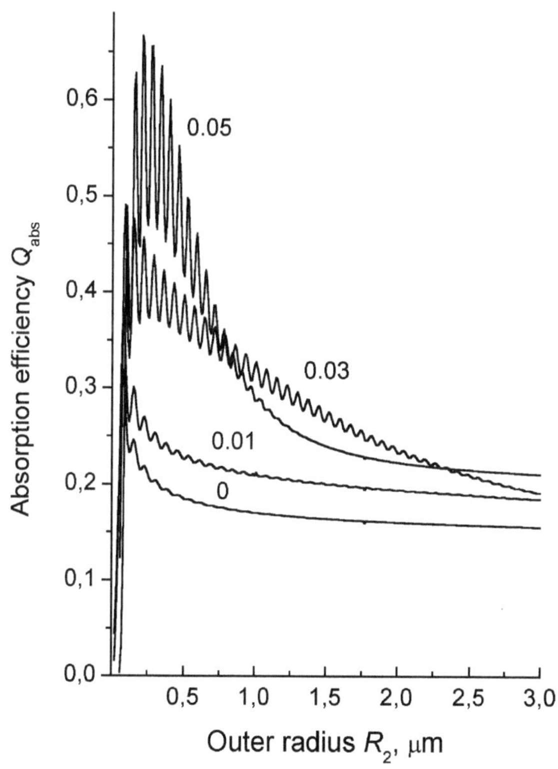

Figure 7.1. Absorption efficiency Q_{abs} of TLPs with an aluminium core and aluminium oxide shell versus the outer radius R_2. Wavelength of incident radiation is $\lambda = 0.53\ \mu m$. The numbers near the curves are the values of the shell thicknesses $h(\mu m)$.

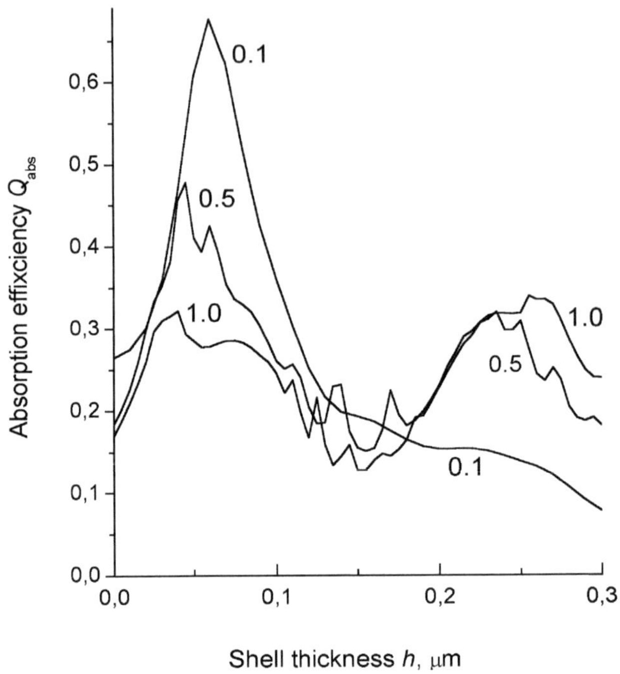

Figure 7.2. Absorption efficiency Q_{abs} of TLPs with an aluminium core and aluminium oxide shell versus the shell thickness h. Wavelength of incident radiation is $\lambda = 0.53\ \mu m$. The numbers near the curves are the radii of the particle core $R_1\,(\mu m)$.

μm the increase of oxide thickness from 0.01 to 0.05 μm leads to an increase in Q_{abs} for TLPs when compared with the aluminum particle of 1.2 and 3 times accordingly. When $R_2 > 0.9\ \mu m$ and shell thickness $h > 0.03$ the TLP absorption decreases but remains substantially more than for the aluminum HP. The increase of the aluminum particles absorption efficiency, when the oxide films appear on their surfaces, is attributable to the focussing of the laser beam transmitting the transparent outer layer of TLPs on the absorbing core. Moreover, the dependencies of the absorption efficiency Q_{abs} of TLPs on the outer layer thickness h for different particle sizes have a pronounced oscillating character.

Such behaviour of the dependencies $Q_{\mathrm{abs}}\,(h)$ may be explained with the help of Figure 7.2. The influence of the transparent outer oxide layer on TPL absorption resembles to some extent the analogous processes which are observed upon the falling of radiation on a parallel-plane transparent film depositing on a plane surface of a plate (Born and Wolf, 1999). Such thicknesses of an aluminum oxide transparent film, which result in the maximum or minimum values of reflectivity, approximately correspond to the location of maxima and minima values of $Q_{\mathrm{abs}}\,(h)$ dependence (illustrated by Figure 7.2).

The characteristic peculiarity of the curves in Figure 7.1 is the presence of regular oscillations. The amplitude of these oscillations essentially increases upon growth of the oxide film. For $h \simeq 0.05\ \mu m$ the oscillation amplitude is the third part of the absorption efficiency factor in the interval $R_2 < \lambda$. The increase of the amplitude oscillation of

Q_{abs} with growing transparent film thickness takes place, to all appearances, because of interference of light within this film. However, these oscillations are impossible to explain by only the interference effects because the oscillations are observed even when the film is absent in spite of strong absorption of the radiation with $\lambda = 0.53\ \mu m$ by the cores of TLPs (Astafyeva *et al.*, 1992). The oscillations can be explained by the fact that, when incident electromagnetic radiation acts on the particles, different orders of electric and magnetic partial waves are exited (Born and Wolf, 1999). It turns out that, in certain spherical particles, intensity of excitement of one of the electric partial waves predominates over the others. This is confirmed by the fact that the first maxima of the electric partial waves are in accordance with the oscillation maxima on the curve $Q_{abs}(R_2)$ for small size parameters. Taking magnetic partial waves into account does not affect this dependence. Increasing the particle radii means that the number of electric vibrations with growing values of l, which carry a significant contribution to the resulting electromagnetic field, also increases. Therefore, as a result of the superposition of a large number of partial waves the extreme structure of the absorption efficiency factor becomes less pronounced, and the oscillations are smoothed out.

The above explanation of the variation of absorptive properties of metallic particles with oxide films on their surface is confirmed by observation of the internal intensity distribution in such particles. The calculations show that intensity distributions inside TLPs are substantially inhomogeneous (Figure 7.3). Practically all the energy is concentrated in the thin spherical layer near the surface of the particle core. A maximum of the energy takes place on the core particle surface at various angles depending of the shell thickness. Upon removal from the core particle surface to its center the released energy decreases. Taking this into account, it is expedient to consider the intensity distribution on the core surface (Figure 7.4). It is seen that the formation and growth of the oxide film on the metallic core leads, as a rule, to an increase of internal intensity. For example, the presence of a thin film ($h = 0.01\ \mu m$) on the metallic core results in increasing the intensity by 20–100%; in this case, the dependence $B(\theta)$ qualitatively resembles that for metallic HP. The growth of shell thickness to 12.5% from the particle radius ($h = 0.05\ \mu m$) is accompanied by the increase of the maximum internal intensity by more than 15 times as compared with metallic HP. Here the effect of repeated paths of rays in the shell and its interaction with the highly absorbed core is apparently confirmed.

It is noteworthy that the energy distribution on the aluminum core surface of TLPs in the case of $h > 0.01\ \mu m$ has a strongly oscillating character (Figure 7.4) with the maximum in the shadow hemisphere. It is apparently connected with the interference effect in the transparent shell of the particle. The concentration of the radiation in the shadow hemisphere is explained by the focused properties of the transparent oxide layer on the metallic particle surface.

The dependencies $Q_{abs}(R_2)$ of the absorption efficiency for TLPs versus their outer radius R_2 at $\lambda = 10.6\ \mu m$ are shown in Figure 7.5. The curves are plotted at the constant values of the ratio of the core radius to outer radius of TLPs $R_1/R_2 = \text{const}$. In the range being considered $0 < R_2 \leq 20\ \mu m$ the absorption efficiency of aluminum HPs is small ($Q_{abs} < 0.06$), that is due to high reflection of radiation from metallic surfaces. As to the absorption of the aluminum oxide HP, the absorption efficiency factor for such a particle

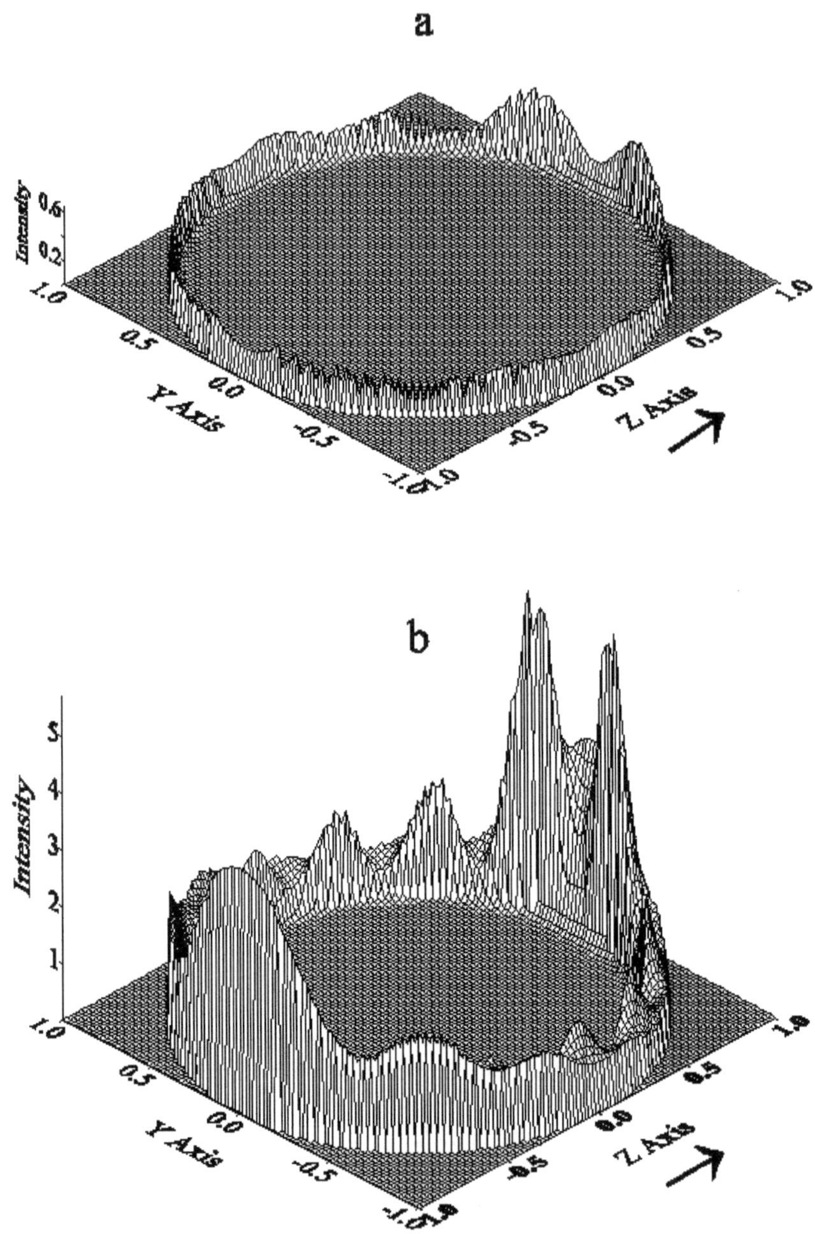

Figure 7.3. Relative internal intensity distribution in TLPs with aluminium cores and aluminium oxide shells for wavelength $\lambda = 0.53\ \mu m$. The core radius is $R_1 = 0.4\ \mu m$ and the shell thicknesses are h =0.01 μm (*a*) and 0.05 μm (*b*). The arrow shows the direction of the propagation of incident radiation.

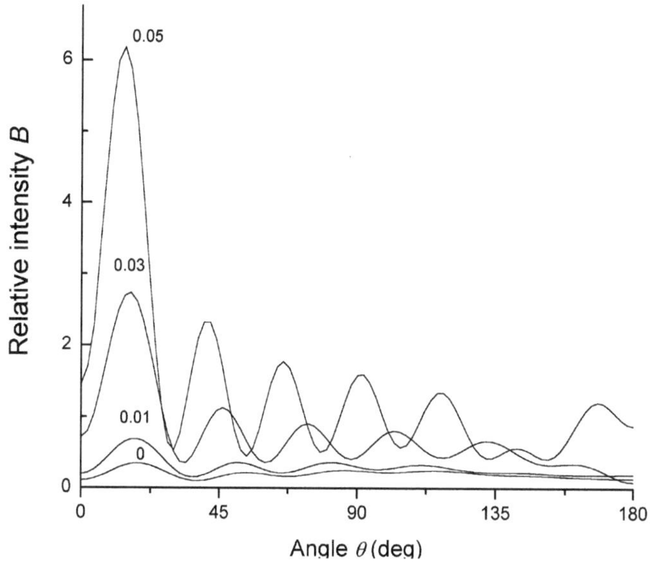

Figure 7.4. Relative intensity distribution over the aluminium core surface of the TLP. The numbers next to the curves are the values of aluminium oxide shell thicknesses in μm. The core radius is $R_1 = 0.4\,\mu m$.

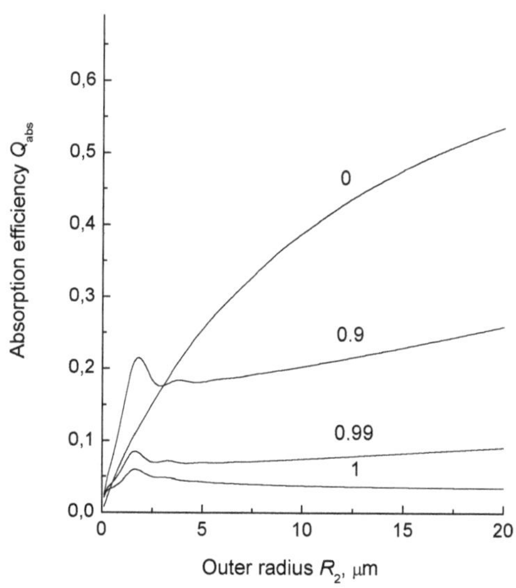

Figure 7.5. Absorption efficiency $Q_{\rm abs}$ of TLPs with aluminium core and aluminium oxide shell versus the outer radius R_2. Wavelength of incident radiation is $\lambda = 10.6\,\mu m$. The numbers near the curves are the values of the ratio R_1/R_2 of the core radius to outer radius.

increases monotonically with increasing size. It follows from Figure 7.5 that in the range of radii $0 < R_2 \leq 20\ \mu m$ the absorption efficiency of the Al_20_3 particle is essentially more than for an aluminum HP of the same radius (for $R_2 > 9\ \mu m$ it is higher by 10 or more times). When increasing the TLP radius the absorption efficiency first increases attaining the maximum value, then decreases. In this case for $R_1/R_2 = 0.99$ and $R_1/R_2 = 0.9$ the absorption efficiency begins to grow at once with a further increase in R_2.

Thus, the presence of an aluminum oxide layer on the aluminum particle leads to a noticeable increase in the effectiveness of energy absorption of such particles as compared with aluminum HPs of the same size. For example, a very thin oxide shell, which is only $\sim$1% from the particle radius ($R_1 = 9.9\ \mu m$, $R_2 = 10\ \mu m$), increases the absorption at this wavelength by a factor of two, and the oxide film, which is approximately 10% from the particle radius ($R_1 = 9\ \mu m$, $R_2 = 10\ \mu m$), increases the absorption more than 5 times. In the range $0.1 \leq R_2 < 3\ \mu m$ the absorption efficiency of a TLP with $R_1/R_2 = 0.9$ is substantially higher than that for the aluminum oxide particles.

The changes of the absorption by the aluminum particles, when oxide films are formed on them, are naturally connected with the energy distribution inside such particles. The intensity distribution inside aluminum particles, which is also typical for other metallic particles, was outlined in Section 6.2. The energy distribution inside the aluminum core of the TLP with the oxide layer resembles that inside the aluminum HP, with B_{max} being in the illuminated hemisphere. However, this B_{max} is somewhat less than that in the corresponding aluminum HP (Figure 7.6). In the shadow hemisphere the largest value B_{max} is also found on the core surface and is also just less than the corresponding B_{max} in the aluminum HP. The maximum of B is defined by the thickness of the oxide layer and decreases with its growth. The reason for this is that by increasing the thickness of the outer layer increasingly more energy is released in the outer shell and progressively less energy is reflected from the core surface.

The distribution of the internal intensity in the aluminum oxide shell is essentially different. Just as in the case of aluminum oxide HPs, the energy distribution inside the oxide layer is characterized by the presence of B maximum shifted from the main diameter to the surface of the circular region with the angular coordinate depending on the particle size and layer thickness.

For example, at the layer thickness $\Delta R = 0.1\ \mu m$ and $R_2 = 3\ \mu m$ the maximum B is situated in the angle range $\theta = 55$–45^0 (in the shadow hemisphere), whereas for $R_2 = 10\ \mu m$ this maximum shifts to angles $\theta \sim 105$–100^0 (in the illuminated hemisphere, see Figure 7.6). Notice that B_{max} are higher inside TLPs with thinner oxide layers. Moreover, the angle, where B_{max} is localized, becomes larger. For example, for the particle with outer radius $R_2 = 10\ \mu m$ and core radius $R_1 = 9\ \mu m$ $B_{\mathrm{max}} \approx 2.2$ ($\theta \approx 109^0$), for $R_2 = 10\ \mu m$ and $R_1 = 9.9\ \mu m$ $B_{\mathrm{max}} \approx 3.6$ ($\theta \approx 101^0$). For the direction $\theta = 90^0$ energy density is high enough compared with the same direction inside HP from Al_2O_3.

In conclusion of this section, consider the internal intensity distribution over the particle surface in the dependence on the angle θ (Figure 7.7). The internal intensity is maximal in the angular region $70^0 < \theta < 130^0$. The thinner the oxide layer, the higher the intensity inside the shell of the TLP.

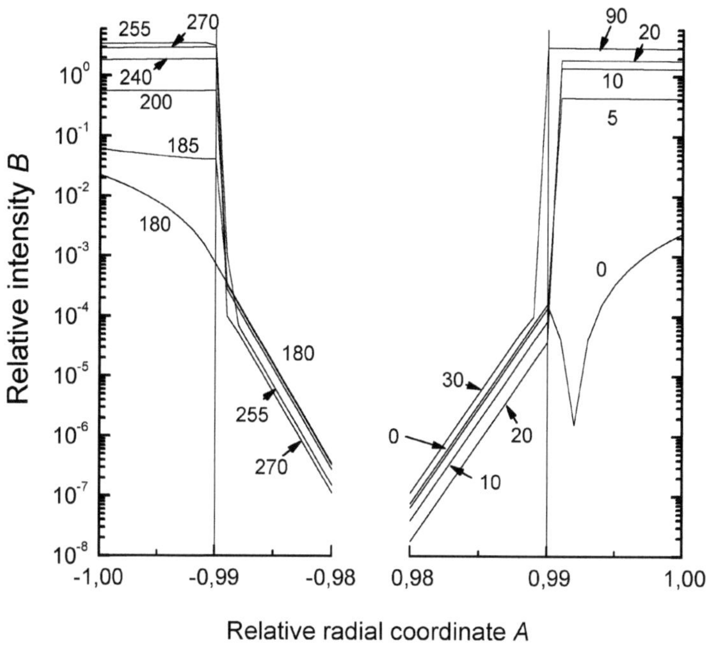

Figure 7.6. Relative intensity distribution along different directions inside TLPs with $R_1 = 9.9$ μm and $R_2 = 10$ μm for wavelength $\lambda = 10.6$ μm. The numbers next to the curves are the values of the angle θ in degrees. Vertical lines correspond to the threshold between the core and the shell.

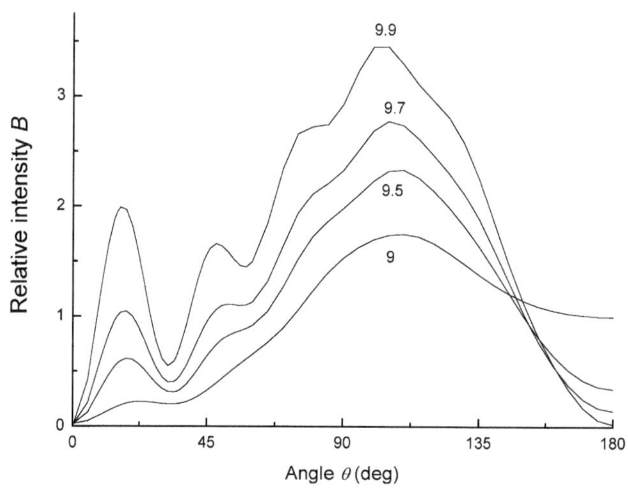

Figure 7.7. Relative internal intensity distribution over the TLP (aluminium/aluminium oxide) surface versus the angle θ for wavelength $\lambda = 10.6$ μm. The numbers near the curves are the values of the core radius R_1 in μm.

Thus, the appearance of even a thin oxide layer on the metallic particle surface leads to substantial change of the internal intensity distribution in this TLP and to the corresponding substantial growth of the absorption of the incident radiation.

7.2 ABSORPTION OF IR RADIATION AND OPTICAL FIELDS INSIDE HOLLOW METALLIC AND METALLIZED PARTICLES

In nature (Cadle, 1976; Green and Lane, 1972), as well as in different technological processes (Kudinov and Ivanov, 1981; Kudinov *et al.*, 1981) all kinds of hollow particles (HOPs) and two-layered particles (TLPs) are often encountered. The specific features of the absorption of radiation by metallic HOPs and TLPs with metallic shells and also the optical fields within such particles are discussed by Astafyeva and Prishivalko (1990*a*). As a example, aluminum was selected as the metal, and the material of the type of polyethylene with $N_1 = 1.52 + i0$ (Speranskaya and Tarutina, 1976) was selected as the core of a TLP. It is assumed that the particles are exposed to radiation with wavelength $\lambda = 10.6$ μm. The optical constants of aluminum were obtained with the help of Drude theory with correction, taking into account the model of limitation of conductivity electron mean free path in the shell volume (Kreibig and von Fragstein, 1969).

In Figure 7.8 the dependencies of the absorption efficiency factors of HOPs and spherical TLPs on their outer radii R_2 are shown. The curves are plotted holding constant the ratio of the core radii to the outer particle radii R_1/R_2. The greater is the ratio R_1/R_2, the thinner is the shell of the HOP and metallized TLP. With increasing the outer radius R_2 the absorption of the particles grows to a certain maximum. The value of this maximum depends on the thickness and the optical constants of the shell. More intense absorption takes place in the case of the thinner outer metallic layer. For example, for HOPs with radii $R_2 = 3$ μm and $R_2 = 10$ μm the decrease in the shell thickness by a factor of 10 results in an increase of the absorption efficiency by a factor of 9 and 5, respectively. Notice that the lower curve in Figure 7.8, corresponding to the ratio $R_1/R_2 = 0.99$, completely coincides with the curve, describing the absorption of the homogeneous aluminum particle (i.e., such a cavity inside TLPs does not influence its optical properties).

So, these calculations show that for $R_1/R_2 < 0.995$ the absorption efficiency of a TLP is still independent of the core material, but if $R_1/R_2 > 0.995$, then the presence of a polymer core increases the absorption of a TLP as compared with an HOP in the radius range $R_2 < 3$ μm.

The dependence $Q_{\text{abs}}(R_2)$ oscillates when the outer particle radius is increased. It seems likely that such a tendency in the absorptive properties of HOPs and TLPs while decreasing the thickness of the metallic shells is due to the following. Obviously, if the shell is thick then radiation is mainly reflected by the metal particle surface and is absorbed very little. For thin shells radiation begins to penetrate inside the particle and its greater part is absorbed in the shell as a result of multiple passage of beams within it. To confirm such an explanation, the energy distribution inside aluminum HOPs and metallized TLPs was considered (Figure 7.9). This figure illustrates that the internal intensity maximum is

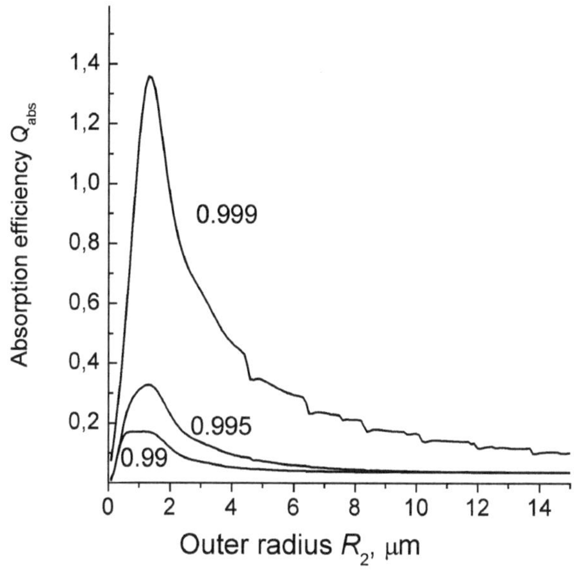

Figure 7.8. Absorption efficiency Q_{abs} for the aluminium HOP versus the outer radius R_2 for wavelength $\lambda = 10.6\ \mu m$. The numbers near the curves are the values of the ratio R_1/R_2 of the core radius to outer radius.

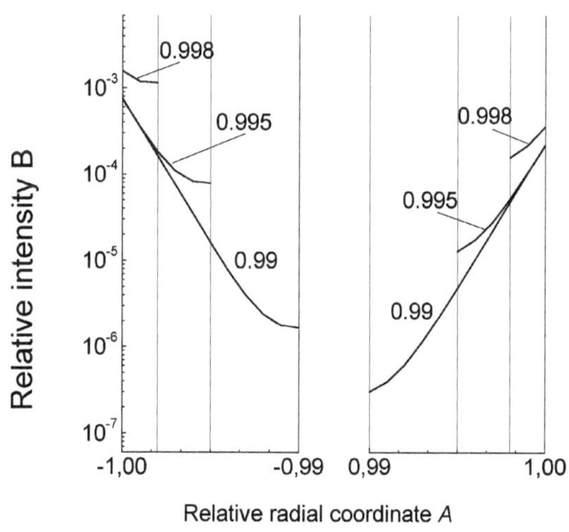

Figure 7.9. Relative intensity distribution inside the shell ($R_2 = 10\ \mu m$) of the aluminium HOP along the particle diameter coincident with the incident radiation (wavelength $\lambda = 10.6\ \mu m$) direction. The numbers near the curves are the values of the ratio R_1/R_2 of the core radius to outer radius. Vertical lines correspond to the threshold between the core and the shell.

located on the illuminated surface of the HOP. Further from the surface, to the particle center, the intensity B decreases both in the illuminated and in the shadow parts of shell. The maximum inhomogeneity of energy release is usually observed along the main particle diameter. For particles with the thick shells ($\Delta R = 0.1\ \mu m$) the energy density in the illuminated hemisphere sharply decreases (on entering a shell $B \sim 7 \times 10^{-4}$, near the core $B \sim 2 \times 10^{-6}$). In the shadow part of the shell the intensity increases from $B \sim 3 \times 10^{-7}$ to $B \sim 2 \times 10^{-4}$. With decreasing TLP shell thickness ($\Delta R = 0.05\ \mu m$) the energy density in the illuminated hemisphere of the shell also decreases, though not so sharply, from $B \sim 7 \times 10^{-4}$ to $B \sim 8 \times 10^{-5}$, in the shadow hemisphere it grows almost by a factor of 20. Inside the particles with the thinnest shells ($\Delta R \sim 0.001\ \mu m$) the energy density in the shell increases substantially in magnitude ($B \sim 10^{-2}$ in the illuminated hemisphere and $B \sim 5 \times 10^{-3}$ in the shadow hemisphere) as compared with thicker shells. The value of B changes weakly along the radial coordinate, slightly decreasing from the particle surface to the core.

Moreover, in the case of the thinnest shells ($\Delta R \sim 0.001\ \mu m$) the energy density in the illuminated and shadow parts of the shell is higher for a TLP with polymer core than for an HOP (as a result of which the total absorption of such TLPs with polymer cores is larger). The distribution of the energy within polyethylene particles covered with an aluminum shell is basically the same as that inside HOPs. The internal intensity has substantial magnitude only near the surface of metallic shells.

Thus, it was shown that the energy deposition within an aluminum shell depends on its thickness. As the shell thickness decreases from $\Delta R = 0.05\ \mu m$ to $\Delta r = 0.02\ \mu m$ for $R_2 = 10\ \mu m$ the maximum value of the internal intensity increases by a factor of 2.1. Beginning at a certain shell thickness the intensity distribution inside HOPs changes weakly and then increases further. This shell thickness weakly depends on the value of the outer particle radius and is usually $\Delta R \sim 0.07 \div 0.08\ \mu m$. Decreasing the shell thickness, the incident energy penetrates into the particle increasing the energy density in the shell by several orders of magnitude and changing the distribution of the internal intensity there.

7.3 ABSORPTION RESONANCES FOR HOPs

One of the topical problems of propagation of radiation in a disperse medium is scattering of light by inhomogeneous particles including HOPs, often occur in hail and are formed as a result of volcano eruptions. HOPs may be obtained from solid materials in the volatile liquids. When the surfaces of such particles are spherical, morphology dependent resonances (MDRs) can be observed in extinction, scattering, and absorption by these particles.

In this section, the conditions of the resonant absorption by spherical HOPs and also peculiarities of the electrical field intensity distributions inside them are discussed (Astafyeva *et al.*, 1986). Calculations were fulfilled for HOPs with the following optical constants: $N_1 = 1 + i0$ in the cavity and $N_2 = 1.6 + i10^{-7}$ in the shell. The cavity and outer size parameters are changed in the range $0 < x_1 \leq 50$ and $0 < x_2 \leq 55$ ($x_i = 2\pi R_i / \lambda\ (i = 1, 2)$, R_1 is the radius of the cavity, and R_2 is the outer radius of

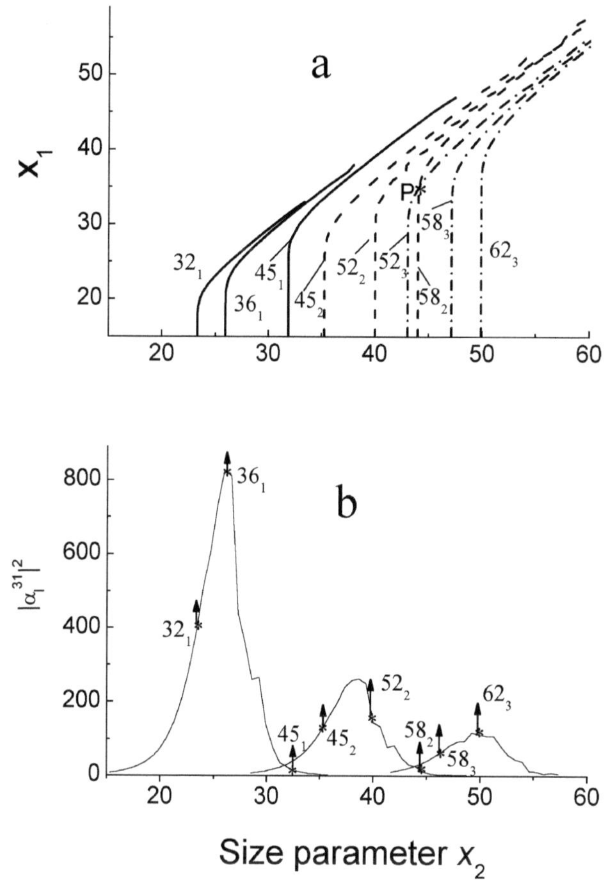

Figure 7.10. Loci of points with the coordinates x_1 and x_2, corresponding to the MDRs of the magnetic partial waves of HOPs with $N_2 = 1.6 + i10^{-7}$ (*a*), and dependences of squares of moduli of the amplitude coefficients of the magnetic resonant modes on the size parameter x_2 for MDRs of the first three orders: $s = 1, 2, 3$ (*b*). The asterisks indicate positions of the resonant modes; the figures near curves (a) and near asterisks (b) are the numbers of MDRs with the order l_s.

HOP). Consideration is given to partial waves with the numbers $l = 30 \div 60$ and MDRs of the first three orders $s = 1 \div 3$.

As is seen from Figure 7.10, formation of the internal cavity in the center of the resonant spherical HP and its growth to some definite size does not break the resonance conditions for mode with number l_s at constant outer particle radius. The equality of the resonant size parameters for HOPs and HPs for several modes of the first three orders is shown by the arrows in Figure 7.11*b*. When the cavity becomes higher than the indicated size, the conditions of MDRs for modes with the same number l_s are fulfilled for a somewhat increased outer radius of HOP. The shell thickness continues to decrease (Figure 7.11*a*).

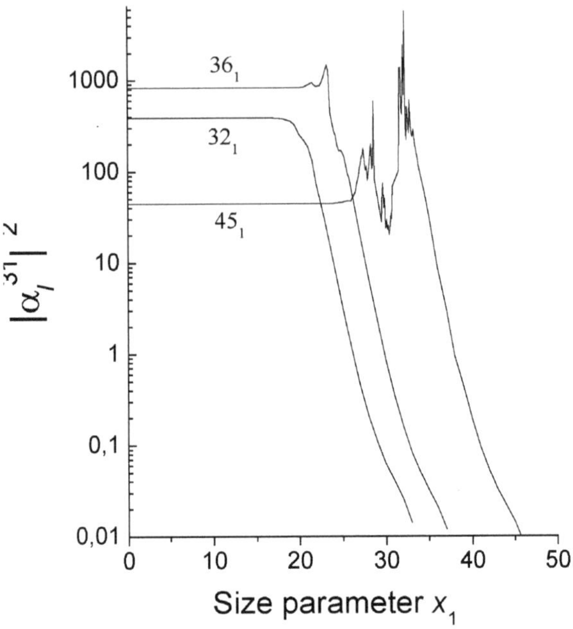

Figure 7.11. Squares of moduli of the amplitude coefficients of the magnetic resonant modes versus the size parameter x_1 for $N_2 = 1.6 + i10^{-7}$. The figures near the curves are the numbers of MDRs with the order l_s.

The change of squared moduli of the amplitude coefficients $|\alpha_l^{31}|^2$ in going from HP to HOP for four MDRs of the first order is shown in Figure 7.11. This figure shows that at first the values of $|\alpha_l^{31}|^2$ are practically not changed by increasing the cavity. In this case, as mentioned above, the outer size of HOPs, for which the resonance conditions are valid, is also not changed. Calculations show that here the energy distribution inside the cavity remains, which is typical for the magnetic MDR of the first order in HP (i.e., there are two regions of the energy concentration in the illuminated and shadow hemispheres near the particle surface). The energy density in these regions is higher by 3–4 orders of magnitude than in the rest of the particle. Similar regularities also take place for MDRs of the second and third orders; the corresponding structure of the internal field in HPs are described by Prishivalko *et al.* (1984*c*).

As the cavity further increases (with appropriate increase of the resonant size parameter x_2 of HOPs), the amplitude coefficients of resonant modes are changed simultaneously. Increase of the cavity size results in different resonant values of $|\alpha_l^{31}|^2$, depending on the resonant mode number. Figure 7.11 shows that for the resonant modes with $l_1 \leq 36_1$ (i.e., for modes corresponding to the left slope and maximum of Figure 7.10*b)* the growth of the cavity leads to an essential decrease of $|\alpha_l^{31}|^2$. Here lowering the energy density inside the particle takes place up to the destruction of the resonant structure of the internal field. For the resonant modes with $l_1 > 36_1$, corresponding to the right slope of the curve in Figure 7.10*b*, the increase of the cavity size at first leads to growth of the amplitude coefficients of resonant modes and only subsequently to a decrease. As this takes place, the maximum value of $|\alpha_l^{31}|^2$ may substantially exceed that for the same resonance magnetic mode in HPs. For example, for $l_1 = 45_1$ the maximum value of $|\alpha_l^{31}|^2$ of HOPs, attained at $x_1 = 31.7$, $x_2 = 33.886689691$, is $|\alpha_l^{31}|^2 \approx 1362.7$; in the case of HPs the resonant value of $|\alpha_l^{31}|^2$ of the same mode is much less ($|\alpha_l^{31}|^2 \approx 44.5$). This small value of the 45th magnetic (first order) amplitude coefficient in HPs implies that the internal field is non-resonant. Such a non-resonant structure of the field remains inside HOPs at a resonance of 45th magnetic mode with increasing cavity radius in the rather wide interval to $x_1 \approx 27$. Further increase of the cavity radius (more than $x_1 \approx 27$) changes the internal intensity distribution in HOPs for magnetic MDRs with $l_1 = 45$ and leads to the appearance of a typical resonant structure (Figure 7.12). But further growth of the cavity must be accompanied by the destruction of the resonant structure of the field.

The calculations show that in contrast to HPs two resonant modes may be observed in HOPs for definite size parameters x_1 and x_2, but the internal intensity distributions are different for various modes. For example, in HOP with the outer radius $x_2 = 44.12171$ and the cavity size $x_1 = 35.142595$ there are two resonant magnetic modes with numbers

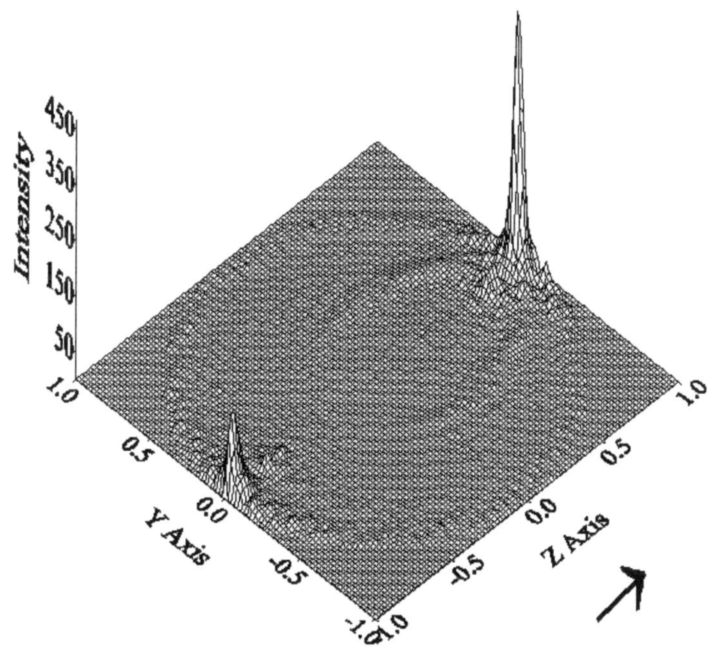

Figure 7.12. Relative intensity distribution inside the HOP ($x_1 = 31.7$, $x_2 = 33.88668969$, and $N_2 = 1.6 + i10^{-7}$) for the magnetic resonant mode with number $l = 45$ and order $s = 1$. The arrow shows the direction of the propagation of incident radiation.

$l_s = 52_3$ and $l_s = 58_2$ (point P in Figure 7.10a). The intensity distribution for the mode with $l_s = 52_3$ corresponds to the resonant structure of the electric field of the third order, and with $l_s = 58_2$ corresponds to that of the second order. Notice that maximum energy density for the mode with $l_s = 58_2$ is almost higher by two orders, than for the mode with $l_s = 52_3$.

From the preceding text it follows that MDRs in HOPs have typical peculiarities, which may essentially change the internal electric field intensity distribution in them. Inside HOPs the energy density has the same high values as inside HPs, and in some cases may be even higher.

8

Light scattering by radially inhomogeneous particles

In the previous chapters we discussed the light scattering by particles with the step-like inhomogeneity (two-, three- and many-layered particles). Here we consider the second variant of the spherically symmetric model that is a sphere with a smooth variation of optical properties in the radial direction. Particles of this type we call radially inhomogeneous particles (RIP). The formal solution of the problem of light scattering by RIPs and the analytical solutions for several inhomogeneity profiles are considered in Section 8.1 along with the review of appropriate works. The comparison of results calculated for the individual RIPs, homogeneous particles (HPs), and two-layered particles (TLPs) and for the polydisperse systems of these particles are made in Section 8.2 and 8.3, respectively. In particular situations it is appropriate to introduce combined models (e.g., the coated sphere with the homogeneous core and the radially inhomogeneous shell (Section 8.4)). The power profile of inhomogeneity for such a combined model is considered in detail. This profile is used for development of the asymptotic optical model of fractal cluster (Section 8.5). The calculated spectral dependencies of the specific extinction and scattering by the fractal cluster comprising of Ag nanoparticles are also presented at the end of the chapter.

8.1 SOLUTION OF THE PROBLEM OF SCATTERING BY SPHERICAL PARTICLES WITH SMOOTH RADIAL INHOMOGENEITY

Consider the scattering of a plane electromagnetic wave on a spherical RIP. As in the previous chapters we shall assume, that the incident wave ($e^{-i\omega t}$) propagates in the positive direction of axis z, and its electric vector oscillates along axis x. Let us assume, that the particle relative refractive index $N(r)$ is the continuous and differentiable function of distance r from the sphere centre ($N = N(r), 0 \leq r \leq R$, where R is the radius of sphere) and the relative magnetic conductivity is equal to unity. From Maxwell's equations (1.1.1–4) it is possible to receive the vector wave equations for the fields outside and inside the sphere with indicated properties. Outside the sphere the usual wave equation (1.1.5) is valid, and inside the sphere the electrical and magnetic fields satisfy to the equations:

$$\nabla \times \nabla \times \mathbf{E}(\mathbf{r}) - k_0^2 N^2(r)\,\mathbf{E}(\mathbf{r}) = 0 \qquad r \leq R, \tag{8.1.1}$$

$$\nabla \times \nabla \times \mathbf{H}(\mathbf{r}) - \frac{1}{N^2(r)} \nabla N(r) \times [\nabla \times \mathbf{H}(\mathbf{r})]\, k_0^2 N^2(r)\,\mathbf{H}(\mathbf{r}) = 0 \tag{8.1.2}$$

The solutions of the vector wave equation (1.1.5) for external area are the vector wave functions for the homogeneous medium $\mathbf{M}_{\sigma ml}(k_0\mathbf{r})$ and $\mathbf{N}_{\sigma ml}(k_0\mathbf{r})$ defined by Eqs.(1.3.1)–(1.3.5). Inside the sphere the electrical wave functions (we shall designate them as ${}^{(e)}\mathfrak{M}$ and ${}^{(e)}\mathfrak{N}$), satisfying Eq.(8.1.1), differ from the magnetic wave functions (designate them as ${}^{(h)}\mathfrak{M}$ and ${}^{(h)}\mathfrak{N}$), satisfying Eq.(8.1.2). Function ${}^{(e)}\mathfrak{M}$ together with ${}^{(h)}\mathfrak{N}$ represent the so-called TE-wave, and function ${}^{(h)}\mathfrak{M}$ together with ${}^{(e)}\mathfrak{N}$ represent the TM-wave (Stratton, 1941). The analytical expressions of the vector wave functions for radially inhomogeneous (RI) media look like (Zich, 1969):

$$\begin{aligned}
{}^{(e)}\mathfrak{M}_{\sigma ml}(\mathbf{r}) &= \frac{V_l(k_0 r)}{k_0 r}\mathbf{m}_{\sigma ml}(\theta,\varphi) \qquad {}^{(h)}\mathfrak{M}_{\sigma ml}(\mathbf{r}) = \frac{W_l(k_0 r)}{k_0 r}\mathbf{m}_{\sigma ml}(\theta,\varphi) \\
{}^{(e)}\mathfrak{N}_{\sigma ml}(\mathbf{r}) &= \frac{1}{N^2}\left[\frac{W_l(k_0 r)}{k_0^2 r^2}\mathbf{l}_{\sigma ml}(\theta,\varphi) + \frac{W_l'(k_0 r)}{k_0 r}\mathbf{n}_{\sigma ml}(\theta,\varphi)\right] \\
{}^{(h)}\mathfrak{N}_{\sigma ml}(\mathbf{r}) &= \frac{V_l(k_0 r)}{k_0^2 r^2}\mathbf{l}_{\sigma ml}(\theta,\varphi) + \frac{V_l(k_0 r)}{k_0 r}\mathbf{n}_{\sigma ml}(\theta,\varphi)
\end{aligned} \tag{8.1.3}$$

$$\begin{aligned}
{}^{(e)}\mathfrak{N}_{\sigma ml}(\mathbf{r}) &= \frac{1}{N^2 k_0}\nabla \times {}^{(h)}\mathfrak{M}_{\sigma ml}(\mathbf{r}) \qquad {}^{(h)}\mathfrak{N}_{\sigma ml}(\mathbf{r}) = \frac{1}{k_0}\nabla \times {}^{(e)}\mathfrak{M}_{\sigma ml}(\mathbf{r}) \\
{}^{(e)}\mathfrak{M}_{\sigma ml}(\mathbf{r}) &= \frac{1}{N^2 k_0}\nabla \times {}^{(h)}\mathfrak{N}_{\sigma ml}(\mathbf{r}) \qquad {}^{(h)}\mathfrak{M}_{\sigma ml}(\mathbf{r}) = \frac{1}{k_0}\nabla \times {}^{(e)}\mathfrak{N}_{\sigma ml}(\mathbf{r})
\end{aligned} \tag{8.1.4}$$

where $\mathbf{m}_{\sigma ml}$, $\mathbf{n}_{\sigma ml}$, and $\mathbf{l}_{\sigma ml}$ are the angular vector functions, defined by Eqs.(1.3.2) and (1.3.6), and functions $V_l(k_0 r)$ and $W_l(k_0 r)$ are the solutions of the linear differential equations of second kind (termed as the radial equations), regular in the coordinate origin:

$$V_l(\rho) + \left[N^2(\rho) - \frac{l(l+1)}{\rho^2}\right] V_l(\rho) = 0 \tag{8.1.5}$$

$$W_l''(\rho) - \frac{d}{d\rho}\left[\ln N^2(\rho)\right] W_l'(\rho) + \left[N^2(\rho) - \frac{l(l+1)}{\rho^2}\right] W_l(\rho) = 0 \tag{8.1.6}$$

where $\rho = k_0 r$. For reasons of convenience, in this chapter we shall designate the size parameter as $\rho_0 = k_0 R$. The solution of Eqs.(8.1.5) and (8.1.6) is the basic difficulty of the theory of scattering by the spherical RIPs.

The electric field inside the particle is represented as the expansion:

$$\mathbf{E}_{\text{int}}(\mathbf{r}) = E_I \sum_{l=1}^{\infty} \mu_l \left[\alpha_l\, {}^{(e)}\mathfrak{M}_{o1l}^{(1)}(\mathbf{r}) - i\beta_l\, {}^{(e)}\mathfrak{N}_{e1l}^{(1)}(\mathbf{r})\right] \tag{8.1.7}$$

and the incident and scattered fields are determined by expansions (3.1.4) and (3.1.4) accordingly. The magnetic field can be found with the help of Eqs.(1.1.7) and (8.1.4). Having substituted these expansions in the boundary conditions on the sphere surface (at $\rho = \rho_0$), we shall receive the set of linear equations for expansion coefficients of scattered (a_l, b_l) and internal (α_l, β_l) fields, solving it for these coefficients we have:

$$a_l = \frac{\psi_l(\rho)}{\xi_l(\rho)} \left\{ 1 - \left[i\psi_l(\rho)\xi_l(\rho) \left(\frac{V_l'(\rho)}{V_l(\rho)} - G_l(\rho) \right) \right]^{-1} \right\} \Bigg|_{\rho=\rho_0} \tag{8.1.8}$$

$$b_l = -\frac{\psi_l(\rho)}{\xi_l(\rho)} \left\{ 1 - \left[iN^{-2}(\rho)\psi_l(\rho)\xi_l(\rho) \left(\frac{W_l'(\rho)}{W_l(\rho)} - N^2(\rho)G_l(\rho) \right)^{-1} \right\} \Bigg|_{\rho=\rho_0} \tag{8.1.9}$$

$$\alpha_l = \left\{ -i\xi_l(\rho) \left[V_l'(\rho) - G_l(\rho)V_l(\rho) \right] \right\}^{-1} \Big|_{\rho=\rho_0} \tag{8.1.10}$$

$$\beta_l = \left\{ iN^{-2}(\rho)\xi_l(\rho) \left[W_l'(\rho) - N^2(\rho)G_l(\rho)W_l(\rho) \right] \right\}^{-1} \Big|_{\rho=\rho_0} \tag{8.1.11}$$

where $G_l(\rho)$ is the logarithmic derivative of function $\xi_l(\rho)$ defined in Section 1.5. Inasmuch as the structure of expressions for fields outside and inside the RIPs is completely identical to structure of expressions for fields in the case of HPs (see Eq.(2.3.26)), for calculation of light scattering characteristics of RIPs it is possible to use the formulae shown in Section 1.4. The difference, naturally, is that in the case of HPs the amplitude coefficients (1.3.14) enter in these formulae, and in the case of RIPs – the more complex coefficients (8.1.8) and (8.1.9).

Thus, now the only unanswered question is regarding the functions, which yield the solutions of the radial equations (8.1.5) and (8.1.6). Before going to its solution, we shall analyse literature concerning the problem of scattering by RI objects of a diversified physical nature.

The problem of the electromagnetic radiation scattering by RI spherical objects has attracted attention for the first time in the 1940s in connection with discussion about the idea of an opportunity to use the spherically symmetric dielectric lenses (lenses of Luneburg, Eaton, and some others) as the radar-tracking mirrors for a short wave range. In the initial stage of study the geometric optics method, grounded on consideration of ray-tracing inside a lens, was widely applied. However, it was rapidly found out, that this elementary consideration can give explicitly incorrect results even at lens sizes very greater than the wavelength. The inadequacy of the geometric optics approach has stimulated the occurrence of works, in which the problem began to be put and solved as the boundary value problem for Maxwell's equations. The first systematic and detailed solution is carried out by Tai (1958). The works of Garbacz (1962), Alexopoulos and Uslenghi (1969), Uslenghi (1969), Uslenghi and Weston (1970), Shafai (1972), Laxpati and Uslenghi (1973) were published some time later. In these, the lenses with enough simple inhomogeneity (as a rule, with one parameter) were considered, for which the radial equations are solved analytically. However, the numerical results were few. Attention was also attracted to the problem of scattering by the two-layered lenses with a homogeneous core and non-uniform shell (Alexopoulos, 1969; Alexopoulos and Barcilon, 1972; Alexopoulos and

Huebner, 1973). The review of the works which have been carried out in the given direction up to the end of the 1960s, is given in the book of Kerker (1969). Some interest is being shown in electrodynamic calculation of such lenses in recent years (e.g., Brenner and Singer, 1993; Hardaker and Holt, 1994; Golubyatnikov and Katsenelenbaum, 1997).

In the paper of Wyatt (1962) an optical model of atom cores appeared. The approach, advanced in it, does not differ conceptually from the approach used in works on dielectric lenses. Later it was applied to another object – the low-temperature plasma sphere with a diffuse boundary (Wyatt, 1963, 1965). In this case the originality of the problem is that the dielectric function can accept negative values, and also be converted to zero on some surface $r = r_0$ (so-called critical point). This circumstance complicates the problem, as the critical point is an additional singular point of the radial equations (8.1.5) and (8.1.6). The results of Wyatt's works were used by some authors (see, e.g., Lynch, 1963; Arnush, 1964; Gutman and Chesnokov, 1969) for investigation of scattering by spherical plasma regions in a backward direction depending on the radiation frequency.

As it was specified above, the solution of the differential equations (8.1.5) and (8.1.6), defining the behaviour of radial functions for a smooth RI profile, are complicated enough. To bypass this difficulty, the idea to replace the RI sphere by a sphere divided on separate concentric layers with constant optical properties in the limits of each layer (a smooth profile of RI approximated by a step-like one) was proposed. To simplify calculations for multilayered particles, the analogy between the theory of a spherical transmission line and theory of electromagnetic wave scattering by a sphere was utilized. In so doing, the necessity to use the unwieldy matrices, originating with traditional treatment of multilayered particles, disappears. Such an approach was theoretically advanced in the work of Wait (1963). However, this method, despite the relative simplicity and attractiveness, has an essential deficiency. As calculations have shown (Mikulski and Murphy, 1963), the scattered field, especially in the backward direction, strongly depends on the number of layers, that is the rather slow convergence to the asymptotic value occurs with the increase of the number of layers. Apparently, influence of reflections on layer boundaries are exhibited here. The necessity to estimate functions on each layer boundary very strongly complicates the calculations. This method justified itself at calculations of the scattering characteristics of actual multilayered objects, but the advisability of its application to a particle with smoothly varying RI was doubtful for a long time. However, the situation changed drastically in recent years. Great progress in computing makes it possible to calculate relatively easy the scattering characteristics for a layered sphere with a very large number of concentric layers (up to several thousands), that is quite sufficient for modeling practically any RI profile. Our iterative algorithm for calculation of scattered and internal fields for a sphere with an arbitrary number of concentric layers was described above (Section 3.5) with some references on this issue being cited.

Furthemore, the RI sphere may be replaced by a multilayered sphere with a smooth profile of the refractive index within the limits of each layer. It is common to use the linear profile, that is, the radial inhomogeneity is approximated by the piecewise-linear one. A similar approach was elaborated in the paper of Permyakov (1968). Such a model, despite the sharp complication of field calculations in each layer, can give, apparently, some advantages in view of the decreasing number of layers. The formal solution for the multilayered sphere, where the RI profile suffers discontinuity on layer boundaries, is also constructed by Perelman (1979, 1996). The synthesis of these two approaches is possible: in the limits of some layers the optical properties vary under the linear law (e.g., in the neighborhood of a critical point), but the optical properties of remaining layers being constant. This model was utilized by Krepak *et al.* (1973) for investigation of the influence of the plasma sphere critical layer on backscattering intensity.

Except for the above mentioned strict methods of the solution for problem of scattering of a plane electromagnetic wave by the RI spherical object, attempts were made at different times to use the approximate approaches. In particular, the known Rayleigh theory on scattering by small particles was extended to RIP (Tai, 1963). In this case, the problem is reduced to the solution of the simplified scalar wave equation, however, even in this simple variant the estimation of radial functions is concerned with calculation of hyper-geometric series. The computer realization of the method to solve the scattering problem by expansion of the scattering characteristics in a series in powers of the ratio – the scatterer size/wavelength – is similarly complicated (von Roos, 1971). Though this method formally allows us to consider the particles of the arbitrary shape and with arbitrary inhomogeneity, in practice it can be applied to small particles only.

The Rayleigh–Gans approximation ("soft" particles with small phase shift) regarding the non-uniform spherical particles has long been known (see, e.g., (Bohren and Huffman, 1983)). Some improvement of this theory is carried out by Farone (1965). Similar "soft" particle approximations were used by Albini (1962) and Shiff (1962).

In the paper of Gutman (1965) the method of cross sections, widely used in the theory of irregular wave-guides and resonators with slowly varying parameters, was applied to the problem of the plane wave diffraction on a sphere with arbitrary (not necessarily radial) inhomogeneity. The solution is gained as an infinite set of equations. This system is generally unsolvable. However, if the sphere parameters in azimuthal and meridional directions vary slowly enough, then it is possible to truncate the set and to receive the approximate solutions quite applicable for practice. In a case, when the inhomogeneity is radially symmetric, the infinite equation set splits into the simple subsets and the final relations coincide with the formulae obtained by the variable separation method.

In the work of Reilly (1973) the Galerkin variation method is applied to the problem of diffraction of the plane electromagnetic wave on a non-spherical non-uniform object. This approach is actually some generalization of the method of expansion in terms of spherical harmonics and reduces to it precisely, if the object has radial symmetry.

The approach on the basis of integral equations (Hizal and Tosun, 1973) is also possible. The obtained set of integral equations can be reduced to the set of the ordinary differential equations, whose solutions are derived with iterative procedure. This method is rather exact, however, its application does not give advantages in comparison with the usual method of variable separation due to great expenditures of computer time.

It is necessary also to mark the work of Latham (1968), where the so-called "invariant imbedding" is applied. This method gives rise to the necessity to solve some linear differential equation. However, it is much more complicated than Eqs.(8.1.4) and (8.1.5), appearing at the usual variable separation, and leaves room for numerical solution only. Choi and Brock (1997) developed the hybrid numerical method based on the boundary element approach.

Notice also that the model of the RI sphere was widely used to describe the scattering by aggregates and clusters (Borghese *et al.*, 1987; Chen, 1991; Fucile *et al.*, 1993). At such modeling, the aggregate of the large number of very small subparticles is replaced by one great sphere with some RI profile. This issue is discussed in Section 8.4 in relation to the fractal cluster. The questions on specific morphology dependent resonances (MDRs) in RIP (Chowdhury et al., 1991), on scattering by the evaporating RI fuel droplets at high temperature (Schneider et al., 1993) and on the inverse problem of construction of the RI profile (Ludlow and Everitt, 2000) also received attention.

The analysis of literary data allows us to conclude, that the approach to solving the problem on scattering of the plane electromagnetic wave by RI spherical objects, founded on the direct separation of variables in spherical coordinates and described in the beginning of this section, is the most simple and convenient for calculations. The application of other methods would not be appropriate, as they either complicate the problem unjustifiably, or give insufficiently precise results.

Come back now to the radial equations Eqs.(8.1.5) and (8.1.6). The concrete form of these equations and, hence, of their solutions depends on selection of the refractive index profile $N = N(\rho)$. It is necessary to mark, that the equations of this type arise also in problems of electromagnetic radiation (in particular, dipole radiation) propagation in the non-uniform Earth's atmosphere (see, e.g., Nomura and Takaku, 1955; Bremmer, 1960; Gould and Burman, 1964). In these works the analytical solutions of the Eqs.(8.1.5) and (8.1.6) for some RI profiles of the atmosphere's optical properties are obtained. The results of the above and similar works have remained, apparently, unknown for the investigators engaging in the study of scattering by the RI sphere, that in some cases results in repeated derivation of the analytical solutions for some profiles.

In principle, a procedure of the direct numerical evaluation may be applied to the radial equations. Such an approach was utilized in the papers of Garbacz (1962), Wyatt (1962, 1963), Bisbing (1966), and Buley (1967), however it has rather essential deficiencies. At first, for each set of input parameters it is necessary to solve some tens of radial equations (for each l). Secondly, the functions W_l and V_l very strongly oscillate and standard numerical methods of differential equation integration are not suitable. For these reasons the concrete calculations on the basis of this method are rather inconvenient. The approach described by Shafai (1972) is much more convenient. It is grounded on the introduction of the so-called amplitude and phase functions, which are weakly oscillating. A number of authors (Lynch, 1963; Arnush, 1964; Margulies and Scarf, 1964) for the numerical solution of the radial equations have utilized the analogy between these equations and wave equations, describing scattering of particles in a field of central force. In so doing, the approximate methods, developed to solve the Schrödinger equation, in particular the WKB-approximation, were applied.

In our opinion, deriving the analytical solutions of the radial equations is most preferable, as only they provide the possibility of systematic analysis and derivation of approximations. In this regard, it is necessary to mark the papers of Westcott (1968*a*, 1968*b*), who has considered the radial equations (8.1.5) and (8.1.6) in terms of the general theory of the differential equations, and has yielded the complete list of RI profiles, when these equations have the analytical solutions as functions not more complicated than the hyper-geometric ones. Acquaintance with this list would facilitate considerably the search for the analytical solutions of the problem of diffraction by non-uniform sphere, but Westcott's publications did not come into the view of the investigators, that was much promoted by their formal-mathematical character.

As shown by Westcott (1968*b*), the Eqs.(8.1.5) and (8.1.6) are simultaneously solved through hyper-geometric functions or their degenerate forms in six cases:

$$\begin{aligned} &A)\ N(\rho) = a\rho^b \\ &B)\ N(\rho) = a\,(\rho \ln b\rho)^{-1} \\ &C)\ N(\rho) = a\left(\rho\sqrt{\ln b\rho}\right)^{-1} \\ &D)\ N(\rho) = a\left[\rho(1+b\rho^{c-1})\right]^{-1} \\ &E)\ N(\rho) = a\rho^{c/2-1}(1+b\rho^2)^{-1} \\ &F)\ N(\rho) = a\left(\rho\sqrt{1+b\rho^c}\right)^{-1} \end{aligned} \tag{8.1.12}$$

where a, b, and c are the arbitrary constants. For the profile *A* both radial equations are solved through the Bessel functions; for the profile *B* the electrical radial equation (8.1.5) is solved through the Bessel functions, and magnetic radial equation (8.1.6) through the Whittaker functions; for the profile *C* both equations are solved through the Whittaker functions, and for profiles *D, E,* and *F* , through the hyper-geometric functions. The two-parameter profiles *A, B,* and *C* are unsuitable for description of the non-uniform particle scattering properties, as they have zero or singular points in the origin of coordinates. They can, however, be applied to the TLP with homogeneous core and non-uniform shell (Levine and Kerker, 1963; Kerker *et al.*, 1966; Alexopoulos and Barcilon, 1972; Alexopoulos and Huebner, 1973; see also Section 8.4). The three-parametric profiles *D, E,* and *F* do have no zero and singular points inside a particle only at values $c = 1(D)$, $2(E)$, and $2(F)$. Thus, there are only three two-parameter profiles, suitable for description of atmospheric particle inhomogeneity:

$$\begin{aligned} &a)\ N(\rho) = N_0(1+b_1\rho)^{-1} \\ &b)\ N(\rho) = N_0(1+b_2\rho^2)^{-1} \\ &c)\ N(\rho) = N_0(1+b_3\rho^2)^{-1/2} \end{aligned} \tag{8.1.13}$$

where N_0 is the refractive index at the particle centre and b_1, b_2, and b_3 are the constant coefficients. The form of these profiles is shown in Figure 8.1*I*.

In our works (Prishivalko *et al.*, 1975; Babenko and Prishivalko, 1973; Babenko, 1975; Babenko *et al.*, 1974, 1975) the solutions of the radial equations (8.1.5) and (8.1.6) for profiles (8.1.13) are studied. They express through the hyper-geometric functions. For example, for the profile b:

$$W_l(\rho) = \rho^{l+1}\left|b_2\rho^2 - 1\right|^{-0.5\left(\sqrt{1+N_0^2/b_2}+1\right)} F_{1l}(\sigma_l, \chi_l, \gamma_l; -b_2\rho^2) \tag{8.1.14}$$

$$V_l(\rho) = \rho^{l+1}\left|b_2\rho^2 - 1\right|^{-0.5\left(\sqrt{1+N_0^2/b_2}+1\right)} F_{2l}(\delta_l, \nu_l, \gamma_l; -b_2\rho^2) \tag{8.1.15}$$

Having substituted these solutions into expressions (8.1.8) and (8.1.9) for amplitude coefficients for the scattered field, we have:

$$a_l = \frac{\psi_l(\rho)}{\xi_l(\rho)}\left\{1 + \frac{i}{\psi_l(\rho)\xi_l(\rho)\left[\dfrac{l+1}{\rho} - G_l(\rho) + f_{2l}\right]}\right\}\Bigg|_{\rho=\rho_0} \tag{8.1.16}$$

$$b_l = -\frac{\psi_l(\rho)}{\xi_l(\rho)}\left\{1 + \frac{i}{\psi_l(\rho)\xi_l(\rho)\left[\dfrac{l+1}{\rho} - N^2(\rho)G_l(\rho) + f_{1l}\right]}\right\}\Bigg|_{\rho=\rho_0} \tag{8.1.17}$$

where

$$f_{1l} = -\rho\frac{b_2 + \sqrt{b_2\left(b_2 + N_0^2\right)}}{b_2\rho_0^2 + 1} - \rho\frac{F_{3l}}{F_{1l}}\left[\frac{b_2(2l+1)}{2l+3} - b_2\left(b_2 + N_0^2\right) + \frac{N_0^2}{2l+3}\right]$$

$$f_{2l} = -\rho\frac{b_2 - \sqrt{b_2\left(b_2 + N_0^2\right)}}{b_2\rho_0^2 + 1} - \rho\frac{F_{4l}}{F_{2l}}\left[b_2 - \sqrt{b_2\left(b_2 + N_0^2\right)} + \frac{N_0^2}{2l+3}\right]$$

$$b_2 = \frac{N_0 - N_1}{\rho_0^2 N_1}$$

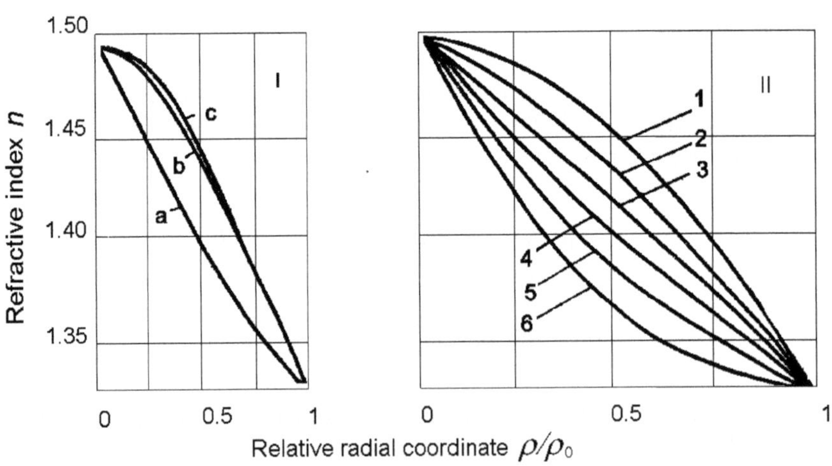

Figure 8.1. Refractive index n versus the relative radial coordinate at $n_0 = 1.5$ and $n_1 = 1.33$: *I* – on Eq.(8.1.13); *II* – on Eq.(8.1.19) at $q' = 0(1)$, 0.1(2), 0.3(3), 0.4(4), 0.5(5), and 0.5439(6).

N_1 is the refractive index on the particle boundary and:

$$F_{1l} = F(\sigma_l, \chi_l, \gamma_l; -b_2\rho^2) \qquad F_{2l} = F(\delta_l, \nu_l, \gamma_l; -b_2\rho^2)$$
$$F_{3l} = F(\sigma_l + 1, \chi_l + 1, \gamma_l + 1; -b_2\rho^2)$$
$$F_{4l} = F(\delta_l + 1, \nu_l + 1, \gamma_l + 1; -b_2\rho^2)$$

are hyper-geometric functions with parameters:

$$\left.\begin{matrix}\sigma_l \\ \chi_l\end{matrix}\right\} = \frac{1}{2}\left[l + \frac{3}{2} - \sqrt{1 + N_0^2/b_2} \pm \sqrt{l^2 + l + 9/4}\right] \qquad \gamma_l = l + \frac{3}{2}$$
$$\delta_l = l + 1 - 0.5\sqrt{1 + N_0^2/b_2} \qquad \nu_l = 0.5\left(l - \sqrt{1 + N_0^2/b_2}\right)$$

For profiles (8.1.13) a, b, and c the refractive index values in the particle centre N_0 and on its boundary N_1 are given. The refractive index run between them is rigidly fixed, that restricts the opportunity to apply these profiles to concrete objects. In this connection we considered the more general profile including profiles a and c as special cases (Babenko, 1975). It can be written as:

$$n(\rho) = n_0\left(1 + a_1\rho + a_2\rho^2\right)^\alpha \tag{8.1.18}$$

where a_1, a_2, and α are the real constants. In the given consideration we shall be restricted to real refractive indexes n, assuming the absorption index κ is negligibly small. In this connection choose the root arithmetical value at the non-integer α value.

The profile (8.1.18) can be written in the normalized form:

$$n(x) = n_0\left\{1 + q'z + \left[\left(\frac{n_1}{n_0}\right)^{1/\alpha} - 1 - q'\right]z^2\right\}^\alpha \tag{8.1.19}$$

where $z = \rho/\rho_0$ and $q' = a_1/\rho_0$. By selection of parameter q' it is possible to vary a run of a refractive index profile between the extreme values n_0 and n_1 (Figure 8.1,*II*). Generally speaking, there are different types of profiles (monotonic and with extremes) depending on the parameter q' value and ratio n_1/n_0. Varying $q'$and α, it is possible to describe practically any actual dependence of $n(z)$.

It is possible to show (see (Babenko, 1975)), that the monotonic run of $n(z)$ takes place in the following cases:

- If $\alpha > 0$ and $n_1/n_0 < 1$, or $\alpha < 0$ and $n_1/n_0 > 1$, then at values q' from the interval $h < q' \leq 0$; and
- If $\alpha > 0$ and $n_1/n_0 > 1$, or $\alpha < 0$ and $n_1/n_0 < 1$, then at values q' from the interval $0 < q' \leq h$, where $h = 2\left[(n_1/n_0)^{1/\alpha} - 1\right]$.

Having substituted Eq.(8.1.18) in Eqs.(8.1.5) and (8.1.6), we have:

$$\frac{d^2W_l}{d\rho^2} - \frac{2\alpha(a_1+2a_2l)}{1+a_1\rho+a_2\rho^2}\frac{dW_l}{d\rho} + \frac{n_0^2\rho^2\left(1+a_1\rho+a_2\rho^2\right)^{2\alpha} - l\,(l+1)}{\rho^2}W_l = 0 \tag{8.1.20}$$

$$\frac{d^2V_l}{d\rho^2} + \frac{n_0^2\rho^2\left(1+a_1\rho+a_2\rho^2\right)^{2\alpha} - l\,(l+1)}{\rho^2}V_l = 0 \tag{8.1.21}$$

In a neighborhood of point $\rho = 0$ the solutions of the differential equations (8.1.20) and (8.1.21) can be presented as a power series in ρ with a non-zero convergence radius:

$$W_l = \rho^{p_l}\sum_{m=0}^{\infty} c_{lm}\rho^m \qquad V_l = \rho^{s_l}\sum_{m=0}^{\infty} d_{lm}\rho^m \tag{8.1.22}$$

where p_l, s_l, c_{lm}, and d_{lm} are constants to be defined. The conditions of the series (8.1.22) and (8.1.23) convergence are explicitly explored by Babenko (1975). It is also shown, that the case $\alpha = -1/2$ is the most simple and convenient for calculations among all possible cases. At this α value $p_l = s_l = l + 1$, and coefficients c_{lm} and d_{lm} are obtained from the recursion relations:

$$c_{lm} = \frac{a_1\left(l-2lm-m^2\right)c_{l,m-1} + \left[a_2(2l+m)(1-m) - n_0^2\right]c_{l,m-2}}{m(2l+m+1)} \tag{8.1.23}$$

$$d_{lm} = \frac{a_1\,(2l+m)(1-m)\,d_{l,m-1} + a_2(2-m)(2l+m-1)d_{l,m-2}}{m(2l+m+1)} \tag{8.1.24}$$

The initial values of the coefficients are equal: $c_{l0} = d_{l0} = 1$, $c_{l1} = a_1/2$, and $d_{l1} = l^2a_1/2l{+}1)$. Thus, the relations (8.1.24) and (8.1.25) in aggregate with Eqs.(8.1.22) and (8.1.23) define the radial functions W_l and V_l, involved in expressions for amplitude coefficients Eq.(8.1.8) and Eq.(8.1.9) in the case of profile (8.1.18). The practical problems, bound with calculations with these formulae, are surveyed in Section 8.2.

8.2 COMPARISON OF LIGHT SCATTERING CHARACTERISTICS OF THE AEROSOL RIPs, HPs, AND TLPs

The detailed information about the character of inhomogeneities of arising in atmospheric aerosol particles is not available now. Our purpose was to study the scattering properties of the RIPs in comparison with HPs and TLPs. The calculations were carried out both for the profile (8.1.13) and the profile (8.1.18). Included in the calculation the Riccati–Bessel, Riccati–Hankel, and angular functions were computed by the procedure described in Section 1.5. The main difficulty in the case of profiles a, b, and c arose with the calculation of the hyper-geometric functions. These functions were calculated as power series in ρ_0, whose convergence and stability degraded with increasing particle size. The calculation accuracy was checked by two methods. On the one hand, at very close n_0 and n_1 the comparison of the scattering characteristics of the RIP with that for the corresponding HP is possible. On the other hand, the algorithm for the profile (8.1.19) includes profiles a and c as special cases, that allowed us to perform the control calculations. Double precision calculations ensured accuracy not less than 6 significant digits in the range $0 < \rho_0 \lesssim 10$, 5 digits in the range $10 \lesssim \rho_0 \lesssim 12$, and 4 digits in the range $12 \lesssim \rho_0 \lesssim 14$. The errors arise because of calculation instability for the sign-alternating power series, when the absolute value of the separate terms in the series is much more than that of the series sum. With the Eqs.(8.1.22) and (8.1.23) the calculation instability is not observed, although the series converge rather slowly. The expenditures of computer time are approximately 2–3 times more, than in similar calculations for profiles a and c.

The scattering efficiencies, phase functions and elements of the reduced scattering matrix for the non-absorbing RIPs with $n_0 = 1.5$ and $n_1 = 1.33$ and parameters $q' =$ 0, 0.1, 0.2, 0.2556, 0.3, 0.4, 0.5, and 0.5439 were calculated. The indicated range of parameter q' envelops all monotonic downward profiles gained from Eq.(8.1.19) (Figure 8.1,*II*). At q' =0 and 0.2556 the profile (8.1.19) coincides with cases b and a accordingly. The calculation of Q_{sc} has shown, that at the chosen n_0 and n_1 the results for profiles b and c practically coincide. For profile a the results are qualitatively rather similar to the corresponding results for other profiles, though the small quantitative distinction, certainly, is present.

The obtained results were compared with similar results for three modifications of HPs: (1) with the refractive index equal to volume-averaged refractive index of the RIP:

$$\bar{n} = \frac{1}{V} \int\limits_V n(\rho) d\rho$$

($n_0 = 1.5$, $n_1 = 1.33$, and $\bar{n} = 1.395$); (2) with $n = 1.5$; and (3) with $n = 1.33$ and for TLPs with the core and shell refractive indexes equal to 1.5 and 1.33 accordingly. In so doing, the ratio of the core and the shell sizes was selected so as the volume-averaged refractive index of TLPs was equal to a similar quantity ($\bar{n} = 1.395$) for the RIPs. It is believed that the particles of all the above types can exist in atmospheric aerosol.

In Figure 8.2 the dependencies $Q_{\text{sc}}(\rho_0)$ for the indicated types of particles are given. From curve comparison it is seen, that the dependence $Q_{\text{sc}}(\rho_0)$ for RIPs (profile b) differs sharply from the corresponding dependencies for HPs with $n = 1.5$ and 1.33. Curves for HPs with $n = 1.395$ and for TLPs lay much closer. In the area $\rho_0 \lesssim 4$ the Q_{sc} for these particles differs from the Q_{sc} for RIPs by no more than 5–6%, but at $\rho_0 \gtrsim 4$ the discrepancy in some cases reaches 30%. From here it follows, that the simulation of the RIP scattering properties with the help of HPs or TLPs with $\bar{n}$ is only possible in the field of small ρ_0 values, in this case at $\rho_0 < 4$. For particles with higher size parameters it is not possible to select such effective $\bar{n}$, at which Q_{sc} of RIPs and HPs with chosen n would differ no more than by 10–15% in a wide enough range of ρ_0. The calculations with the help of the TLP model also do not present such opportunity.

In Figure 8.3 the graphs, illustrating the influence of the n_0 and $n_1$variation (profile b), are depicted. The comparison is carried out only with the run of dependence $Q_{\text{sc}}(\rho_0)$

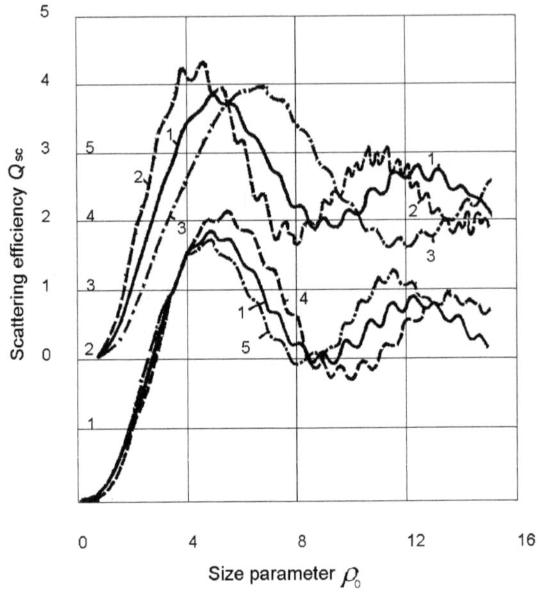

Figure 8.2. Scattering efficiency Q_{sc} versus size parameter ρ_0 for RIPs (*1*) with $n_0 = 1.5$ and $n_1 = 1.33$, for HPs with $n = 1.5(2), 1.33(3)$, and $1.395(4)$, and TLPs (*5*) with $n_1 = 1.5$, $n_2 = 1.33$, and $\bar{n} = 1.395$. The left scale is for the upper group of curves, and the right one is for the lower group.

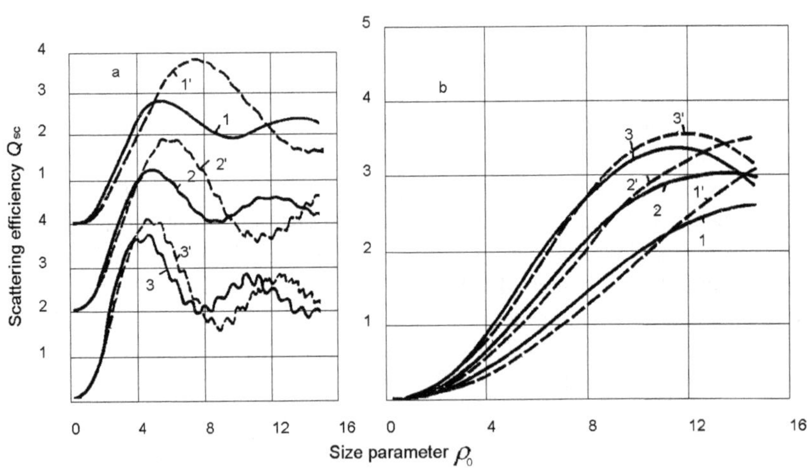

Figure 8.3. Scattering efficiency Q_{sc} versus size parameter ρ_0: *a* – RIPs with $n_0 = 1.6$ and $n_1 = 1.1(1)$, 1.2(*2*), and 1.33(*3*); HPs with $n = 1.269(1')$, 1.341(*2′*), and 1.429(*3′*); *b* – RIPs with $n_0 = 1.2$ and $n_1 = 1.05(1)$, $1.1(2)$, and $1.15(3)$; HPs with $n = 1.107(1')$, $1.139(2')$, and $1.170(3')$. The left upper scale (*a*) is for curves 1,1′, the left lower is for 3, 3′, and the right is for 2,2′.

for HPs with the refractive index equal to the volume-averaged value for the corresponding RIP. From the figure it is seen, that both location and height of maxima and minima of dependence $Q_{sc}(\rho_0)$ are essentially different for HPs and RIPs. The maximum for the RIP is always lower than the one for the corresponding HP, and the first minimum always higher than the corresponding first minimum for HP. In the case of the RIP both extremes are shifted in the direction of smaller ρ_0 in comparison with extremes for the corresponding HP. While increasing the difference $\Delta n = n_0 - n_1$ the distinction between curves enhances, and at constant Δn this distinction increases as n_0 decreases. The difference of Q_{sc} values for RIP and HP with average $\bar{n}$ increases up to 15–20% and more when Δn is equal to 0.3–0.4. Such a great difference occurs also in the field of small ρ_0.

From the run of the curve *1* (Figure 8.3*a*) it is seen most clearly, that for great Δn the simulation of the RIP scattering properties with the help of HP at $\rho_0 > 5$ is generally impossible. For curves coming together it would be necessary to change the effective refractive index of HPs so, that the first maximum was shifted to the left and became lower. The left shift of the maximum can be attained by increasing the effective refractive index, and, by contrast, the maximum lowering can be achieved by reducing the refractive index.

For "soft" particles (Figure 8.3*b*) the interval of ρ_0 values, where the Q_{sc} values for the RIPs and HPs with average n were found to be rather close, is somewhat broader. However, increasing Δn up to 0.15 the distinction of Q_{sc} values in some points reaches 14–16% in this case also. Inasmuch in area $n \approx 1.1$ the Q_{sc} of the HP is very sensitive to changes of n, it is possible to noticeably improve the agreement of curves $Q_{sc}(\rho_0)$ in the field of small ρ_0 by increasing the effective average n value only by 0.003–0.005, but consequently the discrepancy of curves will increase at a great ρ_0 Thus, and in the case of "soft" RIPs the simulation of their scattering properties with the help of HPs is valid only in some restricted ρ_0 interval, basically at $\rho_0 < 10$.

The angular characteristics of light scattering by the RIPs were not considered at all earlier. Some of the results, obtained by us and concerning particles with $n_0 = 1.5$ and $n_1 = 1.33$ are given below. In the text and in the figures the term of the scattered light intensity refers to a quantity $i = i_1 + i_2$, where i_1 and i_2 are defined by Eq. (1.4.10). From Figure 8.4 it is seen, that the phase functions for the RIPs essentially differ from that for HPs with $n = 1.5$ and 1.33. At given ρ_0, the qualitative agreement is observed between the phase functions of HPs with an average refractive index and that for the RIP. As to quantitative intensity values, they noticeably differ in this case also. The greatest distinction takes place for the scattering angle $\theta > 90°$, especially for θ, close to 180°. It is practically impossible to improve the agreement of phase functions by selection of some effective refractive index. If for some angle interval the agreement is improved, for another the degradation necessarily occurs. With increasing the size parameter ρ_0 the agreement is deteriorated both quantitatively and qualitatively.

Similar deductions can be made concerning the angular dependencies of the linear polarization degree of scattered light (Figure 8.5). The curve $P(\theta)$ for HPs with $n = 1.5$ does not bear even qualitative similarities to the corresponding curve for the RIPs. The curve for the particle with $n = 1.33$ lays closer, but at $\theta > 90°$ the distinction becomes very great. For HPs with the volume-averaged n, the greatest diversions are detected at

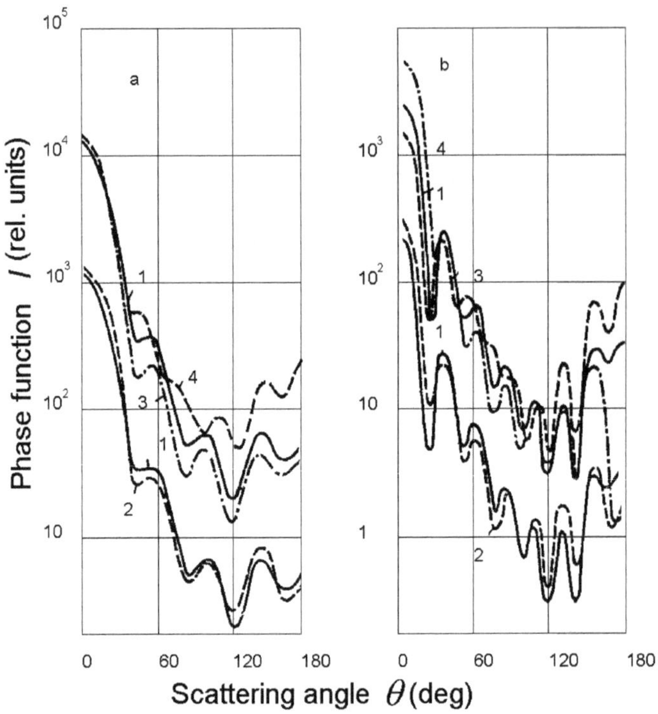

Figure 8.4. Phase functions at size parameter ρ_0 =5.0 (*a*) and 7.5 (*b*) for RIPs with $n_0 = 1.5$ and $n_1 = 1.33$ (*1*) and HPs with $n = 1.395$(*2*), 1.33(*3*), and 1.5(*4*). The right scale is for the upper groups of curves and the left is for the lower groups.

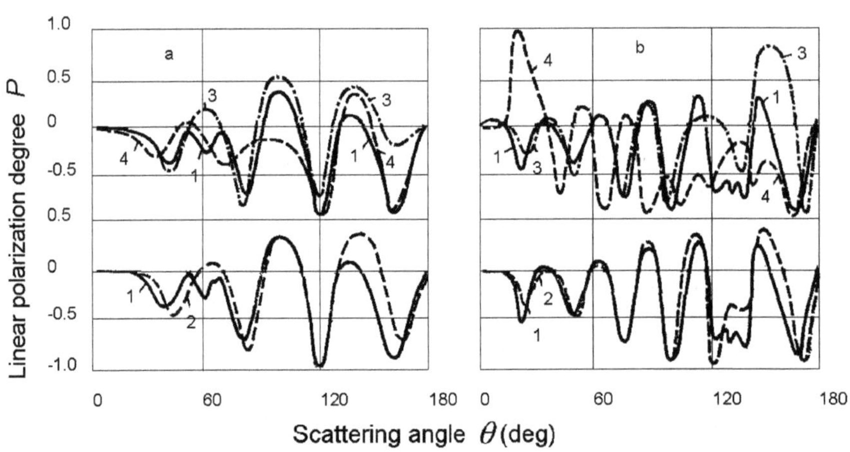

Figure 8.5. Angular dependencies of the linear polarization degree P for particles with size parameters $\rho_0 = 5.0$ (*a*) and 7.5 (*b*). Notation is the same, as in Figure 8.4.

angle $\theta > 120°$ and a little bit smaller at $20° < \theta < 50°$. In other directions there is quite good qualitative and even quantitative agreement at given ρ_0.

At $\theta = 20°$, the polarization degree for HPs and RIPs is practically identical, if $\rho_0 \lesssim 5$ (Figure 8.6). For larger particles the agreement is absent. In the direction $\theta = 90°$ the dependencies $P(\beta)$ are qualitatively rather close to $\rho_0 \approx 14$, thereafter discrepancies occur. Noteworthy also are some relative shift of maxima $P(\rho_0)$. The most essential distinctions are observed in the direction $\theta = 150°$. At $\rho_0 \lesssim 9$ each extreme on the curve for the RIP corresponds to the extreme on curve for HPs with $n = 1.395$, but the numerical P values rather noticeably differ. For the RIP the negative polarization predominates. At $\rho_0 > 9$ both the extreme shapes and their locations are essentially different.

It was already mentioned above, that most considerable diversions of light intensities, scattered by HPs and RIPs, take place in the backward direction, that is at $\theta \approx 180°$. In Figure 8.7*a* the dependencies $I_{\text{bsc}}(\rho_0)$ for HPs with $n = 1.33$, 1.395, and 1.5 and for the

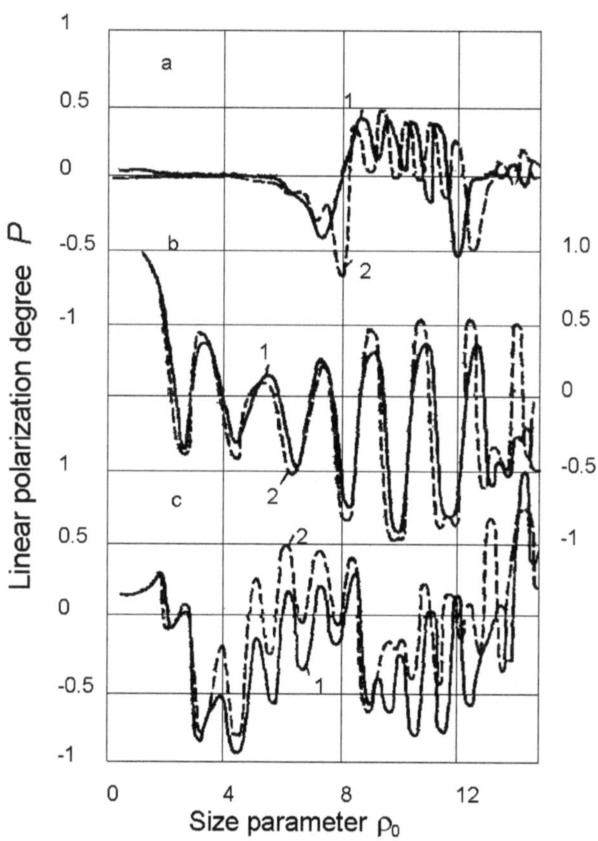

Figure 8.6. Linear polarization degree P versus the size parameter ρ_0 at scattering angles $\theta = 20°(a)$, $90°(b)$, and $150°(c)$ for RIPs with $n_0 = 1.5$ and $n_1 = 1.33$ (*1*) and for HPs with $n = 1.395$ (*2*).

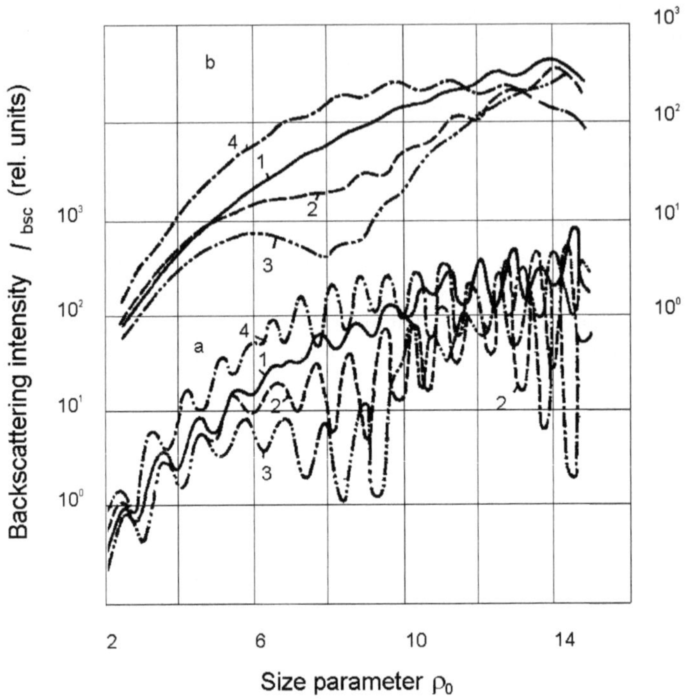

Figure 8.7. Non-averaged (*a*) and averaged (*b*) backscattering intensities versus size parameter. The left scale is for (*a*), and the right is for (*b*). Notation is the same, as in Figure 8.4.

RIPs of the same size are presented. The investigation of these dependencies is of practical interest as regards the laser sensing of the atmosphere.

First of all, it is necessary to mark, that the curve $i_{bsc}(\rho_0)$ for the RIPs is somewhat smoothed in comparison with the same curves for HPs in the entire interval of ρ_0 values. It is clearly exhibited at $\rho_0 > 6$. Nevertheless the comparison of curves becomes inconvenient with increasing ρ_0. Such consideration should be carried out in view of particle polydispersity. However, the characteristics of light scattering by the ensemble of polydisperse particles depend on the great set of quantities, among which a rather essential role is played by parameters of the particle size distribution. Sometimes these parameters can have a defining influence on one or another of the scattering characteristics. Besides, it is hardly probable, that only RIPs would exist in the scattering volume or that their influence would be completely prevailing. Apparently, such particles can exist simultaneously with HPs and affect the overall picture of light scattering by aerosols. Therefore, before one transfers to viewing light scattering by particle ensembles, it is expedient to present and compare the runs of dependencies $i_{bsc}(\rho_0)$ for RIPs and HPs in a more illustrative form.

For the purpose of revealing the pure state of the influence of particle inhomogeneity, smoothing the dependencies $\langle i_{bsc}(\rho_0)\rangle$ for four viewed particle types was carried out under completely identical conditions. The smoothing was implemented by an averaging i_{bsc} on the ensemble of particles in the elementary assumption of their equal probability size distribution. The averaging was executed under the formula:

$$\langle i_{\mathrm{bsc}} \rangle = \frac{1}{2\Delta\rho_0} \int_{\rho_{0i}-\Delta\rho_0}^{\rho_{0i}+\Delta\rho_0} i_{\mathrm{bsc}}(\rho_0) d\rho_0 \tag{8.2.1}$$

where $\Delta\rho_0 = 0.5$ and the integration step over ρ_0 was 0.05. The values of $\langle i_{bsc} \rangle$ were evaluated through 0.5. The results after averaging are presented in Figure 8.7.

The averaged backscattering intensities are rather close (within ~15%) for the RIPs and HPs with $n = 1.395$ at $\rho_0 \lesssim 5$. Further increasing ρ_0 is accompanied by considerable removal of all four curves from each other. On an approximate estimation the $\langle i_{\mathrm{bsc}} \rangle$ values, corresponding to $n = 1.395$, are 2.7–5 times higher, than for $n = 1.33$ at $\rho_0 = 7$–8. In turn, the $\langle i_{bsc} \rangle$ values for RIPs, are 2.2–2.7 times higher, than for $n = 1.395$. The average backscattering intensities for $n = 1.5$ are 2.5–3 times higher, than for RIPs. Notice, that within the interval $12 \leq \rho_0 \leq 14.5$ the curves for all three HPs lay below those of the RIPs, and in range $\rho_0 = 13.25 \pm 0.25$ have rather close values.

In the range $2.5 \leq \rho_0 \leq 14.5$ the run of the curve $\langle i_{\mathrm{bsc}} \rangle (\rho_0)$ for the RIP can be approximated by the smooth curve:

$$\langle i_{\mathrm{bsc}}(\rho_0) \rangle = 0.0159\rho_0^{4.3} \exp(-0.1\rho_0) \tag{8.2.2}$$

The approximating dependence obtained can be useful in revealing the distinctions between light backscattering by HPs and RIPs. These distinctions can be detected for ρ_0 values within the interval from 6 to 12. There is reason to think that a rather considerable part of RIPs in atmospheric aerosol falls into this interval (in the visual wavelength range).

8.3 COMPARISON OF SCATTERING BY SYSTEMS OF ATMOSPHERIC RIPs AND SYSTEMS OF HPs AND TLPs

As was marked in Section 8.2, the characteristics of light scattering by a single RIP is not so essentially different from the similar characteristics of a single HP at small size parameter, but rather noticeably at increasing size parameter. Consider, how these distinctions are exhibited in the case of a polydisperse particle ensemble (Prishivalko and Babenko, 1979). From variants of inhomogeneity treated in Section 8.1 we choose the most general, when the particle optical inhomogeneity is characterized by the profile (8.1.19). The selection of this profile as a basis for the calculation is determined, on the one hand, by the fact that it is general enough due to the presence of three free parameters (n_0, n_1, q'), and on the other hand, by the fact that the radial functions in this case can be obtained with the help of rather simple relations (8.1.24) and (8.1.25).

The values of parameters defining the refractive index radial dependence, are chosen so that it would be possible for comparison with earlier obtained results for systems of HPs and TLPs. The calculations, whose results are given below, are carried out for values $n_0 = 1.65$ and $n_1 = 1.49$ (at $q' = 0.404$), 1.45 ($q' = 0.209$), and 1.43 ($q' = 0.011$). The

q' values are selected so that the volume-averaged refractive index in all three cases is equal $\bar{n} = 1.51$. The obtained dependencies for systems of the indicated RIPs were compared with similar dependencies for systems of HPs with $n = 1.51$ and of TLPs consisting of the spherical core with $n_1 = 1.65$ and concentric salt solution shell having $\bar{n} = 1.51$, that is obtained at the ratio of radii of the water containing particle and its core equal to 1.216. Calculated with Eq.(2.5.13) the shell refractive index will be $n_2 = 1.395$. The refractive index profiles corresponding to all indicated particle types are depicted in Figure 8.8. The absorption was assumed equal to zero. Limits of particle external sizes are identical for all systems.

As follows from Table 8.1, the scattering coefficients are practically insensitive to internal particle structure in the case of the RIP systems also. Its values are determined by the volume-averaged refractive index with an accuracy of 2%. The same is true (with an error no more than 6%) for the forward scattering intensity ($\theta = 0°$). As for the backscattering intensities ($\theta = 180°$), in this case the distinctions between the systems of HPs and RIPs can reach 300–450%, especially for wide distributions.

The phase functions for the RIP systems have much in common with those presented in Figure 4.16. It is possible to mark only, that the greatest discrepancy of phase functions occurs at $\theta \approx 125°$ and $\theta > 165°$. At $\theta \approx 125°$ (where the basic minimum of all calculated phase functions are situated) the highest scattering is observed for TLPs, and the lowest for HPs. For RIPs the intensity of scattering decreases as the q' parameter increases. If for TLPs at $\nu = 2$ the scattering in the direction $\theta = 125°$ is higher by almost 60%, than that

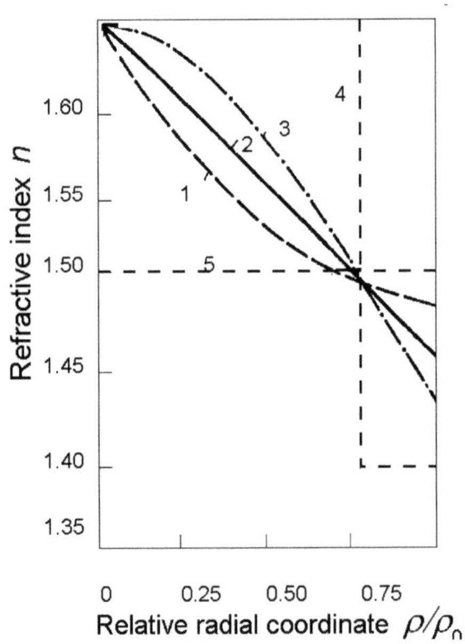

Figure 8.8. Refractive index n versus the relative radial coordinate x for RIPs with: $n_0 = 1.65$ and $n_1 = 1.49$ ($q' = 0.404$) (*1*), $n_1 = 1.46$ ($q' = 0.209$) (*2*), $n_1 = 1.43$ ($q' = 0.11$) (*3*); for TLPs (*4*) with $n_1 = 1,65$; $n_2 = 1.395$, and $\bar{n} = 1.51$; HPs with $\bar{n} = 1.51$ (*5*).

Table 8.1. Ratios of the scattering coefficient and the intensity of forward scattering ($\theta = 0$) for systems of RIPs to those for the corresponding systems of HPs. First line – the ratio of the scattering coefficients, second line – the ratio of forward scattering intensities.

	ν				
q'	2.0	2.5	3.0	3.5	4.0
0.011	0.990	0.992	0.999	1.008	1.018
	0.991	0.981	0.966	0.952	0.944
0.209	0.992	0.994	0.999	1.006	1.012
	0.966	0.964	0.960	0.957	0.956
0.404	0.997	0.997	0.999	1.003	1.006
	0.983	0.981	0.977	0.9974	0.974
TLPs	0.979	0.982	0.995	1.013	1.032
	0.996	0.982	0.958	0.920	0.910

for HPs, and at $\nu = 4$ is higher by almost 20%, then the difference of scattering by RIPs with $q' = 0.209$ is approximately three times less (18 and 6% accordingly). The essential differences in phase function run for different particle types are observed at angles, close to backscattering direction.

In Figure 8.9 the comparative results of angular dependencies of the linear polarization degree P for all five types of particles are given at $\nu = 2.5(0.5)4$. The angular dependence of P for particles with $q' = 0.404$ is closest to HPs within the wide angle interval $\theta < 150°$, although in the area $\theta > 155°$ the distinction at $\nu < 4$ is rather essential. For these scattering directions the curve for HPs transits between the corresponding curves for particles with $q' = 0.209$ and 0.11. For narrow distributions, in which the dominant role is played by small particles, the distinction in polarization degree for the majority of scattering directions (except for the interval $\theta > 150°$) lays within the limits of experimental error. For wide distributions the distinctions are essential at $\theta > 130°$. In this area of the scattering angles, the most deep minima of P occur for the particles with $q' = 0.404$. It is necessary to mark, that only TLPs have the characteristic maximum of linear polarization in the region θ=163–164°.

The angular dependencies of the reduced scattering matrix element f_{43} are rather sensitive to particle internal structure (Figure 8.10). As well as in the previous case, the dependencies $f_{43}(\theta)$ are closest to those for HPs when $q' = 0.404$. However, the difference between them remains considerable even for narrow distributions. The curves for TLPs differs from other curves in Figure 8.10 in that they have the additional narrow maximum at $\theta \approx 167°$. For distributions with $\nu \lesssim 3.5$ there is the small additional maximum in the area $\theta \approx 173°$on curves corresponding to particles with $q' = 0.404$. For distributions with $\nu \leq 3$ the tracks of such maximum are observed on curves corresponding to particles with $q' = 0.209$ also. In this respect the dependencies $f_{43}(\theta)$ for particles with step-like and smoothly varying radial inhomogeneity have some generality.

It is typical, that as parameter q' increases the element f_{43} increases, and the vertex of the positive maximum is shifted in the direction of great angles, coming nearer to the

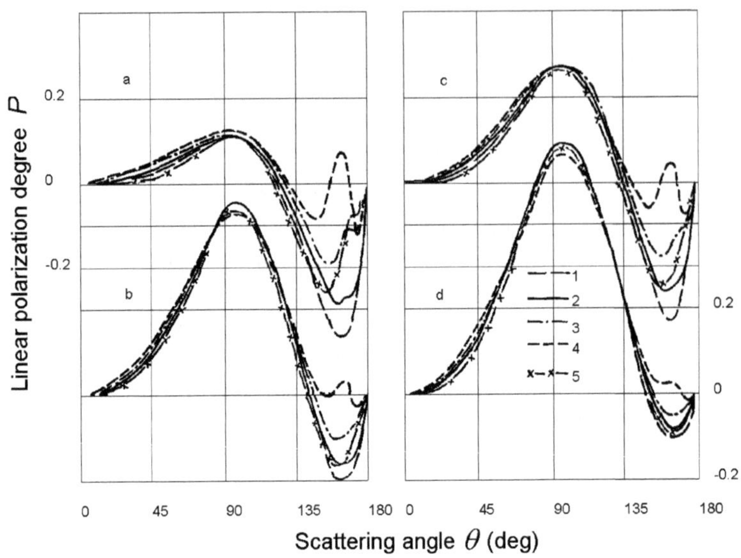

Figure 8.9. Angular dependencies of linear polarization degree P for scattered light with $\lambda = 0.6328\ \mu m$ at $\nu = 2.5(a), 3(b), 3.5(c)$, and $4(d)$. The left scale is for (a, b) and the right scale is for (c, d). Other notation is the same as in Figure 8.8.

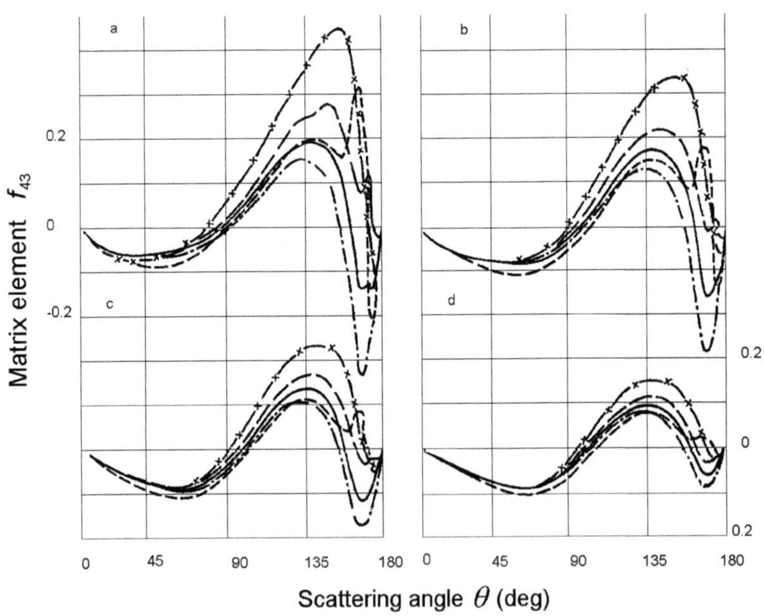

Figure 8.10. Angular dependences of the reduced scattering matrix element f_{43}. Notation is the same as in Figure 8.9.

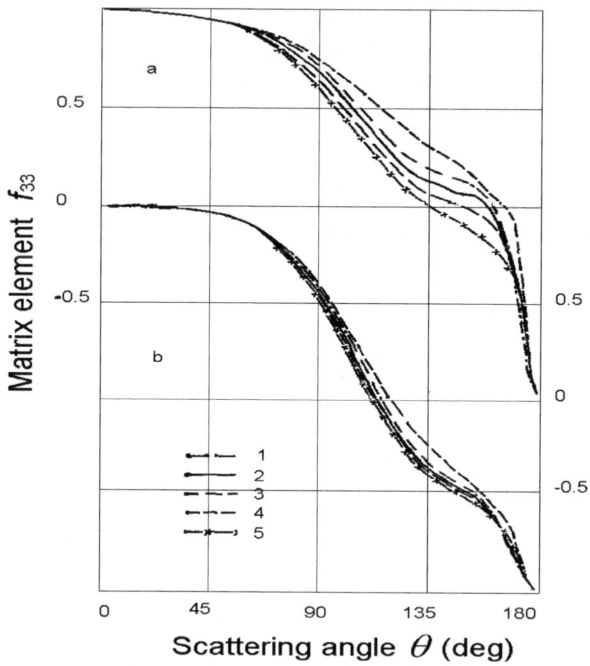

Figure 8.11. Angular dependencies of the reduced scattering matrix element f_{33} at $\nu = 2.5(a)$ and $3.5(b)$. Other notation is the same as in Figure 8.8.

location of the maximum vertex for HPs. Within the scattering angle interval $\theta > 155^\circ$ the angular dependence of the element f_{43} for the RIPs has a rather severe minimum which is especially sharply expressed for wide distributions. The location and depth of this minimum in aggregate with location and height of the positive maximum can serve as the distinctive indicator for light scattering by RIPs compared with scattering by HPs. The presence of the second narrow positive maximum distinguishes light scattering by TLPs.

In view of the similarities of dependencies $f_{33}(\theta)$, they are presented for only two parameter ν values (Figure 8.11). Practically, in all cases, the element f_{33} for TLPs has the highest values, and for HPs the lowest ones. The curves for RIPs are positioned between them in such a manner that the values f_{33} increase with decreasing q'. Closest to HPs appears the dependence $f_{33}(\theta)$ for particles with $q' = 0.404$. Increasing the ν parameter the distinction of f_{33} values for different types of particles rapidly decreases and at $\nu = 4$ does not exceed $\Delta f_{33} = 0.07$.

Thus, the most typical difference between the scattering properties of the RIP systems and those of the HP and TLP systems is the substantially different run of angular dependencies of the linear polarization degree and scattering matrix element f_{43}, especially at scattering angles $\theta > 155^\circ$. The refractive index value averaged over the volume of RIP can be successfully utilized for calculation of such characteristics, as the extinction (scattering) coefficient and intensity of forward scattering. However, by contrast, the angular scattering characteristics at considerable particle humidifying significantly depends on on character of their internal inhomogeneity. This circumstance is necessary to take into account at the solution of inverse problems, connected with revealing the microstructure parameters of aerosol and refractive index of particles, comprising it.

8.4 LIGHT SCATTERING BY SPHERES WITH HOMOGENEOUS CORES AND RI SHELLS

8.4.1 Formal solution

Consider the model of a scattering particle comprising of an homogeneous spherical core (with radius R_1 and complex refractive index N_1) and of an inhomogeneous spherical shell with radius R_2, which complex refractive index N_2 depends on the distance from the particle center: $N_2 = N_2(r)$. For simplicity, the surrounding medium is considered to be a vacuum. Let us introduce Cartesian and spherical coordinate systems with the origin at the particle center. Let the incident wave ($\exp(-i\omega t)$) with wavelength λ and amplitude E_I propagates along the positive direction of the z-axis and its electric vector oscillates along the x-axis.

The vector wave equation takes the following forms for the core region ($r < R_1$):

$$\nabla \times \nabla \times \mathbf{E}(\mathbf{r}) - k_0^2 N_1^2 \mathbf{E}(\mathbf{r}) = 0 \tag{8.4.1}$$

for the region outside the particle ($r > R_2$):

$$\nabla \times \nabla \times \mathbf{E}(\mathbf{r}) - k_0^2 \mathbf{E}(\mathbf{r}) = 0 \tag{8.4.2}$$

and for the RI shell region ($R_1 < r < R_2$):

$$\nabla \times \nabla \times \mathbf{E}(\mathbf{r}) - k_0^2 N_2^2(r) \mathbf{E}(\mathbf{r}) = 0 \tag{8.4.3}$$

$$\nabla \times \nabla \times \mathbf{H}(\mathbf{r}) - \frac{\nabla N_2(r)}{N_2^2(r)} \times [\nabla \times \mathbf{H}(\mathbf{r})] - k_0^2 N_2^2(r) \mathbf{H}(\mathbf{r}) = 0 \tag{8.4.4}$$

where $k_0 = 2\pi/\lambda$ is the wave number for the vacuum. The solutions of the vector wave equations (8.4.1–2) are the ordinary vector wave functions $\mathbf{M}^{(j)}_{\sigma ml}(k\mathbf{r})$ and $\mathbf{N}^{(j)}_{\sigma ml}(k\mathbf{r})$ for an homogeneous medium (see Section 1.3). Because the core ($r < R_1$) includes the null point, we must take $\mathbf{M}^{(1)}_{\sigma ml}(k_1\mathbf{r})$ and $\mathbf{N}^{(1)}_{\sigma ml}(k_1\mathbf{r})$ as the solutions, that is, the field inside the core must be a superposition of these functions. The functions $\mathbf{M}^{(3)}_{\sigma ml}(k_0\mathbf{r})$ and $\mathbf{N}^{(3)}_{\sigma ml}(k_0\mathbf{r})$ will be the solutions for the scattered field. Inside the RI shell ($R_1 < r < R_2$) the vector solutions ${}^{(e)}\mathfrak{M}^{(j)}_{\sigma ml}(\mathbf{r})$ and ${}^{(e)}\mathfrak{N}^{(j)}_{\sigma ml}(\mathbf{r})$ of the wave equation (8.4.3) and the vector solutions ${}^{(h)}\mathfrak{M}^{(j)}_{\sigma ml}(\mathbf{r})$ and ${}^{(h)}\mathfrak{N}^{(j)}_{\sigma ml}(\mathbf{r})$ of the wave equation (8.4.4) are described by Eqs.(8.1.3–6). In so doing, the functions regular at the origin are designated with index $j = 1$, and functions satisfying the radiation conditions in the wave zone are designated with index $j = 3$.

Thus, the elemental vector solutions of the wave equations are constructed for all space regions. Let us apply the known expansion of the incident wave over vector spherical functions (see Section 1.3):

$$\begin{aligned}\mathbf{E}_{\text{inc}} &= E_I \exp(-i\omega t + ik_0 z)\\ &= E_0 \exp(-i\omega t)\sum_{l=1}^{\infty}\mu_l\left[\mathbf{M}_{o1l}^{(1)}(k_0\mathbf{r}) - i\mathbf{N}_{e1l}^{(1)}(k_0\mathbf{r})\right]\\ \mathbf{H}_{\text{inc}} &= -\sqrt{\frac{\varepsilon_0}{\mu_0}}E_I \exp(-i\omega t)\sum_{l=1}^{\infty}\mu_l\left[\mathbf{M}_{e1l}^{(1)}(k_0\mathbf{r}) + i\mathbf{N}_{o1l}^{(1)}(k_0\mathbf{r})\right]\end{aligned} \tag{8.4.5}$$

where $\mu_l = i^l(2l+1)/\left[l(l+1)\right]$. The inessential factor $E_I\exp(-i\omega t)$ will be omitted below. At the sphere external surface $r = R_2$ and at the spherical core–shell interface $r = R_1$ the boundary conditions of the continuity of the tangential field components are:

$$\begin{aligned}\left[\mathbf{r}\times(\mathbf{E}_{\text{inc}}+\mathbf{E}_{\text{sc}})\right]_{r=R_2} &= \left[\mathbf{r}\times\mathbf{E}_{\text{shell}}\right]_{r=R_2}\\ \left[\mathbf{r}\times(\mathbf{H}_{\text{inc}}+\mathbf{H}_{\text{sc}})\right]_{r=R_2} &= \left[\mathbf{r}\times\mathbf{H}_{\text{shell}}\right]_{r=R_2}\\ \left[\mathbf{r}\times\mathbf{E}_{\text{core}}\right]_{r=R_1} &= \left[\mathbf{r}\times\mathbf{E}_{\text{shell}}\right]_{r=R_1}\\ \left[\mathbf{r}\times\mathbf{H}_{\text{core}}\right]_{r=R_1} &= \left[\mathbf{r}\times\mathbf{H}_{\text{shell}}\right]_{r=R_1}\end{aligned} \tag{8.4.6}$$

Considering that these boundary conditions must be linear and identical at all points of the interface, the expressions for the scattered field and fields inside the core and shell must have the same angular structure as the incident field. In this case, the expansion of the field inside the shell must contain both regular ${}^{(e,h)}\mathfrak{M}_{\sigma ml}^{(1)}(\mathbf{r})$ and ${}^{(e,h)}\mathfrak{N}_{\sigma ml}^{(1)}(\mathbf{r})$ and irregular ${}^{(e,h)}\mathfrak{M}_{\sigma ml}^{(3)}(\mathbf{r})$ and ${}^{(e,h)}\mathfrak{N}_{\sigma ml}^{(3)}(\mathbf{r})$functions:

$$\begin{aligned}\mathbf{E}_{\text{core}} &= \sum_{l=1}^{\infty}\mu_l\left[\alpha_{1l}\mathbf{M}_{o1l}^{(1)}(k_1\mathbf{r}) - i\beta_{1l}\mathbf{N}_{e1l}^{(1)}(k_1\mathbf{r})\right]\\ \mathbf{H}_{\text{core}} &= -\sqrt{\frac{\varepsilon_0}{\mu_0}}N_1\sum_{l=1}^{\infty}\mu_l\left[\beta_{1l}\mathbf{M}_{e1l}^{(1)}(k_1\mathbf{r}) + i\alpha_{1l}\mathbf{N}_{o1l}^{(1)}(k_1\mathbf{r})\right]\end{aligned} \tag{8.4.7}$$

$$\begin{aligned}\mathbf{E}_{\text{sc}} &= \sum_{l=1}^{\infty}\mu_l\left[-b_l\mathbf{M}_{o1l}^{(3)}(k_0\mathbf{r}) + ia_l\mathbf{N}_{e1l}^{(3)}(k_0\mathbf{r})\right]\\ \mathbf{H}_{\text{sc}} &= \sqrt{\frac{\varepsilon_0}{\mu_0}}N_1\sum_{l=1}^{\infty}\mu_l\left[a_l\mathbf{M}_{e1l}^{(3)}(k_0\mathbf{r}) + ib_l\mathbf{N}_{o1l}^{(3)}(k_0\mathbf{r})\right]\end{aligned} \tag{8.4.8}$$

$$\begin{aligned}\mathbf{E}_{\text{shell}} &= \sum_{l=1}^{\infty}\mu_l\Big[\alpha_{2l}\,{}^{(e)}\mathfrak{M}_{o1l}^{(1)}(\mathbf{r})+\alpha_{3l}\,{}^{(e)}\mathfrak{M}_{o1l}^{(3)}(\mathbf{r})\alpha_{3l}\\ &\quad -i\beta_{2l}\,{}^{(e)}\mathfrak{N}_{e1l}^{(1)}(\mathbf{r}) - i\beta_{3l}\,{}^{(e)}\mathfrak{N}_{e1l}^{(3)}(\mathbf{r})\alpha_{3l}\Big]\\ \mathbf{H}_{\text{shell}} &= -\sqrt{\frac{\varepsilon_0}{\mu_0}}\sum_{l=1}^{\infty}\mu_l\Big[\beta_{2l}\,{}^{(h)}\mathfrak{M}_{e1l}^{(1)}(\mathbf{r})+\beta_{3l}\,{}^{(h)}\mathfrak{M}_{e1l}^{(3)}(\mathbf{r})\alpha_{3l}\\ &\quad -i\alpha_{2l}\,{}^{(h)}\mathfrak{N}_{o1l}^{(1)}(\mathbf{r}) - i\alpha_{3l}\,{}^{(h)}\mathfrak{N}_{o1l}^{(3)}(\mathbf{r})\alpha_{3l}\Big]\end{aligned} \tag{8.4.9}$$

where a_l, b_l, α_{1l}, α_{2l}, α_{3l}, β_{1l}, β_{2l}, and β_{3l} are the unknown amplitude coefficients. Substituting the expansions (8.4.7-9) into the boundary conditions (8.4.6) we obtain two sets of linear equations in the amplitude coefficients:

$$\begin{aligned} \alpha_{1l}\psi_l(N_1x_1) - N_1\alpha_{2l}V_l^{(1)}(x_1) - N_1\alpha_{3l}V_l^{(3)}(x_1) &= 0 \\ \alpha_{1l}\psi_l'(N_1x_1) - \alpha_{2l}V_l^{(1)\prime}(x_1) - \alpha_{3l}V_l^{(3)\prime}(x_1) &= 0 \\ \alpha_{2l}V_l^{(1)}(x_2) + \alpha_{3l}V_l^{(3)}(x_2) + b_l\xi_l(x_2) &= \psi_l(x_2) \\ \alpha_{2l}V_l^{(1)\prime}(x_2) + \alpha_{3l}V_l^{(3)\prime}(x_2) + b_l\xi_l'(x_2) &= \psi_l'(x_2) \end{aligned} \tag{8.4.10}$$

$$\begin{aligned} \beta_{1l}\psi_l(N_1x_1) - \beta_{2l}W_l^{(1)}(x_1) - \beta_{3l}W_l^{(3)}(x_1) &= 0 \\ N_2^2(x_1)\beta_{1l}\psi_l'(N_1x_1) - N_1\beta_{2l}W_l^{(1)\prime}(x_1) - N_1\beta_{3l}W_l^{(3)\prime}(x_1) &= 0 \\ \beta_{2l}W_l^{(1)}(x_2) + \beta_{3l}W_l^{(3)}(x_2) + a_l\xi_l(x_2) &= \psi_l(x_2) \\ \beta_{2l}W_l^{(1)\prime}(x_2) + \beta_{3l}W_l^{(3)\prime}(x_2) + a_lN_2^2(x_2)\xi_l'(x_2) &= N_2^2(x_2)\psi_l'(x_2) \end{aligned} \tag{8.4.11}$$

Recall that here ψ_l and ξ_l are the Riccati–Bessel and Riccati–Hankel functions respectively (see Section 1.5) and $x_{1,2}$ are the size parameters for core and shell. Recall that W_l and V_l are the radial functions (see Eqs.(8.1.5–6)).

Solving the sets (8.4.10–11) in the amplitude coefficients a_l and b_l of the scattered field we have:

$$\begin{aligned} a_l &= \frac{\psi_l(x_2)}{\xi_l(x_2)}\,\frac{\Phi_l(x_1)\gamma_{1l}\gamma_{2l} - \Phi_l(x_2)\gamma_{3l}\gamma_{4l}}{\Phi_l(x_1)\gamma_{2l}\gamma_{5l} - \Phi_l(x_2)\gamma_{4l}\gamma_{6l}} \\ b_l &= \frac{\psi_l(x_2)}{\xi_l(x_2)}\,\frac{\Omega_l(x_1)\bar{\gamma}_{1l}\bar{\gamma}_{2l} - \Omega_l(x_2)\bar{\gamma}_{3l}\bar{\gamma}_{4l}}{\Omega_l(x_1)\bar{\gamma}_{2l}\bar{\gamma}_{5l} - \Omega_l(x_2)\bar{\gamma}_{4l}\bar{\gamma}_{6l}} \end{aligned} \tag{8.4.12}$$

where

$$\begin{aligned} \begin{pmatrix}\gamma_{1l}\\ \gamma_{3l}\end{pmatrix} &= \begin{pmatrix}U_l^{(3)}(x_2)\\ U_l^{(1)}(x_2)\end{pmatrix} - N_2^2(x_2)D_l(x_2) \\ \begin{pmatrix}\gamma_{2l}\\ \gamma_{4l}\end{pmatrix} &= \frac{N_1}{N_2^2(x_1)}\begin{pmatrix}U_l^{(1)}(x_1)\\ U_l^{(3)}(x_1)\end{pmatrix} - D_l(N_1x_1) \\ \begin{pmatrix}\gamma_{5l}\\ \gamma_{6l}\end{pmatrix} &= \begin{pmatrix}U_l^{(3)}(x_2)\\ U_l^{(1)}(x_2)\end{pmatrix} - N_2^2(x_2)G_l(x_2) \end{aligned} \tag{8.4.13}$$

$$\begin{aligned} \begin{pmatrix}\bar{\gamma}_{1l}\\ \bar{\gamma}_{3l}\end{pmatrix} &= \begin{pmatrix}F_l^{(3)}(x_2)\\ F_l^{(1)}(x_2)\end{pmatrix} - D_l(x_2) \\ \begin{pmatrix}\bar{\gamma}_{2l}\\ \bar{\gamma}_{4l}\end{pmatrix} &= \begin{pmatrix}F_l^{(1)}(x_1)\\ F_l^{(3)}(x_1)\end{pmatrix} - N_1D_l(N_1x_1) \\ \begin{pmatrix}\bar{\gamma}_{5l}\\ \bar{\gamma}_{6l}\end{pmatrix} &= \begin{pmatrix}F_l^{(3)}(x_2)\\ F_l^{(1)}(x_2)\end{pmatrix} - G_l(x_2) \end{aligned} \tag{8.4.14}$$

$D_l(x)$ and $G_l(x)$ are the log-derivatives of functions $\psi_l(x)$ and $\xi_l(x)$ respectively;

$$U_l^{(1,3)}(x) = \frac{W_l^{(1,3)\,\prime}(x)}{W_l^{(1,3)}(x)} \qquad F_l^{(1,3)}(x) = \frac{V_l^{(1,3)\,\prime}(x)}{V_l^{(1,3)}(x)} \tag{8.4.15}$$

are the log-derivatives of the magnetic and electric radial functions respectively; and

$$\Phi_l(x) = \frac{W_l^{(1)}(x)}{W_l^{(3)}(x)} \qquad \Omega_l(x) = \frac{V_l^{(1)}(x)}{V_l^{(3)}(x)} \tag{8.4.16}$$

Let us write the expressions for the amplitude coefficients of the internal fields:

$$\beta_{1l} = -iN_1\frac{N_2^2(x_2)}{N_2^2(x_1)}\frac{W_R(x_1)}{\xi_l(x_2)\psi_l(N_1x_1)W_l^{(1)}(x_2)W_l^{(3)}(x_1)A_l} \tag{8.4.17}$$

$$\beta_{2l} = iN_2^2(x_2)\frac{\gamma_{4l}}{\xi_l(x_2)W_l^{(1)}(x_2)A_l} \qquad \beta_{3l} = -iN_2^2(x_2)\frac{\gamma_{2l}\Phi_l(x_1)}{\xi_l(x_2)W_l^{(1)}(x_2)A_l}$$

where A_l is the denominator of coefficient a_l in Eq.(8.4.12) and W_R is the Wronskian $W_R = W_l^{(1)\,\prime}(\rho)W_l^{(3)}(\rho) - W_l^{(1)}(\rho)W_l^{(3)\,\prime}(\rho)$ of the solutions to Eq.(8.1.6). Analogously:

$$\alpha_{1l} = -iN_1\frac{V_R(x_1)}{\xi_l(x_2)\psi_l(N_1x_1)V_l^{(1)}(x_2)V_l^{(3)}(x_1)B_l} \tag{8.4.18}$$

$$\alpha_{2l} = \frac{i\gamma_{4l}}{\xi_l(x_2)V_l^{(1)}(x_2)B_l} \qquad \alpha_{3l} = -\frac{i\gamma_{2l}\Omega_l(x_1)}{\xi_l(x_2)V_l^{(1)}(x_2)B_l}$$

where B_l is the denominator of coefficient b_l in Eq.(8.4.12); V_R is the Wronskian $V_R = V_l^{(1)\,\prime}(\rho)V_l^{(3)}(\rho) - V_l^{(1)}(\rho)V_l^{(3)\,\prime}(\rho)$ of the solutions to Eq.(8.1.5).

The expressions for the scattering characteristics may be obtained by the usual formulae, and the coordinate representation for the internal electric fields takes the form inside the core:

$$\begin{aligned} E_r^{(\text{core})} &= -\frac{\cos\varphi\sin\theta}{N_1^2\rho^2}\sum_{l=1}^{\infty} i^{l+1}(2l+1)\beta_{1l}\psi_l(N_l\rho)\pi_l(\theta) \\ \begin{pmatrix} E_\theta^{(\text{core})} \\ E_\varphi^{(\text{core})} \end{pmatrix} &= \begin{pmatrix} \cos\varphi \\ -\sin\varphi \end{pmatrix}\frac{1}{N_1\rho}\sum_{l=1}^{\infty}\mu_l\left[\alpha_{1l}\psi_l(N_l\rho)\begin{pmatrix} \pi_l(\theta) \\ \tau_l(\theta) \end{pmatrix}\right. \\ &\quad \left. -i\beta_{1l}\psi_l'(N_l\rho)\begin{pmatrix} \tau_l(\theta) \\ \pi_l(\theta) \end{pmatrix}\right] \end{aligned} \tag{8.4.19}$$

and inside the shell:

$$E_r^{(\text{shell})} = -\frac{\cos\varphi \sin\theta}{N_2^2(\rho)\rho^2} \sum_{l=1}^{\infty} i^{l+1}(2l+1) \left[\beta_{2l} W_l^{(1)}(\rho) + \beta_{3l} W_l^{(3)}(\rho)\right] \pi_l(\theta)$$

$$\begin{pmatrix} E_\theta^{(\text{shell})} \\ E_\varphi^{(\text{shell})} \end{pmatrix} = \begin{pmatrix} \cos\varphi \\ -\sin\varphi \end{pmatrix} \frac{1}{\rho} \sum_{l=1}^{\infty} \mu_l \left\{ \left[\alpha_{2l} V_l^{(1)}(\rho) + \alpha_{3l} V_l^{(3)}(\rho)\right] \begin{pmatrix} \pi_l(\theta) \\ \tau_l(\theta) \end{pmatrix} \right.$$
$$\left. - \frac{i}{N_2^2(\rho)} \left[\beta_{2l} W_l^{(1)\prime}(\rho) + \beta_{3l} W_l^{(3)\prime}(\rho)\right] \begin{pmatrix} \tau_l(\theta) \\ \pi_l(\theta) \end{pmatrix} \right\} \tag{8.4.20}$$

8.4.2 Solution for power profile

The formal scheme of the exact solution to the electrodynamic problem on the scattering of the plane wave by the spherical TLP with the RI shell was constructed in the above section. The main difficulty in computer realization of this solution is the calculation of radial functions. The concrete type of these functions is dependent on the choice of the profile of refractive index $N_2(\rho)$ in the shell. As was demonstrated in Section 8.1, the analytical solutions of the radial equations are possible for a very restricted list of profiles only, and in doing so, the solutions, as a rule, are expressed through hyper-geometric functions which are extremely inconvenient to numerical calculations. The sole profile which enables us to avoid the emergence of the hyper-geometric functions is the power law $N_2(\rho) = A\rho^b$, where A and b are the arbitrary complex constants. After substituting this profile into the radial equations (8.1.5–6) we obtain the differential equations whose solutions are the cylindrical functions:

$$\begin{aligned} V_l^{(1)}(\rho) &= \sqrt{\rho} J_{\zeta_l}\left[X(\rho)\right] & V_l^{(3)}(\rho) &= \sqrt{\rho} H_{\zeta_l}^{(1)}\left[X(\rho)\right] \\ W_l^{(1)}(\rho) &= \rho^{b+1/2} J_{\omega_l}\left[X(\rho)\right] & W_l^{(3)}(\rho) &= \rho^{b+1/2} H_{\omega_l}^{(1)}\left[X(\rho)\right] \end{aligned} \tag{8.4.21}$$

with complex argument

$$X(\rho) = \frac{A}{b+1}\rho^{b+1} = \frac{\rho N_2(\rho)}{b+1} \tag{8.4.22}$$

and indices

$$\zeta_l = \frac{2l+1}{2(b+1)} \qquad \omega_l = \frac{1}{b+1}\sqrt{l(l+1) + \left(b+\frac{1}{2}\right)^2} \tag{8.4.23}$$

It clear that for homogeneous shell ($b = 0$) the argument $X(\rho) \to N_2\rho$ and indices ζ_l and $\omega_l \to l+1/2$, that is, the functions V_l and W_l transform to the Riccati–Bessel and Riccati–Hankel functions respectively with an accuracy of some inessential factor. But if the shell is inhomogeneous ($b \neq 0$) then the indices ζ_l and ω_l are neither integer, half-integer nor even real.

For this profile the log-derivatives of radial functions are:

$$\begin{pmatrix} U_l^{(1)}(\rho) \\ U_l^{(3)}(\rho) \end{pmatrix} = \frac{2b+1}{2\rho} + N_2(\rho) \begin{pmatrix} D_{\omega_l}[X(\rho)] \\ G_{\omega_l}[X(\rho)] \end{pmatrix} \tag{8.4.24}$$

$$\begin{pmatrix} F_l^{(1)}(\rho) \\ F_l^{(3)}(\rho) \end{pmatrix} = \frac{1}{2\rho} + N_2(\rho) \begin{pmatrix} D_{\varsigma_l}[X(\rho)] \\ G_{\varsigma_l}[X(\rho)] \end{pmatrix}$$

where D and G are the log-derivatives of the corresponding cylindrical functions J and $H^{(1)}$. The ratios Φ_l and Ω_l are:

$$\Phi_l(\rho) = \frac{J_{\omega_l}[X(\rho)]}{H_{\omega_l}^{(1)}[X(\rho)]} \qquad \Omega_l(\rho) = \frac{J_{\varsigma_l}[X(\rho)]}{H_{\varsigma_l}^{(1)}[X(\rho)]} \tag{8.4.25}$$

Substituting the solutions (8.4.21) into the expressions for Wronskians we obtain:

$$V_R = -\frac{2i}{\pi}(b+1) \qquad W_R = -\frac{2i}{\pi}(b+1)\rho^{2b} \tag{8.4.26}$$

The interesting special case results when $b = -1$. With this value of b the radial functions simplified to elementary power functions:

$$\begin{pmatrix} W_l^{(1)}(\rho) \\ W_l^{(3)}(\rho) \end{pmatrix} = \rho^{(\pm q_l - 1)/2} \qquad \begin{pmatrix} V_l^{(1)}(\rho) \\ V_l^{(3)}(\rho) \end{pmatrix} = \rho^{(\pm q_l + 1)/2} \tag{8.4.27}$$

where $q_l = 2\sqrt{\left(l+\frac{1}{2}\right)^2 - A^2}$. Hence, in this degenerate case:

$$U_l^{(1)} = -F_l^{(3)} = \frac{q_l - 1}{2\rho} \qquad F_l^{(1)} = -U_l^{(3)} = \frac{q_l + 1}{2\rho} \qquad \frac{\Phi_l(x_1)}{\Phi_l(x_2)} = \frac{\Omega_l(x_1)}{\Omega_l(x_2)} = (x_1 x_2)^{q_l} \tag{8.4.28}$$

This is the astonishing result. The radial functions for the inhomogeneous sphere (though degenerate) are much simpler than for the homogeneous one. It appears to be the only case in the electrodynamic theory of scattering when the solution is expressed through the elementary functions. On the other hand, this profile is inconvenient to use in real situations due to the fact that it has only one free parameter. For example, if we fix the refractive indices at the shell–core interface and on the outer boundary of the particle, the thickness of shell is fixed too. Nevertheless, this profile was intensively used (Shifrin *et al.*, 1985; Kokorin and Shifrin, 1989, 1998) owing to its great simplicity. Moreover, the complex many-layered models were constructed with the use of this profile (Perelman, 1979, 1995, 1996; Perelman and Zinov'eva, 2002).

Let us return to the expressions (8.4.20) for the electric field components inside the shell. Let the incident wave be non-polarized, that is, averaging over angle φ must be done. If we introduce the notations:

$$d_{1l} = \beta_{2l} W_l^{(1)}(\rho) + \beta_{3l} W_l^{(3)}(\rho) \qquad d_{2l} = \alpha_{2l} V_l^{(1)}(\rho) + \alpha_{3l} V_l^{(3)}(\rho) \tag{8.4.29}$$

$$d_{3l} = \beta_{2l} W_l^{(1)\prime}(\rho) + \beta_{3l} W_l^{(3)\prime}(\rho)$$

then Eqs.(8.4.20) for the non-polarized case may be rewritten as:

$$E_r^{(\text{shell})} = -\frac{\sin\theta}{N_2^2(\rho)\rho^2}\sum_{l=1}^{\infty} i^{l+1}(2l+1)d_{1l}\pi_l(\theta) \tag{8.4.30}$$

$$\begin{pmatrix} E_\theta^{(\text{shell})} \\ E_\varphi^{(\text{shell})} \end{pmatrix} = \pm\frac{1}{\rho}\sum_{l=1}^{\infty}\nu_l\left[d_{2l}\begin{pmatrix}\pi_l(\theta)\\ \tau_l(\theta)\end{pmatrix} - \frac{i}{N_2^2(\rho)}d_{3l}\begin{pmatrix}\tau_l(\theta)\\ \pi_l(\theta)\end{pmatrix}\right]$$

where $N_2(\rho) = N_2(x_2)\,(\rho/x_2)^b$. Substitution of solution (8.4.21) into the expressions for amplitude coefficients gives:

$$\begin{pmatrix}\alpha_{2l}\\ \alpha_{3l}\end{pmatrix} = \pm\frac{i}{\sqrt{x_2}B_l}\begin{pmatrix}\bar{\gamma}_{4l}\bar{\lambda}_l\\ \bar{\gamma}_{2l}\bar{\sigma}_l\end{pmatrix} \qquad \begin{pmatrix}\beta_{2l}\\ \beta_{3l}\end{pmatrix} = \mp\frac{iN_2^2(x_2)}{x_2^{b+1/2}A_l}\begin{pmatrix}\gamma_{4l}\lambda_l\\ \gamma_{2l}\sigma_l\end{pmatrix} \tag{8.4.31}$$

where

$$\begin{pmatrix}\lambda_l\\ \bar{\lambda}_l\end{pmatrix} = \xi_l(x_2)\begin{pmatrix}[J_{\omega_l}(X_2)]^{-1}\\ [J_{\zeta_l}(X_2)]^{-1}\end{pmatrix} \qquad \begin{pmatrix}\sigma_l\\ \bar{\sigma}_l\end{pmatrix} = \begin{pmatrix}\Phi_{\omega_l}(X_1)\lambda_l\\ \Omega_{\zeta_l}(X_1)\bar{\lambda}_l\end{pmatrix} \qquad X_{1,2} = X(x_{1,2}) \tag{8.4.32}$$

The factor $x_2^{b+1/2}$ in expressions for β_{2l} and β_{3l} is inconvenient and may cause overflow. So, it is useful to extract it and combine it with the analogous factor depending on ρ and entering in $W_l^{(1)}(\rho)$. Then Eq.(8.4.29) takes the form:

$$d_{1l} = f_{1l} + f_{2l} \qquad d_{2l} = f_{3l} + f_{4l} \qquad d_{3l} = f_{1l}\Lambda_l^{(1)}(\rho) + f_{2l}\Lambda_l^{(3)}(\rho) \tag{8.4.33}$$

where

$$\begin{pmatrix}f_{1l}\\ f_{2l}\end{pmatrix} = \left(\frac{\rho}{x_2}\right)^{b+1/2}\begin{pmatrix}\bar{\beta}_{2l}J_{\omega_l}(X)\\ \bar{\beta}_{3l}H_{\omega_l}^{(1)}(X)\end{pmatrix} \qquad \begin{pmatrix}f_{3l}\\ f_{4l}\end{pmatrix} = \left(\frac{\rho}{x_2}\right)^{1/2}\begin{pmatrix}\bar{\alpha}_{2l}J_{\omega_l}(X)\\ \bar{\alpha}_{3l}H_{\omega_l}^{(1)}(X)\end{pmatrix}$$

$$\begin{pmatrix}\Lambda_l^{(1)}(\rho)\\ \Lambda_l^{(3)}(\rho)\end{pmatrix} = \frac{2b+1}{2\rho} + N_2(\rho)\begin{pmatrix}D_{\omega_l}(X)\\ G_{\omega_l}(X)\end{pmatrix} \qquad \begin{pmatrix}\bar{\alpha}_{2l}\\ \bar{\alpha}_{3l}\end{pmatrix} = \pm\frac{i}{B_l}\begin{pmatrix}\bar{\gamma}_{4l}\bar{\lambda}_l\\ \bar{\gamma}_{2l}\bar{\sigma}_l\end{pmatrix}$$

$$\begin{pmatrix}\bar{\beta}_{2l}\\ \bar{\beta}_{3l}\end{pmatrix} = \mp\frac{iN_2^2(x_2)}{A_l}\begin{pmatrix}\gamma_{4l}\lambda_l\\ \gamma_{2l}\sigma_l\end{pmatrix} \tag{8.4.34}$$

Hence, there is a need to calculate three groups of functions to obtain the amplitude coefficients of the scattered field: (1) the functions $\psi_l(x_2)$ and $\xi_l(x_2)$ and log-derivatives $D_l(x_2)$, $G_l(x_2)$, and $D_l(N_1x_1)$; (2) the log-derivatives $D_{\omega_l}(X_1)$, $D_{\omega_l}(X_2)$, $G_{\omega_l}(X_1)$, and $G_{\omega_l}(X_2)$ and ratios $\Omega_l(X_1)$ and $\Omega_l(X_2)$; and (3) the analogous quantities for indices ζ_l. The computation of functions from the first group presents no special problems (see Section 1.5). Much more complex is the calculation of functions entering in the second and third groups. Clearly, no recurrence over index l is possible here, causing us to calculate functions for each index $l = 1, 2, ..., L$ individually. The fact that we need the ratios of cylindrical functions rather than the functions themselves does not alleviate the situation here. After some attempts we had come to the conclusion that the independent calculation of the function J_ν itself (for simplicity we designate as ν the any index ω_l or ζ_l) is inevitable here. Then the log-derivatives D_ν and $D_{-\nu}$ can be computed through the Lentz's algorithm (see Section 1.5.2.1), which works for an arbitrary complex index. The function $J_{-\nu}$ may be obtained from the relation:

$$J_\nu(z)J_{-\nu}(z) = \frac{2\sin\nu\pi}{\nu z}\left[D_\nu(z) - D_{-\nu}(z)\right] \tag{8.4.35}$$

that follows immediately from the Wronskian, and then the ratio $K_\nu(z) = J_{-\nu}(z)/J_{-\nu}(z)$ may be constructed. Then the required log-derivatives $G_\nu(z)$ and ratio Ω_ν (or Φ_ν) may be calculated using the relations:

$$G_\nu(z) = D_\nu(z) + \frac{K_\nu(z)\left[D_\nu(z) - D_{-\nu}(z)\right]}{L_\nu(z)} \qquad \Omega_\nu(z) = -\frac{if}{2L_\nu(z)} \tag{8.4.36}$$

where $L_\nu(z) = \exp(-i\nu\pi) - K_\nu(z)$ and $f = 2\sin\nu\pi$.

Thus, the central problem here is calculation of the Bessel function $J_\nu(z)$. Bohren and Huffman (1983, p.185) stated that "mere contemplation of complex-order Bessel functions is sufficient to make strong men weep". We fully agree with this opinion. The simplest method consists in using the usual power series (see Eq.(A.2)). But it is known that this series converges extremely slowly (and with catastrophic loss of significant digits) when $|\nu|$ is distinctly less than $|z|$. At $|\nu| \gtrsim |\nu|$ the convergence is better but still remains unsatisfactory. Another method (Lentz, 1976) uses continued fractions for calculation of Bessel functions with arbitrary complex arguments and indices. But the control calculations showed that this method converges no better than the power series. So, we decided to use the adding theorem due to Gegenbauer (Bateman and Erdélyi, 1953*b*):

$$\left(\frac{2}{z}\right)^\nu \sum_{l=0}^{\infty} \frac{(\nu+2l)\Gamma(\nu+1)}{l!} J_{\nu+2l}(z) = 1 \tag{8.4.37}$$

At $\nu = 0$ Eq.(8.4.37) reduces to the well known adding theorem:

$$J_0(z) + 2\sum_{l=1}^{\infty} J_{2l}(z) = 1 \tag{8.4.38}$$

It is easy to see that Eq.(8.4.37) may be rewritten in the form more suitable for numerical calculations:

$$\sum_{l=0}^{\infty} \alpha_l J_{\nu+2l}(z) = 1 \tag{8.4.39}$$

where $\alpha_0 = \left(\frac{2}{z}\right)^\nu \nu\Gamma(\nu) = \left(\frac{2}{z}\right)^\nu \Gamma(\nu+1)$ and subsequent terms may be obtained via the multiplicative upward recursion:

$$\alpha_l = \alpha_{l-1}\frac{(\nu+2l)(\nu+l-1)}{(\nu+2l-2)l} \tag{8.4.40}$$

Let $F_\nu(z) = CJ_\nu(z)$ (where C is the normalizing constant) be the auxiliary function obeying the same recurrence:

$$F_{\nu+l-1}(z) = \frac{2(\nu+1)}{z} F_{\nu+l}(z) - F_{\nu+l+1}(z) \tag{8.4.41}$$

as $J_\nu(z)$. The recursion may be used in the down direction beginning with the initial values $F_{\nu+L-1}(z) = 0 + i0$ and $F_{\nu+L}(z) = 10^{-\delta}(1+i)$ for some large index $l = L$, where δ is some great power (e.g., $\delta = 35$). After summation:

$$\sum_{l=0}^{N/2} \alpha_l F_{\nu+2l}(z) = C \tag{8.4.42}$$

we can recalculate $J_\nu(z)$. The Neimann function:

$$Y_\nu(z) = J_\nu(z) \cot(\nu\pi) - J_{-\nu}(z)/\sin(\nu\pi)$$

(ν is not an integer) may be, in principle, obtained by the similar method using some modification of the Sonine formula:

$$J_{-\nu}(z) = \left(\frac{2}{z}\right)^{2\nu} \Gamma(1-2\nu) \sum_{l=0}^{\infty} \frac{(\nu+2l)\Gamma(\nu+l)}{l!\Gamma(l+1-\nu)\Gamma(1-l-2\nu)} J_{\nu+2l}(z) \tag{8.4.43}$$

Then:

$$Y_\nu(z) = \frac{1}{C} \sum_{l=0}^{N/2} \beta_l F_{\nu+2l}(z) \tag{8.4.44}$$

where the coefficients β_l must meet the recurrence:

$$\begin{aligned} \beta_l &= \frac{(\nu+2l)(\nu+l-1)(2\nu+l-1)}{l(\nu+2l-2)(\nu-1)} \beta_{l-1} \\ \beta_0 &= \cot\ (\nu\pi) - \alpha_0^2/\nu\pi \qquad \beta_1 = 2(\nu+2)\alpha_0^2/\left[\pi(\nu-1)\right] \end{aligned} \tag{8.4.45}$$

The derivative $J'_\nu(z) = \nu J_\nu(z)/z - J_{\nu+1}(z)$ evaluates using $J_\nu, J_{\nu+1}$. The Wronskian $Y_{\nu+1}(z) = \left[J_{\nu+1}(z)Y_\nu(z) - 2/(\pi z)\right]/J_\nu(z)$ gives $Y_{\nu+1}(z)$, and the derivative $Y'_\nu(z) = \frac{\nu}{z}Y_\nu(z) - Y_{\nu+1}(z)$ is obtained.

Thus, the approach gives, in principle, all the required functions. But, the test calculations showed that this method of $Y_\nu(z)$ evaluation is numerically valid at $|\nu| \gg |z|$ only. But in the opposite situation $|\nu| \ll |z|$ (as it usually takes place at summation the series for $Y_\nu(z)$) catastrophic loss of precision occurs. If the index and argument are the values of the same order, then the loss of precision is small, but very slight convergence occurs. Because of this, we used another approach briefly mentioned above which is described in detail in the next section.

8.4.3 Algorithm for power profile

After the preliminary considerations given in the previous section, now we shall describe the algorithm of calculation for the functions D_ν, G_ν, and Φ_ν in detail (Babenko and Leiko, 1991; Babenko and Sinuyk, 1993).

These functions are calculated using the subroutine DIFM (see Appendix D) whose main task is to calculate the function $J_\nu(z)$ for the complex argument z and index $\nu = \nu_1 + i\nu_2$ through the Gegenbauer addition theorem (8.4.37). Its thorough consideration demonstrats that this series converges satisfactorily at $\mathrm{Re}\,\nu > 0$ only. So the functions $J_\nu(z)$ and $J_{-\nu}(z)$ were calculated at $\mathrm{Re}\,\nu \geq 0$ and $\mathrm{Re}\,\nu < 0$, accordingly. In addition, the log-derivative $D_{-\nu}(z)$ was calculated in the first case and $D_\nu(z)$ in the second case using the continued fractions method. It realizes algorithmically as follows: (a) redefining of the index – if $\nu_1 \geq 0$, then $\tilde{\nu} = \nu$ and key KL = 0, if else $(\nu_1 < 0)$, then $\tilde{\nu} = -\nu$ and key KL = 1; and (b) call of the subroutine DLK (see Appendix D) with arguments $(-\tilde{\nu}, z)$ to calculate $D_{-\tilde{\nu}}(z)$. When KL = 0, this function is redesignated as B_ν, and when KL = 1 as D_ν.

The next step is the calculation of the logarithm of the initial coefficient $\bar{\alpha}_0 \equiv \ln \alpha_0 = \tilde{\nu}\ln(2/z) + \ln\Gamma(1+\tilde{\nu})$ using the standard procedure.

Further, the array of auxiliary functions $F_{\tilde{\nu}+l}(z)$ ($l = 0, 1, ..., L_1^*$, definition of L_1^* see below) are calculated using the downward recursion:

$$F_{l-1} = \frac{2(\tilde{\nu}+1)}{z} F_l - F_{l-1} \tag{8.4.46}$$

with initial values $F_{L_1^*+1} = 0$ and $F_{L_1^*} = 10^{-35}(1+i)$. Here and below we use the simplified designation F_l instead of $F_{\tilde{\nu}+l}$. The initial index L_1^* must significantly exceed both $|\tilde{\nu}|$ and $|z|$. We use the following estimate (with rounding off to the nearest odd number):

$$L_1^* = \max\left(1.5\,|z| - |\nu| + 15,\ 30\right) \tag{8.4.47}$$

with the use of which the sequence of $F_l(z)$ converges to the right values through 5–6 recursion steps at most. Hence, the values F_l ($l = 0, 1, ..., 2L_1$, where $2L_1 = L_1^* - 5$) are accepted as correct. Because F_l with even indices only present in Gegenbauer formula, the array obtained must be rearranged: $F_0 \to F(1), F_2 \to F(2), ..., F_{2L_1} \to F(L_1+1)$.

It is advantageous to use the normalization:

$$\tilde{F}(l) = \frac{F(l)}{\mathrm{Re}\,F(1)} \qquad \tilde{\alpha}_l = \frac{\alpha_l}{\alpha_0} \qquad l = 1, 2, ..., L_1 + 1 \tag{8.4.48}$$

Then the sum (8.4.42) may be written as:

$$J_{\tilde{\nu}}(z) = \frac{\tilde{F}(1)}{\alpha_0 C} = \tilde{F}(1)\left[\alpha_0 \sum_{l=0}^{L_1} \tilde{\alpha}_l \tilde{F}(l+1)\right]^{-1} \tag{8.4.49}$$

The next step is the calculation of coefficients through the multiplicative recursion $\tilde{\alpha}_l = c_l \tilde{\alpha}_{l-1}$ $(l = 1, 2, ..., L_1)$, with initial value $\tilde{\alpha}_0 = 1 + i0$, where

$$c_l = \frac{(2l + \tilde{\nu})(l - 1 + \tilde{\nu})}{l(2l - 2 + \tilde{\nu})}$$

These values are combined with $\tilde{F}(l+1)$ for construction of the sum C. Convergence of this sum is checked at each step over l. If some criterion is satisfied, then the summation is terminated. But it is suggested in the algorithm that the estimate (8.4.47) may be not sufficient sometimes. In such a case, when the criterion is not satisfied at $l = L_1$, the calculation of additional terms $\bar{F}_l$ $(l = 2L_1, 2L_1 + 1, ..., L_2^*$, where $L_2^* = 2L_1 + 15)$ takes place using the downward recursion (8.4.46) with initial values $\bar{F}_{L_2^*+1} = 0$ and $\bar{F}_{L_2^*} = 10^{-35}(1 + i)$. The sequence $\bar{F}_l$ and earlier obtained sequence $\tilde{F}_l$ may be "sewn" together at $l = 2L_1$. If we designate the ratio $f = \tilde{F}_{2L_1}/\bar{F}_{2L_1}$, then it is clear that the sequence $\tilde{F}_{2L_1+1} = f\bar{F}_{2L_1+1}$,...,$\tilde{F}_{2L_2} = f\bar{F}_{2L_2}$ (where $2L_2 = L_2^* - 5$) is normalized as well as the initial one. Five terms with even indices are entered in the array $\tilde{F}$: $\tilde{F}(N_2 + 1)$, $\tilde{F}(N_2), ..., \tilde{F}(N_1 + 2)$. The adding of new terms into the sum C takes place. If the convergence is not obtained here, too, then the process must be repeated.

This method allows us to simultaneously find the value of log-derivative $D_{\tilde{\nu}}(z)$ (i.e., $D_\nu(z)$ at KL=0, or $D_{-\nu}(z)$ at KL=1):

$$D_{\tilde{\nu}}(z) = \frac{\tilde{\nu}}{z} - \frac{F_1}{F_0} \tag{8.4.50}$$

Let us go to final expressions on the output of the subroutine DIFM. The arrays of functions $D_{\tilde{\nu}}(z)$, $D_{-\tilde{\nu}}(z)$, and $J_{\tilde{\nu}}(z)$ were computed in the previous stage. Now we must obtain the arrays of functions $\Phi_\nu(z)$ and $G_\nu(z)$. The direct implementation of Eq.(8.4.36) is inconvenient due to a chance to overflow. So, the following scheme was used. The quantities:

$$\begin{aligned}
\bar{f} &= \exp(-|\nu_2|\,\pi) f \\
&= \sin \nu_1 \pi \, [1 + \exp(-2\,|\nu_2|\,\pi)] \pm i \cos \nu_1 \pi \, [1 - \exp(-2\,|\nu_2|\,\pi)] \quad {}^{\nu_2 \geq 0}_{\nu_2 < 0} \\
\bar{K}_\nu &= \ln K_\nu = \pm \left[\ln \frac{\bar{f}}{\pi z (D_\nu - D_{-\nu})} + |\nu_2|\,\pi + 2\tilde{\alpha}_0 + 2 \ln \frac{C}{\tilde{F}(1)} \right] \quad {}^{\nu_1 \geq 0}_{\nu_1 < 0} \\
\bar{\gamma}_\nu &= \ln \gamma_\nu = -i\nu\pi - \bar{K}_\nu
\end{aligned} \tag{8.4.51}$$

were calculated instead of f and K_ν. Then:

$$G_\nu = D_\nu + \frac{D_\nu - D_{-\nu}}{\gamma_\nu - 1} \qquad \Phi_\nu = \bar{\Phi}_\nu \exp(-Q_\nu)$$

where $\bar{\Phi}_\nu = -i\bar{f}\,/\,2(\gamma_\nu - 1)$ and $Q_\nu = \bar{K}_\nu - |\nu_2|\,\pi$. The values $\bar{\Phi}_\nu$ and Q_ν are in the output instead of Φ_ν.

8.5 ASYMPTOTIC MODEL OF A FRACTAL CLUSTER

8.5.1 What is a fractal cluster

The ideas of the mathematical theory of fractal sets (Mandelbrott, 1982) have received wide acceptance in current physics (see Zeldovich and Sokolov, 1985; Smirnov, 1986; Novak and Deway, 1999). More than two decades ago Forrest and Witten (1979) used the fractal concept to describe aggregates of metal smoke. This work provided a basis for developing the Witten and Sander model (1981) imitating the process of stochastic cluster construction. It gave powerful impetus to research in this direction.

The main property of a fractal cluster is the scale invariance, that is, it appears the same when viewed over a range of scales. This concept includes a quantifiable parameter, the Hausdorff or fractal dimension D. The mathematical description of the main structural parameter of the cluster is extremely simple and can be expressed as:

$$N_s \sim \left(\frac{R}{a}\right)^D \tag{8.5.1}$$

where N_s is the number of primary subparticles (monomers) in the aggregate, R is the mean size of the aggregate (e.g., the mean-square-root radius, the radius of gyration), and a is the size of the monomer (for example, the radius of spherical particles). It is clear that $1 \leq D \leq 3$. The value $D = 3$ corresponds to a solid particle, the degree of cluster porosity increases as D decreases. We say that for a fractal aggregate mass scales with the overall size of the D power. Thus, if we experimentally determine the dependence of the subparticle number on the cluster size R, we thereby find the so named internal fractal dimension D_{int}. The external fractal dimension D_{ext} is defined from the power dependence of the pair-correlation function of the subparticle density. Measuring the change of cluster density on microphotographs with the distance r from its center we can find D_{ext}. If a cluster is not too dense then $D_{\text{int}} \approx D_{\text{ext}}$. We shall not distinguish between these quantities in what follows.

Hence, the dimension D is actually determined experimentally or from the computational modelling imitating the physical process of the association of solid subparticles with one or another suppositions. If association proceeds under the principle "cluster-subparticle" (i.e., if subparticles adhere one by one to the available cluster), then the typical value of D is equal to 2.2, which corresponds to a rather closely packed cluster. If, in contrast, the "cluster-cluster" association occurs, then the value of D is near 1.8 and the aggregate is quite porous. In real conditions the two association mechanisms appear to be competitive; as a result the mean values of D at the level 2.0 take place.

It can be shown (Khlebtsov and Melnikov, 1994), that the power law (8.5.1) is rigorously valid at the region of the so named intermediate asymptotic only:

$$a \ll r \ll R \tag{8.5.2}$$

It is clear that the above condition is not valid in the outer region of a cluster. Because of this, the law of descending the pair-correlations of subparticles in this region is important. It it described by the so named cutoff function $h(r/R)$ which enters into the net correlation function $r^{D-3}h(r/R)$ (Khlebtsov, 1996).

The aggregates consisting of the nano-sized metallic subparticles are of special interest in the optics of fractal clusters. The strong specific linear and non-linear effects are observed in it, including the very large local electric fields on the surfaces of individual subparticles. These effects are fully considered in the recent books of Markel and George (2000) and of Shalaev (2000).

Here we are interested in scattering of the electromagnetic wave by three-dimensional fractal cluster. This electrodynamic problem is rather complex, since we must take into account both microscopic interaction of light with subparticles and macroscopic interaction with aggregate as a whole.

The simplest (and clearly the most inexact) approach may be realized with help of the "mean particle" model. In this case the mean filling factor which is the ratio of the mean density of the cluster to the density of subparticle is found on the base of the power dependency (8.5.1). With use of some effective medium theory (EMT) (e.g., with the Maxwell Garnett one) the effective dielectric function ε_{eff} is calculated. The actual aggregate is replaced by a homogeneous spherical particle with the same radius R but with effective optical constants. Such an approach was realized by Hui and Stroud (1986).

More precise methods and models assume the solution of many-particle electrodynamic equations. Methodically the problem may be solved in two ways. The former involves the numerical solution of initial exact equations followed by averaging the results obtained over the realization ensemble (Borghese *et al.,* 1987; Chen *et al.* 1988; Chen *et al.*, 1990). Its main advantage lies in the fact that it can be reconciled with the imitation modeling of the cluster itself. However, this approach is very laborious and unwieldy. Its practical realization is possible only due to the very small size of subparticles. The relatively large computer time restricts the applicability field of the imitation methods but they provide a means of estimating the validity of more simple ones (Nelson, 1989). The Monte Carlo method (Jullien and Botet, 1987, 1989) may be assigned to the imitation ones also. It is appropriate in the case of subparticles which are large compared with the wavelength and totally ignores the coherent scattering.

The second way of solving the problem implies that some physically justified assumptions are made at the stage of obtaining the macroscopical electrodynamic equation that simplifies it significantly. Then the solution of this simplified equation is sought. In particular, the chain of equations for scattered waves may be represented in the approximation of "mean field", where it is assumed that the fields inside and outside the aggregate are formed due to scattering and rescattering of a set of plane waves by the homogeneous system of subparticles (Berry and Percival (1986)). The authors of this often cited paper have solved the diffraction problem for a fractal cluster by the method of successive approximations and obtained, as a result, a variety of useful estimates.

The diffraction problem for the mean field was solved exactly by Kuzmin *et al.* (1988). Since the cluster density and, therefore, its optical properties change only as a function of distance from cluster center, the problem reduces to the problem of scattering by a RIP. Because such a problem for an arbitrary inhomogeneity profile is rather complex, the model of a TLP with homogeneous core and inhomogeneous shell was proposed. This asymptotic model allows us to describe, more or less satisfactorily, the actual profile of optical constants in a cluster.

These three models (mean particle, mean field, and asymptotic) may be uniformly described on the basis of the Dyson equation:

$$\mathbf{E}(\mathbf{r}) = \mathbf{E}_0(\mathbf{r}) + \int\limits_{V'}\int\limits_{V''} \hat{G}(\mathbf{r},\mathbf{r}')M(\mathbf{r}',\mathbf{r}'')\mathbf{E}(\mathbf{r}'')dV'dV'' \qquad (8.5.3)$$

where $\hat{G}(\mathbf{r},\mathbf{r}')$ is the tensor Green function for the free space and $M(\mathbf{r}',\mathbf{r}'')$ is the local mass operator which may be written as the product of Dirac delta-function, two-particle correlation function, and the amplitude function of forward scattering for the cluster subparticles $M(\mathbf{r}',\mathbf{r}'') = \delta(\mathbf{r}'-\mathbf{r}'')\,G_2(\mathbf{r}')\,F$. In spite of the simple spherically symmetric kernel, Eq.(8.5.3) is rather complex to solve. In the mean particle model the kernel is simplified further through factoring out it from the integral according the theorem of mean value. In so doing, the ordinary diffraction problem for a homogeneous sphere is obtained. In the mean field model Eq.(8.5.3) is solved by the method of successive approximations; each iteration is interpreted as the order of scattering. Lastly, in the asymptotic model Eq.(8.5.3) is solved exactly. Thus, it can be considered as some generalization of two preceding methods. At least, the formulae of mean particle and mean field approaches can be derived from the integral equation for the RIP with the help of simple transformations. Numerical tests performed both by the imitation modeling (Nelson, 1989) and in the framework of the asymptotic model have shown that results obtained with the models of mean particle and mean field may have great inaccuracy.

It should be noted that the investigation of optical properties of fractal clusters (especially of metallic and carbon ones) has been progressing very rapidly in recent years. A detailed review can be found in Sorensen (2001).

It should also be borne in mind that the concept of fractality does not cover the entire diversity of aggregates. The non-fractal aggregates are likewise available although such aggregates are rarely the result of some natural process. Moreover, a great deal of investigation has been devoted to scattering and absorption of light by the ensembles comprised of a relatively small number of subparticles when the concept of fractality is not yet applicable (see, e.g., the review of Fuller and Mackowski (2000)).

8.5.2 Model construction

Let us describe briefly the asymptotic model used by us for the calculations of spectral dependencies of the optical properties of the cluster comprising of nano-particles. The characteristic peculiarity of the fractal cluster is the drop of its mean density ρ as a function of distance r from the aggregate center:

$$\rho(r) = \frac{const}{r^{3-D}} \qquad (8.5.4)$$

Assume with some stretch that Eq.(8.5.4) is valid for all r, and not just for r in the region of the intermediate asymptote (8.5.2). Then it is clear that:

$$\rho(r) = \rho_0 f(r) = \rho_0 \left(\frac{a}{r}\right)^{3-D} \qquad (8.5.5)$$

where ρ_0 is the density of the substance of a subparticle with radius a and $f(r)$ is the so-called filling factor. The filling factor $\bar{f}$ averaged from $r = 0$ till some $r = r_0$ is obtained by integration of the cluster inhomogeneity profile over r. If $r_0 \gg a$, then $\bar{f}(r_0) \approx 3f(r_0)/D$. In particular, the filling factor averaged over the entire cluster is:

$$\tilde{f} \approx \frac{3}{D}\left(\frac{a}{R}\right)^{3-D} \tag{8.5.6}$$

It follows that the volume occupied by all subparticles of a cluster is $V = 4\pi a^3 \left(R/a\right)^D/D$, and the number of subparticles in a cluster is $N_s = 3\left(R/a\right)^D/D$.

We used the model of the RI sphere with the relatively dense small homogeneous core (radius r_1, which is nevertheless much more than the subparticle radius a) and great RI shell to describe the scattering properties of a sufficiently great cluster. The value r_1 may be reasonably selected from the condition $f(r_1) = 0.3$, because this level of filling factor defines the percolation threshold in the Bruggeman (BR) theory for subparticles having the dielectric function of Drude type. In other words, it is such a critical value of the subparticle concentration, that the jump change of the properties of system as a whole takes place due to the stochastic sticking of particles and formation of a coherent cluster-like structure in the entire volume of the system. When this threshold is exceeded the system takes new features typical for metals. Clearly this concept is not applicable to pure dielectric subparticles. Experimental values of the percolation threshold vary from $f = 0.2$ to 0.5. In fact, this value $f(r_1)$ is the additional parameter of the model. It follows from Eq.(8.5.5) that the core radius is:

$$r_1 = af(r_1)\exp\left(\frac{1}{D-3}\right) \tag{8.5.7}$$

The optical properties of the cluster change in accordance with the dropping of its density from center to periphery. They can be characterized by the effective absolute complex refractive index $N_{\text{eff}} = n_{\text{eff}} + i\kappa_{\text{eff}}$, that depends on the distance r and differs from the similar quantity for the subparticle matter N_{sub}.

The particles comprising the cluster are, as a rule, very small compared with wavelength λ. For example, the size of metallic subparticles is of the order of some nanometers. In this case, it is appropriate to use one or another EMT to find the constant refractive index of core N_1 and profile of the shell refractive index $N_2(\rho)$. The most popular approximations of this kind are the well known formulae from Maxwell Garnett (MG) theory and BR theory:

$$f\frac{\varepsilon_{\text{sub}} - \varepsilon_{\text{m}}}{\varepsilon_{\text{sub}} + 2\varepsilon_{\text{m}}} = \frac{\varepsilon_{\text{eff}} - \varepsilon_{\text{m}}}{\varepsilon_{\text{eff}} + 2\varepsilon_{\text{m}}} \qquad f\frac{\varepsilon_{\text{sub}} - \varepsilon_{\text{eff}}}{\varepsilon_{\text{sub}} + 2\varepsilon_{\text{eff}}} = (1-f)\frac{\varepsilon_{\text{eff}} - \varepsilon_{\text{m}}}{\varepsilon_{\text{eff}} + 2\varepsilon_{\text{m}}} \tag{8.5.8}$$

where ε_{sub} and ε_{m} are the absolute dielectric functions of the subparticles and medium surrounding them and ε_{eff} is the effective dielectric function.

Equations (8.5.8) are usually obtained from simple physical considerations and can also be derived from the exact many-particle electrodynamic equations in the approximations of the mean T-matrix and coherent potential for the isotropic monodisperse system of spherically symmetric scatterers in the long wave limit, when the subparticle size is much less than the wavelength (Davis and Schwartz, 1985). The generalization of MG theory is the polarization theory of Clausius–Mosotti, the basis for which is the Lorentz relation between the local and external fields and where the correlations of particle positions are considered:

$$\frac{\varepsilon_{\text{eff}} - \varepsilon_{\text{m}}}{\varepsilon_{\text{eff}} + 2\varepsilon_{\text{m}}} = f\frac{\varepsilon_{\text{sub}} - \varepsilon_{\text{m}}}{\varepsilon_{\text{sub}} + 2\varepsilon_{\text{m}}}(1 + Af + Bf^{2}) \tag{8.5.9}$$

where A and B are the correlation-dependent functions. This formula may also be obtained from the exact electrodynamic equations.

MG theory describes the microstructure of the type when particles of different kinds (the particles of one kind in this specific case) are embedded in the homogeneous medium–matrix. In so doing, it is actually suggested that the interaction of particles with the surrounding medium is strong and interaction of particles with one another is weak. By this it is meant that MG theory is most effective in the case of strongly diluted systems, when the particles are randomly distributed in the homogeneous medium. There is the inverse variant also when it is assumed at large filling factor ($f > 0.5$) that the matrix is not the initial surrounding medium but metal (i.e., the components are interchanged). However, the direct and inverse variants of the MG approximation do not match at $f = 0.5$. In this regard the more exact modifications of this quasi-static EMT were developed, in particular, the Sheng (1980) approximation that eliminates the jump at $f = 0.5$.

In BR theory the local field is selected by the selfconsisted way (i.e., assuming that the mean efficiency of scattering by an individual structural element located in the effective medium is zero). The BR theory describes well the microstructure in the form of a random mixture of particles of different kinds, when it is impossible to draw a distinct border between components. The weak interaction of particles with surrounding medium and strong interaction between particles themselves are assumed in BR theory.

In this regard we assume that one can calculate the absolute dielectric function for the core, with a knowledge of its mean filling factor:

$$\bar{f}_{\text{core}} = \left(\frac{a}{r_1}\right)^3 \left\{1 + \frac{3}{D}\left[\left(\frac{r_1}{a}\right)^D - 1\right]\right\} \tag{8.5.10}$$

and utilizing the BR formula, or more precisely, finding a solution to the BR square equation:

$$2\varepsilon^2 + \varepsilon\left[3\bar{f}_{\text{core}}\left(\varepsilon_{\text{m}} - \varepsilon_{\text{sub}}\right) - 3\varepsilon_{\text{m}} + \varepsilon_{\text{sub}}\right] - \varepsilon_{\text{m}}\varepsilon_{\text{sub}} = 0 \tag{8.5.11}$$

that has a positive imaginary part.

It is evident, that MG theory is most suitable in the cluster periphery where $f \ll 1$. In the internal cluster region adjacent to the core it is expedient to use the BR approach.

Many attempts to improve the classical EMTs were undertaken in recent years. In particular, Chýlek *et al.* (2000) developed the approach based on the account of all terms in the Mie series (instead of dipole ones only, as it is done in the MG and BR approaches) and allowing, in principle, to go further towards the greater particles. Regrettably, the final expressions in these papers are erroneous. We rederived the expressions which have the form:

$$\left(N_{\text{eff}}^2\right)_{k+1} = N_{\text{m}}^2 \frac{A_k(1-f) + fB_k}{A_k(1-f) - 2fB_k} \qquad A_k = -i\frac{12\pi^2}{\lambda^3}\left(N_{\text{eff}}^3\right)_k \tag{8.5.12}$$

$$B_k = \frac{3}{4\pi a^3}\sum_{l=1}^{\infty}(2l+1)\left[a_l\left(r, \frac{N_{\text{sub}}}{(N_{\text{eff}})_k}\right) + b_l\left(r, \frac{N_{\text{sub}}}{(N_{\text{eff}})_k}\right)\right]$$

where a_l and $b_l$are the Mie amplitude coefficients for the scattered field. Coefficients A_k and B_k are calculated iteratively where $(m_{\text{eff}})_1$, evaluated by BR formula, is chosen as an initial approximation ($k = 1$).

It was mentioned above that the main idea of our approach is to replace the sufficiently great fractal cluster by the sphere with a small homogeneous core and RI shell, in so doing the profile of the optical inhomogeneity being calculated with the help of some EMT. There are two different approaches used by us to calculate the characteristics of scattering by such a coated sphere.

a) The first one is as follows. It can be shown that all EMTs reduce to the MG theory at small filling factor f. Since $f \ll 1$ at the cluster periphery, MG theory was used by us to evaluate the refractive index on the outer border of the fractal. If we solve the MG equation in ε_{eff} then we have:

$$\varepsilon_{\text{eff}} = \varepsilon_{\text{m}}\left(1 + \frac{3f}{\varepsilon_{12} - f}\right) \qquad \text{where } \varepsilon_{12} = \frac{\varepsilon_{\text{sub}} + 2\varepsilon_{\text{m}}}{\varepsilon_{\text{sub}} - \varepsilon_{\text{m}}} \tag{8.5.13}$$

If assume that $f \ll \varepsilon_{12}$ then $\varepsilon_{\text{eff}}(r) \approx \varepsilon_{\text{m}}\left[1 + 3f(r)/\varepsilon_{12}\right]$ or in the same approximation taking account of Eq.(8.5.5):

$$N_{\text{eff}}(r) \approx N_{\text{m}}\left[1 + \frac{3}{2\varepsilon_{12}}\left(\frac{a}{r}\right)^{3-D}\right] \tag{8.5.14}$$

As was discussed in Section 8.1, the analytical solution to the problem of scattering by RI spheres can be found in the relatively simple form (i.e., by use of known special functions) in very limited cases. Regrettably, it is impossible for profile (8.5.13), but the solution for the power profile $N(r) = Ar^b$ (where A and b are the arbitrary complex constants) which is very close to profile (8.5.13) is relatively simple (see Section 8.4.2). These two profiles actually differ in the additive term only and drop from the cluster center to its periphery. Both profiles are singular in the center but this singularity is eliminated by the presence of a small homogeneous core. That is the main reason that the core is introduced in the model. The profiles can be made coincident in two points (e.g., in point r_1 inside the cluster and in the point R in the exterior). Then the constants A and b can be evaluated. Let us consider it in more detail. If these constants are represented as $A = \gamma\exp(i\Phi)$ and $b = b' + ib''$ then it is evident that:

$$\begin{pmatrix} n(r) \\ \kappa(r) \end{pmatrix} = \gamma r^{b'}\begin{pmatrix} \cos \\ \sin \end{pmatrix}(\Phi + b''\ln r) \tag{8.5.15}$$

Designate the refractive indices at points r_1 and R as $N_{\text{eff}}(r_1) = n_1 + i\kappa_1$ and $N_{\text{eff}}(R) = n_2 + i\kappa_2$. Then:

$$
\begin{aligned}
b' &= \frac{\ln\left[\left(n_2^2+\kappa_2^2\right)/\left(n_1^2+\kappa_1^2\right)\right]}{2\ln(R/r_1)} \qquad a = \frac{\sqrt{n_1^2+\kappa_1^2}}{r_1^{b'}} \\
b'' &= \frac{\arctan\left(\kappa_2/n_2\right) - \arctan\left(\kappa_1/n_1\right)}{\ln(R/r_1)} \\
\Phi &= \frac{\ln R \arctan\left(\kappa_1/n_1\right) - \ln r_1 \arctan\left(\kappa_2/n_2\right)}{\ln(R/r_1)}
\end{aligned} \tag{8.5.16}
$$

Now the scattering characteristics are calculated using the computational scheme for power profile (see Section 8.4.3). Such an approach was initially put forward by Kuzmin *et al.* (1988) and used for calculation of plasmon extinction spectra of fractal aggregates of Ag subparticles embedded in the dielectric (KCl) matrix (Kuzmin and Moroz, 1992) and for investigation of influence of aggregation of atmospheric soot on its optical properties (Kuzmin, 1992).

However, our experience showed that the model needs some refinements. Firstly, it follows from Eq.(8.5.15) that the power profile contains not only the pure power factor but also an oscillating component whose dependence on the radial distance r is rather complex. Test calculations showed that this component sometimes produces the sharp non-physical maxima of n and κ near the center of a cluster consisting of metallic subparticles. But with a core, these maximums are in it and, therefore, are eliminated. Secondly, the rough approximation (8.5.14) resulting from the MG formula and, at first sight, practically analogous to it, may sometimes give (unlike the original MG approximation) non-physical negative n_{eff} at the exterior of a cluster comprising of metallic subparticles. Hence, it is essential to use the original MG formula instead of approximation (8.5.14). Thirdly, using the MG formula for an internal region of a cluster where the filling factor may be sufficiently large is a drawback. As was already stated above, the BR formula is more suitable for the internal region.

This modified variant of the asymptotic model was used for calculation of the specific extinction of light by aggregates of Al subparticles in the visible and IR spectral ranges (Babenko, 1996).

(b) The second variant is based on the spherical model with a relatively small homogeneous core covered by a large number (sufficient to represent the smooth homogeneity profile) of homogeneous concentric layers. The filling factor is calculated in the middle of each layer with Eq.(8.5.5) and the effective complex refractive index of the layer is evaluated through some EMT. The optical constants of the quasi-homogeneous core are the same as in the previous variant, and computed with Eqs.(8.5.10–11). Note that here we fail to avoid a step at the core–shell interface which, however, may be smoothed artificially.

The characteristics of scattering by this model particle were calculated in two ways that complement and control each other: (1) the algorithm of a multi-layered sphere (see Section 3.5); and (2) the algorithm of direct numerical integration of the radial equations. The former requires relatively large computer time and, thus, have become competitive only in recent years. As for the latter – the idea of numerical integration of the radial equation is so evident that it was proposed in one of the first publications on light scattering by RIPs (Wyatt, 1962). However, it was quickly realized that some difficulties are related with such an approach. Firstly, there are several tens of radial equations for each set of input parameters (even for a moderate size parameter x). Secondly, the functions $W_l(\rho)$ and $V_l(\rho)$ (see Eqs.(8.1.5–6)) very quickly oscillate and the standard numerical schemes of differential equation integration will not work for them. The second problem may be eliminated by construction of such combinations of sought functions, that will be more convenient for numerical analysis. For example, it is easy to see that the low oscillating log-derivatives of radial functions, instead of the functions themselves, enter in the expressions (8.1.8–9) for the external amplitude coefficients. However, the corresponding equations for the log-derivatives are non-linear, that is not convenient.

We use other approach (Shafai (1972)), involving the search for solutions to Eqs.(8.15–16) of the form:

$$\begin{aligned} W_l(\rho) &= A_l(\rho)\,[\psi_l(\rho)\cos\delta_l(\rho) + \chi_l(\rho)\sin\delta_l(\rho)] \\ V_l(\rho) &= B_l(\rho)\,[\psi_l(\rho)\cos\varepsilon_l(\rho) + \chi_l(\rho)\sin\varepsilon_l(\rho)] \end{aligned} \qquad (8.5.17)$$

where $\psi_l(\rho)$ and $\chi_l(\rho)$ are the Riccati–Bessel and Riccati–Neiman functions; and $A_l(\rho)$, $B_l(\rho)$, $\delta_l(\rho)$ and $\varepsilon_l(\rho)$ are auxiliary amplitude and phase functions. Thus, the oscillating component "transfers" to Riccati functions, that are well investigated from a computational point of view (see Section 1.5).

In what follows we shall present some results for aggregates of Ag particles in the visible range. The metallic particles are known to be among the most popular objects in the optics of disperse media. This is due to both the extremely wide variety of physical effects in systems of metallic particles (Kerker, 1985; Petrov, 1986; Kreibig and Vollmer, 1995) and the technological importance of metallic nano-structures in the epoch of miniaturization of electronic devices. As early as in the beginning of the last century, it was recognized that colloidal metals occur, as a rule, in the form of aggregates, whose optical properties may be substantially different from ones of isolated particles.

The experimental data of Johnson and Christy (1972) was used as the refractive index of bulk Ag. These values were modified taking account of the influence of subparticle size (so-called size effect) on the subparticle refractive index in the framework of the model of limitation of conductivity electron mean free path in the subparticle volume (Kreibig and von Fragstein, 1969). Note that silver particles (especially in the region of plasmon resonance) are extremely inconvenient for any EMT due to the unusual optical constants of silver in the visible range. In order that the subparticles can be associated into an aggregate their radius a should be quite small. Typical values of a = 2–10 nm were used in calculations. The outer radius of cluster R changes in the very broad range: from 0.25 μm to values of the order of 10 μm. Calculation of spectra at greater R is not appropriate, because, firstly, it is unlikely that so large 3D clusters exist in real situations, and, secondly, the optical properties of structures which are large compared with λ is slightly sensitive to their internal structure. Calculations were carried out for the visual range 0.32–0.80 μm and monodisperse clusters. It is reasonable that a distribution (sometimes very broad) over R occurs in real systems, but it is better to reveal the influence of the cluster structural parameters on its optical properties "in pure state", without considering a distribution.

Also in calculations it is accepted that the clusters of Ag subparticles are embedded in a gelatine matrix. The spectral dependence of the gelatine refractive index was calculated with the dispersion relation.

The extinction spectra shown below were calculated with the model of a multi-layered sphere when the number of layers is 1000. The validity check was carried out by the method of direct numerical integration of radial equations. It was revealed that the two approaches give practically identical results (variance is no more than 0.1%) when the number of layers is quite great. Four variants of EMT described above (due to MG, BR, Sheng, and Chýlek) were used. However, our calculations showed that it is not expedient to use the modified EMTs with regard to the fractal cluster, because they in one way or another reduce to MG or BR theory. For example, the results of calculations with Cheng formulae, practically coincide with MG results, and the results obtained with the Chýlek approach are very close to ones calculated by BR theory. If, in addition, we take into account the abrupt increase of calculation time when the iteration scheme (8.5.12) is in use, then it is believed that such, to some extent, opposite approaches as MG and BR are quite sufficient in calculations.

Some typical spectra are shown in Figures 8.12–8.16. For example, the extinction and scattering spectra for the fractal $(D = 2.0)$ clusters of Ag subparticles $(a = 2, 5$, and 10 $nm)$ in gelatine are plotted in Figures 8.12–8.14. Note, that the values of the efficiency factors are plotted in the graphs hereafter, instead of experimentally determining specific extinction and scattering:

$$\alpha_{\text{ext,sc}} = \frac{3\pi Q_{\text{ext,sc}}}{2\lambda\rho_0 x\bar{f}} \tag{8.5.18}$$

where x is the size parameter for cluster $(x = kR)$.

As seen from the Figures 8.12–8.14, the main parameter determining the spectrum form in this case is the subparticle radius. At extremely small $a = 2$ nm the extinction and scattering spectra contain only the sharp plasmon resonance near $\lambda = 0.43\ \mu m$, its position being somewhat long wave shifted compared with the plasmon peak for solid particles of the same radius and does not practically depend on the external cluster radius R. The maximum height is reduced slightly as R increases. The spectra calculated with EMT due to MG and BR are practically identical, but BR gives slightly conservative $Q_{\text{ext,sc}}$ in the maximum. This difference decreases as R increases.

When the subparticle radius grows up to 5 nm the spectra becomes more complicated and the values $Q_{\text{ext,sc}}$ in the maximum increase $(Q_{\text{ext}} \sim 1.8$–1.9, $Q_{\text{sc}} \sim 0.8)$. The slight broadening of maxima is observed compared with case $a = 2$ nm; the maximum in the scattering spectra becomes double (the first peak is at $\lambda \sim 0.41\ \mu m$, the second one is at $\lambda \sim 0.45\ \mu m$ (MG) and 0.46–$0.48\ \mu m$ (BR)). These tendencies are enhanced by going to $a = 10$ nm. The spectra is quite broad, the splitting of the main maximum becomes well defined both in extinction and scattering. A small long wave shift is also observed.

The spectra in Figures 8.12–8.14 were calculated at fractal dimension $D = 2.0$ typical for clusters with mediate porosity. Analogous spectra for loose $(D = 1.8)$ and quite dense $(D = 2.2)$ clusters are plotted in Figures 8.15–8.16. The single plasmon peak, much more narrow than for $D = 2.0$, manifests itself for loose clusters. In contrast, for $D = 2.2$ both

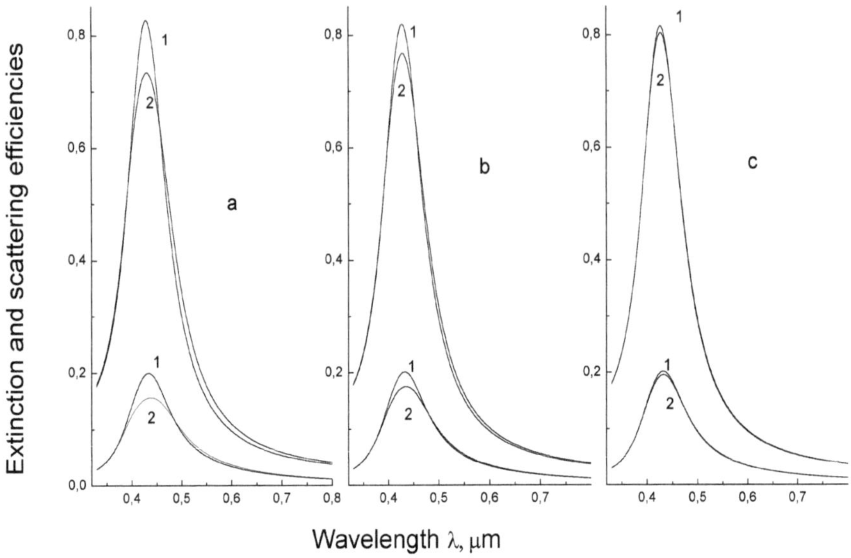

Figure 8.12. Extinction Q_{ext} (upper curves) and scattering Q_{sc} (lower curves) efficiencies versus the wavelength λ for the fractal cluster of Ag subparticles in a gelatine matrix. The fractal dimension is $D = 2$ and the subparticle radius is $a = 2\ nm$. The outer cluster radius is $R = 0.5(a), 1.0(b)$, and $5.0(c)\ \mu m$. Effective medium theory used are MG (*1*) and BR (*2*).

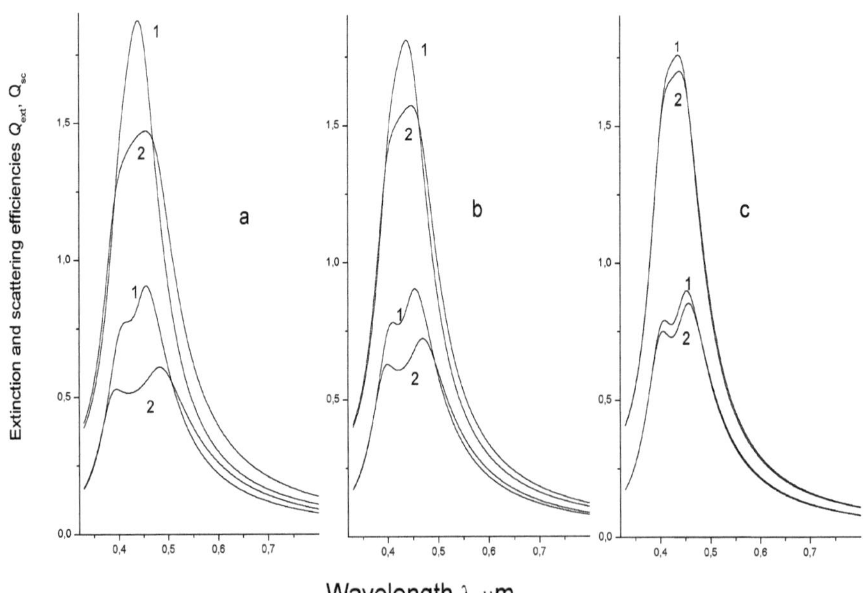

Figure 8.13. The same as in Figure 8.12 but for the subparticle radius $a = 5\ nm$.

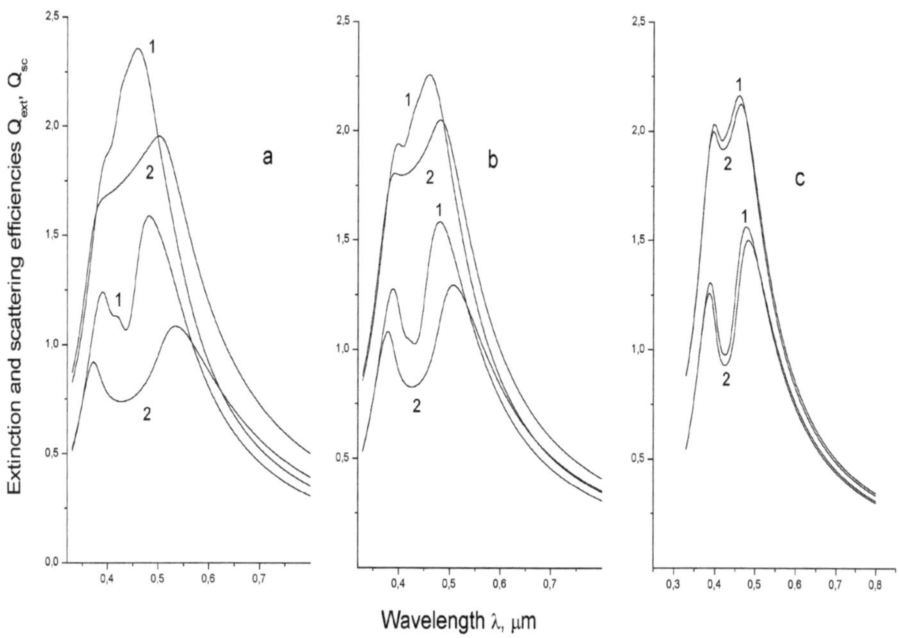

Figure 8.14. The same as in Figure 8.12 but for the subparticle radius $a = 10\ nm$.

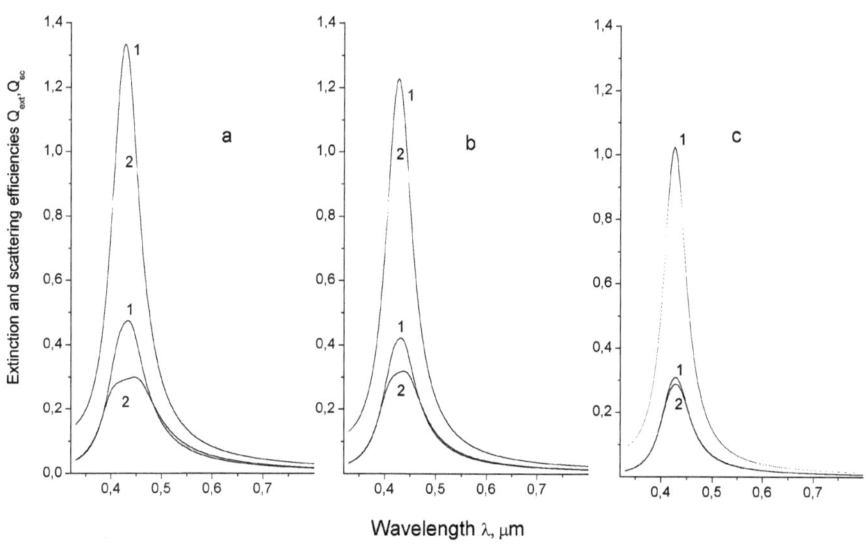

Figure 8.15. The same as in Figure 8.12 but for the subparticle radius $a = 5\ nm$ and fractal dimension $D = 1.8$.

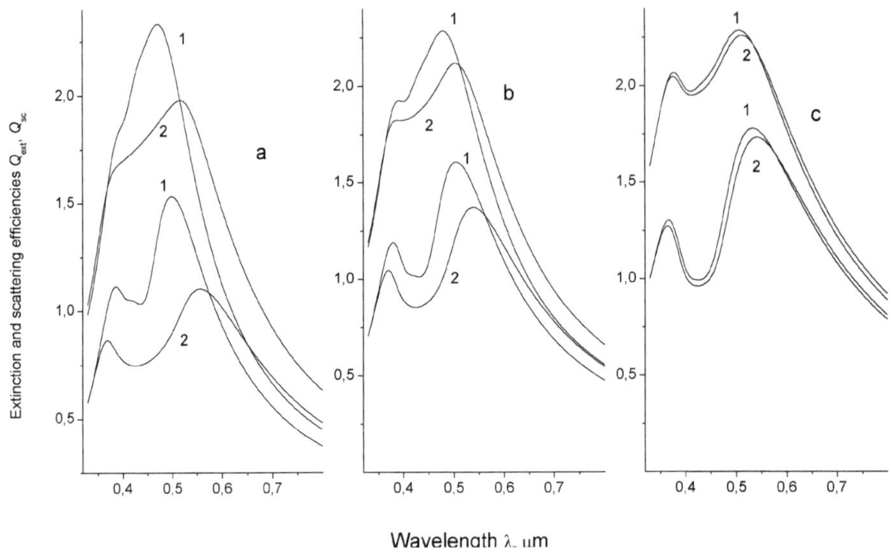

Figure 8.16. The same as in Figure 8.12 but for the subparticle radius $a = 5\ nm$ and fractal dimension $D = 2.2$.

the strong broadening of the spectrum with splitting of the main maximum and long wave shift are observed.

Comparison of results calculated with formulae of MG and BR shows that these approximations give practically identical tendencies for transformation of spectra under variations of the cluster structural parameters, though the absolute values are different to some extent. Since these versions of EMT describe limiting, and actually opposite, situations, one would expect that other intermediate versions will not introduce something new.

It must be noted, that all features of spectra of Ag subparticles in gelatine mentioned above correlate qualitatively with the experimental spectra (Khlebtsov *et al.*, 2000). However, the quantitative comparison with experiment requires the data on the microstructure parameters of the clusters in the specific experiment.

Thus, the asymptotic model is capable of describing scattering by fractal clusters adequately. Nevertheless, this model (even in the improved version) is the first approximation only in describing the optical properties of clusters. Model refinement is strongly hampered by the fact that the RI profile must be obtained beforehand in the framework of some EMT. Applicability of such theories for the internal region of the fractal cluster is open to question.

9

Light scattering by weakly inhomogeneous and weakly anisotropic particles

The morphological analysis of scattering particles shows that they are far from always being spherical, homogeneous, or isotropic. It is, therefore, desirable to search for more adequate models of single scatterers. But even the most precise model will still only be a model . The application of one or another model is motivated by the degree of accuracy within which it is necessary to describe a physical phenomenon. Sometimes, rather rough, approximate models allow us to explain the essence of the phenomenon under observation, and sometimes for this purpose more precise and complex models are required. All depends on the concrete situation.

The diffraction problems, as a rule, are very sensitive to parameters of the body on which diffraction occurs, therefore, of essential value is the question of how small distortions of a model from the actual properties influence the light scattering characteristics. The investigation of the scattering characteristics sensitivity to variations of model parameters allows one to reveal those ranges of values of these characteristics where it is necessary to take into account the inappreciable, and furthermore, the essential features of the scattering centre internal structure. However, for such investigations it is necessary to have computer codes to calculate the light scattering characteristics for models being analysed. The development of such codes often runs into great problems. In this chapter a rather simple and universal enough approach for investigation of sensitivity of the light scattering characteristics to parameters of the model scattering particle is offered. This approach is implemented for weakly inhomogeneous (Section 9.1) and weakly anisotropic (Sections 9.3–9.6) objects. If one takes into account that in the visual wavelength range the anisotropic properties of mineral aerosol component particles are exhibited weakly enough (Section 9.2), it appears possible to carry out actual calculations and to tackle the question of when it is necessary to use the model of anisotropic particles.

9.1 MODIFIED DIFFRACTION EQUATION AND THE BORN DISTORTED WAVE APPROXIMATION FOR RADIALLY INHOMOGENEOUS PARTICLES (RIPs)

Frequently the scattering potential can be divided into two physically not equivalent parts: refractive and absorption, determined part and random part, and so on. The diffracted fields are not quantities additive with respect to such dividing, however, the basic diffraction integral equation (1.1.4) can be changed to separate the term responsible for independent scattering on a chosen part of potential. A similar approach is especially useful when the potential part is given inexactly or when one needs to account for the influence of the known potential approximately. For deriving the modified diffraction equation the scattering potential $k_0^2(N_1^2 - 1)$ of the basic diffraction integral equation (1.1.4) should be divided into two parts:

$$k_0^2(N_1^2 - 1) = k_0^2(N_2^2 - 1) + k_0^2(N_1^2 - N_2^2)$$

Designate as $\mathbf{E}_2(\mathbf{r})$ the field, that would turn out at scattering only on potential $k_0^2(N_2^2-1)$. Then, as well as in the case of Eq.(1.2.36), it is possible to write:

$$\mathbf{E}(\mathbf{r}) = \mathbf{E}_I(\mathbf{r}) + \mathbf{E}_2(\mathbf{r}) + \int\limits_V \hat{G}\left(k_0\mathbf{r}, k_2\mathbf{r}'\right) k_0^2(N_1^2 - N_2^2)\mathbf{E}(\mathbf{r}')d^3\mathbf{r}' \tag{9.1.1}$$

Here $\hat{G}\left(k_0\mathbf{r}, k_2\mathbf{r}'\right)$ is the one-particle Green's function for the particle with refractive index N_2.

Assume now, that the field $\mathbf{E}_2(\mathbf{r})$ and the Green's function $\hat{G}\left(k_0\mathbf{r}, k_2\mathbf{r}'\right)$ are known, and the quantity $(N_1^2 - N_2^2)$ is small enough. Then it is possible to try to solve Eq.(9.1.1) by the method of successive approximations. Such an approach is known in quantum scattering theory as the method of distorted waves (Newton, 1982). In this approach it is obviously important to select the quantity N_2 so that it, on the one hand, would be not so different from the quantity N_1, and on the other hand, that for it the "inoculation" field $\mathbf{E}_2(\mathbf{r})$ and corresponding Green's function would be known. It is clear, that such selection is non-unique, and the convergence rate of the Born series depends on it. For more complex "inoculation" models a smallness of $(N_1^2{-}N_2^2)$ can be achieved that increases the convergence rate, but therewith more complicated functions describing the field $\mathbf{E}_2(\mathbf{r})$ can be obtained that hinder the calculation of the integral term. Therefore, here it is necessary to find a reasonable compromise depending on a actual physical situation.

For the "inoculation" fields, it is possible to use the diffracted fields for homogeneous spherical particles introduced in Section 1.3. In more complex situations the fields for two layered particles (TLPs) (see Section 3.1) or those for RIPs with a profile giving a well-convergent analytical solution (see Section 8.1) can be utilized.

Within the framework of the distorted wave method consider the Born approximation for RIPs. Assume, that the inhomogeneity profile can, with a sufficient degree of accuracy, be approximated by the constant function of radius, that is:

$$\left|N_1^2(r) - N_2^2(r)\right| = |\varepsilon(r)| \ll 1 \tag{9.1.2}$$

where N_2^2 is determined, for example, from the relation:

$$N_2^2 = \frac{1}{R}\int_0^R N_1^2(r)dr$$

Taking into account Eq.(9.1.2), we have the first iteration from Eq.(9.1.1):

$$\mathbf{E}(\mathbf{r}) = \mathbf{E}_I(\mathbf{r}) + \mathbf{E}_{2,\mathrm{out}}(\mathbf{r}) + \int_V \hat{G}\left(k_0\mathbf{r}, k_2\mathbf{r}'\right) k_0^2\varepsilon(r')\mathbf{E}_{2,\mathrm{in}}(\mathbf{r}')d^3\mathbf{r}' \qquad (9.1.3)$$

where $\mathbf{E}_{2,\mathrm{out}}(\mathbf{r})$ and $\mathbf{E}_{2,\mathrm{in}}(\mathbf{r}')$ are the external and internal fields for the particle with the refractive index N_2, respectively. The solution of Eq.(9.1.3) represents the Born approximation in the distorted wave method.

It is necessary to note, that Eq.(9.1.3) gives the solution of Eq.(9.1.1) with accuracy no better than $|\max\varepsilon(r)|^2$. A more precise numerical estimate can be obtained on the basis of the following reasons. Take the maximal value of $\varepsilon(r)$ as a majorant in Eq.(9.1.1). Because the majorant is a constant, Eq.(9.1.1) can be solved exactly, as for a particle with constant refractive index. On the other hand, Eq.(9.1.3) gives the approximate solution. The difference between them will serve as the more precise estimate of the error in the Born approximation.

If the error estimate is greater, than the accuracy with which we need to calculate the light scattering characteristics, then it is necessary to increase the iteration number. However, it can happen that the error estimate will increase as a result (i.e., at such a model choice the Born series does not converge). In this case it is possible to try to approximate the potential $N_1(r)$ by a step function, taking the TLP field as the "inoculation field". If in this case, the required accuracy is not achieved, it is possible to utilize the approximation with the profile function (that is to select $N_2 = f(r)$), having a well-converged solution of the scattering problem.

This hierarchy of "inoculation" fields is stipulated primarily by the complexity of a numerical realization of the solution algorithm of the corresponding problem. However, depending on the actual situation it can be disturbed.

Let us give the formula for the scattered field in the case of RIPs in the Born approximation. In the coordinate system where the z-axis coincides along the wave vector of the plane incident wave and the x-axis along the vector of intensity oscillations in the incident electric field, the scattered radiation field is given by:

$$\mathbf{E}_{\mathrm{sc}}(\mathbf{r}) = E_I \sum_{l=1}^{\infty} \mu_l \left[A_{1l}\mathbf{M}_{o1l}^{(3)}(k_0\mathbf{r}) - iB_{1l}\mathbf{N}_{e1l}^{(3)}(k_0\mathbf{r})\right] \qquad (9.1.4)$$

where each superposition coefficient consists of two terms:

$$A_{1l} = \alpha_l^{33} + A'_{1l} \qquad B_{1l} = \beta_l^{33} + B'_{1l}$$

where the first one represents the amplitude coefficient corresponding to the homogeneous particle with refractive index N_2, and the second one takes into account only the particle inhomogeneity. The additional amplitude coefficients have the form:

$$A'_{1l} = \frac{i}{k_0} \frac{\int\limits_0^{\rho_0} \varepsilon\left(\frac{\rho}{k_0 N_2}\right) \psi_l(N_2\rho)\psi_l(N_2\rho)d\rho}{\xi'_l(\rho_0)\psi_l(N_2\rho_0) - N_2\xi_l(\rho_0)\psi'_l(N_2\rho_0)} \tag{9.1.5}$$

$$B'_{1l} = -\frac{i}{k_0} \frac{\int\limits_0^{\rho_0} \varepsilon\left(\frac{\rho}{k_0 N_2}\right) \left[\psi'^2_l(N_2\rho) + \frac{l(l+1)}{\rho^2}\psi^2_l(N_2\rho)\right] d\rho}{\xi_l(\rho_0)\psi'_l(N_2\rho_0) - N_2\xi'_l(\rho_0)\psi_l(N_2\rho_0)}$$

where $\rho_0 = k_0 R$ is the size parameter. Knowing the scattered radiation field (Eq.(9.1.4)), it is easy to find all requisite light scattering characteristics. For this purpose it is necessary to take advantage of Eqs.(1.3.50), (1.3.57), and (1.3.58), replacing the coefficients α_l^{33} and β_l^{33} with A_{1l} and B_{1l}.

No strong restriction is superimposed on the function $\varepsilon(\rho)$, apart from condition (9.1.2) and the existence conditions for integrals which are included in the numerators of Eqs.(9.1.5). It is quite permissible within the framework of these conditions that the function $\varepsilon(\rho)$ can be random. Physically this situation corresponds to the spherically-symmetric distribution of inhomogeneities inside a particle. The "fluffy" soot particles, agglomerates included in the composition of aerosols and hydrosols and so on are a good example. In this case for an estimation of the scattering characteristics it is necessary to specify correctly the random function $\varepsilon(\rho)$.

9.2 ANISOTROPIC PARTICLES IN THE ATMOSPHERE

In atmospheric aerosols, as it was already mentioned, a considerable proportion is made by mineral particles of soil-erosive origin. Their basic mass (modes A and B) is localized in the surface atmosphere layer at an altitude up to 2 km (Khrgian, 1969). In so doing, the coarse mode B is situated, as a rule, in the lower layers. It is non-stable, and its existence depends in large measure on the presence of local aerosol sources and relevant meteorological conditions. For example, in dust storms the particle concentration increased with the visibility range diminishing to 10–20 m. The mode A can be detected both at small, and high altitudes. Particles of this mode are "long-lived" in the atmosphere. They participate in considerable journey of the air masses above the Earth surface, and their remoteness from sources and distribution are determined basically by air mass circulations.

It is general practice to classify the atmospheric dust sources as either global or local ones. The Sahara region and Asian deserts (Gobi and Takla–Makan) qualify as global sources. The local sources determine a dust content above restricted Earth regions and, as a rule, their power has a periodic seasonal character. To such local sources it is possible to relate the Middle Asia deserts.

The chemical and structural composition of dust particles is widely diversified, and quantitative content of the different minerals within them depends on sources, which form an aerosol background in the given region, though in general it may be suggested that the dust particles are basically the silica–alumina ones, because the compounds of these elements make up 85% of the Earth's crust.

The particles of the coarse fraction (mode B) are particles of primary minerals or minerals formed from melted magma at high temperature and pressure in deep layers of the Earth, which have since risen or given vent on the surface. Under activity of the different natural factors they dispersed and have found their way into the soil composition. The primary minerals are divided into basic four groups (Gorbunov, 1963).

The first group is feldspars. Potash feldspars (orthoclase and microcline), kalium–sodium feldspars and sodium–calcium feldspars (plagioclases) enter into this group. All components of this group were detected in the Sahara aerosol samples. The feldspars have similar crystalline structure. They form monoclinic and triclinic crystals, and the basic element of crystal lattice is the silica–oxygen tetrahedron. In the optical sense they are no simpler than the biaxial crystals (Fedorov, 1958). Let us note, that the spars are the basic source of clay formation .

The group of micas, whose basic representatives are muscovite and serisite, form the monoclinic type crystals. They belong to triplex minerals and consist of two layers of tetrahedrons and one layer of octahedrons.

The quartz group comprise quartz itself and some silicon dioxides, which (e.g., opal) can be amorphous. Quartz exists in two basic modifications. At high temperatures it is α-quartz (mineral of trigonal type). At low temperatures (846K) it transforms into β-quartz (mineral of hexagonal type with one optical axis). On the Earth's surface the β-quartz modification occurs.

The chlorite group are the aqueous silicates of aluminium, magnesium, iron, and chrome. All minerals of this group are monoclinic. It is necessary also to mark, that the chlorites occur both in the coarse dispersion fraction (mode B), and in finely divided fraction (mode A).

In the finely divided fraction of soil-derived aerosol, as has been shown by the samples analyses, the clay particles prevail (Kondratiev and Pozdnyakov, 1981). The clay minerals fall into secondary ones, and are gained from primary minerals under the activity of the different natural factors. They consist of Si, Al, O, OH, and sometimes Fe, Mg, K, and Mn with other elements entering into them (Gorbunov, 1963). These minerals fall into the group of schistose silicates and have original structures. The layers consist of silicon with oxygen combinations, which form tetrahedrons, jointed in a flat grid (thus, the silicon atom is positioned at the centre of the oxygen atom group), and of oxygen–hydroxyl–aluminium groups forming the octahedrons. In so doing the oxygen atoms and the hydroxyl groups in the octahedrons are enclosed by atoms of aluminium, magnesium, iron, and other elements.

The minerals of the montmorillonite and the kaolinite groups are most representative in aerosol samples. Two tetrahedral silicon–oxygen layers and one octahedral alumina–oxygen–hydroxyl layer are peculiar to the montmorillonite group. The minerals of the kaolinite group are of dioctahedron type and have the tetrahedral silicon–oxygen and alumina–oxygen–hydroxyl layer. Both groups fall into the group of the lowest syngony crystal. The distinctive feature of these clays is that they can bind the free water, that results in a refractive index change over a rather wide interval. For example, one of the montmorillonite refractive indices oscillates from 1.485 up to 1.600 for the same wavelength.

Particles of hydromicas (illite) and hydroxides of iron and aluminium occur in smaller amounts. The latter (e.g., haematite) can define to a large extent the absorption index of aerosol particles in the visible spectral range.

Salt minerals are present as impurities in clay minerals. These are chlorides, sulfates and sulfides, carbonates, nitrates, phosphates, and others. The crystals of these compounds fall into different syngonies, but in basic to lowest ones. Many of them are soluble in water, but some are only soluble in acids.

It is characteristic for all minerals that they are optically anisotropic. The quartz particles may be considered as the typical representatives of anisotropic particles in the atmosphere. Really, as seen from Table 2.1, quartz is in almost all the samples. Certainly, the amount of quartz particles is different in different regions. However, quartz may be thought of as comprising approximately 30–50% of the continental aerosol on average (Kondratiev and Pozdnyakov, 1981). The quartz is typical for the atmospheric aerosol in another respect also. The real part of quartz refractive index in the visible range lays in limits, which usually are considered as typical for a mineral aerosol. The imaginary part of the aerosol substance refractive index is determined by other components, though, probably, it is possible to speak about the polluted quartz (haematite films on the particle surface). The small amount of impurity does not essentially change the refractive index real part, but can noticeably affect its imaginary part. So, when speaking about quartz, we shall mean the mineral part of the anisotropic aerosol or dust anisotropic particles and vise versa.

All kinds of processes of anthropogenic and natural character accompanied by soot particle formation may be served as the anisotropic aerosol sources. In such particles, as has been shown by X-ray diffraction analysis, there can be blobs of crystalline carbon (Khrgian, 1969). In some cases the organic aerosol components can also have anisotropic properties.

Though in this book the problems, connected with cloud optics are not considered, it is necessary to mark, that the ice particles of hexagonal structure with the pronounced anisotropic properties can enter into crystalline (cirrus) clouds. The properties of similar clouds are intensively studied by experimental methods. The results of these examinations are reported in works (Dugin *et al.*, 1971, 1976, 1977; Volkovitskiy *et al.*, 1974*a*, 1974*b, 1975*). The bibliography of later research can be found in the review papers of Sassen (2000) and Liou *et al.* (2000).

In the closing part of this section it should be mentioned that the anisotropic particles are frequently also non-spherical, and the non-sphericity of their shapes (when present) is likely to have a stronger effect on the scattering patterns than their anisotropy (Mishchenko *et al.*, 1997; Volten *et al.*, 2001). On the other hand, the joint consideration of anisotropy and non-sphericity is a very difficult task.

9.3 LIGHT SCATTERING BY WEAKLY ANISOTROPIC SPHERICAL PARTICLES

Interest in problems, related to diffraction of electromagnetic radiation by anisotropic particles (APs) – spherulites, ice, and quartz particles in the atmosphere and interstellar space and so on – has increased recently. The difficulties emerging with their solution, are connected, as a rule, that it is impossible to separate the variables in Maxwell's equations. However, there are approaches permitting us to bypass these difficulties and to solve a series of practical problems.

First of all we shall point out problems, related to the diffraction of electromagnetic radiation on infinite gyroelectric and gyromagnetic cylinders and cylindrical shells (Platzman and Ozaki, 1960; Wait, 1961; Ohba,1963; Chen and Cheng, 1964, 1965, 1966; Yeh, 1964; Rusch and Yeh, 1967; Mayhon and Schultz, 1967). The cases of perpendicular and oblique incidence were considered. The scattering object was plasma, magnetized by a constant magnetic field along a symmetry axis. In this case the problem is simplified due to the specific form of the tensor dielectric function in the cylindrical coordinate system. This allows us "to uncouple" equations for the component of the electromagnetic field and to obtain an analytical solution of a problem (Chen and Cheng, 1966).

The formal exact solution of the problem of scattering by a gyrotropic sphere was obtained by Ford and Werner (1978) and Devaty (1988). Solutions of Oganessyan and Baregamyan (1981) for gyrotropic spheres and of Oganessyan (1982) for gyrotropic spheroids appear to be incorrect (see comments of Varadan *et al.* (1989)). Note also the work of Cohoon (1989), who obtained the analytic solution of the scattering problem for multilayered spheres with a special tensor dielectric function, and the works of Varadan *et al.* (1989) and Lakhtakia (1992), where scattering by APs is calculated in the discrete dipole approximation.

The analytical solution was also obtained for isotropic – gyrotropic (chiral) objects of both cylindrical, and spherical configurations (Bohren, 1975; 1978; Pendleton and Rosen, 1998; Kokhanovsky, 2002). Such properties have the particles and shells of organic origin in fluid and amorphous states. The structure of the obtained solutions is determined by a series expansion over the angular and radial functions similar to that in an isotropic case. Naturally, the coefficients of this expansion are much more complicated. The scattering problem was also solved for chiral spheroidal objects on the basis of the T-matrix approach (Lakhtakia *et al.*, 1985), the method of variables separation in spheroidal coordinates (Cooray and Ciric, 1993), the integral equation method (Rojas, 1992; Karonis *et al.*, 1999), and for chiral particles of arbitrary (helix, for the most part) form using the discrete dipole approximation (Singham, 1987, 1988, 1989).

The direct methods of solving the scattering problems reduce, in the main, the diffraction integral equation to an infinite set of linear equations (Sveshnikov and Il'inski, 1971; Barber and Yeh, 1975; Nikol'skiy, 1977; Holt *et al.*, 1978; Sveshnikov and Eremin, 1980). These methods are attractive in that they allow us, within the framework of a uniform procedure, to take into account any changes in structure of the scattering centre, including anisotropic properties of particles (Graglia and Uslenghi, 1984, 1987; Monzon, 1987; Beker *et al.*, 1989; Graglia *et al.*, 1989; Su, 1989; Papadakis *et al.*, 1990; Piller and

Martin, 1998; Kiselev *et al.*, 2002; Doicu, 2003). However, they are complicated in numerical realization, as a series, through which the required functions are represented, converge slowly. Even with a powerful computer it is not always possible to obtain the solution with a sufficient degree of accuracy for single AP, to say nothing of an ensemble of such particles. The convergence of direct methods is determined, as a rule, by the relative refractive index and size of a particle. The greater they are, the lower the achievable accuracy of the solution. However, in a range of these quantities, where the direct methods are easily implemented and yield a high accuracy, the more simple methods frequently exist. In particular, in many works the light scattering characteristics of APs were explored on the basis of approximate methods, suitable for a class of objects, restricted by some or another physical or geometrical condition. Basically, three approximate methods well known in the light scattering theory were used in calculations: approximation of Rayleigh, approximation of Rayleigh–Gans, and approximation of van de Hulst.

Approximation of Rayleigh was utilized for investigation of light scattering by ice APs and shells supposedly entering into the composition of some interstellar dust clouds (Dolginov and Silant'ev, 1976; Dolginov *et al.*, 1995). A necessary condition of applicability of this approximation is a small particle size compared with the wavelength of incident radiation. It is suggested that this condition is valid for interstellar grains. Approximation of Rayleigh–Gans (Goldstein, 1956; Baranov, 1966; Kotov and Novikov, 1967; Picot *et al.*, 1968; Stein *et al.*, 1968; Sundararajan and Marchessault, 1970; Holoubek, 1972, 1991; Misra and Stein, 1972; Turner, 1973; Khlebtsov, 1979*a, 1979b*; Zumer and Doane, 1986; Piazza *et al.*, 1989; Sano, 1989; Zumer *et al.*, 1989; Khlebtsov and Melnikov, 1990) is grounded on representation of an internal field of a particle as a unperturbed incident wave. For correctness of investigations grounded on this approximation, the condition of proximity of particle optical properties to properties of a surrounding medium should be satisfied. Such properties, for example, have the spherulitic aggregates, which are formed at a polymerization of films (Kotov and Novikov, 1967). For spherulitic aggregates an exact solution of the scattering problem is obtained (Roth and Dignam, 1973), but this solution is complicated for numerical realization; in contrast to this, the Rayleigh–Gans approximation is easily implemented, and, as shown by Sundararajan and Marchessault (1970), these calculations correlate well with experiment.

The van de Hulst approximation can be used for great particles, whose refractive index does not differ strongly from the refractive index of a surrounding medium. This approximation enables us to calculate light scattering characteristics, near to the forward scattering direction. Therefore, it was utilized for an estimation of effective refractive indices of suspensions consisting of APs (Meeten, 1980*a,b,c;* Zumer, 1988; Yakovlev and Afonin, 1997).

None of the above methods can be utilized to study the optical properties of the mineral component of atmospheric aerosol. Really, as was noted in Chapter 2, that it contains a great enough number of particles with high refractive indices. But, the distinguishing feature of the majority of crystalline substances is that the components of their tensor dielectric function differ little in the visible spectral range. This fact allows us to convert to the Born approximation of a distorted wave method surveyed in Section 9.1. Why it is so, will be shown below. Beforehand we shall note, that the given approach for the problem of scattering by anisotropic objects of atmospheric optics was proposed to use in several works (Kuzmin, 1980, 1981*a*, 1981*b*; Kuzmin and Babenko, 1981; 1982). It is necessary to also note other works (Cherkasova *et al.*, 1977; Cherkasova and Khijnyak, 1978), closely connected with the works mentioned above. In these the scattering of an electromagnetic wave by a ferrite sphere situated in a magnetized field (that results in magnetic gyrotropy) was considered.

Let us assume, that the optical properties of APs are described by a symmetric complex dielectric tensor:

$$\hat{\varepsilon}' = \varepsilon' + \frac{4\pi i\sigma'}{\omega} \tag{9.3.1}$$

In general, the position of the principal axes of tensors $\varepsilon'$and σ' in the absorbing crystals is arbitrary, that gives 12 independent parameters, but only 9 of them carry an invariant character: six principal values of tensors and three angles defining the position of the principal axes of one tensor with respect to principal axes of another. The remaining parameters characterize the position of axes in the chosen coordinate system (Fedorov, 1958).

The complex dielectric tensor can be presented as the sum:

$$\hat{\varepsilon}' = N^2\hat{I} + \hat{\varepsilon} \tag{9.3.2}$$

where $\hat{I}$ is the unit tensor. The first item we can term as an isotropic part of the dielectric tensor, and second item as the anisotropic part. The selection of a constant N^2 is dictated by the concrete physical situation or reasons of convenience. In particular, when the components of the dielectric tensor, or its main values differ a little from each other, the isotropic part of expression (9.3.2) can be taken as the arithmetic mean from principal values of real and imaginary parts of the dielectric tensor. Then the anisotropic part is small and it is possible to take advantage of the Born approximation of the distorted wave method, i.e., the solution of the scattering problem is possible to note as:

$$\mathbf{E}_{\rm sc}(\mathbf{r}) = \mathbf{E}_{\rm out}(\mathbf{r}) + \int_V \hat{G}_{31}\left(k_0\mathbf{r}, k\mathbf{r}'\right) k_0^2\hat{\varepsilon}\mathbf{E}_{\rm in}(\mathbf{r}')d^3\mathbf{r}' \tag{9.3.3}$$

If the incident plane wave is such that $\{\mathbf{e_z} \upuparrows \mathbf{k_0},\ \mathbf{e_x} \upuparrows \mathbf{E}_I\}$ then the values of the exterior $\mathbf{E}_{\rm out}(\mathbf{r})$ and internal $\mathbf{E}_{\rm in}(\mathbf{r}')$ fields are yielded by relations (1.3.32), which will be utilized for calculations further. The Green's function is picked according to Eq.(1.3.11).

The structure of expression (9.3.3) is rather complicated because of the presence of the tensor $\hat{\varepsilon}$ in integrand, "intermixing" components. The procedure to simplify Eq.(9.9.3) and reduce it to a view, convenient for investigations, is simple in principle, but cumbersome. It is reduced to a calculation of triple integrals, whose integrands contain combinations of Legendre polynomials, Bessel, and trigonometrical functions. If we shall substitute expansions (1.3.11) and (1.3.41) into Eq.(9.3.3), integrate over angle φ, repeatedly apply recursion relations for Legendre polynomials and recursion and integral relations for Bessel functions, then we come to a final expression for an electric field outside the scatterer (in this and subsequent expressions we shall omit the argument kr of functions **M** and **N**):

$$\mathbf{E}_{\text{sc}}(\mathbf{r}) = \mathbf{E}_{\text{out}}(\mathbf{r}) + E_I \sum_{l=1}{}' \left\{ i^{l+1} \left[A_l \mathbf{M}^{(3)}_{o1l} + B_l \mathbf{M}^{(3)}_{o3l} + \mathcal{E}_l \mathbf{M}^{(3)}_{o2l} + C_l \mathbf{M}^{(3)}_{e1l} \right. \right.$$
$$\left. + D_l \mathbf{M}^{(3)}_{e3l} + F_l \mathbf{M}^{(3)}_{e2l} \right] + i^l \left[G_l \mathbf{N}^{(3)}_{e1l} + H_l \mathbf{N}^{(3)}_{e3l} + K_l \mathbf{N}^{(3)}_{e2l} \right. \tag{9.3.4}$$
$$\left. \left. + I_l \mathbf{N}^{(3)}_{o1l} + J_l \mathbf{N}^{(3)}_{o3l} + L_l \mathbf{N}^{(3)}_{o2l} \right] \right\}$$

Here $\sum_{l=1}{}'$ means, that the terms with $m > l$ are eliminated from the sum and A_l, B_l, $\mathcal{E}_l$, C_l, D_l, F_l are the supplemented amplitude coefficients ($\varepsilon_{xx}, \varepsilon_{xy}, \ldots$, are the components of tensor $\hat{\varepsilon}$) for **M**-part of the field:

$$A_l = f_l \left\{ \frac{2l+1}{l(l+1)} \alpha_l^{31} \left[(3l^2 + 3l - 2)\varepsilon_{xx} + (l-1)(l+2)\varepsilon_{yy} + 4\varepsilon_{zz} \right] \right.$$
$$+ \frac{l-1}{l} \beta_{l-1}^{31} \left[(l-2)\varepsilon_{xx} - (l+2)\varepsilon_{yy} + 4\varepsilon_{zz} \right]$$
$$\left. + \frac{l+2}{l-1} \beta_{l+1}^{31} \left[(l+3)\varepsilon_{xx} - (l-1)\varepsilon_{yy} - 4\varepsilon_{zz} \right] \right\} \tag{9.3.5}$$
$$\frac{B_l}{f_l(\varepsilon_{xx} - \varepsilon_{yy})} = -\frac{D_l}{2 f_l \varepsilon_{xy}} = \frac{2l+1}{l(l+1)} \alpha_l^{31} - \frac{1}{l} \beta_{l-1}^{31} - \frac{1}{l+1} \beta_{l+1}^{31}$$
$$C_l = 2 f_l \varepsilon_{xy} \left[-(2l+1)\, \alpha_l^{31} - (l+2)\beta_{l+1}^{31} - (l-1)\beta_{l-1}^{31} \right]$$
$$-\frac{\mathcal{E}_l}{2 f_l \varepsilon_{xz}} = \frac{F_l}{2 f_l \varepsilon_{zy}} = \frac{3\,(2l+1)}{l(l+1)} \alpha_l^{31} + \frac{l-3}{l} \beta_{l-1}^{31} - \frac{l+4}{l+1} \beta_{l+1}^{31}$$

where

$$f_l = \frac{\rho_0 \alpha_l^{31}}{8l(l+1)N^3} \left[\psi_l^2(N\rho_0) + \psi_{l-1}(N\rho_0)\psi_{l+1}(N\rho_0) \right]$$

and G_l, H_l, J_l, K_l, L_l, I_l are the supplemented amplitude coefficients for the **N**-part of the field:

$$G_l = g_l \left\{ \frac{1}{l+1} \alpha_{l+1}^{31} \left[(l-1)\varepsilon_{xx} - (l+3)\varepsilon_{yy} + 4\varepsilon_{zz} \right] + \frac{l+3}{2l+3} \beta_{l+2}^{31} \left(3\varepsilon_{xx} + \varepsilon_{yy} - 4\varepsilon_{zz} \right) \right.$$
$$\left. + \frac{1}{(l+1)\,(2l+3)} \beta_l^{31} \left[(3l^2 + 7l + 6)\varepsilon_{xx} + (l+3)(l+2)\varepsilon_{yy} + 4l(l+2)\varepsilon_{zz} \right] \right\}$$
$$+ h_l \left\{ \frac{1}{l} \alpha_{l-1}^{31} \left[(l+2)\varepsilon_{xx} - (l-2)\varepsilon_{yy} - 4\varepsilon_{zz} \right] + \frac{l-2}{2l-1} \beta_{l-2}^{31} \left(3\varepsilon_{xx} + \varepsilon_{yy} - 4\varepsilon_{zz} \right) \right.$$
$$\left. + \frac{1}{l\,(2l-1)} \beta_l^{31} \left[(3l^2 - l + 2)\varepsilon_{xx} + (l-1)(l-2)\varepsilon_{yy} + 4\,(l+1)\,(l-1)\varepsilon_{zz} \right] \right\} \tag{9.3.6}$$

$$\frac{H_l}{\varepsilon_{xx}-\varepsilon_{yy}} = \frac{J_l}{2\varepsilon_{xy}} = g_l\left[\frac{\alpha^{31}_{l+1}}{(l+2)(l+1)} - \frac{\beta^{31}_{l+2}}{(2l+3)\,(l+2)} - \frac{\beta^{31}_{l}}{(2l+3)\,(l+1)}\right]$$
$$+\,h_l\left[\frac{\alpha^{31}_{l-1}}{l(l-1)} - \frac{\beta^{31}_{l-2}}{(2l-1)\,(l-1)} - \frac{\beta^{31}_{l}}{(2l-1)\,l}\right]$$

$$\frac{K_l}{2\varepsilon_{xz}} = \frac{L_l}{2\varepsilon_{yz}} = g_l\left[-\frac{(l+4)\alpha^{31}_{l+1}}{(l+2)(l-1)} + \frac{2(l+3)\beta^{31}_{l+2}}{(2l+3)\,(l+2)} + \frac{\beta^{31}_{l}}{(2l+3)\,(l+1)}\right]$$
$$+\,h_l\left[\frac{(l-3)\,\alpha^{31}_{l-1}}{l(l-1)} - \frac{2(l-2)\beta^{31}_{l-2}}{(2l-1)\,(l-1)} + \frac{3\beta^{31}_{l}}{(2l-1)\,l}\right]$$

$$I_l = 2\varepsilon_{xy}\left\{g_l\left[\alpha^{31}_{l+1} + \frac{(l+3)\beta^{31}_{l+2}}{2l+3} + \frac{l\beta^{31}_{l}}{2l+3}\right]\right.$$
$$\left.+h_l\left[\alpha^{31}_{l-1} + \frac{(l-2)\beta^{31}_{l-2}}{2l-1} + \frac{(l+1)\beta^{31}_{l}}{(2l-1)\,l}\right]\right\}$$

where
$$g_l = \frac{\rho_0\beta^{31}_l}{8(l+1)N^3}\left[\psi^2_{l+1}(N\rho_0) - \psi_l(N\rho)\psi_{l+2}(N\rho_0)\right]$$
$$h_l = \frac{\rho_0\beta^{31}_l}{8lN^3}\left[\psi^2_{l-1}(N\rho_0) - \psi_l(N\rho_0)\psi_{l-2}(N\rho_0)\right]$$

Thus, we have received an expression for the intensity of the electromagnetic field, scattered by spherical APs. The influence of anisotropy is characterized, at first, by the occurrence of the additions to the amplitude coefficients of Mie theory (these additions are submitted by the terms $A_l\mathbf{M}^{(3)}_{o1l}$ and $G_l\mathbf{N}^{(3)}_{e1l}$ in Eq.(9.3.4)) and, second, by the presence of the terms with completely other angular structures in the expressions for scattered fields. How the presence of these terms can change the scattering characteristics of APs in comparison with the characteristics of isotropic particles of the same size, will be discussed in Section 9.4.

It is seen from the above expressions that the calculation of coefficients A_l, B_l, C_l... is simple, despite their unwieldness, because they are combinations of elementary functions and Riccati–Bessel functions, the procedure of which calculation is well known (see Section 1.5). It is possible to consider the scattered wave as a transverse one in the wave zone, that is, it is possible to neglect r-components in comparison with θ and φ-components. Besides this, replace the Riccati–Hankel function $\xi_l(y)$ and its derivative $\xi_l'(y)$, appearing in the expressions for $\mathbf{M}^{(3)}_{\sigma 1l}$ and $\mathbf{N}^{(3)}_{\sigma 1l}$, by the asymptotic values (at $y \gg l$):

$$\xi_l(y) \approx (-i)^{l+1}e^{iy} \qquad \xi_l'(y) \approx (-i)^l e^{iy}$$

According to this, we have in the wave zone:

$$\mathbf{M}^{(3)}_{\substack{e\\o}ml}(k_0\mathbf{r}) \approx \frac{(-i)^{l+1}}{k_0 r}e^{ik_0 r}\mathbf{m}_{\substack{e\\o}ml}(\theta,\varphi) \qquad \mathbf{N}^{(3)}_{\substack{e\\o}ml}(k_0\mathbf{r}) \approx \frac{(-i)^{l}}{k_0 r}e^{ik_0 r}\mathbf{n}_{\substack{e\\o}ml}(\theta,\varphi) \tag{9.3.7}$$

(see Eqs.(1.3.2) and (1.3.4) for the definition of $\mathbf{m}$ and $\mathbf{n}$-functions).

In the obtained expressions there are the angular functions $\pi_l^{(1),(2),(3)}(\theta)$ and $\tau_l^{(1),(2),(3)}(\theta)$ (at upper index $m = 1$ they coincide with the angular functions in Mie theory). The calculation of these functions can be carried out with known recursion relations (see Section 1.5.3).

Thus, according to Eq.(9.3.7) we have:

$$E_{\text{sc},\theta} = \frac{E_I}{k_0 r} e^{ik_0 r} \sum_{l=1}^{\infty}{}' \left[-i \left(\alpha_l^{33} \pi_l^{(1)} + \beta_l^{33} \tau_l^{(1)} \right) \cos\varphi + R_l' \right] \tag{9.3.8}$$

$$E_{\text{sc},\varphi} = \frac{E_I}{k_0 r} e^{ik_0 r} \sum_{l=1}^{\infty}{}' \left[i \left(\alpha_l^{33} \tau_l^{(1)} + \beta_l^{33} \pi_l^{(1)} \right) \sin\varphi + T_l' \right]$$

where

$$\begin{aligned} T_l' = {} & \left(A_l' \pi_l^{(1)} + G_l' \tau_l^{(1)} \right) \sin\varphi + \left(I_l \tau_l^{(1)} - C_l \pi_l^{(1)} \right) \cos\varphi \\ & + \left(2\mathcal{E}_l \pi_l^{(2)} + K_l \tau_l^{(2)} \right) \sin 2\varphi - \left(L_l \tau_l^{(2)} - 2F_l \pi_l^{(2)} \right) \cos 2\varphi \\ & + \left(H_l \tau_l^{(3)} + 3B_l \pi_l^{(3)} \right) \sin 3\varphi - \left(J_l \tau_l^{(3)} - 3D_l \pi_l^{(3)} \right) \cos 3\varphi \end{aligned} \tag{9.3.10}$$

$$\begin{aligned} R_l' = {} & \left(A_l' \tau_l^{(1)} + G_l' \pi_l^{(1)} \right) \cos\varphi + \left(C_l \tau_l^{(1)} - I_l \pi_l^{(1)} \right) \sin\varphi \\ & + \left(\mathcal{E}_l \tau_l^{(2)} + 2K_l \pi_l^{(2)} \right) \cos 2\varphi + \left(2L_l \pi_l^{(2)} - F_l \tau_l^{(2)} \right) \sin 2\varphi \\ & + \left(B_l \tau_l^{(3)} + 3H_l \pi_l^{(3)} \right) \cos 3\varphi + \left(3J_l \pi_l^{(3)} - D_l \tau_l^{(3)} \right) \sin 3\varphi \end{aligned} \tag{9.3.11}$$

It is possible to see that when we compare expression (9.3.4) with the similar expression (1.3.32) for isotropic particles, that Eq.(9.3.4) has a more complex structure, and, hence, in this case the amplitude matrix of scattering cannot be found by analogy with Mie theory, as it could done in the previous sections.

9.4 AMPLITUDE SCATTERING MATRIX FOR SPHERICAL APs

In Section 1.3 the complete analysis of scattering matrix in coordinate and momentum representations was carried out on the example of the spherical isotropic particle. However, for investigation of equation set (1.2.18) it is quite enough to know the more simple matrix – the amplitude scattering matrix, which is given by Eq.(1.2.11). The comparative analysis of Eqs.(1.2.11) and (1.3.32) shows, that the amplitude scattering matrix can be found from the asymptotes of scattered fields. We can invoke it to find the amplitude scattering matrix in the case of spherical APs (Kuzmin, 1980).

Before going to the arbitrarily chosen coordinate system, use the solution of the problem in a special reference frame, connected with the direction of radiation propagation. For this purpose enter a Cartesian coordinate system, whose z-axis coincides with the wave vector of the incident plane monochromatic wave. Axes x and y we shall choose arbitrarily. Then the electric field vector of the incident wave can be written as an expansion in two mutually perpendicular directions: $\mathbf{E}_I = E_{xI}\hat{\mathbf{e}}_x + E_{yI}\hat{\mathbf{e}}_y$. In a far-field region the scattered field expresses itself through an amplitude matrix $\hat{A}$ as follows:

$$\mathbf{E}_{\rm sc}(\mathbf{r}) = \frac{e^{ik_0 r}}{r}\hat{A}(\mathbf{k}_0', \mathbf{k}_0)\mathbf{E}_I$$

from which it follows, that this matrix can be written as:

$$\hat{A}(\mathbf{k}_0', \mathbf{k}_0) = \frac{e^{ik_0 r}}{r}\left[\mathbf{E}_{{\rm sc},x} \otimes \hat{\mathbf{e}}_x + \mathbf{E}_{{\rm sc},y} \otimes \hat{\mathbf{e}}_y\right] \tag{9.4.1}$$

where $\mathbf{E}_{{\rm sc},x}$ and $\mathbf{E}_{{\rm sc},y}$ are the fields forming in the far-field region as the result of scattering of the electromagnetic wave with unit amplitude and oscillations of the electric vector along the orths $\hat{\mathbf{e}}_x$ and $\hat{\mathbf{e}}_y$, accordingly. Apparently, it is possible to express the field $\mathbf{E}_{{\rm sc},x}$ through the asymptotic ($k_0 r \gg 1$) value of the scattered field $\mathbf{E}_{\rm sc}(\mathbf{r})$, noted in the coordinate system, where the z-axis coincides with the propagation direction, and the incident wave electric vector oscillates along the x-axis:

$$\mathbf{E}_{{\rm sc},x} \approx E_{xI}^{-1}\mathbf{E}_{\rm sc}(\mathbf{r})$$

The expression for this field intensity is given by Eq.(9.3.4).

For finding of $\mathbf{E}_{{\rm sc},y}$ one can take the problem solution for APs, already obtained. For this purpose enter a new coordinate system so that its x-axis coincides with the y-axis of the old coordinate system. In this new coordinate system the dielectric tensor will be expressed through components of the dielectric tensor in the old coordinate system (rotation around the z-axis by 90°) as follows:

$$\hat{\varepsilon}^{\rm new} = \begin{pmatrix} \varepsilon_{yy} & -\varepsilon_{yx} & -\varepsilon_{yz} \\ -\varepsilon_{xy} & \varepsilon_{xx} & \varepsilon_{xz} \\ -\varepsilon_{zy} & \varepsilon_{xz} & \varepsilon_{zz} \end{pmatrix}$$

Then the solution in the new coordinate system will look like Eq.(9.3.4) with the corresponding replacement of components of tensor $\hat{\varepsilon}$ by $\hat{\varepsilon}^{\rm new}$. The obtained vector field at transition to the old coordinate system is transformed with the help of the Wigner D-functions (Varshalovich *et al.*, 1988):

$$\mathbf{M}^{(3)}_{\sigma ml}(k_0\mathbf{r}') = \sum_{m'=0}^{l}\sum_{\sigma'=e,o} D^{l(\sigma\sigma')}_{mm'}(\pi/2, 0, 0)\mathbf{M}^{(3)}_{\sigma' m' l'}(k_0\mathbf{r})$$

$$\mathbf{N}^{(3)}_{\sigma ml}(k_0\mathbf{r}') = \sum_{m'=0}^{l}\sum_{\sigma'=e,o} D^{l(\sigma\sigma')}_{mm'}(\pi/2, 0, 0)\mathbf{N}^{(3)}_{\sigma' m' l'}(k_0\mathbf{r})$$

$$D^{l(\sigma\sigma')}_{mm'}(\gamma_1, \gamma_2, \gamma_3) = \left[\frac{(l+m')!\,(l-m)!}{(l-m')!\,(l+m)!}\right]^{1/2} \times \begin{array}{cc|l} \sigma=e & \sigma=o & \\ \mathrm{Re}\left[D^l_{mm'}(\gamma_1,\gamma_2,\gamma_3)\right] & \mathrm{Im}\left[D^l_{mm'}(\gamma_1,\gamma_2,\gamma_3)\right] & \sigma'=e \\ -\,\mathrm{Im}\left[D^l_{mm'}(\gamma_1,\gamma_2,\gamma_3)\right] & \mathrm{Re}\left[D^l_{mm'}(\gamma_1,\gamma_2,\gamma_3)\right] & \sigma'=o \end{array}$$

where $D^l_{mm'}(\gamma_1, \gamma_2, \gamma_3)$ is the Wigner function.

Having done the indicated transformations, we shall obtain the solution to the problem of diffraction by the spherical AP when the electric vector of incident radiation oscillates along the y-axis in the originally selected coordinate system. On the strength of tensor symmetry the final output takes the form:

$$E^1_{\text{sc}} = E_{yI} \sum_{l=1}^{\infty}{}' \left[\frac{2l+1}{l(l+1)} i^l \left(\alpha_l^{33} \mathbf{M}^{(3)}_{e1l} - \beta_l^{33} \mathbf{N}^{(3)}_{o1l} \right) \right.$$
$$+ i^{l+1} \left(-C_l \mathbf{M}^{(3)}_{o1l} + D_l \mathbf{M}^{(3)}_{o3l} + F_l \mathbf{M}^{(3)}_{o2l} - A'_l \mathbf{M}^{(3)}_{e1l} - B_l \mathbf{M}^{(3)}_{e3l} - E_l \mathbf{M}^{(3)}_{e2l} \right)$$
$$\left. + i^l \left(I_l \mathbf{N}^{(3)}_{o1l} - J_l \mathbf{N}^{(3)}_{o3l} - L_l \mathbf{N}^{(3)}_{o2l} + G'_l \mathbf{N}^{(3)}_{e1l} + H_l \mathbf{N}^{(3)}_{e3l} + K_l \mathbf{N}^{(3)}_{e2l} \right) \right] \tag{9.4.2}$$

where

$$A'_l = f_l \left\{ \frac{2l+1}{l(l+1)} \alpha_l^{31} \left[(3l^2+3l-2)\varepsilon_{yy} + (l-1)(l+2)\varepsilon_{xx} + 4\varepsilon_{zz} \right] \right.$$
$$+ \frac{l-1}{l} \beta_{l-1}^{31} \left[(l-2)\varepsilon_{yy} - (l+2)\varepsilon_{xx} + 4\varepsilon_{zz} \right]$$
$$\left. + \frac{l+2}{l-1} \beta_{l+1}^{31} \left[(l+3)\varepsilon_{yy} - (l-1)\varepsilon_{xx} - 4\varepsilon_{zz} \right] \right\}$$

$$G'_l = g_l \left\{ \frac{1}{l+1} \alpha_{l+1}^{31} \left[(l-1)\varepsilon_{yy} - (l+3)\varepsilon_{xx} + 4\varepsilon_{zz} \right] + \frac{l+3}{2l+3} \beta_{l+2}^{31} \left(3\varepsilon_{yy} + \varepsilon_{xx} - 4\varepsilon_{zz} \right) \right.$$
$$\left. + \frac{1}{(l+1)(2l+3)} \beta_l^{31} \left[(3l^2+7l+6)\varepsilon_{yy} + (l+3)(l+2)\varepsilon_{xx} + 4l(l+2)\varepsilon_{zz} \right] \right\}$$
$$+ h_l \frac{1}{l} \alpha_{l-1}^{31} \left[(l+2)\varepsilon_{yy} - (l-2)\varepsilon_{xx} - 4\varepsilon_{zz} \right] + \frac{l-2}{2l-1} \beta_{l-2}^{31} \left(3\varepsilon_{yy} + \varepsilon_{xx} - 4\varepsilon_{zz} \right)$$
$$\left. + \frac{1}{l(2l-1)} \beta_l^{31} \left[(3l^2-l+2)\varepsilon_{yy} + (l-1)(l-2)\varepsilon_{xx} + 4(l+1)(l-1)\varepsilon_{zz} \right] \right\}$$

In the compact form the expressions (9.3.4) and (9.4.2) can be written as:

$$\mathbf{E}_{\text{sc}}(\mathbf{r}) = E_{xI} \sum_{\sigma ml} \left[\alpha^{\sigma}_{ml} \mathbf{M}^{(3)}_{\sigma ml}(k_0 \mathbf{r}) - i \beta^{\sigma}_{ml} \mathbf{N}^{(3)}_{\sigma ml}(k_0 \mathbf{r}) \right] \tag{9.4.3}$$

$$\mathbf{E}^1_{\text{sc}}(\mathbf{r}) = E_{yI} \sum_{\sigma ml} \left[\delta^{\sigma}_{ml} \mathbf{M}^{(3)}_{\sigma ml}(k_0 \mathbf{r}) - i \gamma^{\sigma}_{ml} \mathbf{N}^{(3)}_{\sigma ml}(k_0 \mathbf{r}) \right] \tag{9.4.4}$$

Using the asymptotic expansions (9.3.6), it is possible to obtain from Eqs.(9.4.1), (9.4.3), and (9.4.4) the amplitude scattering tensor in the coordinate system, related to the wave vector of incident radiation:

$$\hat{A}(\mathbf{k}'_0, 0) = k_0^{-1} \sum_{\sigma ml} (-i)^{l+1} \left[\alpha^{\sigma}_{ml} \mathbf{m}_{\sigma ml}(\theta, \varphi) + \beta^{\sigma}_{ml} \mathbf{n}_{\sigma ml}(\theta, \varphi) \right] \otimes \hat{\mathbf{e}}_x$$
$$+ \left[\delta^{\sigma}_{ml} \mathbf{m}_{\sigma ml}(\theta, \varphi) + \gamma^{\sigma}_{ml} \mathbf{n}_{\sigma ml}(\theta, \varphi) \right] \otimes \hat{\mathbf{e}}_y \tag{9.4.5}$$

Rewrite Eq.(9.4.5) in a more convenient form. For this purpose take into account the relations:

$$\hat{\mathbf{e}}_x = \frac{2}{l(l+1)}\mathbf{m}_{o1l}(0,0) = \frac{2}{l(l+1)}\mathbf{n}_{e1l}(0,0)$$
$$\hat{\mathbf{e}}_y = \frac{2}{l(l+1)}\mathbf{n}_{o1l}(0,0) = -\frac{2}{l(l+1)}\mathbf{m}_{e1l}(0,0) \qquad (9.4.6)$$

From Eq.(9.4.6) it follows:

$$\hat{A}(\mathbf{k}_0',0) = k_0^{-1}\sum_{\sigma ml}\sum_{\sigma' m'}\lambda_{m'l}\left[\alpha_{ml}^{\sigma\sigma'}\mathbf{m}_{\sigma ml}(\theta,\varphi)\otimes\mathbf{m}_{\sigma' m' l}(0,0)\right.$$
$$\left.+\beta_{ml}^{\sigma\sigma'}\mathbf{n}_{\sigma ml}(\theta,\varphi)\otimes\mathbf{n}_{\sigma' m' l}(0,0)\right]$$

where

$$\alpha_{ml}^{\sigma o} = (-i)^{l+1}\frac{\pi l(l+1)}{2l+1}\alpha_{ml}^{\sigma} \qquad \alpha_{ml}^{\sigma e} = (-i)^{l+1}\frac{\pi l(l+1)}{2l+1}\delta_{ml}^{\sigma}$$
$$\beta_{ml}^{\sigma o} = (-i)^{l+1}\frac{\pi l(l+1)}{2l+1}\gamma_{ml}^{\sigma} \qquad \beta_{ml}^{\sigma e} = (-i)^{l+1}\frac{\pi l(l+1)}{2l+1}\beta_{ml}^{\sigma}$$

The function arguments (θ,φ) and $(0,0)$ correspond to scattering angles and incidence angles in the entered coordinate system. To receive the scattering amplitude matrix in an arbitrary coordinate system, rotate the initial coordinate system L around the y-axis through angle γ_2, and then around the z'-axis, obtained after the first rotation of system L, through the angle γ_3. At such coordinate system transformation the vector angular wave functions are transformed with the following formulae:

$$\mathbf{m}_{\sigma ml}(\theta',\varphi') = \sum_{\sigma' m'} D_{m'm}^{l(\sigma\sigma')}(0_1,\gamma_2,\gamma_3)\mathbf{m}_{\sigma' m' l}(\theta,\varphi) \qquad (9.4.7)$$
$$\mathbf{n}_{\sigma ml}(\theta',\varphi') = \sum_{\sigma' m'} D_{m'm}^{l(\sigma\sigma')}(0_1,\gamma_2,\gamma_3)\mathbf{n}_{\sigma' m' l}(\theta,\varphi)$$

Applying Eq.(9.4.7), we shall obtain an expression for the amplitude scattering tensor in the arbitrarily chosen coordinate system:

$$\hat{A}(\mathbf{k}_0',\mathbf{k}_0) = k_0^{-1}\sum_{l}\sum_{\sigma' m'}\sum_{\sigma'' m''}\left[K_{m'm''}^{l(\sigma'\sigma'')}(\gamma_2,\gamma_3)\mathbf{m}_{\sigma' m' l}(\theta',\varphi')\otimes\mathbf{m}_{\sigma'' m'' l}(\gamma_2,\gamma_3)\right.$$
$$\left.+L_{m'm''}^{l(\sigma'\sigma'')}(\gamma_2,\gamma_3)\mathbf{n}_{\sigma' m' l}(\theta',\varphi')\otimes\mathbf{n}_{\sigma'' m'' l}(\gamma_2,\gamma_3)\right] \qquad (9.4.8)$$

where

$$K_{m_1m_2}^{l(\sigma_1\sigma_2)} = \sum_{m_3m_4}\sum_{\sigma_3\sigma_4}\lambda_{m_4l}\alpha_{m_3l}^{\sigma_3\sigma_4}\left[D_{m_3m_1}^{l(\sigma_3\sigma_1)}(0,\gamma_2,\gamma_3)\right]_T D_{m_4m_2}^{l(\sigma_4\sigma_2)}(0,\gamma_2,\gamma_3)$$
$$L_{m_1m_2}^{l(\sigma_1\sigma_2)} = \sum_{m_3m_4}\sum_{\sigma_3\sigma_4}\lambda_{m_4l}\beta_{m_3l}^{\sigma_3\sigma_4}\left[D_{m_3m_1}^{l(\sigma_3\sigma_1)}(0,\gamma_2,\gamma_3)\right]_T D_{m_4m_2}^{l(\sigma_4\sigma_2)}(0,\gamma_2,\gamma_3)$$

This tensor is more complex, than the tensor of the electromagnetic wave scattering by an isotropic particle, that is exhibited in dependence of the expansion coefficients $K^{l(\sigma_1\sigma_2)}_{m_1m_2}$ and $L^{l(\sigma_1\sigma_2)}_{m_1m_2}$ on the direction of incident radiation propagation. As expected, in the general case it is not invariant concerning rotations in three-dimensional space. It, in particular, results in the appearance of the off-diagonal terms in the extinction matrix (1.2.33). Really, the amplitude forward scattering tensor does not degenerate into scalar, as it does for isotropic particles. In the coordinate system, related to the direction of radiation propagation, it, as follows from Eqs.(9.4.5) and (9.4.6), has three different components:

$$\begin{aligned} A_{11}(0) &= -\frac{4\pi i}{k_0}\sum_{l=1}^{\infty}\left[(2l+1)\left(\alpha_l^{33}+\beta_l^{33}\right)+il(l+1)(A_l+G_l)\right] \\ A_{12}(0) = A_{21}(0) &= \frac{4\pi}{k_0}\sum_{l=1}^{\infty} l(l+1)(I_l-C_l) \\ A_{22}(0) &= -\frac{4\pi i}{k_0}\sum_{l=1}^{\infty}\left[(2l+1)\left(\alpha_l^{33}+\beta_l^{33}\right)+il(l+1)(A_l'+G_l')\right] \end{aligned} \tag{9.4.9}$$

Hence, there are 5 independent components in the extinction matrix. The remaining ones are turned to zero due to equality of the $A_{12}(0)$ and $A_{21}(0)$ components in the first approximation.

As to the scattering cross section, it can be noted regardless of Stokes parameters only for linearly polarized radiation:

$$\sigma_{\text{sc}}^{x} = \frac{2\pi}{k_0^2}\sum_{l=1}^{\infty}\left[(2l+1)\left(\left|\alpha_l^{33}\right|^2+\left|\beta_l^{33}\right|^2\right)+2l(l+1)\,\text{Im}(A_l^*\alpha_l^{33}+G_l^*\beta_l^{33})\right] \tag{9.4.10}$$

$$\sigma_{\text{sc}}^{y} = \frac{2\pi}{k_0^2}\sum_{l=1}^{\infty}\left[(2l+1)\left(\left|\alpha_l^{33}\right|^2+\left|\beta_l^{33}\right|^2\right)+2l(l+1)\,\text{Im}(A_l'^*\alpha_l^{33}+G_l'^*\beta_l^{33})\right] \tag{9.4.11}$$

and for natural or unpolarized radiation:

$$\begin{aligned} \sigma_{\text{sc}} = \frac{2\pi}{k_0^2}\sum_{l=1}^{\infty}&\Big\{(2l+1)\left(\left|\alpha_l^{33}\right|^2+\left|\beta_l^{33}\right|^2\right) \\ &+2l(l+1)\,\text{Im}\,\left[\left(A_l^*+A_l'^*\right)\alpha_l^{33}+\left(G_l^*+G_l'^*\right)\beta_l^{33}\right]\Big\} \end{aligned} \tag{9.4.12}$$

Notice, that in Eqs.(9.4.10)–(9.4.12) the terms proportional to $\hat{\varepsilon}^2$ are not taken into account.

At transition to the ensemble of APs it is necessary to average not only over the particle sizes, as it was done in the previous chapters, but also over their space orientation. The orientation of spherical APs is intimately associated in this case with the presence of the chosen directions – the AP optical axes. Let us stop and consider it in more detail.

As is known, it is possible to express the components of the real dielectric tensor in the coordinate system, which has been turned around the coordinate system of the main dielectric axes, through principal values of the dielectric tensor and the Eulers angles (Varshalovich *et al.*, 1988). These relations look like:

$$\hat{\varepsilon} = \hat{Q}\hat{\alpha}\hat{Q}_T \qquad \hat{\alpha} = \begin{pmatrix} \alpha^{(1)} & 0 & 0 \\ 0 & \alpha^{(2)} & 0 \\ 0 & 0 & \alpha^{(3)} \end{pmatrix}$$

where $\alpha^{(i)}$ are the eigenvalues of the dielectric tensor and $\hat{Q}$ is a matrix ($\hat{Q}_T$ is the transpose of the matrix), which has the following components:

$$\begin{aligned} Q_{xx} &= \cos\gamma_1 \cos\gamma_2 \cos\gamma_3 - \sin\gamma_1 \sin\gamma_3 \\ Q_{xy} &= -\cos\gamma_1 \cos\gamma_2 \sin\gamma_3 - \sin\gamma_1 \cos\gamma_3 \\ Q_{xz} &= \cos\gamma_1 \sin\gamma_2 \qquad Q_{yx} = \sin\gamma_1 \cos\gamma_2 \cos\gamma_3 + \cos\gamma_1 \sin\gamma_3, \\ Q_{yy} &= -\sin\gamma_1 \cos\gamma_2 \sin\gamma_3 + \cos\gamma_1 \cos\gamma_3 \qquad Q_{yz} = \sin\gamma_1 \sin\gamma_2, \\ Q_{zx} &= -\sin\gamma_2 \cos\gamma_3 \qquad Q_{zy} = \sin\gamma_2 \sin\gamma_3 \qquad Q_{zz} = \cos\gamma_2 \end{aligned} \tag{9.4.13}$$

In relations (9.4.13) γ_1, γ_2, and γ_3 are the Eulers angles uniquely describing rotation of some coordinate system around the coordinate system, coinciding with directions of main dielectric axes. Thus, to each AP in the coordinate system, related to the direction of incident radiation propagation, it is possible to assign some orientation parameter or the triple of Eulers angles. Let us term this triple of angles as the orientation vector and denote it as $\gamma = \{\gamma_1, \gamma_2, \gamma_3\}$. Here we shall note the following circumstance. If the dielectric function is the complex tensor, then the diagonal representation is generally impossible, as the real and imaginary parts can have their own coordinate systems, turned relative to each other. In this case the orientation vector can be entered both on the basis of the coordinate system for the real part of the dielectric tensor, and on the basis of that for imaginary part. The selection of this or that orientation parameter does not play a key role by virtue of linearity of the Eulers angles transformation at coordinate system rotation.

So, to carry out the orientational averaging, it is necessary to average the light scattering characteristics over Eulers angles. In our case it is possible to utilize one simplifying circumstance. The dependence on the Eulers angles completely enters into components of tensor $\hat{\varepsilon}$, through which, in turn, the amplitude coefficients of the scattered field expansion are linearly expressed. Moreover, these components linearly enter in the squared characteristics, such as the matrix of Stokes parameter transformation and the extinction matrix, because the square terms are small and not taken into account. Hence, having substituted into light scattering characteristics the average values of the tensor $\hat{\varepsilon}$ components instead of their unaveraged values, we shall obtain the required averaged characteristics.

Now we shall consider the case of AP random orientation in space (i.e., when all orientation directions have equal probability). Integrating the components of tensor $\hat{\varepsilon}$ with account of Eq.(9.4.13), we gain, that only the averaged diagonal components are not equal to zero. The straightforward substitution of the obtained expressions results in an expression simpler than Eq.(9.4.4):

$$\mathbf{E}_{sc}(\mathbf{r}) = E_{xI} \sum_{l=1}^{\infty} \frac{(2l+1)}{l(l+1)} i^l \left[\alpha_l \mathbf{M}^{(3)}_{o1l}(k_0\mathbf{r}) - i\beta_l \mathbf{N}^{(3)}_{e1l}(k_0\mathbf{r}) \right] \tag{9.4.14}$$

where

$$\alpha_l \equiv \alpha_l^{33} + \bar{A}_l = \alpha_l^{33} - \frac{il(l+1)}{2(2l+1)} f_l \left\{ \frac{2l+1}{l(l+1)} \alpha_l^{31} \left[l(l+1)\left(3\alpha + 2\alpha^{(3)}\right) - \beta \right] \right.$$
$$\left. +\beta \left(\frac{l+2}{l+1} \beta_{l+1}^{31} - \frac{l-1}{l} \beta_{l-1}^{31} \right) \right\} \tag{9.4.15}$$

$$\beta_l \equiv \beta_l^{33} + \bar{G}_l = \alpha_l^{33} - \frac{l(l+1)}{2(2l+1)} g_l \left\{ \frac{1}{(l+1)(2l+3)} \beta_l^{31} \right. \tag{9.4.16}$$
$$\left. \times \left[(5l^2 + 13l + 9)\alpha + (3l^2 + 7l + 3)\alpha^{(3)} \right] + \beta \left(\frac{l+3}{2l+3} \beta_{l+2}^{31} - \frac{1}{l+1} \alpha_{l+1}^{31} \right) \right\}$$
$$+ \frac{l(l+1)}{2(2l+1)} h_l \left\{ \frac{1}{l(2l-1)} \beta_l^{31} \left[(15l^2 - 3l + 1)\alpha + (3l^2 - l - 1)\alpha^{(3)} \right] \right.$$
$$\left. + \beta \left(\frac{l-2}{2l-1} \beta_{l-2}^{31} - \frac{1}{l} \alpha_{l-1}^{31} \right) \right\}$$

where $\alpha = \alpha^{(1)} + \alpha^{(2)}$ and $\beta = \alpha - 2\alpha^{(3)}$ ($\alpha^{(1)}, \alpha^{(2)}$, and $\alpha^{(3)}$ are the principal values of the dielectric tensor $\hat{\varepsilon}$).

It follows at once, that in this case the structure of the amplitude scattering matrix is absolutely similar to the structure of the matrix for the system of isotropic particles and, hence, the extinction matrix is converted into scalar. But the amplitude coefficients entering into the corresponding extinctions matrices for anisotropic and isotropic particles are different.

Considering the computational aspects of the Born approximation in the distorted wave method for APs, it is possible to note, that the calculation formulae are simple enough and are easily transformed into algorithms implemented on the basis of codes for isotropic spheres.

9.5 INTEGRAL SCATTERING CHARACTERISTICS FOR ENSEMBLES OF APs

Characteristics of light scattering by a unit volume of dispersion medium consisting of anisotropic scatterers, shall be considered for the example of quartz dust APs. The choice of quartz particles as an object for investigation is determined by the fact that they are the typical representatives of crystalline substances in an atmosphere. Quartz, or silicon dioxide, is the most stable silicon compound. Under usual conditions it is a colorless high-melting substance. From three silicon dioxide modifications (monocrystal, polycrystal, and fused quartz) only the first is of interest in our investigation. The quartz crystal is the uniaxial positive one. It does not absorb radiation in the visual and near-IR spectral ranges ($\lambda = 0.3$–$5\ \mu m$). The optical constants of quartz are widely represented in the literature (see, e.g., Sosman, 1927; Laikin, 1961; Voronkova *et al.*, 1965). The dust, in which the composition quartz enters, has particle diameters up to 20 μm, but at heights of 1–2 km the particles with diameter 0.7–2 μm predominate (Khrgian, 1969). The AP size distribution can be described by the gamma or log-normal distribution functions.

Consider the behaviour of the effective refractive index of crystalline quartz AP systems in air for the case, when their preferred orientation takes place (Kuzmin, 1981). We mention in passing that the results of this section can be useful also for the study of refraction in colloidal systems.

Generally speaking, the complex dielectric tensor cannot be represented in a diagonal form (Fedorov, 1958). However, it can be made in that spectral range, where the quartz does not absorb radiation. In the coordinate system, connected with the principal axes of the dielectric tensor, it looks like:

$$\hat{\varepsilon}' = \begin{pmatrix} N^2 + \alpha^{(1)} & 0 & 0 \\ 0 & N^2 + \alpha^{(2)} & 0 \\ 0 & 0 & N^2 + \alpha^{(3)} \end{pmatrix}$$

where $\alpha^{(1)}, \alpha^{(2)}$, and $\alpha^{(3)}$ are the diagonal components of tensor $\hat{\varepsilon}$. Since quartz is the positive uniaxial crystal, the following relations are fulfilled:

$$\alpha^{(1)} = \alpha^{(2)} \equiv \alpha_{\text{ord}} \qquad \alpha^{(3)} = \alpha_{\text{eord}} \qquad \alpha_{\text{ord}} < \alpha_{\text{eord}} \tag{9.5.1}$$

where α_{ord} and α_{eord} are related to the ordinary and extraordinary refractive indices:

$$n_{\text{ord}} = \sqrt{N^2 + \alpha_{\text{ord}}} \qquad n_{\text{eord}} = \sqrt{N^2 + \alpha_{\text{eord}}} \tag{9.5.2}$$

Thus, in the coordinate system, connected with incident radiation, the dielectric tensor is completely determined by two values $\alpha_{\text{ord,eord}}$ and two Euler angles α and β, describing the direction of the main dielectric axes.

If the concentration of scattering centres is small enough, then without the account of density fluctuations the refractive index can be expressed as follows:

$$\hat{N}_{\text{eff}} = 1 + \frac{2\pi}{k_0^2} \left\langle \hat{A}(\hat{\varepsilon}, 0) \right\rangle C_N \tag{9.5.3}$$

where $\left\langle \hat{A}(\hat{\varepsilon}, 0) \right\rangle$ is the orientationally averaged amplitude forward scattering matrix. Generally it depends both on the propagation direction, and on the direction of the electric vector oscillations in the incident wave. In the case of random space orientation of dielectric axes it degenerates into a scalar and does not depend on the Euler angles. Thus, the complex refractive index generally is a complex tensor quantity, and, hence, such a system will have a linear dichroism. Really, the imaginary part of the refractive index (9.5.3) is related to the extinction coefficient for a system of scattering particles:

$$\operatorname{Im} \hat{N}_{\text{eff}} = \frac{1}{2k_0} K_{\text{ext}} \tag{9.5.4}$$

Such dichroism is connected not with radiation absorption, but with removing some part of the radiation from the beam, therefore it is frequently termed as the conservative dichroism (Stoimenova *et al.*, 1980).

For the plane homogeneous incident wave it is enough to know the extinction efficiencies for two independent components of the wave. In one of them the electric vector oscillates along the x-axis , in the second, along the y-axis. The expressions for them follow from Eqs.(9.4.9):

$$Q^x_{\text{ext}} = Q^1_{\text{ext}} - \frac{2}{\rho_0^2}\,\text{Im}\left[\sum_{l=1}^{\infty} l(l+1)(A_l + G_l)\right] \tag{9.5.5}$$

$$Q^y_{\text{ext}} = Q^1_{\text{ext}} - \frac{2}{\rho_0^2}\,\text{Im}\left[\sum_{l=1}^{\infty} l(l+1)(A'_l + G'_l)\right]$$

where Q^1_{ext} is the extinction efficiency for the "average" isotropic particle, that is for the isotropic particle with a refractive index equal to the arithmetic mean of refractive indices of ordinary and extraordinary waves.

Following the work of Kuzmin (1981*a*), consider the spectral dependence of linear dichroism of the quartz APs in air. The calculations were carried out on the basis of Eqs.(9.5.4) and (9.5.5) for APs with the size $R = 1\ \mu m$ in the wavelength range 0.48–0.61 μm using the data from Table 9.1. The refractive indices for intermediate wavelengths missing in this table were built by linear interpolation. The calculations were performed for a monodisperse AP system. For elimination of the so-called ripple structure in studied dependence it was flattened by averaging over some wavelength interval. It was supposed, that AP optical axes are oriented in one direction, and $\gamma_1 = 0$, $\gamma_2 = \pi/2$, and $\gamma_3 = 0$.

In Figure 9.1 the spectral dependence is depicted for the relative difference of refractive index imaginary parts of waves with different direction of the electrical vector oscillations, defined as:

$$\Delta m' = (\text{Im}\, N^{yy}_{\text{eff}} - \text{Im}\, N^{xx}_{\text{eff}})\,/\,\text{Im}\, N^{xx}_{\text{eff}} \tag{9.5.6}$$

and similar dependence for the relative difference $\Delta m''$ of the real parts, describing the birefringence in the system of quartz APs in air. The quantity $|\Delta m'|$ reaches a maximum in the area of $\lambda \approx 0.55\ \mu m$ and converts to zero in the area of $\lambda \approx 0.49\ \mu m$. The calculations in the wider spectral range show, that the overall run of spectral dependence of quantity $\Delta m'$ looks like the oscillations about the zero value. The maximal value $|\Delta m'|$ in the field $\lambda \approx 0.55\ \mu m$ achieves 3% (i.e., it is a value of the same magnitude as a relative difference of real parts of refractive indices of ordinary and extraordinary waves).

Table 9.1. The refractive index of crystalline quartz.

$\lambda(\mu m)$	n_{ord}	n_{eord}	$\lambda(\mu m)$	n_{ord}	n_{eord}
0.3404	1.5675	1.5774	0.5791	1.5447	1.5538
0.3587	1.5639	1.5737	0.5893	1.5442	1.5534
0.4341	1.5540	1.5634	0.6278	1.5428	1.5519
0.4678	1.5510	1.5604	0.6563	1.5419	1.5509
0.4800	1.5501	1.5594	1.600	1.5270	1.5355
0.4862	1.5497	1.5590	1.609	1.5261	1.5352
0.5086	1.5482	1.5582	1.615	1.5268	1.5342
0.5184	1.5477	1.5569	1.682	1.5258	1.5332
0.5461	1.5462	1.5554			

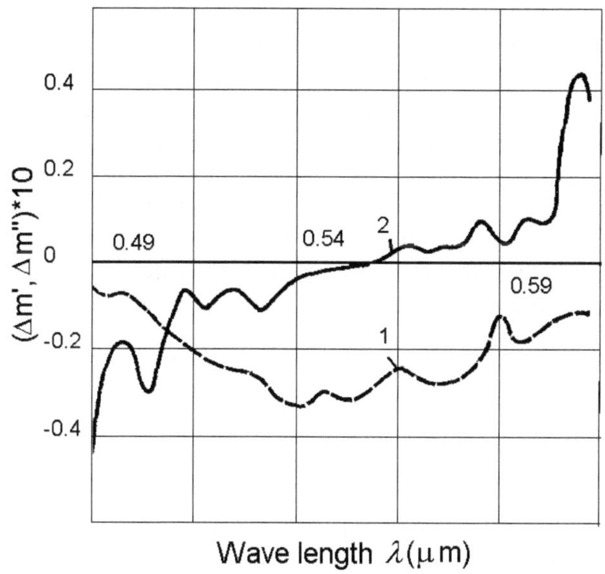

Figure 9.1. Spectral dependencies of the relative linear dichroism $\Delta m'(1)$ and birefringence $\Delta m''(2)$ of the scattering system consisting of the oriented quartz APs with radius $R = 1\ \mu m$.

In the work of Meeten (1980*a*) the approximate empirical rule for colloid systems from the oriented APs is given, according to which the linear dichroism of such a system is maximal by absolute value for those wavelengths, where the birefringence is equal to zero and vise versa. This rule was theoretically justified for great "soft" particles. The agreement between experiment and theory is observed, as shown by Kuzmin (1981*a*), and also in other cases. If taking into account, that the APs in the studied case are not great and "soft", it is clear from Figure 9.1, that, on the one hand, the theoretical calculations are confirmed qualitatively by experimental material, and, on the other hand, can serve as a theoretical basis of the above empirical rule in the field of strong diffraction interaction of radiation with a substance.

Consider now the influence of anisotropy of the particle optical properties on the medium spectral transparence, which in the single scattering approximation is immediately related to spectral dependence of the extinction cross section. If the APs in the system are oriented identically, then the extinction of linearly polarized radiation can be described by Eqs.(9.4.10) and (9.4.11). If the statistical distribution of axes in space takes place, these formulae are necessary to average over all possible orientations of axes. In case of complete random distribution of axes the coefficients A_l, A_l', G_l, and G_l' in Eqs.(9.4.10) and (9.4.11) need to be replaced by the coefficients $\bar{A}_l$ and $\bar{G}_l$ defined by Eq.(9.4.15).

As a concrete example we present here the calculated extinction efficiencies for monodisperse system of crystalline quartz APs with radius $R = 7\ \mu m$ in the wavelength range $\lambda = 0.4400$–$0.6328\ \mu m$ (Kuzmin, 1981*b*). The average refractive index appearing while dividing the dielectric tensor into isotropic and anisotropic parts, is chosen equal to the arithmetic mean from the refractive indices of ordinary and extraordinary waves.

For investigation of the deviation of the spectral transparency of an APs system from that of a system of model isotropic particles with the average refractive index, we enter the following relative characteristics: $\Delta_x = \delta_x/Q^1_{\text{ext}}$ and $\Delta_{xy} = \delta_{xy}/Q^x_{\text{ext}}$, where $\delta_x = Q^x_{\text{ext}} - Q^1_{\text{ext}}$ and $\delta_{xy} = Q^y_{\text{ext}} - Q^x_{\text{ext}}$. The quantity Δ_x describes the above mentioned deviation in the case of incident electric vector oscillations along the x-axis, when the particle system has strictly fixed space orientation of the main dielectric axes. The quantity Δ_{xy} describes the difference in the spectral transparency of an AP system, when the incident electric vector oscillates along the x-axis and along the y-axis. The fixed AP orientation is defined by the Euler angles $\{\gamma_1 = 0,\ \gamma_2 = \pi/2,\ \text{and}\ \gamma_3 = 0\}$.

In Figure 9.2*a* the spectral dependencies $\Delta_x(\lambda)$ and $\Delta_{xy}(\lambda)$ are given for the above mentioned monodisperse system of APs in the indicated spectral range. The range being studied envelops the size parameter values from 69.5 (at $\lambda = 0.6328\ \mu m$) to 100 (at $\lambda = 0.44\ \mu m$). As the figure indicates, the dependencies $\Delta_x(\lambda)$ and $\Delta_{xy}(\lambda)$ periodically oscillate with approximately identical frequencies and shifting on a half-period from each other. The number of basic extremes on curves $\Delta_x(\lambda)$ and $\Delta_{xy}(\lambda)$ coincides with that for curve $Q^1_{\text{ext}}(\lambda)$ (Figure 9.2*b*). The maximal absolute values of $|\Delta_{xy}(\lambda)|$ are more than those of $|\Delta_x(\lambda)|$ at a fixed wavelength. Notice also, that $\Delta_x(\lambda)$ and $\Delta_{xy}(\lambda)$ are close to zero for those wavelengths, where the spectral transparency $Q^1_{\text{ext}}(\lambda)$ reaches extreme values. Maximal values $|\Delta_{xy}(\lambda)|$ and $|\Delta_x(\lambda)|$ do not exceed 0.03 and 0.02 accordingly. Behaviour of dependencies $\Delta_x(\lambda)$ and $\Delta_{xy}(\lambda)$ for other fixed γ_3 values is qualitatively

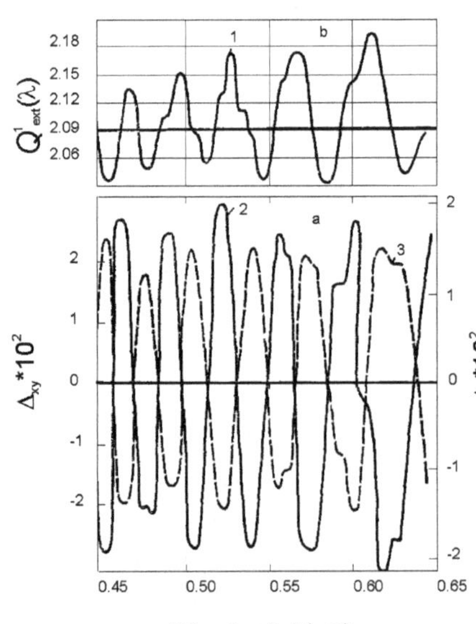

Figure 9.2. Spectral dependencies of an extinction efficiency $Q^1_{\text{ext}}(\lambda)(1)$ and diversions $\Delta_{xy}(2)$ and $\Delta_x(3)$ for the monodisperse system of quartz APs at $R = 7\ \mu m$, $\gamma_1 = 0$, $\gamma_2 = \pi/2$, and $\gamma_3 = 0$.

similar. Maximal values of $|\Delta_{xy}(\lambda)|$ and $|\Delta_x(\lambda)|$ remain small. The study of similar deviation for unpolarized radiation and systems with random space distribution of AP axes shows, that in these cases the extreme values of deviation are less, than when the polarized radiation incidents upon the system of APs with fixed orientation of optical axes.

This implies, that in the visual wavelength range it is possible to simulate the spectral transparency for systems consisting of quartz APs with sizes, characteristic for atmospheric dust ($R = 7\ \mu m$), with the help of isotropic particle system with an average refractive index. For systems with space random distribution of main dielectric axes the error of such a model does not exceed 1%, and for systems with fixed orientation at incidence of linearly polarized radiation – 3%. The deduction about low sensitivity of the radiation extinction to anisotropy concerns not only the quartz AP in the visible spectral range. It is confirmed by the spectral dependencies $\Delta_x(\lambda)$ and $\Delta_{xy}(\lambda)$ for a series of other situations, in particular for APs in the resonance region of size parameter ($\rho_0 \approx 5$), where at first sight it would be possible to expect the great values of $|\Delta_{xy}(\lambda)|$ and $|\Delta_x(\lambda)|$. In Figure 9.3 the results of calculations for model APs with an average refractive index $n = 1.30$ and anisotropy parameters $\alpha^{(1)} = \alpha^{(2)} = -0.01$ and $\alpha^{(3)} = 0.01$ are given. The AP orientation is determined by angles $\gamma_1 = 0$, $\gamma_2 = \pi/2$, and $\gamma_3 = 0$. Apparently, the maximal values $|\Delta_{xy}|$ and $|\Delta_x|$ grow with size parameter, however, they remain small and do not exceed $|\Delta_x| \sim 1\%$ and $|\Delta_{xy}| \sim 3\%$.

Thus, the spectral dependence of transparency of the dispersion medium consisting of weakly anisotropic spherical particles can be simulated (with an accuracy of 2–5 percent) in a wide interval of the size parameter by the similar dependence $Q^1_{\text{ext}}(\lambda)$ for the system of isotropic particles with an average refractive index. Presence of the averaging factors (non-polarized incident radiation, chaotic distribution of optic axes, polydispersity) increases the accuracy.

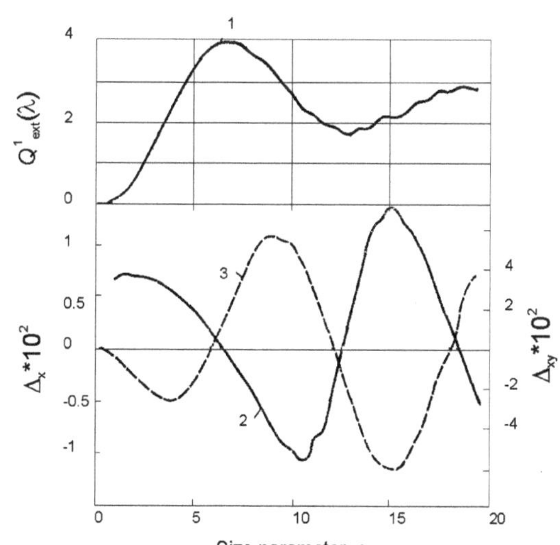

Figure 9.3. Extinction efficiency $Q^1_{\text{ext}}(1)$ and diversions $\Delta_{xy}(2)$ and $\Delta_x(3)$ versus size parameter ρ_0 of the model APs with $n = 1.30$.

9.6 ANGULAR CHARACTERISTICS OF LIGHT SCATTERING BY SYSTEMS OF APs

The matrix of Stokes parameter transformation (the Mueller matrix (MM)) appears in the vector radiative transfer equation and, hence, is one of the basic characteristics of unit volume when solving the direct problems of radiative transfer. On the other hand, under conditions of single scattering the elements of this matrix are measurable quantities and can serve as initial experimental material for the solution of a wide circle of inverse problems on revealing the microstructure of disperse systems and optical constants of particles making up these systems. Thus, the investigation of angular dependencies of MM elements for APs is of practical interest both for direct, and inverse problems.

Upon dividing the tensor $\hat{\varepsilon}'$ into isotropic and anisotropic parts the amplitude scattering matrix can also be divided (accurate to $\hat{\varepsilon}^2$) into components:

$$\hat{A}\left(\hat{\varepsilon}', \theta, \varphi\right) = \hat{A}_1\left(N, \theta, \varphi\right) + \hat{A}_2\left(\hat{\varepsilon}, \theta, \varphi\right) \tag{9.6.1}$$

where the first describes the scattering by the "average" isotropic particle with refractive index N, and the second is stipulated by departure of particle properties from the isotropic. Included in Eq.(9.6.1) the scattering angles θ and φ are defined in the spherical coordinate system, connected with the direction of the incident wave propagation. Expression for $\hat{A}\left(\hat{\varepsilon}', \theta, \varphi\right)$ is given by Eq.(9.4.5), which in the expanded form notes as:

$$\hat{A}_{11}\left(\hat{\varepsilon}', \theta, \varphi\right) = \frac{1}{k_0}\sum_{l=1}^{\infty}{}' \left[\frac{2l+1}{l(l+1)} i\left(-\alpha_l^{33} Q_l^{(1)} - \beta_l^{33} S_l^{(1)}\right)\cos\varphi + R_l\right]$$

$$\hat{A}_{21}\left(\hat{\varepsilon}', \theta, \varphi\right) = \frac{1}{k_0}\sum_{l=1}^{\infty}{}' \left[i\frac{2l+1}{l(l+1)} i\left(\alpha_l^{33} S_l^{(1)} + \beta_l^{33} Q_l^{(1)}\right)\sin\varphi + T_l\right]$$

$$\hat{A}_{12}\left(\hat{\varepsilon}', \theta, \varphi\right) = \frac{1}{k_0}\sum_{l=1}^{\infty}{}' \left[\frac{2l+1}{l(l+1)} i\left(-\alpha_l^{33} Q_l^{(1)} - \beta_l^{33} S_l^{(1)}\right)\sin\varphi + T_l'\right]$$

$$\hat{A}_{22}\left(\hat{\varepsilon}', \theta, \varphi\right) = \frac{1}{k_0}\sum_{l=1}^{\infty}{}' \left[\frac{2l+1}{l(l+1)} i\left(-\alpha_l^{33} S_l^{(1)} - \beta_l^{33} Q_l^{(1)}\right)\cos\varphi + R_l'\right]$$

where R_l and T_l are determined by Eqs.(9.3.10), and T_l' and R_l' take the form:

$$\begin{aligned} T_l' = {} & \left(A_l' Q_l^{(1)} + G_l' S_l^{(1)}\right)\sin\varphi + \left(I_l S_l^{(1)} - C_l Q_l^{(1)}\right)\cos\varphi \\ & + \left(2\mathcal{E}_l Q_l^{(2)} + K_l S_l^{(2)}\right)\sin 2\varphi - \left(L_l S_l^{(2)} - 2F_l Q_l^{(2)}\right)\cos 2\varphi \\ & + \left(H_l S_l^{(3)} + 3B_l Q_l^{(3)}\right)\sin 3\varphi - \left(J_l S_l^{(3)} - 3D_l Q_l^{(3)}\right)\cos 3\varphi \end{aligned} \tag{9.6.2}$$

$$\begin{aligned} R_l' = {} & \left(A_l' S_l^{(1)} + G_l' Q_l^{(1)}\right)\cos\varphi + \left(C_l S_l^{(1)} - I_l Q_l^{(1)}\right)\sin\varphi \\ & + \left(\mathcal{E}_l S_l^{(2)} + 2K_l Q_l^{(2)}\right)\cos 2\varphi + \left(2L_l Q_l^{(2)} - F_l S_l^{(2)}\right)\sin 2\varphi \\ & + \left(B_l S_l^{(3)} + 3H_l Q_l^{(3)}\right)\cos 3\varphi + \left(3J_l Q_l^{(3)} - D_l S_l^{(3)}\right)\sin 3\varphi \end{aligned} \tag{9.6.3}$$

The functional relation of the terms included in Eq. (9.6.2) and (9.6,3), is yielded by Eqs.(9.3.5), (9.3.7), and (9.3.10). Notice that R_l,T_l, T_l',R_l' depend on the Euler angles γ_1, γ_2, and γ_3, and in case of uniaxial crystal – on γ_1 and γ_2.

The MM elements are determined (accurate to $\hat{\varepsilon}^2$) by expression:

$$D_{\lambda\mu}(\theta,\varphi) = \frac{C_N}{2}\sum_{ijkl=1}^{2}\sigma_{ik}^{\lambda}\left\{\langle A_{1ij}(\theta,\varphi)A_{1lk}^{*}(\theta,\varphi)\rangle\right. \\ \left. + \langle A_{2ij}(\theta,\varphi)A_{1lk}^{*}(\theta,\varphi)\rangle + \langle A_{1ij}(\theta,\varphi)A_{2lk}^{*}(\theta,\varphi)\rangle\right\}\sigma_{jl}^{\nu} \tag{9.6.4}$$

For ensemble of APs having a statistical space distribution of the principal dielectric axes, Eq.(9.6.4) is necessary to average over this distribution. Taking into account, that $\hat{A}_1$and $\hat{\sigma}^{\lambda}$ do not depend on AP orientation, we shall note the MM for ensemble of APs with completely random (that is, with equal probability of all directions) orientation is:

$$D_{\lambda\mu}(\theta,\varphi) = \frac{C_N}{2}\sum_{ijkl=1}^{2}\sigma_{ik}^{\lambda}\left\{A_{1ij}(\theta,\varphi)A_{1lk}^{*}(\theta,\varphi)\right. \\ \left. + \langle A_{2ij}(\theta,\varphi)\rangle A_{1lk}^{*}(\theta,\varphi) + A_{1ij}(\theta,\varphi)\langle A_{2lk}^{*}(\theta,\varphi)\rangle\right\}\sigma_{jl}^{\nu} \tag{9.6.5}$$

where $< ... >$ means an orientational averaging. As the functional structure of the averaged matrix $\hat{A}_2$ is similar to the matrix structure in the case of isotropic particles, the MM in the scattering plane ($\varphi = 0$) looks like:

$$\hat{D}(\theta,\varphi) = \begin{bmatrix} D_{11}(\theta,\varphi) & D_{12}(\theta,\varphi) & 0 & 0 \\ D_{12}(\theta,\varphi) & D_{11}(\theta,\varphi) & 0 & 0 \\ 0 & 0 & D_{33}(\theta,\varphi) & -D_{43}(\theta,\varphi) \\ 0 & 0 & D_{43}(\theta,\varphi) & D_{33}(\theta,\varphi) \end{bmatrix} \tag{9.6.6}$$

Apparently, the structure of the matrix (9.6.6) coincides with structure of a similar matrix for isotropic particles.

Notice some features of the amplitude scattering matrix of APs with the chosen space orientation, immediately following from its analytical form (Kuzmin and Babenko, 1982). The elements of this matrix depend on orientation of principal dielectric axes relative to the coordinate system, connected with the direction of incident wave propagation. This orientation is uniquely characterized by the triple of Euler angles. Thus, for APs the scattering matrix and, hence, the MM are not invariant concerning AP rotations. At arbitrary orientation $(\gamma_1, \gamma_2, \gamma_3)$ the amplitude scattering matrix does not diagonalize in the coordinate system, connected with the scattering plane, as it takes place for isotropic particles. From here it follows, that those elements of the MM, which are equal to zero for isotropic particles $(D_{13}, D_{14}, D_{23}, D_{24}, D_{42}, D_{41}, D_{31}, D_{32})$, accept non-zero values for APs. Symmetry relations $D_{11} = D_{22}$, $D_{12} = D_{21}$, $D_{43} = -D_{34}$, and $D_{33} = D_{44}$, fulfilled for isotropic particles, will be fulfilled (with accuracy $\hat{\varepsilon}^2$) for APs, however, any other symmetry relations are not present. Notice also, that for APs the D_{11} value depends on the azimuthal angle φ.

Behaviour of angular dependencies of the MM elements we shall consider on an example of the system of quartz APs at incident radiation wavelength $\lambda = 0.6328\ \mu m$

(Kuzmin and Babenko, 1982). The refractive indices of ordinary and extraordinary waves in quartz at this wavelength were obtained from Table 9.1. The calculations were performed for the range of the typical dust particle sizes in the surface layer of atmosphere (Khrgian, 1969). For a monodisperse system of APs, with great enough values of size parameter and refractive index, the angular dependencies of the MM elements strongly oscillate, which impedes the analysis. For flattening the oscillations enter a narrow particle size distribution (gamma distribution with half-width parameter $\mu = 100$). We shall carry out consideration for systems with modal radius $R_0 = 0.4; 1;$ and $3\ \mu m$, which correspond to a modal size parameter $\rho_0 = 4; 10;$ and 29.9.

It was assumed, that all APs are oriented identically (optic axes of uniaxial quartz crystals are oriented perpendicularly to incident wave propagation). The coordinate system was selected so that the z-axis coincides with the incident wave vector, and the x-axis was oriented along optical axis. The triple of Euler angles $(0,\ \pi/2,\ 0)$ corresponds to such geometry. The scattered fields were explored in the plane xz (that is at $\varphi = 0$), as in this case the number of independent elements of MM is minimal and is equal to four, as in the case of isotropic particles.

Compare the obtained angular dependencies of the MM elements for a system of APs with similar dependencies for systems of isotropic particles, whose refractive index was chosen as the arithmetic mean from refractive indices of quartz for ordinary and extraordinary waves at $\lambda = 0.6328\ \mu m$. We shall make a comparison for elements of a reduced MM. The difference of the elements is used as a measure of discrepancy:

$$\Delta f_{\lambda\nu}(\theta) = f_{\lambda\nu}(\theta) - f^{\,\text{isot}}_{\lambda\nu}(\theta) \tag{9.6.7}$$

The analysis of dependencies $\Delta f_{\lambda\nu}(\theta)$ shows, that this discrepancy increases with an increase of the particle size (Figure 9.4). If at $R_0 = 0.4\ \mu m$ maximal diversion makes 0.025, at $R_0 = 1\ \mu m$ it is equal 0.055, and at $R_0 = 3\ \mu m$ makes 0.13. The greatest departures are exhibited at scattering angles in the back hemisphere $\theta \sim$140–175°. This feature is characteristic for all MM elements, that is, the indicated angle range may be thought of as the most sensitive to particle anisotropic structure. Figure 9.5 illustrates sited features for the example of the $D_{11}(\theta)$ component.

For calculation of scattering characteristics the approximate method was utilized. It, naturally, introduces some error, which under analysis can be accepted for the development of anisotropy of a particle substance. As is known, in perturbation theory the convergence and error depend on the quantity of a small parameter (in our case it is $\hat{\varepsilon}$) and the greater the parameter, the greater is the error. The estimate of the greatest possible error of the solution was carried out as follows. For a given tensor $\hat{\varepsilon}$ the principal values of $\alpha^{(i)}$ were obtained and calculations were performed with the approximate theory, developed for APs, and in so doing it was assumed, that the "isotropic" part of the dielectric function is equal $\left(N^2 + \alpha^{(i)}_{\max}\right)$, where $\alpha^{(i)}_{\max}$ is the maximal principal value of tensor $\hat{\varepsilon}$, and the "anisotropic" part is determined by tensor $\hat{\varepsilon}$, for which all principal values are identical and are equal to $(-\alpha^{(i)}_{\max})$. It is seen, that at such a selection of the dielectric tensor it degenerates into scalar N, that is, it is possible to perform a parallel calculation of the scattering characteristics with the exact Mie theory for isotropic particles. The difference of the solutions obtained by this method it can serve as an error estimate of the approxima-

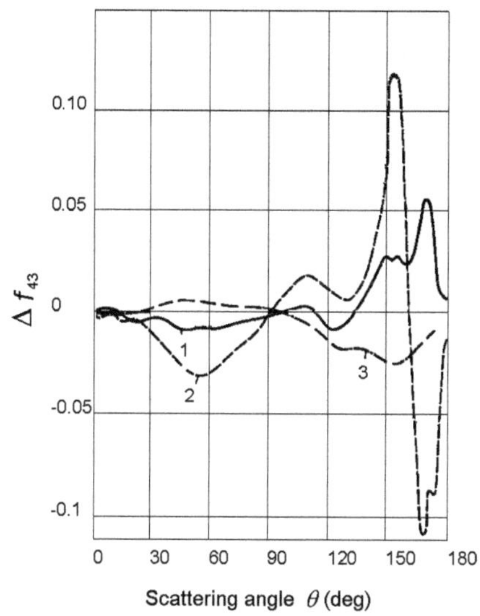

Figure 9.4. Angular dependencies for the difference of the element f_{43} for quartz APs from values f_{43}^{is} for isotropic particles with an average refractive index. The particle modal size is $R_0 = 1$ (*1*), 3 (*2*), and 0.4 μm (*3*). Curve 3 is given with an opposite sign.

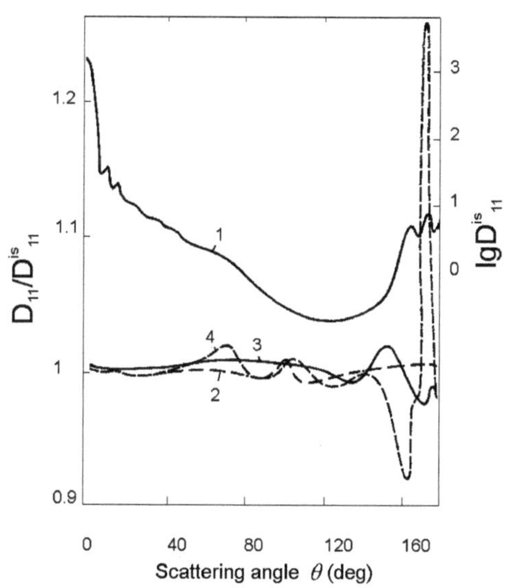

Figure 9.5. Angular dependencies of the element $D_{11}^{\text{is}}(\theta)$ for isotropic particles with an average refractive index (*1*) and radius $R_0 = 3\ \mu m$ (right scale) and ratio of elements $D_{11}(\theta)$ of quartz APs to $D_{11}^{\text{is}}(\theta)$ for radii $R_0 = 0.4(2), 1(3)$, and 3 μm (*4*) (left scale) at the wavelength $\lambda = 0.6328$ μm.

te computational method. Such assessment is rather rough and, apparently, somewhat overestimates the real maximal error.

By way of illustration of such an approach to error estimation, the angular dependencies of the MM element $f_{33}(\theta)$ for AP systems with $R_0 = 3\ \mu m$ are given in Figure 9.6. The difference in the curves between the angle range 0–150° is insignificant and does not exceed 5%. As can be seen from Figure 9.6, the solution error is comparable with the difference $\Delta f_{33}(\theta)$ for APs and isotropic "average" particles, that is, in this angle region here is difficulty in deciding, whether this difference is due to particle anisotropy or to computational method error. However, in the angle range $\theta \approx 150$–$170°$ the difference is great and achieves 25% at $\theta \approx 160°$, which essentially exceeds the calculation error, that is, in this range the influence of particle substance anisotropy on element f_{33} is great enough.

We will describe briefly the behavior of differences $\Delta f_{\lambda\nu}(\theta)$ for other MM elements. From Figure 9.5 it is seen, that at $\theta < 150°$ the $D_{11}(\theta)$ values are comparable with calculation error, however, at $\theta > 150°$ they exceed the error and attain 25% (an error of about 10%) in the glory region (0–175°). The values Δf_{12} also exceed the solution error at $\theta > 150°$, however, the absolute value of difference Δf_{12} is less (5–10%). In the same angle interval the value Δf_{43} makes 10–15% (see Figure 9.4), however, the error estimate gives the comparable value. Probably, a more exact error estimation is necessary in this case.

All results presented above deal with the case, when the angle γ_2 between the optical axis of quartz APs and the direction of incident wave propagation is fixed and equal to 90°. While at the angle γ_2 variation (for a fixed scattering angle θ), the values of MM elements

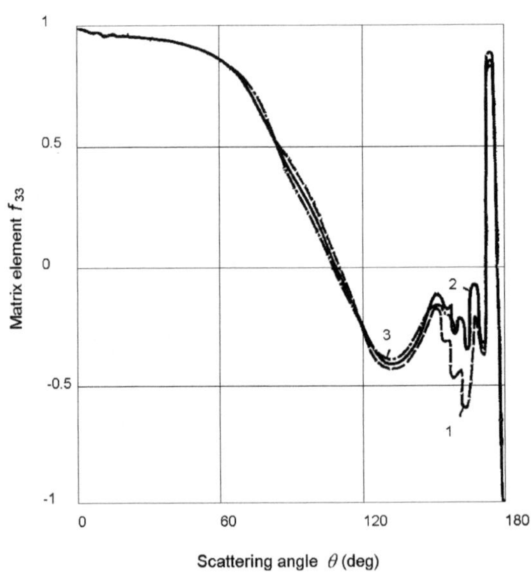

Figure 9.6. Angular dependencies of the element f_{33} for systems of particles with $R_0 = 3\ \mu m$: quartz APs (*1*); isotropic particles with the average refractive index, calculation under the exact formulae (*2*); and calculation under the approximate formulae (*3*).

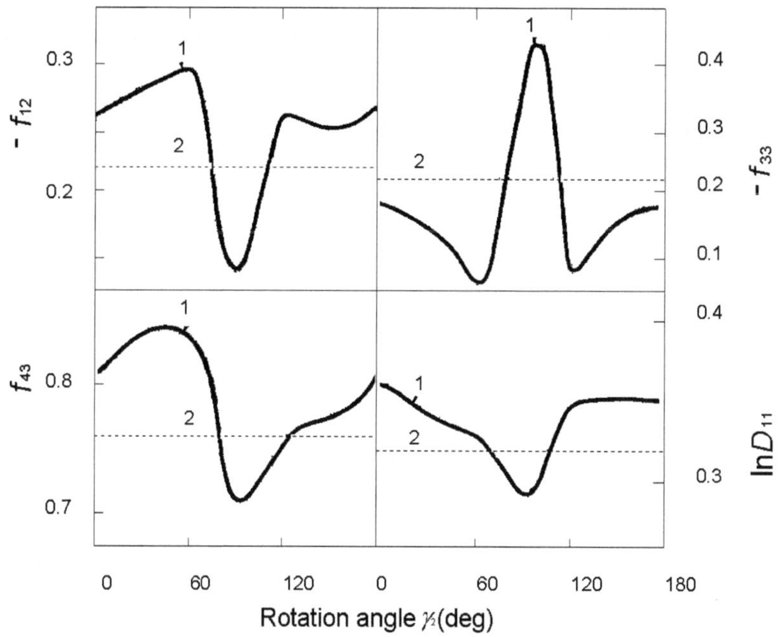

Figure 9.7. MM elements versus the rotation angle γ_2 at the scattering angle $\theta = 160°$: quartz APs (*1*) and isotropic particles with the average refractive index (*2*) at $R_0 = 3\ \mu m$.

will vary. This change, as it was already noted, is characteristic for the MM of APs. In Figure 9.7 the dependencies of the MM elements on angle γ_2 are given at a fixed value $\theta = 160°$ (it corresponds to light source rotation in the plane xz, or rotation of AP itself). It is seen, that at a γ_2 change the MM elements oscillate about the corresponding values for the "average" isotropic particle, and the extremes of these oscillations for all elements are implemented at angles $\gamma_2 = 60$, 90, and 120°. The range of variation of the MM elements according a γ_2 change is most significant for $f_{33}(\gamma_2)$ (0.06–0.42); and less so for $f_{12}(\gamma_2)$ (0.14–0.30) and $f_{43}(\gamma_2)$ (0.72–0.83). Such changes (e.g., at transition from angle $\gamma_2 = 60°$ to angle $\gamma_2 = 90°$) it is quite possible to detect during experimental investigations.

The above analysis of the angular characteristics is carried out for a system of quasi-monodisperse particles. However, as calculations show, the increasing distribution function half-width does not result in an essential change of the marked regularities. Therefore, for such a variable object, as aerosol, it is possible to formulate the following recommendations for the account of anisotropic properties of particles (at any rate for the visual wavelength range). Anisotropic properties need not be taken into account in the calculations of: (1) the diagonal terms of the extinction matrix; (2) the scattering characteristics for systems of the chaotically oriented particles and particles of submicron aerosol fraction, oriented along some direction; and (3) the angular scattering characteristics for fractions A and B at scattering angles $\theta < 90°$. To the contrary, the anisotropic properties should be taken into account in the case of the oriented particle systems for calculation of off-diagonal components of the extinction matrix and angular light scattering characteristics in the angle range $\theta > 90°$ for fractions A and B.

In conclusion of this chapter it is worth noting once again that the possible non-sphericity of anisotropic particles is likely to have a stronger effect on the scattering patterns than their anisotropy.

10

Some practical aspects of light scattering by inhomogeneous and anisotropic particles

A large share of the results in the preceding chapters were concerned with atmospheric optics. However, as repeatedly indicated earlier, the methods of light scattering are applicable to the extremely broad range of other disperse media. The examples of such applications were discussed in Chapters 6 and 7 and Sections 8.4–8.5. In addition, in this chapter several interesting practical applications of the theory of light scattering by the inhomogeneous and anisotropic particles are considered. In particular, the method of detecting and sizing the super-thin layer on the surface of a particle, based on comparison of the theoretical calculations with the results of laboratory experiments, is described in Section 10.1. The further three sections of this chapter are devoted to different aspects of development of the dispersion light filters with the use of two-layered particles (TLPs) (Section 10.2), quasi-homogeneous particles (Section 10.3), and anisotropic particles (APs) (Section 10.4). Use of such particles as an alternative to commonly employed homogeneous particles has opened up fresh opportunities for a smooth spectral shift of the transmission band, and for regulation of its half-width.

10.1 DETECTION OF ADSORBED LAYERS ON PARTICLE SURFACES AND ESTIMATION OF THEIR THICKNESS

It was noticed by Dudo (1973) during the development of a method to define the refractive index of a disperse substance from the measured values of the scattering coefficient that for polystyrene latex particles with radii $R < 0.2\ \mu m$ the dependence of obtained n values on particle size is observed in the visible spectral range (the refractive index decreases with decreasing R). Prishivalko and Dudo (1975) analysed the reasons for such behaviour and clarified the opportunities of its practical usage.

As is known, the emulsifiers, forming the very thin film on the particle surface are used to prevent latex particles sticking (Figurovski and Futran, 1947; Slonim *et al.*, 1960; Mehta *et al.*, 1971; Singh and Singh, 1973; Urgiris and Costaschuk, 1973). For complete elimination of sticking, it is good practice to add a small amount of dilute ammonia solution to the latex water suspension before taking optical measurements (Slonim *et al.*, 1960). In so doing, some ammonia turns to a gaseous state, forming the thinnest gas layer on latex particles.

There are rather contradictory literature data on the emulsifier shell width resulting on latex: from one nanometer up to a tenth of a micrometer and more. The wide range of shell thickness, indicated by the different authors, is caused, apparently, by the distinction of particle properties itself and emulsifiers used and also by a lack of sufficiently reliable methods to estimate the shell size.

With the purpose of estimating the influence of thin surface layers the measurements of the scattering coefficient K_{sc} in the wavelength interval $0.35 \leq \lambda \leq 0.8\ \mu m$ were carried out on 11 samples of polystyrene latex with particle radius $0.045 \leq R \leq 0.39\ \mu m$ (Prishivalko and Dudo, 1975). The particle size was determined with the help of an electron microscope. The measurements of the scattering coefficient were performed with the help of the special device attached to spectrophotometer. Latex samples with emulsifier were diluted in distilled water, and then ammonia was added to the same samples in such a quantity, that the dispersion medium would represent the 0.1, 0.5, and 1% ammonia solution. The measurements have shown, that the adding ammonia results in lowering the scattering coefficient of suspension, if the radius of its particles is less than some defined value. The change in ammonia concentration within the indicated limits does not influence the scattering coefficient.

From Table 10.1 it is seen, that for samples, consisting of particles with a small radius, the lowering of measured K_{sc} values observed after adding ammonia considerably exceeds the experimental error. As shown by Dudo (1973), for these samples the error comprises 1–1.5%.

In Figure 10.1 the results of evaluation of the relative refractive index n' of polystyrene latex particles in water on the basis of the scattering coefficient values measured with the help of the method described by Dudo (1973) are represented. It is seen, that for the smallest particles the refractive index values obtained for samples without ammonia is lower on the average by 0.007, and for a sample with ammonia is lower by 0.024, than in the case of larger particles with $R > 0.35\ \mu m$. The error of the n' definition itself does

Table 10.1. The experimental values of the scattering coefficient (normalized to a unit volume and unit mass concentration) for samples without ammonia $K_{\text{sc}}^{(1)}$ and with ammonia $K_{\text{sc}}^{(2)}$ (cm^2/mg) and ratio $\delta = \left(K_{\text{sc}}^{(1)} - K_{\text{sc}}^{(2)}\right)/K_{\text{sc}}^{(1)}$.

	$\lambda = 0.45\ \mu m$			$\lambda = 0.55\ \mu m$			$\lambda = 0.65\ \mu m$		
$R(\mu m)$	$K_{\text{sc}}^{(1)}$	$K_{\text{sc}}^{(2)}$	$\delta(\%)$	$K_{\text{sc}}^{(1)}$	$K_{\text{sc}}^{(2)}$	$\delta(\%)$	$K_{\text{sc}}^{(1)}$	$K_{\text{sc}}^{(2)}$	$\delta(\%)$
0.045	2.5	2.6	11.0	1.8	1.6	8.0	1.2	1.1	10.0
0.055	3.7	3.3	11.0	1.9	1.7	10.0	1.6	1.5	11.0
0.085	12.0	10.8	10.0	5.7	5.3	8.0	3.3	3.0	8.0
0.110	16.9	16.0	5.5	8.9	8.2	8.0	5.7	5.2	8.3
0.135	27.3	26.0	5.0	14.0	13.0	7.0	8.5	7.8	8.1
0.205	39.0	37.0	5.0	23.2	22.5	3.0	15.7	15.3	2.6
0.235	46.3	45.0	3.0	28.9	28.9	0.0	19.5	19.5	0.0

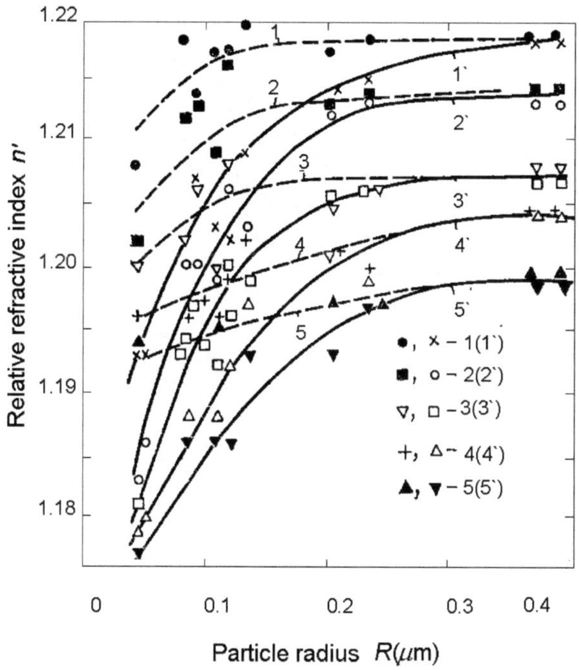

Figure 10.1. Relative refractive index of polystyrene latex in water (Dudo, 1973) for the different particle sizes at wavelength λ_0 =0.35 (*1*, *1'*), 0.4 (*2*, *2'*), 0.5 (*3*, *3'*), 0.6 (*4*, *4'*), and 0.7 μm (*5*, *5'*). The curves *1–5* correspond to samples without ammonia, *1'–5'* correspond to samples with ammonia.

not exceed ± 0.003. The similar dependence of the refractive index values on the particle size was marked also by Smart and Wills (1967).

It is possible to assume, that the dependence $n'(R)$, observed in the case of small particles, is due to the presence of a fluid emulsifier thin film and layer of gaseous ammonia. To test the validity of such a guess, the calculations of scattering cross sections for homogeneous (HP) and TLPs with refractive indices of a core and shell equal or close to having place in experiment were performed.

In Figure 10.2 the results of correlation between the calculated values of scattering cross section, for polystyrene HPs and TLPs with the polystyrene core and different shells, are given. Shell radius R_2, and accordingly size parameter $x_2 = 2\pi R_2 n_w / \lambda_0$, were varied over a wide range (n_w is the water refractive index). The polystyrene refractive index relative to water at $\lambda_0 = 0.54\ \mu m$ was accepted equal to $n_1' = 1.207$. The value $n_2' = 0.75$ corresponds to the gaseous ammonia refractive index relative to water. It was supposed, that the fluid emulsifier refractive index is close to or slightly below the polystyrene one. To take into account the emulsifier influence, in view of absence the more precise information, the calculations were carried out for several values of the shell relative refractive index.

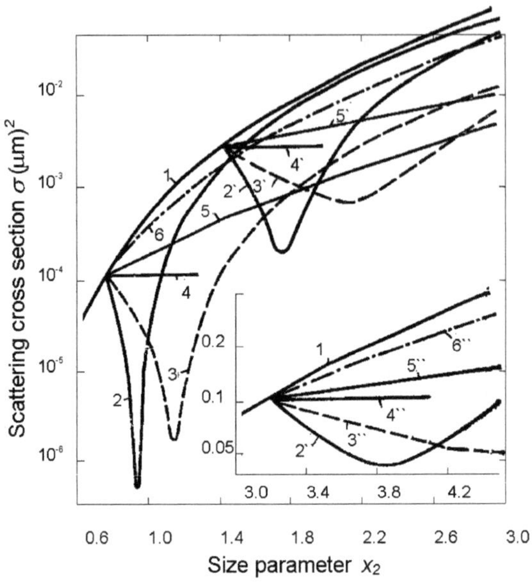

Figure 10.2. Scattering cross sections versus the external particle size at $\lambda_0 = 0.54\ \mu m$ for HPs with n' =1.207 (*1*) and TLPs with $n'_1 = 1.207$, $n'_2 = 0.75$ $(2, 2', 2'')$, 0.9 $(3, 3', 3'')$, 1 $(4, 4', 4'')$, 1.05 $(5, 5', 5'')$, and 1.15 $(6, 6'')$; $R_1 = 0.05$ (*2–6*), 0.0925 (*2′–5′*), and 0.205 μm (*2″–6″*).

In the studied range of size parameter x_2 the scattering cross section $\sigma_{\rm sc}^{(2)}$ of TLPs, for which the shell refractive index is lower than that for the core, is smaller, than $\sigma_{\rm sc}^{\rm hom}$ for HPs from the core substance with the same external size. The formation of the shell, with the relative refractive index, smaller than unity, on HPs in the beginning results in a sharp lowering of $\sigma_{\rm sc}^{(2)}$. At some shell width $\Delta R^{\rm min}$, corresponding to $x_2^{\rm min}$, the minimal value $\sigma_{\rm sc}^{\rm min}$ is reached. Further shell increasing is accompanied by rapid growth of $\sigma_{\rm sc}^{(2)}$.

It is evident from Table 10.2, in surveyed cases the minimal values $\sigma_{\rm sc}^{\rm min}$ are reached, when the shell width is 20–22% from the core radius at $n'_2 = 0.75$ and 45–48% at $n'_2 = 0.9$. At a given value of the shell refractive index minimal scattering takes place at some ratio of shell and core volumes that practically is almost independent of the particle size. However, depth of minima $\sigma_{\rm sc}(x_2)$ very rapidly decreases with increasing core size.

It would naturally appear that the scattering by a particle is minimal, when the refractive index averaged over particle volume (see Eq.(2.5.12)) is equal to the surrounding medium refractive index $\bar{n} = n_{\rm med}$. This condition is executed, when:

$$\frac{\Delta R}{R_1} = \sqrt[3]{\frac{n'_1 - n'_2}{1 - n'_2}} - 1$$

Having substituted $n'_1 = 1.207$ and $n'_2 = 0.75$, we find, that the scattering should be minimal, when the shell reaches 22.3% from the core radius, and for $n'_2 = 0.9 - 45.3\%$,

Table 10.2. Calculated values of $x_2^{\min}$, $\sigma_{\rm sc}^{\min}$, and $\Delta R^{\min}$ for $n'_1 = 1.207$ at different core radius in comparison with $\sigma_{\rm sc}^{\rm core}$, $\sigma_{\rm sc}^{\rm hom}$, and R_1. Asterisk denotes that the values of $\sigma_{\rm sc}^{\rm hom}$ are obtained at $x = x_2^{\min}$ and $n' = 1.207$.

R_1 (μm)	$x_2^{\min}$	$\sigma_{\rm sc}^{\min}$ (μm^2)	$\Delta R^{\min}$ (μm)	$\sigma_{\rm sc}^{\rm hom}$ (μm^2)*	$\frac{\sigma_{\rm sc}^{\rm core}}{\sigma_{\rm sc}^{\min}}$	$\frac{\sigma_{\rm sc}^{\rm hom}}{\sigma_{\rm sc}^{\min}}$	$\frac{\Delta R^{\min}}{R_1}$
			$n'_2 = 0.75$				
0.05	0.94	5.1–7	0.011	3.3–4	230	660	0.22
0.0925	1.75	2.1–4	0.021	7.4–3	14.8	37	0.22
0.205	3.83	4.5–2	0.043	2.2–1	2.31	4.9	0.21
0.370	6.83	5.95–1	0.072	1.8	1.67	3.0	0.195
			$n'_2 = 0.9$				
0.05	1.14	1.8–6	0.024	9.8–4	63.9	544	0.48
0.0925	2.11	6.9–4	0.044	1.3–2	4.3	18.8	0.475
0.205	4.60	5.3–2	0.092	0.45	2.0	8.5	0.45

that correlates well with Table 10.2 data. It is necessary once again to note, that the finer the TLP, the better its scattering properties can be described by the HP model with the volume-averaged refractive index and vise versa (Prishivalko and Babenko, 1975).

The calculations show, that the formation of the gaseous ammonia layer on the polymer particle should give an "enlightenment" of the latex water suspension. In particular (at $\lambda = 0.54\ \mu m$, $n'_1 = 1.207$, and n'_2=0.75, the core radius $R_1 = 0.05\ \mu m$, and shell width ΔR =0.5 nm) σ_{sc} reduced by 11.5%, at $\Delta R = 1$ nm by 21.6%, and at $\Delta R = 1.5$ nm by 30.7%. In experiments this effect is attenuated by the presence of the fluid emulsifier. Lack of data for the emulsifier refractive index means there is no way to evaluate precisely this influence. The rough estimated calculations have shown, that it is rather insignificant and can be neglected in the first approximation, assuming, that the main enlightenment effect is stipulated by gaseous ammonia.

The experiment showed, that at $R_1 \approx 0.205\ \mu m$ the influence of shells, forming on particles, is small, and at $R_1 \geq 0.235\ \mu m$ it practically is undetected. Therefore, it is possible to consider, that the refractive index value of particles of these sizes corresponds to that of a polymer. Thus, measuring the scattering coefficient for fine particles and using the shell refractive index and spectral dependence for the polystyrene refractive index, obtained from experiments on large particles, we can estimate the shell width on fine particles.

For this purpose the nomograms of $\sigma_{\rm sc}$ dependence on the core size parameter x_1 for TLPs with several given ammonia layer thickness ($\Delta R = 0.5$; 1.0; 1.5, and 2.0 nm) and n'_1 =1.18 (0.01) 1.23 were calculated and constructed. Values of n'_1were determined by entering the experimental values of $\sigma_{\rm sc}$ on the nomogram and by the interpolating between curves corresponding to different n'_1 values. It is natural, that the same experimental value of $\sigma_{\rm sc}$ corresponds to different n'_1 on nomograms for different ΔR.

The resulting n'_1 values were plotted on diagrams such as those presented in Figure 10.3. Despite some dispersion of points, it is possible to draw the average curves. As is seen from the figure, the evaluation of n'_1 by nomograms for ΔR =1.5 and 2 nm has given

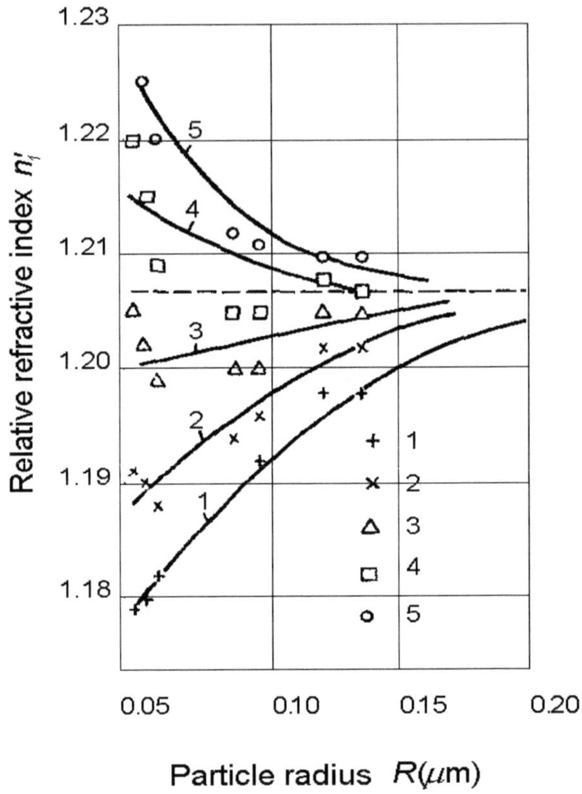

Figure 10.3. Core relative refractive index of TLPs retrieved from nomograms for ΔR=0 (*1*), 0.5 (*2*), 1 (*3*), 1.5 (*4*), and 2 nm (*5*) at $\lambda_0 = 0.54\ \mu m$.

the overestimated refractive index values, and by nomograms for $\Delta R = 0.$,0.5 and 1 nm, underestimated ones. The value $n_1' = 1.207$, to which all curves converge (fan-shaped) as particle size increases corresponds to an ammonia layer width $\Delta R \approx 1.2-1.3$ nm.

The value $\Delta R \approx 1.2$–1.3 nm is obtained as a result of the estimate of aggregate activity of the emulsifier film and ammonia layer. It appears to be practically independent of the particle diameter. Apparently, both fluid emulsifier and gaseous ammonia form almost monomolecular layers on its surface. Notice, that the rather close value of the emulsifier layer width $\Delta R = 1$ nm (irrespective of its concentration) is obtained by the Doppler spectroscopy method (Urgiris and Costaschuk, 1973).

The volume of the shell with width 1.2 nm accounts for about 7% of the core volume with $R_1 = 0.05\ \mu m$, approximately 4% of that with $R_1 = 0.0925\ \mu m$ and about 0.55% at $R_1 = 0.205\ \mu m$. It was shown above, that with core size increasing the influence of shell sharply decreases even when maintaining the constant ratio between volumes of the core and shell. If the core volume share decreases, then the enlightenment effect disappears even faster.

For our samples the influence of emulsifier and ammonia ceases to be detectable at $R = 0.235\ \mu m$. For other substances the minimal value of R (starting with it, it is possible to neglect the influence of surface films) can vary in between rather wide limits depending on the refractive index and layer width of the surface substance. If the cores and shell refractive indices are essentially distinct, the film formation can be confidently identified, and its width can be evaluated at the scattering coefficient measurements.

10.2 DISPERSION LIGHT FILTERS WITH TLPs

In Charters 3 and 4 it was indicated, that the smaller the non-homogeneous particle, the better its scattering properties can be described by the model of HPs, whose refractive index is equal to that averaged over the volume of the corresponding non-homogeneous particle. Such an approximate model allows us, in some cases, to considerably simplify the solution of some scattering problems for non-homogeneous particles. In the previous section this representation was used to determine the conditions of minimal light scattering by polystyrene latex particles with gas shells suspended in water. Experiment (Prishivalko and Dudo, 1975) has confirmed "enlightenment" of such suspensions at the formation of the gaseous ammonia shell on latex particles.

Subsequently, some papers (Kerker *et al.*, 1975; Kerker, 1975, 1977; Chew and Kerker, 1976) have appeared, in which the opportunity of a sharp (by some orders of magnitude) decrease of the scattering (extinction) by small non-absorbing dielectric particles enclosed by an air shell, and embedded in non-absorbing medium is investigated. This idea was illustrated by the scattering (at λ=0.546 μm) calculation for the white pigment (titanium dioxide, $N_1 = 2.97 + i0$) particle with an air shell embedded in resin ($N_{\text{med}} = 1.51 + i0$). In this case the core relative refractive index of the formed TLP is equal to $n_1' = 1.97$, and the refractive index for the microcavity enveloping this core is $n' = 0.66$. At the size parameter $x_2 = 0.3$ and ratio $q = R_1/R_2 = \sqrt[3]{0.416}$, the scattering cross section decreases by almost 5–6 orders of magnitude in comparison with that for the particle of the same size consisting only of pigment or representing the microcavity. Thus, the microcavity with a strongly refractive particle of particular size, contained at its centre, becomes invisible. Under the appropriate relations between refractive indices and sizes of the particle core and shell the rather great particle with the size parameter $x_2 \approx 1.3 - 1.6$ can be almost invisible (scattering cross section at $q = \sqrt[3]{0.3}$ drops more than 100 times in comparison with the pigment HP and almost 20 times in comparison with the microcavity of the same size). The comparable effect is attained if we interchange cavity and pigment (i.e., a pigment particle with microhollow inside in the resin medium) only in this case for $x_2 = 0.3$ the ratio of core and shell radii should satisfy the condition $q = \sqrt[3]{0.782}$. This phenomenon was recommended for practical use in the paint industry.

The device was developed for monochromatization of IR radiation using TLPs (Prishivalko and Dudo, 1983). Two points form the basis for this device: (1) the enlightenment of the latex suspension detected experimentally (Dudo, 1973; Prishivalko and Dudo, 1975); and (2) the existence of the narrow intervals of size parameter x_2, within which the rather sharp decrease of scattering is observed (see Figure 10.2). These results are of special interest, because due to considerable expansion of investigations and practical use of the IR wavelength range the problem of monochromatization in this range has gained special urgency.

The dispersion filters consisting of connecting medium with particles of other substances embedded in it are widely applied to monochromatization of radiation in the IR spectral range in last several years (Borisevich *et al.*, 1971). If dispersion curves $n(\lambda)$ for the medium and the particle substance intersect at some wavelength λ_0, the radiation with this wavelength transmits through the filter without scattering. For radiation with wavelengths $\lambda < \lambda_0$ and $\lambda > \lambda_0$ there is some difference Δn between refractive indices of the medium and the particle. The radiation with such wavelengths is scattered by the filter more strongly, the greater this difference. As a result, the width of a filter transmission band the less, the less the intersection angle for the dispersion curves of the filter components. The dispersion filters of the crystal–crystal and crystal–air types are especially widespread. They are applied in astrophysical and space investigations, in pyrometry and other fields of science and technique. However, in developing the dispersion light filters separating the narrow spectral interval with a high enough transmission, there are fundamental difficulties. As the basic condition of dispersion light filter work is the crossing of dispersion curves for connecting medium and particles, embedded in it, under a great enough angle the restrictions are superimposed first of all by nature itself. It does not always happen that it is possible to select the filter components with an intersection of dispersion curves at the necessary wavelength, and if such intersection does occur, the angle of the mutual decline of these curves can be insufficient. Some substances cannot be utilized as filter components due to technological reasons.

The spectral shift of the maximum of the dispersion filter transmission band is achieved by: (1) replacement of one filter component by a substance with another run of the dispersion curve; (2) change of the spectral dependence of the refractive index of one filter component under the influence of the external factors; or (3) dilution of immersion liquid in the crystal–fluid type filter by another fluid. In some cases it is possible to receive the desired result, applying the so-called mixed crystals, whose refractive indices it is possible to change smoothly within limits of values for its components. However here again there are restrictions, defined by the known Goldshcmidt rule.

In the IR spectral range the intersection of dispersion curves usually takes place near to the absorption band of one filter component. The presence of even small absorption causes a shift of the filter transmission maximum and noticeable drop. For all these reasons, it is highly questionable to receive a set of filters with high transmission, whose narrow spectral bands, located enough close to each other, would fill a given spectral interval.

Further we shall show, that the possibilities of making dispersion light filters can be considerably enlarged, if we use TLPs, consisting of a spherical core and concentric shell. In doing so, in the filter transmission maximum the medium refractive index should have the intermediate value between that for core and shell. If the particles are small enough ($R_2 < 0.1\lambda_0$), their scattering properties are determined by the volume-averaged refractive index:

$$\bar{n} = n_2 + (n_1 - n_2)\,(R_1/R_2)^3 \tag{10.2.1}$$

For ease of calculations, on the first investigation phase we shall assume that absorption both in the particle core and shell is negligibly small.

The filter transmission maximum will take place under condition of equality of the particle volume-averaged refractive index to the refractive index of medium $\bar{n} = n_{\text{med}}$. At given n_1, n_2, and n_{med} we have the relation between the core and shell sizes satisfying the condition $\bar{n} = n_{\text{med}}$:

$$q^3 = (n_{\text{med}} - n_2)/(n_1 - n_2). \tag{10.2.2}$$

The shift of the maximum of the filter pass band is achieved by: (1) selection of the proper core and shell substances (rough shift of the transmission band); or (2) change of the ratio of the core and shell sizes (smooth shift). The change of maximum transmission can be reached both by change of particle concentration, and by some variation of the core and shell sizes at their invariable ratio.

From the above the important technological requirement follows: the particle cores should be monodisperse, and the shells should have identical thicknesses. Otherwise the transmission band is strongly broadened, and the maximum transmission will decrease.

The calculations which have been carried out with the equations of Section 3.1 have shown, that the condition (10.2.2) is approximate. It yields good results for small particles at slightly distinguishing n_1 and n_2. At deviation from these requirements the location of transmission maximum is determined from the condition $\bar{n} = n_{\text{med}}$ with some error.

Guttler (1952) obtained the expression for the light scattering efficiency for small TLPs as a power series in size parameter:

$$Q_{sc} = 6\,|p_1'|^2\, x_2^4 - 12\,|p_1'|^2 \cos\delta \cdot x_2^6 - 12\,|p_1'|^2\, \mathrm{Im}\, p_1' x_2^7 + O(x_2^8) \tag{10.2.3}$$

where

$$p_1' = \frac{1}{3}\frac{\xi_1 + 2\eta_1}{\xi_1 - \eta_1},\ \xi_1 = 20\left[\left(n_1^2 + 2n_2^2\right) - \left(n_1^2 - n_2^2\right) q^3\right]$$
$$\eta_1 = -10 n_2^2 \left[\left(n_1^2 + 2n_2^2\right) + 2\left(n_1^2 - n_2^2\right) q^3\right] \tag{10.2.4}$$

From Eq.(10.2.3) it is easily seen, that if $|p_1'|^2$ =0, then in the expansion the terms containing x_2^8 and higher powers of size parameter will remain. The series terms corresponding to the electrical and magnetic dipole scattering reduce to zero. There are non-zero terms corresponding to electrical quadrupole and higher orders of scattering. Then at $x_2 < 0.6$ scattering is practically absent.

From Eqs.(10.2.4) in a fashion similar to that in Kerker's book (1969, Section 5), we gain in explicit form the expression for $|p_1'|^2$ through refractive indices of core, shell, and medium as:

$$|p_1'|^2 = \frac{4}{9}\left|\frac{\left(n_2^2 - n_{\text{med}}^2\right)\left(n_1^2 + 2n_2^2\right) + q^3\left(2n_2^2 + n_{\text{med}}^2\right)\left(n_1^2 - n_2^2\right)}{\left(n_2^2 + 2n_{\text{med}}^2\right)\left(n_1^2 + 2n_2^2\right) + 2q^3\left(n_2^2 - n_{\text{med}}^2\right)\left(n_1^2 - n_2^2\right)}\right|^2 \tag{10.2.5}$$

It is clear, that $|p_1'|^2$ is converted to zero only at a particular relation between the sizes of the particle core and shell and three indicated refractive indices n_1, n_2, and n_{med}. Equating

(10.2.5) to zero, we obtain the following conditions:

$$q^3 = 1 - \frac{3n_2^2(n_{\mathrm{med}}^2 - n_1^2)}{(n_{\mathrm{med}}^2 + 2n_2^2)(n_2^2 - n_1^2)} \qquad n_{\mathrm{med}}^2 = n_2^2 \frac{(n_1^2 + 2n_2^2) + 2q^3(n_1^2 - n_2^2)}{(n_1^2 + 2n_2^2) - q^3(n_1^2 - n_2^2)} \tag{10.2.6}$$

At fulfillment of these conditions and $x_2 < 0.6$ the media containing particles is practically transparent. It is possible also to find the minimality condition for the series (10.2.3) term with x_2^8. However, it has no essential value, and for the practical choice of q, n_1, n_2, and n_{med} it is enough to use Eqs.(10.2.6). With their help for any set of R_1, R_2, n_1, and n_2 it is possible to find such n_{med}, and for given values n_1, n_2, and n_{med} – such a ratio $q = R_1/R_2$, that the scattering practically disappears. If the dispersion curves $n_1(\lambda), n_2(\lambda)$ and, $n_{\mathrm{med}}(\lambda)$ are known, the problem is reduced to definition of the ratio q, satisfying Eq.(10.2.6) at given wavelength λ_0 and corresponding to it $n_1(\lambda_0), n_1(\lambda_0)$, and $n_{\mathrm{med}}(\lambda_0)$. Naturally, the obtained q value can differ a little from the approximate one, found with Eq.(10.2.2). If the dispersion curve of the core substance $n_1(\lambda)$ intercrosses at some wavelength λ_0 the dispersion curve of medium $n_{\mathrm{med}}(\lambda)$, then the shift of the transmission maximum in the direction of greater wavelength is achieved through coating on these cores with the shell from a substance with a refractive index $n_2 > n_1$, and in the direction of smaller wavelength, with the shell with $n_2 < n_1$. The shift quantity is regulated by shell thickness.

The essential advantage, stipulated by application of TLPs, implies that with their help it is possible to receive the dispersion filter when there is no way to select a pair of substances with intersecting dispersion curves in the given spectral range. A necessary condition required is presence of a relative decline of the dispersion curve of one filter component in relation to that for medium refractive index.

Consider some concrete examples. In Figure 10.4 the spectral dependencies of the refractive index of substances frequently used as dispersion filter components in the IR spectral range are given: KBr, MgO, KRS-5, and LiF (Voronkova *et al.*, 1965; Borisevich *et al.*, 1971). As is known, at the molding of magnesium oxide particles with potassium bromide powder the filter is obtained with transmission maximum close to $\lambda_0 = 7\ \mu m$. For comparison the dependencies of effective refractive index $\bar{n}(\lambda)$ calculated with Eq.(10.2.1) for TLPs are shown. Notice at once, that in Figure 10.4 the results of model calculations for substances with well known dependencies $n(\lambda)$ are given. They only illustrate opportunities, opening up for the use of TLPs as components of dispersion filters. In so doing, the problem of an opportunity to obtain the actual TLPs from concrete substances, utilized in calculations was not considered.

In Figure 10.5*a* the spectral dependencies of the radiation extinction efficiency for the MgO HPs introduced in KBr medium and for the TLPs corresponding to effective dispersion curves in Figure 10.4 are given. In Figure 10.5*b* the similar dependencies for TLPs (core – MgO, shell – KRS-5) embedded in KBr medium are depicted, where the ratio of the core and shell sizes is constant ($R_1/R_2 = 0.96$), but the particle size vary. These dependencies correspond to the effective dispersion curve 4 in Figure 10.4. As it is seen, the value of the extinction efficiency in minimum essentially depends on the particle size. The location of minimum of $Q_{\mathrm{ext}}(\lambda)$ is slightly displaced with decreasing particle

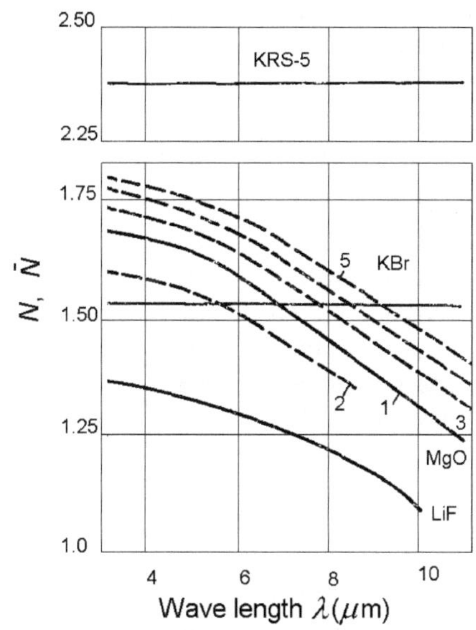

Figure 10.4. Dispersion curves of KBr, MgO (*1*), LiF, and KRS-5 and calculated with Eq.(10.2.1) the dependencies $n(\lambda)$ (*2–5*) for TLPs (core – MgO, shell – LiF (*2*) and KRS-5 (*3–5*)). The sizes of cores and shells here and in Figures 10.5, 10.6, and 10.8 are indicated in Table 10.3.

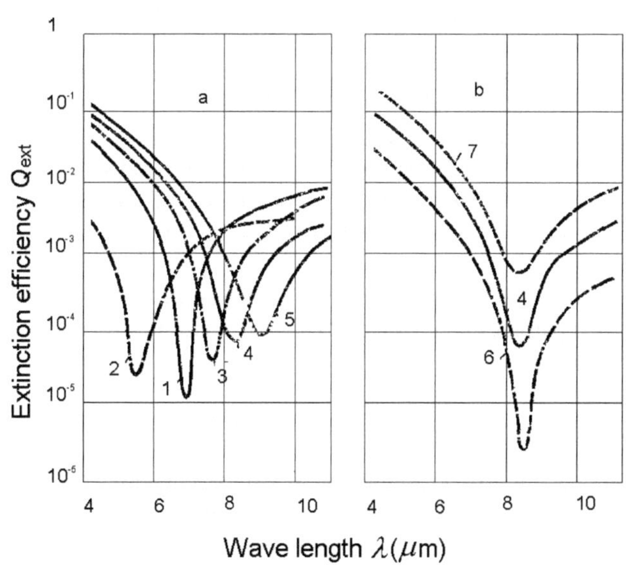

Figure 10.5. Spectral dependencies of the extinction efficiency for MgO HPs (*1*), TLPs (core – MgO, shell – LiF) (*2*), and KRS-5 particles (*3–7*)). All particles are embedded in KBr medium. Notation of curves 2–5 is the same as in Figure 10.4.

Table 10.3. Values of R_1 and R_2 used in the calculations and the wavelength in the transmission maximum (HOP = hollow particles).

No. Figure	Particles	Core	Shell	Medium	No. curve	R_1	R_2	λ_0
						(μm)		
10.4	HPs	MgO	–	KBr	1	0.85	0.85	6.94
10.5	TLPs	MgO	LiF	KBr	2			5.5
10.6	TLPs	MgO	KRS-5	KBr	3	0.85	0.87	7.7
	TLPs	MgO	KRS-5	KBr	4	0.85	0.885	8.4
	TLPs	MgO	KRS-5	KBr	5	0.85	0.90	9.0
	TLPs	MgO	KRS-5	KBr	6	0.55	0.573	8.5
	TLPs	MgO	KRS-5	KBr	7	1.15	1.198	8.25
	TLPs	KRS-5	LiF	KBr	8,11	0.5	0.716	8.55
	TLPs	KRS-5	LiF	KBr	9,12	0.5	0.776	7.15
	TLPs	KRS-5	LiF	KBr	10	0.5	0.832	6.5
10.8	HOPs	air	KBr	LiF	1	0.5	0.656	6.1
	HOPs	air	KBr	LiF	2	0.5	0.596	8.1
	HOPs	air	KRS-5	MgO	3	0.5	0.600	6.2
	HOPs	air	KRS-5	MgO	4	0.5	0.572	7.85
	HOPs	air	KBr	MgO	5	0.5	0.968	8.2
	HPs	MgO	–	KBr	6	0.85	0.85	6.94

size, coming nearer to that predicted on the basis of Eq.(10.2.6).

Essential dependence of the extinction efficiency minimum value and of all curves $Q_{\text{ext}}(\lambda)$ on the particle size derives from the fact that for $x_2 \approx 0.2 - 1.0$ the Q_{ext} values lay on the uprising branch of the curve $Q_{\text{ext}}(x_2)$; therefore, in the given range of the particle size and wavelength of radiation being filtered it is very important to comply with the core monodispersity and with persistence of particle shell thickness. The manufacture of such particles runs into some problems, however, they rapidly decrease with advance in the far IR region, as the manufacture of larger TLPs is much easier. Just in the far IR the filters of the offered type should have the greatest advantages over the usual dispersion filters.

In Figure 10.6*a* the transmission spectral dependencies for filters manufactured from KBr with embedded particles are depicted. The transmission was estimated with the help of the relation:

$$T = \frac{I}{I_0} = \exp(-\pi R_2^2 k_0 C_N l)$$

where I and I_0 are the intensities of radiation, passed through the filter and incidenting on it, C_N is the particle numerical concentration, and l is the layer thickness. Comparison of the transmission of different filters was made at constant $C_N l$. This constant value gives the transmission $T = 90\%$ for the filter from MgO HPs in KBr medium at wavelength $\lambda = 7\ \mu m$. It is evident, that the transmission of filters with TLPs compares favourably with transmission of filters with HPs. However, with increasing particle size the transmission rapidly impinges.

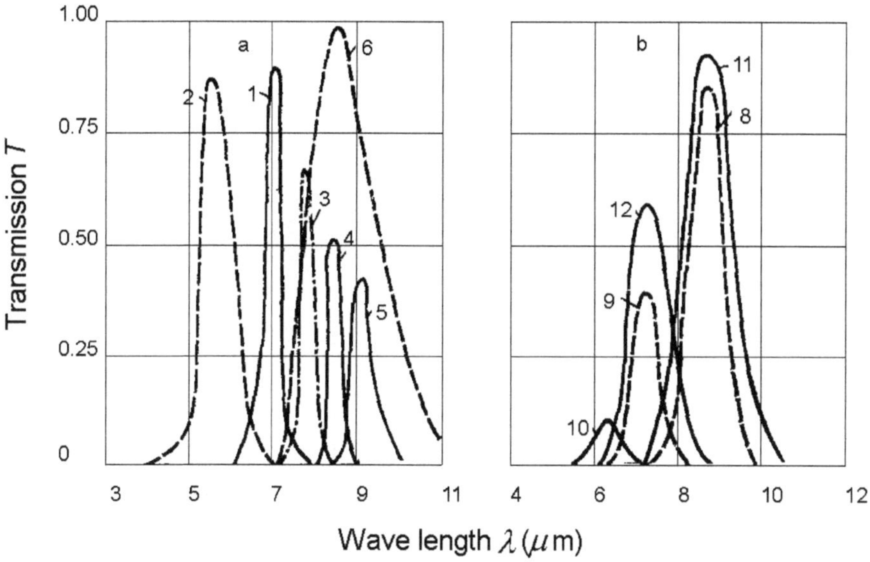

Figure 10.6. Spectral transmission dependencies of filters from KBr with different filling components: *a* – monodisperse particles, whose extinction efficiencies are shown in Figure 10.5 (curve notation is the same), *b* – monodisperse particles (core – KRS-5, shell – LiF).

Up to this point the cases considered were, when dispersion curves of core and media intersect, that is, a shell is only meant for a shift of the filter transmission maximum. In Figure 10.6*b*, the spectral transmission dependencies are given which were calculated in a similar manner for filters comprising of substances with non-intersecting dispersion curves in a given spectral range.

The dashed curves in Figure 10.6*b* refer to filters with particles of the above mentioned composition at the same value of $C_N l$. In this case for particles with $R_1 = 0.5\ \mu m$ and $= 0.832\ \mu m$ transmission at wavelength $\lambda = 6.5\ \mu m$ was less than 1%. In order that the maximum transmission grows to 10% for these R_1 and R_2 the calculations were made with reduced value of $C_N l$ (curve 10). This reduced value was also used in cases, corresponding to curves 11 and12.

One additional comment regarding the special feature of such filters is necessary. As the small particles have the phase function, close to a Rayleigh one, the receiver reception angle should be small enough. Otherwise, the filtering effect will be strongly attenuated by superimposition of radiation, scattered in directions distinct from zero.

If the intersection point of the effective dispersion curves is moved by applying the shells on particles, the intersection angle of these curves can decrease, that will give a broadening of the filter pass band (Figure 10.7). Therefore the substances with a steep run of the dispersion curve are preferable as a shell material.

The application of dispersion filters with TLPs allows us to considerably expand the spectral range of filtered radiation in the direction of far IR, for which the filters are not practically available at present. Having a set from three substances, from which at least

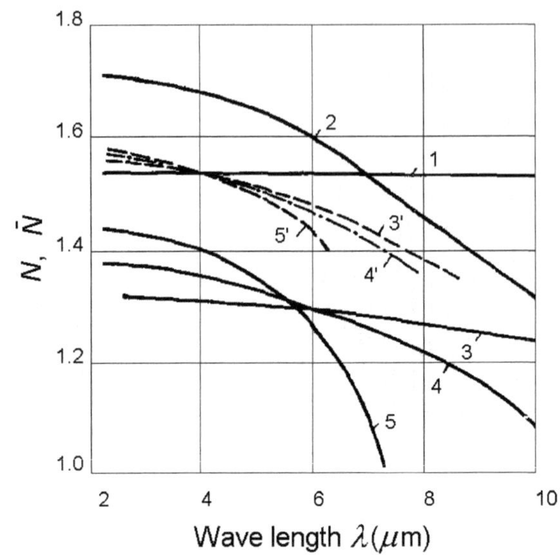

Figure 10.7. Dispersion curves for KBr(*1*), MgO(*2*), NaF(*3*), LiF(*4*), and fused quartz (*5*) and TLPs (core – MgO, shell – NaF(*3′*), LiF(*4′*), and SiO_2 (*5′*)).

one has an increasing or decreasing run of the dispersion curve (intersection of $n(\lambda)$ curves is by no means needed), it is possible, by varying the ratio q, to receive a number of filters for setting aside the transmission bands with any predetermined location of their maxima.

Similarly, it is possible to supplement the list of existing nowadays filters for the wavelength range $\lambda \approx 1 - 25\ \mu m$, having ensured monochromatization in those parts of the spectrum, where it is impossible to select pairs of substances with dispersion curves intersecting at the necessary wavelength.

The HOPs can be used as the dispersion filter component in some cases. Then we have the approximate relations:

$$n = n_2 - (n_2 - 1)\,q^3 = n_{\text{med}} \qquad q^3 = (n_2 - n_{\text{med}})/(n_2 - 1) \tag{10.2.1$'$}$$

and, accordingly, the more precise ones:

$$q^3 = 1 - \frac{3n_2^2(n_{\text{med}} - 1)}{(n_{\text{med}} + 2n_2^2)\,(n_2^2 - 1)} \qquad n_{\text{med}} = n_2^2\frac{(2n_2^2 n_1^2 + 1) - 2q^3\,(n_2^2 - 1)}{(2n_2^2 + 1) + q^3\,(n_2^2 - 1)} \tag{10.2.6$'$}$$

We shall indicate the opportunities represented by HOPs, on the example of the same substances (see Figure 10.4). In Figure 10.8*a* the spectral dependencies of the extinction efficiency for HOPs from different substances in different media are given. The spectral transmission dependencies for filters (Figure 10.8*b*) were calculated in the same manner as in the previous case. For comparability of the results the product of particle concentration

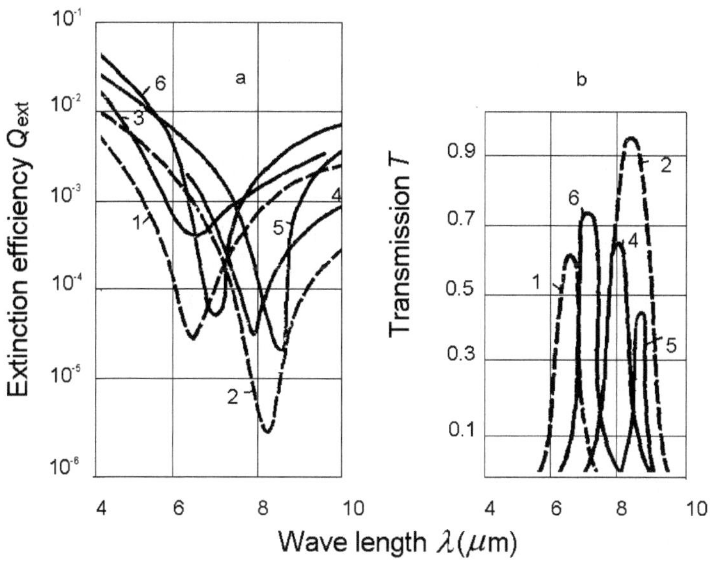

Figure 10.8. Spectral dependencies of the extinction efficiency (*a*) and transmission (*b*) for HOPs from KBr in LiF medium (*1*, *2*), from KRS-5 in MgO medium (*3*, *4*), from KBr in MgO medium (*5*) and for MgO HPs in KBr medium (*6*).

by the filter thickness $C_N l$ was identical in all cases and equal to its value, which for the filter from MgO HPs in KBr medium ensures $T = 75\%$ at a wavelength $\lambda = 6.94\ \mu m$.

Thus, having in hand only two substances, from which one has an increasing or decreasing run of the dispersion curve (intersection of curves $n(\lambda)$ is not required), it is possible, by small variations of the ratio of HOP internal radius to the exterior one, to obtain filters for selecting the transmission bands with practically any predetermined location of their maxima. It is only necessary, that for the wavelength corresponding to the filter transmission maximum, the medium refractive index was less than that of the HOP substance.

Consider now the influence of absorption on transmission of filters with the TLPs. As is marked by Borisevich *et al.* (1971), in usual dispersion filters the absorption is exhibited both as a drop of transmission maximum, and some shift of it aside from the absorption band. From Table 10.4 it is seen, that the quartz absorbs in this spectral range, and absorption increases at $\lambda \geq 5\ \mu m$ sharply. Here MgO slightly absorbs also. The core and shell sizes used in calculations, were selected (without the account of absorption) so that the transmission maximum ($\bar{n} = n_{\rm med}$) was at $\lambda_0 = 5\ \mu m$.

As suggested in Figure 10.9*a,* in response to absorption the long-wave values of $Q_{\rm ext}$ are much more than those in the short-wave region. This tendency is opposite to the one in Figures 10.5*a* and 10.8*a*. According to this, the right side of the transmission bands in Figure 10.9*b* (transmission in maximum is $T = 0.5$) are more steep. The transmission maximum is shifted to the left from $\lambda = 5\ \mu m$. As would be expected, the absorption influence is stronger, the smaller the particles. At the identical particle external size the half-width of the transmission band is less in the event that the core absorbs, instead of

Table 10.4. Optical constants of MgO, SiO_2, and KBr (Voronkova *et al.*, 1965).

	MgO		SiO_2		KBr
$\lambda(\mu m)$	n	κ	n	κ	n
3.0	1.69	–	1.42	8.2–6	1.535
3.5	1.68	–	1.406	6.1–6	1.534
4.0	1.67	–	1.389	6.4–5	1.533
4.5	1.66	–	1.366	2.4–4	1.533
4.8	1.65	–	1.35	7.6–4	1.532
5.0	1.64	–	1.34	1.0–3	1.532
5.2	1.63	1.2–6	1.327	1.9–3	1.531
5.6	1.61	5.0–6	1.297	3.7–3	1.531
6.0	1.59	1.3–5	1.258	5.5–3	1.530
7.0	1.525	4.6–5	1.098	1.0–2	1.530

shell. The transmission band shift can be considerable, and it is necessary to take this into account at a selection of substances for dispersion filters. One of the advantages gained with the use of TLPs, lies in the fact that it is almost always possible to select the core and shell materials so that their absorption bands are not superimposed on the filter pass band. From this point of view the TLPs are more preferable than HOPs, as they represent more opportunities to combine the substances.

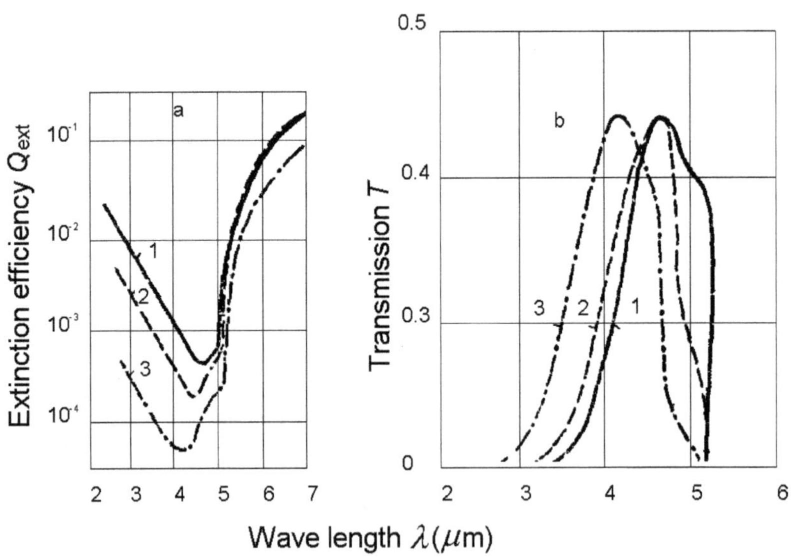

Figure 10.9. Spectral dependencies of extinction efficiency (*a*) and transmission (*b*) for TLPs with cores from MgO and shells from SiO_2 (*1*, *3*), and also with cores from SiO_2 and shells from MgO (*2*) in KBr medium: $R_1 = 0.4\ \mu m$ and $R_2 = 0.461\ \mu m$ (*1*), $R_1 = 0.327\ \mu m$ and $R_2 = 0.461\ \mu m$ (*2*), $R_1 = 0.174\ \mu m$ and $R_2 = 0.2\ \mu m$ (*3*).

The practical embodying of TLPs and HOPs becomes simpler with increasing size. Therefore filters with such particles are easier to make for the far IR spectral range. It is important to note here, that there is the well laid down optical procedure of microparticle sorting (Ashkin, 1970), and also the procedure to obtain and sort automatically (both on diameter, and on shell thickness) the HOPs with sizes of about tens of micrometers, used for the manufacture of spherical laser targets for experiments on the laser controllable thermonuclear fusion (Isakov *et al.*, 1976).

Summing up, it is necessary to underline once again, that use of the TLPs and HOPs in dispersion filters, despite some technological difficulties at their manufacture, has opened up fresh opportunities for: (1) smooth spectral shift of the transmission band; (2) regulation of its half-width at given λ_0 (through the change of decline of effective dispersion curves by applying the shells from different substances); and (3) selecting the transmission bands so as not to overlap with absorption bands of core or shell material. All these advantages can also be implemented in those spectral ranges, where the available natural or artificial materials have no intercrossing of dispersion curves. In the following section the modification of this new type of dispersion light filter will be described, for which the more simple manufacture technology for the filling component particles can be used.

10.3 QUASI-HOMOGENEITY OF SCATTERING MEDIA AND IR RADIATION MONOCHROMATIZATION

The integral characteristics of light scattering by small particles with a layered structure or with smoothly varying radial inhomogeneity, as was shown earlier, practically coincide with the corresponding characteristics of HPs of the same size with some effective values of optical constants. This statement is, apparently, valid and for small particles with an arbitrary character of internal inhomogeneity, of which the typical example are the particles of the submicron fraction of atmospheric aerosol. This is supported by the fact, that values of optical constants of dry aerosol substance obtained on the basis of a two-fractional model of aerosol particle composition (Hänel, 1968; Volz, 1972) and on the basis of the multi-fractional (synthetic) model of aerosol particles (Ivlev and Popova, 1973; Kondratiev *et al.*, 1973), are in satisfactory agreement. Used in calculations, they yield quite a good fit to experimental data (Jakobi *et al.*, 1952). This statement is also confirmed by the results of the calculations carried out on the basis of the HP model with some effective optical constants, which give rather good agreement with experimental data in some cases (Gorchakov and Emilenko, 1974).

Thus, notwithstanding that the particles of the atmospheric aerosol submicron fraction represent a mechanical mixture (agglomerate of different substances) in many optical experiments they exhibit themselves as optically homogeneous. It is possible to term this property of particles with composite internal structure as a quasi-homogeneity.

The concept of the optical quasi-homogeneity was used by Rozenberg (1958) in studies of optical properties of thin metal films and analysis of applicability to them of the Maxwell Garnett's theory of the 3D colloid. As is known, for descriptions of electrodynamic properties of medium with a great number of small inhomogeneities Maxwell

Garnett has entered the concept of the effective field inside medium and effective dielectric function. Its quantity with the account only of dipole scattering on micro-inhomogeneities is determined with the help of the Lorentz–Lorenz formula.

In works of Kuzmin and Prishivalko (1980) and Prishivalko *et al.* (1980) the optical properties of the scattering medium were studied with due regard for multipole scattering, and the conditions were found, at fulfilment of which it is possible to consider such medium as quasi-homogeneous one. From comparison of these quantities including in the equation for the average field in the medium and in the radiative transfer equation, obtained on the basis of the statistical–electrodynamic theory of multiple scattering (Babrabanenkov and Finkelberg, 1971; Barabanenkov, 1975; Kravtsov *et al.*, 1975; Rytov *et al.*, 1978), the criteria for quasi-homogeneity are found. It was shown that there is the principal opportunity to manufacture the quasi-homogeneous materials, suitable for obtaining the particles of filling component when producing the dispersion filters with a given location of the transmission band.

As was shown in Chapter 1, the property of the medium containing a great number of particles, can be described within the framework of statistical electrodynamics. For description of light propagation in such medium it is necessary to know at least the first and the second moments of the electromagnetic field. The more inhomogeneous the medium in the statistical sense, the greater the number of statistical moments required to describe the process of radiation propagation. The number of the required statistical moments decreases as the medium homogeneity increases. In the extreme case it is enough to know the first moment or the average field. The macroscopic Maxwell's equations yield an example of complete description of a statistically homogeneous medium.

Examination the transition from the statistically non-homogeneous medium to the homogeneous medium requires taking account of two aspects. At first, the statistical properties of the medium as a whole should be simple enough. This can be achieved by increasing the concentration of scattering particles. Secondly, even the small distinctions of medium properties from that for a statistically homogeneous medium can nevertheless result in great fluctuations of the average field. Therefore it is necessary to reduce the efficiency of scattering by a single particle, decreasing its size. The process of the particle concentration increasing with simultaneous decreasing of their sizes will eventually produce the completely homogeneous medium, except for the fluctuations at molecular level.

Let us trace the change of medium scattering properties stipulated by increasing concentration. Media obtaining at considerable increasing concentration, we shall term quasi-homogeneous. Consider, what will be the consequences of such the transition to quasi-homogeneous medium for optics.

Increasing of particle concentration first of all gives an infringement of conditions (1.2.16), that does not allow us to utilize in the transfer equation the characteristic of elementary volume, defined by Eqs.(1.2.33) and (1.2.41). The correctness of the transfer equation itself in this case was already partially considered in Chapter1 and more detailed consideration can be found in works of Rozenberg (1982; 1977). Estimates of the characteristics of elementary volume (see Eqs.(1.2.33) and(1.2.41)) at increasing concentration can be improved within the framework of the optical theorem (1.2.40). To ensure that the optical theorem was valid, we have rejected part of the terms in the expressions for the elements of the Stokes parameter transformation matrix, as at small concentrations they are small. Now we shall take into account these earlier rejected terms, and "renormalize" the extinction matrix to maintain the validity of the optical theorem. For this purpose we shall redefine the effective refractive index as follows:

$$N'_{\text{eff}} = N_0 + \frac{2\pi}{k^2 N_{\text{eff}}} A\left(\frac{N}{N_{\text{eff}}}, 0\right) C_N \tag{10.3.1}$$

Here N_0 is the refractive index of the medium in which the particles are suspended. Such replacement gives the relation:

$$\frac{4\pi}{k} \operatorname{Im} \frac{1}{N_{\text{eff}}} A\left(\frac{N}{N_{\text{eff}}}, 0\right) = \frac{1}{4\pi} \int Sp\left[A^2\left(\frac{N}{N_{\text{eff}}}, \Omega, \Omega'\right)\right] d\Omega\, d\Omega' \tag{10.3.2}$$

which in this case does express the optical theorem (1.2.40). The right part of Eq.(10.3.2) represents the scattering cross section per particle, which we shall denote as $\sigma_{\text{sc}}^{\text{eff}}$. From Eq.(10.3.2) it is seen, that its quantity is determined not only by the single particle properties, but also by the mutual influence of neighboring particles, that is by properties of medium as a whole. If the particle concentration in the medium is small,$\sigma_{\text{sc}}^{\text{eff}}$ is equal to the scattering cross section of the single particle taking place in the medium with a refractive index N_0. Increasing the number of scattering centres, the effective refractive index of the medium varies: at $C_N \to \infty, N_{\text{eff}} \to N$ and $\sigma_{\text{sc}}^{\text{eff}} \to 0$. Decreasing $\sigma_{\text{sc}}^{\text{eff}}$ testifies that the medium homogeneity degree grows. At $\sigma_{\text{sc}}^{\text{eff}} \to 0$ the scattering medium may be considered quasi-homogeneous and radiation propagation in it can be described, as in the case of continuous homogeneous medium.

As the quantitative characteristic of concentration influence on the degree of scattering medium homogeneity we shall enter the coefficient of inhomogeneity:

$$\gamma = \frac{\|I(\mathbf{r}, \Omega) - I_{\text{aver}}(\mathbf{r}, \Omega)\|}{\|I(\mathbf{r}, \Omega)\|} \tag{10.3.3}$$

where $I_{\text{aver}} = |\mathbf{E}_{\text{aver}}|^2$ is the average field intensity, and integration over all spatial angles is denoted by double direct brackets. The coefficient of inhomogeneity varies from 0 up to 1, and the less it differs from zero, the more homogeneous is medium.

For estimation of quantity γ write the scalar radiative transfer equation as the integral equation:

$$I(\tau, \Omega) = I_{\text{aver}} + e^{-\tau} \int_0^{\tau} e^{\tau'} \int_{(4\pi)} \frac{1}{K_{\text{ext}}} \Gamma\left(\frac{N}{N_{\text{eff}}}, \Omega, \Omega'\right) I(\tau', \Omega') d\Omega\, d\Omega' d\tau' \tag{10.3.4}$$

where $\tau = K_{\text{ext}} L$ (L is the light path length). Utilizing Eq.(10.3.4), for non-absorbing particles we have:

$$\gamma \leq \left\| \exp\left(-\tau_0 \frac{\sigma_{\text{sc}}^{\text{eff}}}{\sigma_{\text{sc}}}\right) - 1 \right\| \frac{\max I(\tau, \Omega)}{\|I(\tau, \Omega)\|} \qquad \tau_0 = \sigma_{\text{sc}} C_N L \tag{10.3.5}$$

Inasmuch as the ratio $\max I(\tau,\Omega)/\left\|I(\tau,\Omega)\right\|$ varies slightly with the concentration, its influence on the homogeneity degree of the scattering medium can be characterized by the ratio $\sigma_{\text{sc}}^{\text{eff}}/\sigma_{\text{sc}}$.

For model medium consisting of spherical MgO particles, embedded in fused quartz medium, the calculations of the ratio $\sigma_{\text{sc}}^{\text{eff}}/\sigma_{\text{sc}}$ were performed at $\lambda = 4.8\ \mu m$ under Eqs.(10.3.1) and (10.3.2) with the use of Mie theory. The refractive and absorption indices of MgO and fused quartz for this wavelength are given in Table. 10.4. The size of MgO particles varied from 0.2 to 3 μm.

In Figure 10.10 the dependencies of $\sigma_{\text{sc}}^{\text{eff}}/\sigma_{\text{sc}}$ versus the volume concentration of MgO particles are depicted. As might be expected, the effective scattering cross section decreases in comparison with σ_{sc} when the volume concentration increases. Different steepness of curves $\sigma_{\text{sc}}^{\text{eff}}/\sigma_{\text{sc}} = f(C_V)$ testifies that the medium homogeneity degree depends on the particle size. The fewer particles, the smaller the concentration needed to decrease $\sigma_{\text{sc}}^{\text{eff}}$ by an identical factor. For particles with radius $R = 0.1\ \mu m$ $\sigma_{\text{sc}}^{\text{eff}}$ is an order of magnitude less than σ_{sc} at $C_V = 0.356$, and for particles with $R = 1.25\ \mu m$ – at $C_V = 0.456$. It follows from above, that under certain conditions ($\sigma_{\text{sc}}^{\text{eff}} \ll \sigma_{\text{sc}}$) the scattering medium with particles contained in it, comes close, with its optical properties, to an optically homogeneous medium with an effective refractive index, defined from Eq.(10.3.1).

Mixing powdery substances in different proportions and obeying the quasi-homogeneity conditions, it is possible to create a material with such run of dispersion curve, that none of the known substances has. The particles from such materials can be utilized as a filling component at the manufacture of dispersion filters with the necessary location of the tran-

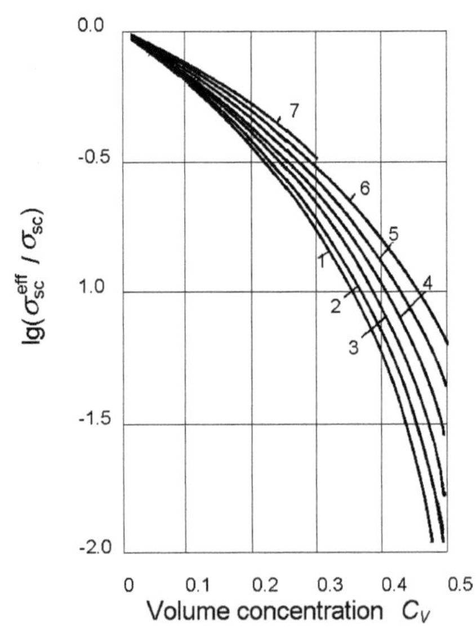

Figure 10.10. Ratio $\sigma_{\text{sc}}^{\text{eff}}/\sigma_{\text{sc}}$ versus the volume concentration of the MgO monodispersed particles with radius $R = 0.1(1), 0.25(2), 0.5(3), 0.75(4), 1(5), 1.25(6)$, and $1.5\ \mu m(7)$ in fused quartz.

smission band. Naturally, the technology of the quasi-homogeneous medium manufacture should ensure the optical contact between constituent components, and eliminate the presence of air, which in this case will manifest itself as a separate component.

For estimated calculations it is possible to take advantage of approximate relations (Heller, 1965), according to which the effective refractive index of a medium consisting of several components, is determined as:

$$\bar{n}(\lambda) = \frac{\sum\limits_i V_i n_i(\lambda)}{\sum\limits_i V_i} \tag{10.3.6}$$

where $n_i(\lambda)$ are the refractive indices of separate components and V_i are their volumes.

Transferring from volumetric relations to mass ones, we have:

$$\bar{n}(\lambda) = \frac{\sum\limits_i M_i n_i(\lambda)/\rho_i}{\sum\limits_i M_i/\rho_i} \tag{10.3.7}$$

where M_i, ρ_i are the mass and density of the i-component. If we use the relative mass concentrations of separate component $m_j = M_j / \sum\limits_i M_i$, then Eq.(10.3.7) transforms to:

$$\bar{n}(\lambda) = \frac{\sum\limits_i m_i n_i(\lambda)/\rho_i}{\sum\limits_i m_i/\rho_i} \tag{10.3.8}$$

The simplest and easiest adjustable case refers to particles from the mixture of two substances. To receive the filter with a maximal transmission at wavelength λ_0, it is necessary to select components of composite particles and connecting medium of the filter, so that the known condition was fulfilled:

$$\bar{n}(\lambda_0) = n_{med}(\lambda_0) \tag{10.3.9}$$

With the account of Eq.(10.3.7) it is easy to obtain a necessary relation of mass concentrations for components comprising particles:

$$\frac{m_1}{m_2} = \frac{n_{med}(\lambda_0) - n_2(\lambda_0)}{n_1(\lambda_0) - n_{med}(\lambda_0)} \cdot \frac{x_1}{x_2} \tag{10.3.10}$$

The opportunity to obtain the filters with quasi-homogeneous particles shall be considered using a concrete example. Filters of the crystal–crystal type (connecting medium – KCl, filling component – the quartz particles) have maximum of the transmission band at $\lambda = 4.16\ \mu m$. The filters with the same connecting medium and filling component from MgO particles have the transmission maximum at $\lambda = 8.15\ \mu m$ (see book of Borisevich *et al.*, 1971). In this book a great deal of substances ($CaCO_3$, Al_2O_3, $BaSO_4$, NaCl, KJ *etc.*) were required to manufacture the filters for the range 4.16–8.15 μm. When employing the quasi-homogeneous particles with the different content of two substances only (MgO and quartz) we can ensure monochromatization of radiation with practically any wavelength from the indicated spectral interval (Prishivalko *et al.*, 1981).

For manufacture of the filling component particles the powders of MgO and SiO_2 were mixed in the proportion 0.717:0.283 (the first sample) and 0.512:0.488 (the second sample). The corresponding volume concentration of MgO powder equals $C_{V_1} = 0.65$ and $C_{V_2} = 0.44$. The MgO particles were less than 0.5 μm, the quartz particles – no more than 3.5 μm. The estimated calculations with Eqs.(10.3.1) and (10.3.12) have shown, that it is possible to obtain the quasi-homogeneous material with $\sigma_{\rm sc}^{\rm eff} \ll \sigma_{\rm sc}$ from these particles.

The prepared powder mixtures after careful intermixing were heated in the vacuum furnace at a pressure of 1.33 Pa and temperature 2070 K to complete fusion of quartz. Then the melt was cooled to room temperature. The cooled samples were ground, the fraction with the particle size of 30–50 μm was selected. These particles were mixed in the proportion 1:10 with the KCl powder exposed to molding under high pressure (Prishivalko *et al.*, 1983).

In Figure 10.11 the transmission bands of filters obtained by the above method are reproduced (spectrograph was used to register these bands). Use of particles from the quasi-homogeneous material allowed us to create filters with transmission bands in the middle of the interval $4.16 \le \lambda \le 8.15$ μm (maximal transmission at $\lambda = 6.29$ μm (Figure 10.11.a) and at $\lambda = 5.84$ μm (Figure 10.11.b). The indicated filters are highly competitive with earlier known ones regarding the maximal transmission and the transmission band half-width. This is clearly confirmed by the data in Table 10.5.

The reported data confirms the practical opportunity to obtain the quasi-homogeneous media and use it for manufacture of dispersion filters. This opens up new ways for monochromatization of radiation in the IR spectral range.

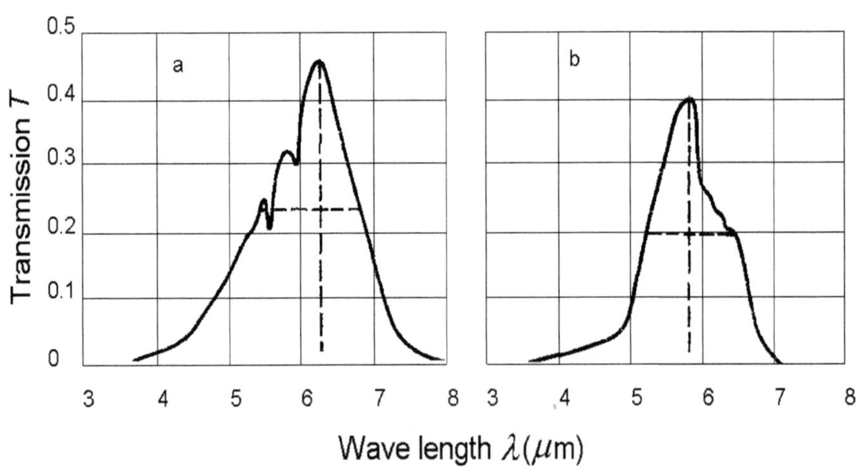

Figure 10.11. Spectral transmission of experimental samples of dispersion filters with quasi-homogeneous particles of fused quartz ($C_V^{\rm MgO} = 0.65(a), 0.44(b)$).

Table 10.5. Characteristics of dispersive crystal-crystal type filters (Borisevich *et al.*, 1971) and filters (denoted by *) with quasi-homogeneous particles (Prishivalko *et al.,* 1981): the wavelength at the maximum of transmission band, band halfwidth, and transmission in maximum .

Filter components	$\lambda_0(\mu m)$	$\Delta\lambda(\mu m)$	$T_{\max}(\%)$
SiO_2-KCl	4.16	1.22	40
$SrSO_4$-KCl	5.85	0.8	41
$BaSO_4$-NaCl	5.85	1.0	54
$(0.44MgO+0.56SiO_2)$-KCl*	5.88	1.24	41
$BaSO_4$-KCl	6.06	1.1	66
Al_2O_3-KBr	6.31	1.4	35
$(0.65MgO+0.35SiO_2)$-KCl*	6.29	1.4	46
Al_2O_3-NaCl	6.36	0.85	47
Al_2O_3-KCl	6.67	1.4	43
MgO-KCl	8.15	1.0	53

The estimated calculations show, that the dispersion filters with quasi-homogeneous particles yield wider transmission bands, than the filters with TLPs or HOPs, but technology of their manufacture is much easier, that is a rather essential advantage.

10.4 CHRISTIANSEN EFFECT IN DISPERSE MEDIA WITH APs

Let us assume, that the non-monochromatic light beam with almost plane phase front incidents on the system of APs having some preferred orientation of optic axes in space. On the strength of dependencies of the dielectric function of the medium, in which the APs are suspended, and components of the dielectric tensor of an AP itself on the radiation wavelength, the considerable decreasing of the light scattering efficiency can be observed in some spectral range at fulfillment of particular conditions. It would appear reasonable that it will happen in the wavelength range, where the values of the dielectric function of medium are close to values of the AP dielectric tensor components. Consider the characteristics of radiation scattering by a such system on some arbitrary, but fixed wavelength within a given interval.

Assume, that three conditions are fulfilled:

$$\operatorname{Im}\alpha^{(i)} = 0 \qquad i = 1,2,3 \tag{10.4.1}$$

$$\left|\frac{1}{3}Sp\left[\hat{\varepsilon}'(\lambda)\right] - \varepsilon_0(\lambda)\right| \ll 1 \tag{10.4.2.}$$

$$\left|\alpha^{(i)} - \alpha^{(j)}\right| \ll 1 \qquad i \neq j \tag{10.4.3}$$

where $\alpha^{(i)}$, as well as in Section 9.3 are the principal values of the tensor dielectric function $\hat{\varepsilon}$. It follow from the condition (10.4.1), that the case of non-absorbing APs will be considered only. The conditions (10.4.2) and (10.4.3), similar to conditions of "softness" for isotropic particles, are necessary to apply the Rayleigh–Gans approximation, or the Born approximation of the basic integral equation (1.1.4), within the framework of which further consideration will be carried out.

As zero approximation we shall choose the electric field of incident radiation $\mathbf{E}_I(\mathbf{r}) = \mathbf{E}_I e^{i\mathbf{k}_0\cdot\mathbf{r}}$. Then the scattered field intensity can be found from relation:

$$\mathbf{E}_{sc}(\mathbf{r}) = \int_V G^0(\mathbf{r},\mathbf{r}') k_0^2 \left[\hat{\varepsilon}'(\lambda) - \varepsilon_0(\lambda)\right] \mathbf{E}_I e^{i\mathbf{k}_0\cdot\mathbf{r}'} d^3\mathbf{r}' \tag{10.4.4}$$

where $\varepsilon_0(\lambda)$ is the dielectric function of the medium, in which the APs are suspended. Assume, that the medium does not absorb in the investigated spectral range. Let us enter the coordinate system, connected with the direction of incident radiation propagation, that is, the z-axis is along the wave vector $\mathbf{k}_0$, and the axes x and y are in the plane, perpendicular to z-axis. Due to transversality of the plane wave in non-absorbing medium the unit orths $\hat{\mathbf{e}}_x$ and $\hat{\mathbf{e}}_y$ completely describe the oscillations of the electrical intensity vector $\mathbf{E}_I$.

Generally speaking, the direction of axes x and y in the given plane can be selected arbitrarily. But, as is known, when the plane wave propagates in non-absorbing crystal there are only two chosen directions, along which the electrical vector can oscillate. Along these directions we direct the axes x and y. It is simple to show, that in this case the expression for $\hat{\varepsilon}'(\lambda)$ is transformed to:

$$\hat{\varepsilon}'(\lambda) = \begin{pmatrix} \varepsilon_1(\lambda) & 0 & \varepsilon_{xz}(\lambda) \\ 0 & \varepsilon_2(\lambda) & \varepsilon_{yz}(\lambda) \\ \varepsilon_{zx}(\lambda) & \varepsilon_{zy}(\lambda) & \varepsilon_{zz}(\lambda) \end{pmatrix} \tag{10.4.5}$$

Substituting Eq.(10.4.5) in Eq.(10.4.4), it is uneasy to receive the amplitude scattering matrix in the far-field region, which in the spherical coordinate system is noted as (arguments λ are omitted here for brevity):

$$\hat{A}(\theta,\varphi) = F(\theta) \begin{pmatrix} (\varepsilon_1-\varepsilon_0)\cos\theta\cos\varphi - \varepsilon_{zx}\sin\theta & (\varepsilon_2-\varepsilon_0)\cos\theta\sin\varphi - \varepsilon_{zy}\sin\theta \\ -(\varepsilon_1-\varepsilon_0)\sin\varphi & (\varepsilon_2-\varepsilon_0)\cos\varphi \end{pmatrix} \tag{10.4.6}$$

where $F(\theta)$ is the scattering form factor having the same form, as in the case of isotropic particles:

$$F(\theta) = k_0^2 \int_V \exp\left[i\left(\mathbf{k}_0' - \mathbf{k}_0\right)\cdot\mathbf{r}\right] d^3\mathbf{r} \tag{10.4.7}$$

Apparently, that amplitude forward scattering matrix, on which basis the extinction matrix is constructed, has the diagonal form with the following components:

$$A_{11}(0) = \left[\varepsilon_1(\lambda) - \varepsilon_0(\lambda)\right] F(0) \qquad A_{22}(0) = \left[\varepsilon_2(\lambda) - \varepsilon_0(\lambda)\right] F(0) \tag{10.4.8}$$

At real $\varepsilon_1(\lambda), \varepsilon_0(\lambda)$, and $\varepsilon_2(\lambda)$ in the Rayleigh–Gans approximation Eq.(10.4.8) yields only real part of the amplitude forward scattering matrix. The imaginary part is in line with the optical theorem (Newton, 1982):

$$\frac{4\pi}{k_0}\,\mathrm{Im}\,[A_{11}(0)] = \int\limits_{(4\pi)} d\Omega\;[A_{11}(\theta,\varphi)\,A_{11}^{*}(\theta,\varphi) + A_{12}(\theta,\varphi)\,A_{21}^{*}(\theta,\varphi)] \qquad (10.4.9)$$

$$\frac{4\pi}{k_0}\,\mathrm{Im}\,[A_{22}(0)] = \int\limits_{(4\pi)} d\Omega\;[A_{22}(\theta,\varphi)\,A_{22}^{*}(\theta,\varphi) + A_{21}(\theta,\varphi)\,A_{12}^{*}(\theta,\varphi)]$$

Having substituted in Eq.(10.4.9) the amplitude matrix components from Eq. (10.4.6), we shall receive imaginary parts of the required components:

$$\mathrm{Im}\,[A_{11}(0)] = \frac{k_0}{2}\int\limits_0^{\pi} d\theta\sin\theta\cdot F^2(\theta)\left\{[\varepsilon_1(\lambda)-\varepsilon_0(\lambda)]\cos^2\theta + \varepsilon_{zx}^2(\lambda)\sin^2\theta - \right.$$
$$\left. - [\varepsilon_1(\lambda)-\varepsilon_0(\lambda)]\,[\varepsilon_2(\lambda)-\varepsilon_0(\lambda)]\cos\theta\right\} \qquad (10.4.10)$$

$$\mathrm{Im}\,[A_{22}(0)] = \frac{k_0\,[\varepsilon_2(\lambda)-\varepsilon_0(\lambda)]}{2}\int\limits_0^{\pi} d\theta\sin\theta\cdot F^2(\theta)\left\{[\varepsilon_2(\lambda)-\varepsilon_0(\lambda)]\right.$$
$$\left. - [\varepsilon_1(\lambda)-\varepsilon_0(\lambda)]\cos\theta\right\}$$

For lack of the off-diagonal components in the forward scattering amplitude matrix the extinction matrix has the block-diagonal form (the arguments (0) are omitted here for brevity):

$$\alpha = \frac{2\pi C_N}{k_0} = \begin{pmatrix} \mathrm{Im}\,[A_{11}+A_{22}] & \mathrm{Im}\,[A_{11}-A_{22}] & 0 & 0 \\ \mathrm{Im}\,[A_{11}-A_{22}] & \mathrm{Im}\,[A_{11}+A_{22}] & 0 & 0 \\ 0 & 0 & \mathrm{Im}\,[A_{11}+A_{22}] & -\mathrm{Re}\,[A_{11}-A_{22}] \\ 0 & 0 & \mathrm{Re}\,[A_{11}-A_{22}] & \mathrm{Im}\,[A_{11}+A_{22}] \end{pmatrix} \qquad (10.4.11)$$

Such a form of the extinction matrix considerably simplifies the calculation of the coherent component of the scattered field. At asymptotic estimates of directly transmitted radiation the incoherent component can be neglected, as for it the effective optical trajectory is always more than for coherent one. Then, calculating the coherent component and going to asymptotes, we shall receive the characteristics of radiation after its transmission through the optically thick enough layer. As follows from Eq.(10.4.11) the system of four equations for the coherent component is divided into two systems of two equations:

$$\frac{dS_1}{dz} = -\alpha_{11}S_1 - \alpha_{12}S_2 \qquad \frac{dS_3}{dz} = -\alpha_{11}S_3 - \alpha_{43}S_4$$
$$\frac{dS_2}{dz} = -\alpha_{12}S_1 - \alpha_{11}S_2 \qquad \frac{dS_4}{dz} = -\alpha_{43}S_3 - \alpha_{11}S_4$$

The solutions of these two systems are easily found and look like:

$$S_{1,2} = \frac{1}{2}\left(S_{10} + S_{20}\right) e^{-(\alpha_{11}+\alpha_{12})z} \pm \frac{1}{2}\left(S_{10} - S_{20}\right) e^{-(\alpha_{11}-\alpha_{12})z}$$

$$S_3 = \frac{1}{2}\,\mathrm{Re}\left[\left(S_{30} + iS_{40}\right) e^{-(\alpha_{11}+i\alpha_{43})z} + \left(S_{30} - iS_{40}\right) e^{-(\alpha_{11}-i\alpha_{43})z}\right]$$

$$S_4 = \frac{1}{2}\,\mathrm{Re}\,\frac{1}{i}\left[\left(S_{30} + iS_{40}\right) e^{-(\alpha_{11}+i\alpha_{43})z} - \left(S_{30} - iS_{40}\right) e^{-(\alpha_{11}-i\alpha_{43})z}\right] \qquad (10.4.12)$$

where S_{10}, S_{20}, S_{30}, and S_{40} are the Stokes parameters of radiation incidenting at system of APs. With increasing of the optical thickness of a layer $L \longrightarrow \infty$, all Stokes parameters of emergent radiation turn to zero. The sole exception provided by one case deserves detailed consideration. It follows from Eq.(10.4.11):

$$\alpha_{11} + \alpha_{12} = \frac{4\pi C_N}{k_0}\,\mathrm{Im}\,\left[A_{11}(0)\right] \qquad \alpha_{11} - \alpha_{12} = \frac{4\pi C_N}{k_0}\,\mathrm{Im}\,\left[A_{22}(0)\right] \qquad (10.4.13)$$

It results from Eq.(10.4.10), that the second expression in Eq.(10.4.13) is converted to zero at fulfillment of equality $\varepsilon_2(\lambda) = \varepsilon_0(\lambda)$, that is, at intercrossing the dispersion curves corresponding to the refractive index of the ordinary wave in a particle and that of a medium. In this case, as follows from Eq.(10.4.12), the Stokes parameters for radiation, coming out of a scattering medium, looks like:

$$S_1 = \frac{1}{2}\left(S_{10} - S_{20}\right) \qquad S_2 = -\frac{1}{2}\left(S_{10} - S_{20}\right), \quad S_3 = S_4 = 0 \qquad (10.4.14)$$

It is seen from Eq.(10.4.14), that the incident radiation component, whose electric vector is perpendicular to the propagation direction and to one optical axis (ordinary wave), transits through the dispersion medium without scattering and the light becomes linearly polarized at output. Thus, the system of the oriented APs under the corresponding conditions operates as element polarizing and filtering radiation. Notice, that such a system is sensitive to a change of incident radiation direction, that is exhibited both in wavelength change of transmitted light, and in its polarization direction.

Consider one case that is simpler, than previous. Let the incident wave propagate along the direction of any AP dielectric axis. Then the component ε_{zx} which is included in Eq.(10.4.10), is converted to zero and there is an additional spectral interval of transmission by the system stipulated by equality $\varepsilon_1(\lambda) = \varepsilon_0(\lambda)$. The radiation in this spectral interval is linearly polarized, and polarization direction is perpendicular to polarization of the ordinary wave in crystal. This case is similar to the case of phase plates, where the directions of propagation for ordinary and extra-ordinary waves in crystal coincide. The indicated situation is simply interpreted on the basis of the dispersion curve diagrams (Figure 10.12). The points of dispersion curve intersection give the wavelength values λ_1 and λ_2, where the disperse system has a maximal transmission. If λ_1 and λ_2 are close enough to each other, then two transmission bands combine, forming, depending on quantity $(\lambda_1 - \lambda_2)$, the transmission curve, typical for bandpass filters, which are energetically more favorable, than the usual filters (Borisevich *et al.*, 1971), or, alternatively, the trans-

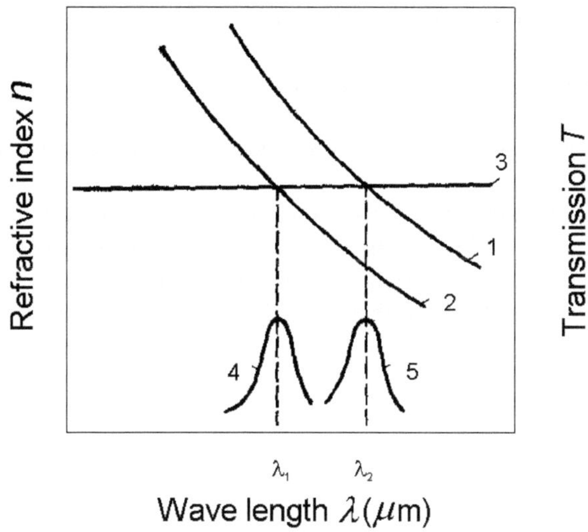

Figure 10.12. Schematic interpretation of the Christiansen effect in dispersion media comprising of APs: the dispersion curves for the ordinary wave (*1*), extra-ordinary wave (*2*), and medium, in which the particles are suspended (*3*); *4*, *5* – schematic transmission curves $T(\lambda)$ for the extra-ordinary and ordinary waves.

mission curve of the usual almost triangular form. If λ_1 and λ_2 are far from each other and form two transmission bands, then, it is possible, nevertheless, to receive at the output the transmission curve, typical for a bandpass filter by governing the radiation angular structure.

Apparently, in actual systems it is impossible to achieve the rigorous orientation of all APs in one direction. Some dispersion in AP orientation is always due to thermal oscillations. It results in "slurring over" the effects studied here in the pure state. As a consequence, the linear polarization degree of output radiation, and its spectral distribution, and so on, will be changed. In the extreme case, when the AP optical axes distribute in space by an equal-probability fashion, the usual Christiansen effect in isotropic dispersion media will be observed.

Here we survey a case of natural APs. Similar features will be observed for systems of artificial APs under the action of magnetic or electrical fields.

It is necessary to stress in closing, that the present work is not intended to cover exhaustively all of the numerous problems concerning scattering and absorption of light by non-homogeneous and anisotropic particles. The same is true of the quoted literature list. The aim of the book is to summarize some of the investigation stages and to give a focused presentation of major aspects of this important problem. The atmospheric aerosol was the basic object, used to study and illustrate the regularities obtained. But we argue that it is wise to further expand the range of objects under investigation, which can result in a refinement of the facts presented, and possibly, in the discovery of new essential ones.

An additional point to emphasize, once again, is that this book completely disregards the issue of particle non-sphericity, which is in some cases more important than particle inhomogeneity and/or anisotropy. The influence of particle non-sphericity on electromagnetic scattering has been very much investigated in recent years. Interested readers can find much information on this subject in the recently published books of Mishchenko *et al.* (2000*a*, 2002*a*).

It seems likely that the present day computers and the latest advances in light scattering theory will allow us to construct and calculate the models of scatterers taking into account simultaneously non-sphericity and inhomogeneity and/or anisotropy. But this is the theme of other books. We hope that they will be published in the not so remote future.

Appendix A

The Riccati–Bessel functions

A.1 THE DEFINITION OF BESSEL FUNCTIONS

The Bessel functions are the solutions of the Bessel differential equation:

$$u\frac{d}{du}(u\frac{dw}{du}) + (u^2 - \nu^2)w = 0 \qquad \text{(A.1)}$$

In general, the order ν and the independent variable u may be any complex number. This equation has the regular singular point at $u = 0$ the non-regular singular point at $u = \infty$. The independent solutions of Eq.(A.1) are:

$$J_\nu(u) = \sum_{m=0}^{\infty} \frac{(-1)^m (u/2)^{2m+\nu}}{m!\Gamma(m+\nu+1)} \qquad \text{(A.2)}$$

and $J_{-\nu}(u)$. The first solution $J_\nu(u)$ is referred to as the Bessel function of the first kind. The linear combinations:

$$\begin{aligned} Y_\nu(u) &= \frac{1}{\sin\nu\pi}\left[\cos\nu\pi J_\nu(u) - J_{-\nu}(u)\right] \qquad \nu \neq 0, \pm1, \pm2, \ldots \qquad \text{(A.3)} \\ H_\nu^{(1)}(u) &= J_\nu(u) + iY_\nu(u) = \frac{1}{i\sin\nu\pi}\left[J_{-\nu}(u) - J_\nu(u)e^{-i\nu\pi}\right] \\ H_\nu^{(2)}(u) &= J_\nu(u) - iY_\nu(u) = \frac{1}{i\sin\nu\pi}\left[J_\nu(u)e^{i\nu\pi} - J_{-\nu}(u)\right] \end{aligned}$$

are the solutions of Eq.(A.1) also. The function $Y_\nu(u)$ is referred to as the Bessel function of the second kind, or the Neuman function; the functions $H_\nu^{(1),(2)}(u)$ are the Bessel functions of the third kind, or the first and the second Hankel functions.

If ν is the integer then the right-hand parts of Eqs.(A.3) are indeterminate. But the limits of these parts at $\nu \to l$ (integer) exist (Bateman and Erdélyi, 1953*a*).

The Bessel functions may be expressed as some linear combinations of the elementary functions if and only if $\nu = l + 1/2$. The functions with half-integer order are common for the scattering problems with spherical symmetry. In this situation it is common to use the so-called named spherical Bessel functions:

$$z_l^{(1)}(u) = \sqrt{\frac{\pi}{2u}} J_{l+1/2}(u) \qquad z_l^{(2)}(u) = \sqrt{\frac{\pi}{2u}} Y_{l+1/2}(u) \qquad z_l^{(3)}(u) = \sqrt{\frac{\pi}{2u}} H_{l+1/2}^{(1,2)}(u) \tag{A.4}$$

or, which is more convenient, the Riccati–Bessel functions:

$$\begin{aligned} \psi_l(u) &= u z_l^{(1)}(u) = \sqrt{\frac{\pi u}{2}} J_{l+1/2}(u) \\ \chi_l(u) &= -u z_l^{(2)}(u) = -\sqrt{\frac{\pi u}{2}} Y_{l+1/2}(u) = (-1)^l \sqrt{\frac{\pi u}{2}} J_{-(l+1/2)}(u) \\ \xi_l(u) &= \sqrt{\frac{\pi u}{2}} H_{l+1/2}^{(1)}(u) \qquad \eta_l(u) = \sqrt{\frac{\pi u}{2}} H_{l+1/2}^{(2)}(u) \end{aligned} \tag{A.5}$$

It is evident, that:

$$\begin{aligned} \xi_l(u) &= \psi_l(u) - i\chi_l(u) \qquad \eta_l(u) = \psi_l(u) + i\chi_l(u) \\ \psi_l(u) &= \frac{1}{2}\left[\xi_l(u) + \eta_l(u)\right] = \frac{1}{2}\left[\xi_l(u) + \xi_l^*(u)\right] \\ \chi_l(u) &= \frac{1}{2i}\left[\eta_l(u) - \xi_l(u)\right] \end{aligned} \tag{A.6}$$

The Wronskians of the Riccati–Bessel functions are as follows:

$$\begin{aligned} \xi_l\psi_l' - \xi_l'\psi_l &= -i \qquad \eta_l\psi_l' - \eta_l'\psi_l = i \\ \xi_l\chi_l' - \xi_l'\chi_l &= 1 \qquad \eta_l\chi_l' - \chi_l'\psi_l = -1 \\ \psi_l\chi_l' - \psi_l'\chi_l &= -1 \qquad \xi_l\eta_l' - \xi_l'\eta_l = -2i \end{aligned} \tag{A.7}$$

Let us introduce notation for the log-derivatives:

$$D_l(u) = \frac{\psi_l'(u)}{\psi_l(u)} \qquad G_l^{(1)}(u) = \frac{\xi_l'(u)}{\xi_l(u)} \qquad G_l^{(2)}(u) = \frac{\eta_l'(u)}{\eta_l(u)} \tag{A.8}$$

A.2 RECURSIONS AND ASYMPTOTES

The standard recursions for the Bessel functions are:

$$\begin{aligned} \frac{2\nu}{u} J_\nu(u) &= J_{\nu-1}(u) + J_{\nu+1}(u) \qquad 2J_\nu'(u) = J_{\nu-1}(u) - J_{\nu+1}(u) \\ uJ_\nu'(u) + \nu J_\nu(u) &= uJ_{\nu-1}(u) \qquad uJ_\nu'(u) - \nu J_\nu(u) = -uJ_{\nu+1}(u) \end{aligned} \tag{A.9}$$

which for the spherical Bessel functions (argument u is omitted below) become:

$$\begin{aligned} z_l &= \frac{u}{2l+1}\left(z_{l-1}+z_{l+1}\right) \\ z_l' &= -\frac{l+1}{u}z_l + z_{l-1} = \frac{l}{u}z_l - z_{l+1} \\ &= \frac{1}{2}\left(-\frac{1}{u}z_l + z_{l-1} - z_{l+1}\right) \\ &= \frac{1}{2l+1}\left[lz_{l-1} - (l+1)z_{l+1}\right] \end{aligned} \tag{A.10}$$

and for the Riccati-Bessel functions become:

$$\begin{aligned} \psi_l &= \frac{u}{2l+1}\left(\psi_{l-1}+\psi_{l+1}\right) \\ \psi_l' &= -\frac{l}{u}\psi_l + \psi_{l-1} = \frac{l+1}{u}\psi_l - \psi_{l+1} \\ &= \frac{1}{2}\left(\frac{1}{u}\psi_l + \psi_{l-1} - \psi_{l+1}\right) \\ &= \frac{1}{2l+1}\left[(l+1)\,\psi_{l-1} - l\psi_{l+1}\right] \end{aligned} \tag{A.11}$$

with initial values:

$$\begin{aligned} \psi_0 &= \sin u \qquad \psi_1 = \frac{\sin u}{u} - \cos u \\ \psi_2 &= \left(\frac{3}{u^2}-1\right)\sin u - \frac{3}{u}\cos u \end{aligned} \tag{A.12}$$

The recursions for χ_l, ξ_l, and η_l are the same but the initial values are:

$$\begin{aligned} &\chi_0 = \cos u \quad \chi_1 = \sin u + \frac{\cos}{u} \qquad \chi_2 = \left(\frac{3}{u^2}-1\right)\cos u + \frac{3}{u}\sin u \\ &\begin{pmatrix}\xi_0\\ \eta_0\end{pmatrix} = \mp i\exp(\pm iu) \qquad \begin{pmatrix}\xi_1\\ \eta_1\end{pmatrix} = -\exp(\pm iu)\left(1\pm\frac{i}{u}\right) \\ &\begin{pmatrix}\xi_2\\ \eta_2\end{pmatrix} = -\exp(\pm iu)\left(1\pm\frac{3i}{u}-\frac{3}{u^2}\right) \end{aligned} \tag{A.13}$$

The recursion for the log-derivative of any Riccati–Bessel function is:

$$D_{l-1} = \frac{l}{u} - \left(D_l + \frac{l}{u}\right)^{-1} \tag{A.14}$$

Note some useful additional relations for the Riccati–Bessel functions:

$$\begin{aligned}
\psi_l'' &= \left[\frac{l(l+1)}{u^2} - 1\right]\psi_l \\
\psi_{l+1}' &= \frac{l+1}{u}\psi_{l-1} + \left[1 - \frac{(2l+1)(l+1)}{u^2}\right]\psi_l \\
&= \frac{u}{2l+1}\psi_{l-1} + \left(-\frac{l+1}{u} + \frac{u}{2l+1}\right)\psi_{l+1} \\
\psi_{l-1}' &= -\frac{l}{u}\psi_{l+1} - \left[1 - \frac{(2l+1)l}{u^2}\right]\psi_l \\
&= \frac{u}{2l+1}\psi_{l+1} + \left(\frac{l}{u} - \frac{u}{2l+1}\right)\psi_{l-1} = \frac{l}{u}\psi_l' + \left(\frac{l^2}{u^2} - 1\right)\psi_l \\
\psi_l' &= \frac{l}{u}\psi_{l-1}' + \left(1 - \frac{l^2}{u^2}\right)\psi_{l-1} \\
&= -\frac{(l+1)u}{u^2 - l(2l+1)}\psi_{l-1}' - \left[1 + \frac{l(l+1)}{u^2 - l(2l+1)}\right]\psi_{l+1}
\end{aligned} \tag{A.15}$$

$$\begin{aligned}
\psi_{l-1} &= \frac{1}{u^2 - (l+1)^2}\left\{lu\psi_{l+1}' + \left[u^2 - (l+1)(2l+1)\right]\right\}\psi_l' \\
&= \frac{u}{l}\psi_{l-1}'\left[1 + \frac{l+1}{(2l+1)}\frac{u^2}{l(l+1) - u^2}\right] - \frac{u^3}{2l+1}\frac{1}{l(l+1) - u^2}\psi_{l+1}' \\
\left[\frac{l(l+1)}{u^2} - 1\right]\psi_l &= \frac{1}{2l+1}\left[(l+1)\psi_{l-1}' - l\psi_{l+1}'\right] \\
\psi_l' &= \frac{u}{2l+1}\left\{\left[1 + \frac{l+1}{l(l+1) - u^2}\right]\psi_{l-1}' + \left[1 - \frac{l}{l(l+1) - u^2}\right]\psi_{l+1}'\right\} \\
1 &= \frac{l(l+2)\psi_{l-2} - (l-1)(l+1)\psi_{l+2}}{(l-1)(2l+3)\psi_{l+1}' + (l+2)(2l-1)\psi_{l-1}'} \\
\psi_l' &= \frac{u^2 - l(2l-1)}{u(2l-1)}\psi_l + \frac{u}{2l-1}\psi_{l-2}
\end{aligned} \tag{A.16}$$

$$\begin{aligned}
2(2l+1)u\psi_l\psi_l' &= (2l+1)\psi_l^2 + u^2\left(\psi_{l-1}^2 - \psi_{l+1}^2\right) \\
-\frac{2}{l+1}\frac{\psi_l\psi_l'}{u} + \frac{\psi_l^2}{u^2} + \frac{\left(\psi_l'\right)^2}{(l+1)^2} &= \frac{\psi_{l+1}^2}{(l+1)^2} \\
\frac{2}{l}\frac{\psi_l\psi_l'}{u} + \frac{\psi_l^2}{u^2} + \frac{\left(\psi_l'\right)^2}{l^2} &= \frac{\psi_{l-1}^2}{l^2}
\end{aligned} \tag{A.17}$$

$$\begin{aligned}
-\frac{\psi_l\psi_{l-2}'}{u} + (l-1)\frac{\psi_l\psi_{l-2}}{u^2} + \frac{l-1}{l}\frac{\psi_l'\psi_{l-2}}{u} - \frac{\psi_l'\psi_{l-2}'}{l} &= \frac{\psi_{l-1}^2}{l} \\
\frac{\psi_l\psi_{l+2}'}{u} + (l+2)\frac{\psi_l\psi_{l+2}}{u^2} - \frac{l+2}{l+1}\frac{\psi_l'\psi_{l+2}}{u} - \frac{\psi_l'\psi_{l+2}'}{l-1} &= \frac{\psi_{l+1}^2}{l+1}
\end{aligned} \tag{A.18}$$

$$\frac{3l+2}{(l+1)^2}\psi_{l+1}^2 + \frac{3l+1}{l^2}\psi_{l-1}^2 = (2l+1)\left[\frac{2}{l(l+1)}\frac{\psi_l\psi_l'}{u} + \frac{3\psi_l^2}{u^2} + \frac{3l^2+3l+1}{l^2(l+1)^2}\left(\psi_l'\right)^2\right]$$
$$\frac{l}{(l+1)^2}\psi_{l+1}^2 + \frac{l+1}{l^2}\psi_{l-1}^2 = (2l+1)\left[\frac{2}{l(l+1)}\frac{\psi_l\psi_l'}{u} + \frac{\psi_l^2}{u^2} + \frac{l^2+l+1}{l^2(l+1)^2}\left(\psi_l'\right)^2\right] \tag{A.19}$$

Note several asymptotic relations for the Riccati–Bessel functions. If order is much more than argument ($|u| \ll l$), then:

$$\psi_l(u) \sim \frac{u^{l+1}}{(2l+1)!!} \qquad \chi_l(u) \sim \frac{(2l-1)!!}{u^l} \tag{A.20}$$
$$\begin{pmatrix}\xi_l(u)\\ \eta_l(u)\end{pmatrix} \sim \mp i(2l-1)!!\exp\left(\pm iu\right)u^{-l}$$

If, on the contrary, $|u| \gg l$, then:

$$\begin{pmatrix}\psi_l(u)\\ \chi_l(u)\end{pmatrix} \sim \begin{matrix}\sin\\ \cos\end{matrix}(u - l\frac{\pi}{2}) \qquad \begin{pmatrix}\xi_l(u)\\ \eta_l(u)\end{pmatrix} \sim (\mp i)^{l+1}\exp\left(\pm iu\right) \tag{A.21}$$

A.3 QUADRATIC COMBINATIONS

Let us introduce the notation of the quadratic combinations of the Riccati–Bessel functions of the first kind ψ_l from arguments x and Nx:

$$\left(\psi_k\psi_{k_0}\right)_l \equiv \psi_l'(x)\psi_l(Nx) - N\psi_l(x)\psi_l'(Nx) \tag{A.22}$$
$$\left(\psi_{k_0}\psi_k\right)_l \equiv \psi_l(x)\psi_l'(Nx) - N\psi_l'(x)\psi_l(Nx)$$

where x is the size parameter and N is the complex refractive index. For other kinds of Riccati–Bessel functions one can introduce the analogous notation. It can be shown that:

$$\left(\psi_k\xi_{k_0}\right)_l\left(\xi_k\psi_{k_0}\right)_l - \left(\xi_k\xi_{k_0}\right)_l\left(\psi_k\psi_{k_0}\right)_l = N \tag{A.23}$$
$$\left(\psi_{k_0}\xi_k\right)_l\left(\xi_{k_0}\psi_k\right)_l - \left(\psi_{k_0}\psi_k\right)_l\left(\xi_{k_0}\xi_k\right)_l = N$$

Useful additional expressions are as follows:

$$\frac{d}{dx}\left(\psi_k\psi_{k_0}\right)_l = (N^2-1)\psi_l(x)\psi_l(Nx)$$
$$\left(\psi_k\psi_{k_0}\right)_{l+1} = -\left(\psi_{k_0}\psi_k\right)_l - \frac{l+1}{Nx}\frac{d}{dx}\left(\psi_k\psi_{k_0}\right)_l$$
$$\left(\psi_k\psi_{k_0}\right)_{l-1} = -\left(\psi_{k_0}\psi_k\right)_l + \frac{l}{Nx}\frac{d}{dx} \tag{A.24}$$
$$\left(\psi_k\psi_{k_0}\right)_{l-1} - \left(\psi_k\psi_{k_0}\right)_{l+1} = \frac{2l+1}{Nx}\frac{d}{dx}\left(\psi_k\psi_{k_0}\right)_l$$

$$
\begin{aligned}
\left(\psi_{k_0}\psi_k\right)_l + N\left(\psi_k\psi_{k_0}\right)_l &= -(N^2-1)\psi_l(x)\psi_l'(Nx)\\
N\left(\psi_{k_0}\psi_k\right)_l + \left(\psi_k\psi_{k_0}\right)_l &= -(N^2-1)\psi_l'(x)\psi_l(Nx)\\
l\left(\psi_k\psi_{k_0}\right)_{l+1} + (l+1)\left(\psi_k\psi_{k_0}\right)_{l-1} &= -(2l+1)\left(\psi_{k_0}\psi_k\right)_l
\end{aligned}
$$

The expressions (1.3.18–1.3.20) for the amplitude coefficients may be rewritten using the notation (A.22) in a concise form:

$$
\begin{aligned}
\alpha_l^{33} &= -\frac{\left(\psi_k\psi_{k_0}\right)_l}{\left(\psi_k\xi_{k_0}\right)_l} \qquad \beta_l^{33} = -\frac{\left(\psi_{k_0}\psi_k\right)_l}{\left(\xi_{k_0}\psi_k\right)_l}\\
\alpha_l^{11} &= -\frac{\left(\xi_k\xi_{k_0}\right)_l}{\left(\psi_k\xi_{k_0}\right)_l} \qquad \beta_l^{11} = -\frac{\left(\xi_{k_0}\xi_k\right)_l}{\left(\xi_{k_0}\psi_k\right)_l}\\
\alpha_l^{31} &= N\alpha_l^{13} = \frac{iN}{\left(\psi_k\xi_{k_0}\right)_l} \qquad \beta_l^{31} = N\beta_l^{13} = -\frac{iN}{\left(\xi_{k_0}\psi_k\right)_l}\\
\frac{l}{\alpha_{l+1}^{31}} &+ \frac{l+1}{\alpha_{l-1}^{31}} = (2l+1)\frac{1}{\beta_{l+1}^{31}}
\end{aligned}
\tag{A.25}
$$

Introducing notation:

$$
\begin{aligned}
\left(\psi_{k_1}\psi_{k_2}\right)_l &\equiv \psi_l(N_1x)\psi_l'(N_2x) - \frac{N_1}{N_2}\psi_l'(N_1x)\psi_l(N_2x)\\
\left(\psi_{k_2}\psi_{k_1}\right)_l &\equiv \psi_l'(N_1x)\psi_l(N_2x) - \frac{N_1}{N_2}\psi_l(N_1x)\psi_l'(N_2x)
\end{aligned}
\tag{A.26}
$$

one can obtain the following expressions:

$$
\begin{aligned}
\left(\psi_{k_1}\psi_{k_2}\right)_l\left(\xi_{k_1}\xi_{k_0}\right)_l - \left(\xi_{k_1}\psi_{k_2}\right)_l\left(\psi_{k_1}\xi_{k_0}\right)_l &= i\frac{N_1}{N_2}\left(\psi_{k_2}\xi_{k_0}\right)_l\\
\left(\psi_{k_2}\psi_{k_0}\right)_l\left(\xi_{k_1}\xi_{k_0}\right)_l - \left(\xi_{k_1}\psi_{k_0}\right)_l\left(\psi_{k_2}\xi_{k_0}\right)_l &= -iN_2\left(\xi_{k_1}\psi_{k_2}\right)_l\\
\frac{1}{iN_2}\left[\left(\psi_{k_1}\psi_{k_0}\right)_l\left(\psi_{k_2}\xi_{k_0}\right)_l - \left(\psi_{k_2}\psi_{k_0}\right)_l\left(\psi_{k_1}\xi_{k_0}\right)_l\right] &= \left(\psi_{k_1}\psi_{k_2}\right)_l
\end{aligned}
\tag{A.27}
$$

A.4 INDEFINITE INTEGRALS

We know three basic indefinite integrals over the quadratic combinations of the Riccati–Bessel functions. Note that in this subsection $\psi_l^{(i)}$ $(i = 1,2,3,4)$ denotes any kind of the Riccati–Bessel function (that is $\psi_l^{(1)} \equiv \psi_l$, $\psi_l^{(2)} \equiv \chi_l$, $\psi_l^{(3)} \equiv \xi_l$, $\psi_l^{(4)} \equiv \eta_l$). The *first* basic integral has the form:

$$
\begin{aligned}
&\int\left[(\alpha^2-\beta^2) - \frac{(n-l)(n+l+1)}{r^2}\right]\psi_n^{(i)}(\alpha r)\psi_l^{(j)}(\beta r)dr\\
&\quad = \beta\psi_n^{(i)}(\alpha r)\psi_{l-1}^{(j)}(\beta r) - \alpha\psi_{n-1}^{(i)}(\alpha r)\psi_l^{(j)}(\beta r) - \frac{n-l}{r}\psi_n^{(i)}(\alpha r)\psi_l^{(j)}(\beta r)
\end{aligned}
\tag{A.28}
$$

The special case of Eq.(A.28) at equal orders $n = l$:

$$\int \psi_l^{(i)}(\alpha r)\psi_l^{(j)}(\beta r)dr = \frac{1}{\alpha^2-\beta^2}\left[\alpha\psi_{l+1}^{(i)}(\alpha r)\psi_l^{(j)}(\beta r) - \beta\psi_l^{(i)}(\alpha r)\psi_{l+1}^{(j)}(\beta r)\right]$$
$$= \frac{1}{\alpha^2-\beta^2}\left[\beta\psi_l^{(i)}(\alpha r)\psi_{l-1}^{(j)}(\beta r) - \alpha\psi_{l-1}^{(i)}(\alpha r)\psi_l^{(j)}(\beta r)\right]$$
$$= \frac{1}{\alpha^2-\beta^2}\left[\beta\psi_l^{(i)}(\alpha r)\psi_l^{(j)\prime}(\beta r) - \alpha\psi_l^{(i)\prime}(\alpha r)\psi_l^{(j)}(\beta r)\right] \tag{A.29}$$

Using designation:

$$^{(1)}I_l^{ij}(\alpha,\beta) \equiv \frac{1}{\alpha\beta}\int \psi_l^{(i)}(\alpha r)\psi_l^{(j)}(\beta r)dr \tag{A.30}$$

we can write:

$$\begin{aligned}
{}^{(1)}I_l^{31}(k,k_0)\Big|_0^R &= -\,{}^{(1)}I_l^{31}(k,k_0)\Big|_R^\infty = \frac{(\xi_k\psi_{k_0})_l}{k_0^3N(N^2-1)}\\
{}^{(1)}I_l^{13}(k,k_0)\Big|_0^R &= -\,{}^{(1)}I_l^{13}(k,k_0)\Big|_R^\infty = \frac{(\psi_k\xi_{k_0})_l}{k_0^3N(N^2-1)}\\
{}^{(1)}I_l^{31}(k_1,k_2)\Big|_0^R &= -\,{}^{(1)}I_l^{31}(k_1,k_2)\Big|_R^\infty = \frac{(\xi_{k_1}\psi_{k_2})_l}{k_0^3N_1(N_1^2-N_2^2)}\\
{}^{(1)}I_l^{13}(k_1,k_2)\Big|_0^R &= -\,{}^{(1)}I_l^{13}(k_1,k_2)\Big|_R^\infty = \frac{(\psi_{k_1}\xi_{k_2})_l}{k_0^3N_1(N_1^2-N_2^2)}\\
{}^{(1)}I_l^{11}(k,k_0)\Big|_0^R &= -\,{}^{(1)}I_l^{11}(k,k_0)\Big|_R^\infty = \frac{(\psi_k\psi_{k_0})_l}{k_0^3N(N^2-1)}
\end{aligned} \tag{A.31}$$

$$\begin{aligned}
{}^{(1)}I_l^{33}(k,k_0)\Big|_0^R &= -\,{}^{(1)}I_l^{33}(k,k_0)\Big|_R^\infty = \frac{(\xi_k\xi_{k_0})_l}{k_0^3N(N^2-1)}\\
{}^{(1)}I_l^{11}(k_1,k_2)\Big|_0^R &= -\,{}^{(1)}I_l^{11}(k_1,k_2)\Big|_R^\infty = \frac{(\psi_{k_1}\psi_{k_2})_l}{k_0^3N_1(N_1^2-N_2^2)}\\
{}^{(1)}I_l^{33}(k_1,k_2)\Big|_0^R &= -\,{}^{(1)}I_l^{33}(k_1,k_2)\Big|_R^\infty = \frac{(\xi_{k_1}\xi_{k_2})_l}{k_0^3N_1(N_1^2-N_2^2)}
\end{aligned} \tag{A.32}$$

$$\begin{aligned}
\alpha_l^{11}\,{}^{(1)}I_l^{11}(k,k_0)\Big|_0^R + {}^{(1)}I_l^{31}(k,k_0)\Big|_0^R &= \frac{\alpha_l^{31}}{ik_0^3N(N^2-1)}\\
\alpha_l^{33}\,{}^{(1)}I_l^{33}(k,k_0)\Big|_a^\infty + {}^{(1)}I_l^{31}(k,k_0)\Big|_a^\infty &= -\frac{\alpha_l^{31}}{ik_0^3N(N^2-1)}\\
ik_0^3(N^2-1)\alpha_l^{31}\,{}^{(1)}I_l^{11}(k,k_0)\Big|_0^R &= \alpha_l^{33}
\end{aligned} \tag{A.33}$$

The special case of Eq.(A.29) at equal arguments and kinds ($\alpha = \beta,\ i = j$) is obtained using the limit $\alpha \to \beta$:

$$\int \left[\psi_l^{(i)}(\alpha r)\right]^2 dr = \frac{r}{2}\left\{\left[\psi_l^{(i)}(\alpha r)\right]^2 - \psi_{l-1}^{(i)}(\alpha r)\psi_{l+1}^{(i)}(\alpha r)\right\} \tag{A.34}$$

Clearly:

$$\int \left[\psi_{l+1}^{(i)}(x)\right]^2 dx = \int \left[\psi_l^{(i)}(x)\right]^2 dx - \frac{2}{x}\psi_l^{(i)}(x)\psi_{l+1}^{(i)}(x) \tag{A.35}$$

$$\int \left[\psi_{l-1}^{(i)}(x)\right]^2 dx = \int \left[\psi_l^{(i)}(x)\right]^2 dx + \frac{2}{x}\psi_l^{(i)}(x)\psi_{l-1}^{(i)}(x)$$

$$^{(1)}I_l^{11}(k,k) = \frac{r}{2k^2}\left[\psi_l^2(kr) - \psi_{l-1}(kr)\psi_{l+1}(kr)\right] \tag{A.36}$$

$$^{(1)}I_l^{33}(k,k) = \frac{r}{2k^2}\left[\xi_l^2(kr) - \xi_{l-1}(kr)\xi_{l+1}(kr)\right]$$

The special case of Eq.(A.29) at equal arguments but different kinds ($\alpha = \beta,\ i \neq j$) does not follows readily from this equation:

$$\begin{aligned}&\int \psi_l^{(i)}(\alpha r)\psi_l^{(j)}(\alpha r)dr \\ &\quad= \frac{r}{4}\left[\psi_l^{(i)}(\alpha r)\psi_l^{(j)}(\alpha r) - \psi_{l+1}^{(i)}(\alpha r)\psi_{l-1}^{(j)}(\alpha r) - \psi_{l-1}^{(i)}(\alpha r)\psi_{l+1}^{(j)}(\alpha r)\right]\end{aligned} \tag{A.37}$$

The special case of Eq.(A.28) at equal arguments $\alpha = \beta$:

$$\begin{aligned}&\int \frac{1}{r^2}\psi_n^{(i)}(\alpha r)\psi_l^{(j)}(\alpha r)dr \\ &\quad= \frac{\alpha}{(n-l)(n+l+1)}\left[\psi_n^{(i)\prime}(\alpha r)\psi_l^{(j)}(\alpha r) - \psi_n^{(i)}(\alpha r)\psi_l^{(j)\prime}(\alpha r)\right]\end{aligned} \tag{A.38}$$

The special case of Eq.(A.38) at equal orders ($n = l$):

$$\begin{aligned}&\int \frac{1}{r^2}\psi_l^{(i)}(\alpha r)\psi_l^{(j)}(\alpha r)dr \\ &\quad= \frac{1}{2lr}\psi_l^{(i)}(\alpha r)\psi_l^{(j)}(\alpha r) + \frac{\alpha}{2l}\left[\psi_{l+1}^{(i)}(\alpha r)\frac{\partial\psi_l^{(j)}(\alpha r)}{\partial l} - \psi_l^{(i)}(\alpha r)\frac{\partial\psi_{l+1}^{(j)}(\alpha r)}{\partial l}\right]\end{aligned} \tag{A.39}$$

The *second* basic indefinite integral is:

$$\begin{aligned}^{(2)}I_l^{ij}(\alpha,\beta) &\equiv \int \left[\psi_l^{(i)\prime}(\alpha r)\psi_l^{(j)\prime}(\beta r) + \frac{l(l+1)}{r^2\alpha\beta}\psi_l^{(i)}(\alpha r)\psi_l^{(j)}(\beta r)\right] dr \\ &= \frac{1}{\alpha^2-\beta^2}\left[\alpha\psi_l^{(i)}(\alpha r)\psi_l^{(j)\prime}(\beta r) - \beta\psi_l^{(i)\prime}(\alpha r)\psi_l^{(j)}(\beta r)\right]\end{aligned} \tag{A.40}$$

It can be found that:

$$
\begin{aligned}
-\,{}^{(2)}I_l^{31}(k,k_0)\Big|_0^R &= {}^{(2)}I_l^{31}(k,k_0)\Big|_R^\infty = -\frac{(\psi_{k_0}\xi_k)_l}{k_0(N^2-1)} \\
-\,{}^{(2)}I_l^{13}(k,k_0)\Big|_0^R &= {}^{(2)}I_l^{13}(k,k_0)\Big|_R^\infty = \frac{(\xi_{k_0}\psi_k)_l}{k_0(N^2-1)} \\
-\,{}^{(2)}I_l^{31}(k_1,k_2)\Big|_0^R &= {}^{(2)}I_l^{31}(k_1,k_2)\Big|_R^\infty = \frac{N_2\,(\psi_{k_2}\xi_{k_1})_l}{k_0N_1(N_1^2-N_2^2)} \\
-\,{}^{(2)}I_l^{13}(k_1,k_2)\Big|_0^R &= {}^{(2)}I_l^{13}(k_1,k_2)\Big|_R^\infty = \frac{N_2\,(\xi_{k_2}\psi_{k_1})_l}{k_0N_1(N_1^2-N_2^2)}
\end{aligned}
\tag{A.41}
$$

$$
\begin{aligned}
{}^{(2)}I_l^{31}(k,k_0)\Big|_0^R + \beta_l^{11}\,{}^{(2)}I_l^{11}(k,k_0)\Big|_0^R &= \frac{\beta_l^{31}}{ik_0(N^2-1)} \\
{}^{(2)}I_l^{31}(k,k_0)\Big|_0^R + \beta_l^{33}\,{}^{(2)}I_l^{33}(k,k_0)\Big|_0^R &= -\frac{\beta_l^{31}}{ik_0(N^2-1)}
\end{aligned}
\tag{A.42}
$$

$$
\begin{aligned}
-\,{}^{(2)}I_l^{11}(k,k_0)\Big|_0^R &= {}^{(2)}I_l^{11}(k,k_0)\Big|_R^\infty = \frac{(\psi_{k_0}\psi_k)_l}{k_0(N^2-1)} \\
-\,{}^{(2)}I_l^{33}(k,k_0)\Big|_0^R &= {}^{(2)}I_l^{11}(k,k_0)\Big|_R^\infty = \frac{(\xi_{k_0}\xi_k)_l}{k_0(N^2-1)} \\
-\,{}^{(2)}I_l^{11}(k_1,k_2)\Big|_0^R &= {}^{(2)}I_l^{11}(k_1,k_2)\Big|_R^\infty = \frac{N_2\,(\psi_{k_2}\psi_{k_1})_l}{k_0(N_1^2-N_2^2)} \\
-\,{}^{(2)}I_l^{33}(k_1,k_2)\Big|_0^R &= {}^{(2)}I_l^{33}(k_1,k_2)\Big|_R^\infty = \frac{N_2\,(\xi_{k_2}\xi_{k_1})_l}{k_0(N_1^2-N_2^2)} \\
ik_0\frac{N-1}{N}\beta_l^{31}\,{}^{(2)}I_l^{11}(k,k_0)\Big|_0^R &= \beta_l^{33}
\end{aligned}
\tag{A.43}
$$

The special case of Eq.(A.40) at equal arguments ($\alpha=\beta$):

$$
\begin{aligned}
{}^{(2)}I_l^{ij}(\alpha,\alpha) &\equiv \int\left[\psi_l^{(i)}{}'(\alpha r)\psi_l^{(j)}{}'(\alpha r) + \frac{l(l+1)}{r^2\alpha^2}\psi_l^{(i)}(\alpha r)\psi_l^{(j)}(\alpha r)\right]dr \\
&= \frac{1}{2\alpha}\left[\psi_l^{(i)}(\alpha r)\psi_l^{(j)}{}'(\alpha r) + \alpha\psi_l^{(i)}{}'(\alpha r)\psi_l^{(j)}{}'(\alpha r) - \alpha\psi_l^{(i)}{}''(\alpha r)\psi_l^{(j)}(\alpha r)\right]
\end{aligned}
\tag{A.44}
$$

The special case of Eq.(A.44) at equal kinds ($i = j$):

$$
\begin{aligned}
{}^{(2)}I_l^{ii}(\alpha,\alpha) &\equiv \int \left\{ \left[\psi_l^{(i)\,\prime}(\alpha r)\right]^2 + \frac{l(l+1)}{r^2\alpha^2}\left[\psi_l^{(i)}(\alpha r)\right]^2 \right\} dr \qquad \text{(A.45)} \\
&= \frac{1}{2\alpha}\left[\psi_l^{(i)}(\alpha r)\psi_l^{(i)\,\prime}(\alpha r) + \alpha\left(\psi_l^{(i)\,\prime}(\alpha r)\right)^2 - \alpha\psi_l^{(i)\,\prime\prime}(\alpha r)\psi_l^{(i)}(\alpha r)\right] \\
&= \left(\psi_l^{(i)}(\alpha r)\right)^2\left(\frac{r}{2} - \frac{l}{\alpha^2 r}\right) - \frac{r}{2}\psi_{l-1}^{(i)}(\alpha r)\psi_{l+1}^{(i)}(\alpha r) + \frac{1}{\alpha}\psi_l^{(i)}(\alpha r)\psi_{l-1}^{(i)}(\alpha r) \\
&= \frac{r}{2}\left(\psi_l^{(i)\,\prime}(\alpha r)\right)^2 + \frac{1}{2\alpha}\psi_l^{(i)}(\alpha r)\psi_l^{(i)\,\prime}(\alpha r) + \frac{r}{2}(\psi_l^{(i)}(\alpha r))^2\left[1 - \frac{l(l+1)}{\alpha^2 r^2}\right]
\end{aligned}
$$

The *third* basic integral has the form:

$$
\begin{aligned}
{}^{(3)}I_l^{ij}(\alpha,\beta) &\equiv \int \left[\frac{\psi_l^{(i)}(\alpha r)\psi_l^{(j)\,\prime}(\beta r)}{\alpha r} + \frac{\psi_l^{(i)\,\prime}(\alpha r)\psi_l^{(j)}(\beta r)}{\beta r} - \frac{\psi_l^{(i)}(\alpha r)\psi_l^{(j)}(\beta r)}{\alpha\beta r^2}\right] dr \\
&= \frac{1}{\alpha\beta r}\psi_l^{(i)}(\alpha r)\psi_l^{(j)}(\beta r). \qquad \text{(A.46)}
\end{aligned}
$$

Several additional integrals:

$$
\begin{aligned}
&\int \left[(l+1)\left(\psi_{l-1}^{(i)}(\alpha r)\right)^2 + l\left(\psi_{l+1}^{(i)}(\alpha r)\right)^2\right] dr \\
&= \frac{2l+1}{2}r\left[\left(\psi_{l-1}^{(i)}(\alpha r)\right)^2 - \frac{2l-1}{\alpha r}\psi_l^{(i)}(\alpha r)\psi_{l-1}^{(i)}(\alpha r) + \left(1 - \frac{2l}{\alpha^2 r^2}\right)\left(\psi_l^{(i)}(\alpha r)\right)^2\right] \qquad \text{(A.47)}
\end{aligned}
$$

$$
\int r^{l+1}\psi_l^{(i)}(\alpha r)dr = \frac{r^{l+1}}{\alpha}\psi_{l+1}^{(i)}(\alpha r) \qquad \text{(A.48)}
$$

$$
\int r^{-l}\psi_l^{(i)}(\alpha r)dr = -\frac{r^{-l}}{\alpha}\psi_{l-1}^{(i)}(\alpha r)
$$

$$
\int_0^x \left|\psi_l^{(i)}(\alpha r)\right|^2 dr = \frac{\mathrm{Im}\left[\alpha\psi_l^{(i)\,*}(\alpha x)\psi_{l+1}^{(i)}(\alpha x)\right]}{\mathrm{Im}(\alpha^2)} \qquad \text{(A.49)}
$$

Appendix B

Basic expressions on associated Legendre polynomials

B.1 DEFINITION OF THE ASSOCIATED LEGENDRE POLYNOMIALS AND THE ANGULAR FUNCTIONS

1. The associated Legendre polynomial $P_l^{(m)}(\cos\theta)$ $(m = 0, 1, 2, ..; l = m, m+1, ...)$ of degree l and order m with a real argument $\cos\theta$ is a solution of the differential equation finite at $\theta = 0$ and $\theta = \pi$:

$$\frac{d}{d\theta}\left[\sin\theta\frac{dP_l^{(m)}(\cos\theta)}{d\theta}\right] - m^2\frac{P_l^{(m)}(\cos\theta)}{\sin\theta} + l(l+1)\sin\theta\, P_l^{(m)}(\cos\theta) = 0 \quad \text{(B.1)}$$

This is the classical form of this equation. After the change of argument $\mu = \cos\theta$ it becomes:

$$\frac{d}{d\mu}\left[A^2\frac{dP_l^{(m)}(\mu)}{d\mu}\right] + \left[l(l+1) - \frac{m^2}{A^2}\right]P_l^{(m)}(\mu) = 0 \quad \text{(B.2)}$$

where $A = \sqrt{1-\mu^2}$. The other way to introduce the associated polynomials consists of using the ordinary Legendre polynomials:

$$P_l^{(0)}(\mu) = \frac{1}{2^l l!}\frac{d^l}{dx^l}A^{2l} \qquad |\mu| \leq 1 \quad \text{(B.3)}$$

as the generic functions:

$$P_l^{(m)}(\mu) = A^m\frac{d^m P_l^{(0)}(\mu)}{d\mu^m} \qquad m \leqslant l \quad \text{(B.4)}$$

We should be bear in mind that in many mathematical books Eq.(B.4) contains the additional factor $(-1)^m$.

2. In the light scattering theory it is much more convenience to work with the so-called angular functions $\pi_l^{(m)}(\mu)$ and $\tau_l^{(m)}(\mu)$, which are closely related with the associated Legendre polynomials:

$$\pi_l^{(m)}(\mu) = \frac{P_l^{(m)}(\cos\theta)}{\sin\theta} = \frac{P_l^{(m)}(\mu)}{A} \tag{B.5}$$

$$\tau_l^{(m)}(\mu) = \frac{dP_l^{(m)}(\cos\theta)}{d\theta} = -\frac{dP_l^{(m)}(\mu)}{d\mu}A \tag{B.6}$$

The functions $\pi_l^{(1)}$ and $\tau_l^{(1)}$ appear in Mie theory (upper index $^{(1)}$ is generally omitted), and functions with other m-values are used in scattering problems for more complex particles (e.g., spheroids. In terms of angular functions the differential equation (B.2) takes the form:

$$\frac{d\tau_l^{(m)}(\mu)}{d\mu} = l(l+1)\pi_l^{(m)}(\mu) + \frac{1}{A^2}\left[\mu\tau_l^{(m)}(\mu) - m^2\pi_l^{(m)}(\mu)\right] \tag{B.7}$$

It immediately follows from definitions (B.5) that:

$$\frac{d\pi_l^{(m)}(\mu)}{d\mu} = \frac{1}{A^2}\left[\mu\pi_l^{(m)}(\mu) - \tau_l^{(m)}(\mu)\right] \tag{B.8}$$

Notice that Eq.(B.4), where the associated Legendre polynomial is constructed by multiple differentiation of the generic function, may be reformulated in terms of angular functions:

$$\pi_l^{(m)}(\mu) = A^{m-1}\frac{d^m}{d\mu^m}\left[\frac{\pi_l^{(0)}(\mu)}{A}\right] \tag{B.9}$$

In particular:

$$\pi_l(\mu) = \frac{dP_l^{(0)}}{d\mu} \tag{B.10}$$

B.2 SYMMETRY OVER ARGUMENT AND VALUES FOR SOME PARTICULAR ARGUMENTS

1. It is easy to obtain the expression regarding the symmetry of argument:

$$\pi_l^{(m)}(-\mu) = (-1)^{l+m}\pi_l^{(m)}(\mu) \qquad \tau_l^{(m)}(-\mu) = (-1)^{l+m-1}\pi_l^{(m)}(\mu) \tag{B.11}$$

In particular, at $m = 1$:

$$\pi_l(-\mu) = (-1)^{l+1}\pi_l(\mu), \;\; \tau_l(-\mu) = (-1)^{l}\pi_l(\mu) \tag{B.12}$$

2. At boundary value $\mu = 1$:

$$\pi_l^{(m)}(1) = \tau_l^{(m)}(1) = \delta_{m1}\frac{l(l+1)}{2} \qquad m \neq 0 \tag{B.13}$$

Some difficulty arises here at $m = 0$. The functions $\pi_l^{(0)}(\pm 1) \to \infty$ due to factor A in the denominator. But for the normalized functions $\bar{\pi}_l^{(0)} = A\pi_l^{(0)}$:

$$\bar{\pi}_l^{(0)}(1) = 1, \quad \bar{\pi}_l^{(0)}(-1) = (-1)^l \tag{B.13a}$$

3. At boundary value $\mu = 0$:

$$\pi_l^{(m)}(0) = \frac{(l+m-1)!!}{(l-m)!!}(-1)^{(l-m)2}\delta_{1,(-1)^{l+m}} \tag{B.14}$$

$$\tau_l^{(m)}(0) = \frac{(l+m)!!}{(l-m-1)!!}(-1)^{(l-m+1)2}\delta_{1,(-1)^{l+m+1}} \tag{B.15}$$

4. Sometimes the following expression for functions with equal upper and lower indices may be useful:

$$\pi_l^{(l)}(\mu) = (2l-1)A\pi_{l-1}^{(l-1)}(\mu) = A^{l-1}(2l-1)!! \tag{B.16}$$

Furthermore:

$$\begin{aligned}
\pi_l^{(l-1)}(\mu) &= \mu A^{l-2}(2l-1)!! \\
\pi_l^{(l-2)}(\mu) &= A^{l-3}\left[(2l-1)\mu^2 - 1\right]\frac{(2l-3)!}{2!} \\
\pi_l^{(l-3)}(\mu) &= \mu A^{l-4}\left[(2l-1)\mu^2 - 3\right]\frac{(2l-3)!!}{3!} \\
\pi_l^{(l-4)}(\mu) &= A^{l-5}\left[(2l-1)(2l-3)\mu^4 - 6(2l-3)\mu^2 + 3\right]\frac{(2l-5)!!}{4!}
\end{aligned} \tag{B.17}$$

and so on.

5. Below are the explicit expressions for the first values of $\pi_l^{(m)}(\mu)$, where the notation $A = \sqrt{1-\mu^2}$ is used:

$$\begin{aligned}
&\pi_0^{(0)} = A^{-1} \quad \pi_1^{(0)} = \mu A^{-1} \quad \pi_2^{(0)} = (2A)^{-1}\left(3\mu^2 - 1\right) \\
&\pi_3^{(0)} = (2A)^{-1}\left(5\mu^3 - 3\mu\right) \quad \pi_4^{(0)} = (8A)^{-1}\left(35\mu^4 - 30\mu^2 + 3\right) \\
&\pi_5^{(0)} = (8A)^{-1}\left(63\mu^5 - 70\mu^3 + 15\mu\right) \\
&\pi_6^{(0)} = (16A)^{-1}\left(231\mu^6 - 315\mu^4 + 105\mu^2 - 5\right) \\
&\pi_7^{(0)} = (16A)^{-1}\left(429\mu^7 - 693\mu^5 + 315\mu^3 - 35\right) \ \ldots
\end{aligned}$$

$$
\begin{aligned}
\pi_1^{(1)} &= 1 \qquad \pi_2^{(1)} = 3\mu \qquad \pi_3^{(1)} = \frac{3}{2}\left(5\mu^2-1\right) \qquad \pi_4^{(1)} = \frac{5}{2}\mu\left(7\mu^2-3\right)\\
\pi_5^{(1)} &= \frac{15}{8}\left(21\mu^4-14\mu^2+1\right) \qquad \pi_6^{(1)} = \frac{21}{8}\mu\left(33\mu^4-30\mu^2+5\right)\\
\pi_7^{(1)} &= \frac{7}{16}\left(429\mu^6-495\mu^4+135\mu^2-5\right) \ldots\\
\pi_2^{(2)} &= 3A \qquad \pi_3^{(2)} = 15\mu A \qquad \pi_4^{(2)} = \frac{15}{2}A\left(7\mu^2-1\right)\\
\pi_5^{(2)} &= \frac{105}{2}\mu A\left(3\mu^2-1\right) \qquad \pi_6^{(2)} = \frac{105}{8}A\left(33\mu^4-18\mu^2+1\right)\\
\pi_7^{(2)} &= \frac{63}{8}\mu A\left(143\mu^4-110\mu^2+15\right) \ldots\\
\pi_3^{(3)} &= 15A^2 \qquad \pi_4^{(3)} = 105\mu A^2 \qquad \pi_5^{(3)} = \frac{105}{2}A^2\left(9\mu^2-1\right)\\
\pi_6^{(3)} &= \frac{315}{2}\mu A^2\left(11\mu^2-3\right) \qquad \pi_7^{(3)} = \frac{315}{8}A^2\left(143\mu^4-66\mu^2+3\right)\\
\pi_4^{(4)} &= 105A^3 \qquad \pi_5^{(4)} = 945\mu A^3 \qquad \pi_6^{(4)} = \frac{945}{2}A^3\left(11\mu^2-1\right) \qquad \text{(B.18)}\\
\pi_7^{(4)} &= \frac{3465\mu}{2}A^3\left(13\mu^2-3\right) \ldots\\
\pi_5^{(5)} &= 9!!A^4 \qquad \pi_6^{(5)} = 10395\mu A^4 \qquad \pi_7^{(5)} = \frac{10395}{2}A^4\left(13\mu^2-1\right) \ldots\\
\pi_6^{(6)} &= 11!!A^5 \quad \pi_7^{(6)} = 13!!\mu A^5 \ \ldots \ \pi_7^{(7)} = 13!!A^6 \ \ldots.
\end{aligned}
$$

The explicit expressions for the first values of $\tau_l^{(m)}(\mu)$:

$$
\begin{aligned}
\tau_0^{(0)} &= 0 \quad \tau_1^{(0)} = -A \quad \tau_2^{(0)} = -3\mu A \quad \tau_3^{(0)} = \frac{3}{2}A(1-\mu^2)\\
\tau_4^{(0)} &= \frac{5}{2}\mu A(3-7\mu^2) \quad \tau_5^{(0)} = -\frac{15}{8}A(21\mu^4-14\mu^2+1)\\
\tau_6^{(0)} &= -\frac{21}{8}\mu A(33\mu^4-30\mu^2+5)\\
\tau_7^{(0)} &= -\frac{7}{16}A(429\mu^6-495\mu^4+135\mu^2-5) \ldots\\
\tau_1^{(1)} &= \mu \quad \tau_2^{(1)} = 6\mu^2-3 \quad \tau_3^{(1)} = \frac{3}{2}\mu(15\mu^2-11)\\
\tau_4^{(1)} &= \frac{5}{2}(28\mu^4-27\mu^2+3) \quad \tau_5^{(1)} = \frac{15}{8}\mu(105\mu^4-126\mu^2+29)\\
\tau_6^{(1)} &= \frac{21}{8}(198\mu^6-285\mu^4+100\mu^2-5)\\
\tau_7^{(1)} &= \frac{7}{16}\mu(3003\mu^6-5049\mu^4+2385\mu^2-275) \ldots\\
\tau_2^{(2)} &= 6\mu A \quad \tau_3^{(2)} = 15A(3\mu^2-1) \quad \tau_4^{(2)} = 30\mu A(7\mu^2-4)\\
\tau_5^{(2)} &= \frac{105}{2}A(15\mu^4-12\mu^2+1) \quad \tau_6^{(2)} = \frac{105}{4}\mu A(99\mu^4-102\mu^2+19)
\end{aligned}
$$

$$
\begin{aligned}
\tau_7^{(2)} &= \frac{63}{8} A(1001\mu^6 - 1265\mu^4 + 375\mu^2 - 15) \ldots \\
\tau_3^{(3)} &= 45\mu A^2 \quad \tau_4^{(3)} = 105A^2(4\mu^2 - 1) \quad \tau_5^{(3)} = \frac{315}{2}\mu A^2(15\mu^2 - 7) \\
\tau_6^{(3)} &= \frac{945}{2} A^2(22\mu^4 - 15\mu^2 + 1) \quad \tau_7^{(3)} = \frac{315}{8}\mu A^2(1001\mu^4 - 902\mu^2 + 141) \ldots \\
\tau_4^{(4)} &= 420\mu A^3 \quad \tau_5^{(4)} = 945A^3(5\mu^2 - 1) \quad \tau_6^{(4)} = 945\mu A^3(33\mu^2 - 13) \qquad \text{(B.19)} \\
\tau_7^{(4)} &= \frac{3465}{2} A^3(91\mu^4 - 54\mu^2 + 3) \ldots \\
\tau_5^{(5)} &= 4725\mu A^4 \quad \tau_6^{(5)} = 10395A^4(6\mu^2 - 1) \quad \tau_7^{(5)} = \frac{10395}{2}\mu A^4(91\mu^2 - 31) \ldots \\
\tau_6^{(6)} &= 62376\mu A^5 \quad \tau_7^{(6)} = 13!!A^5(7\mu^2 - 1) \ldots \quad \tau_7^{(7)} = 7 \cdot 13!!\mu A^6 \ \ldots
\end{aligned}
$$

6. Sometimes we have to deal with the polynomials with negative order $P_l^{(-m)}(\cos\theta)$ $(m = 1, 2, \ldots)$. The expression:

$$
\pi_l^{(-m)}(\mu) = \frac{(l-m)!}{(l+m)!}\pi_l^{(m)}(\mu), \; m = 1, 2, 3, \ldots \quad l \geq |m| \qquad \text{(B.20)}
$$

is useful in this case.

7. One can see from Eqs.(B.18–19) that the functions $\pi_l^{(m)}(\mu)$ and $\tau_l^{(m)}(\mu)$ may be expressed as the finite power series in μ:

$$
\pi_l^{(m)}(\mu) = \gamma_1 \sum_{j=l_1}^{l_2} f_{lj}^{(m)} \mu^j \qquad \tau_l^{(m)}(\mu) = \gamma_2 \sum_{j=l_3}^{l_4} f_{lj}^{(m)} \mu^j \qquad \text{(B.21)}
$$

where the summation is over even indices at l_1(or l_3) $= 0$ and over odd indices at l_1(or l_3) $= 1$ and:

$$
\gamma_1 = \begin{cases} 1 & m \text{ odd} \\ A & m \text{ even and } 0 \end{cases} \qquad
\gamma_2 = \begin{cases} 1/A & m{=}0 \\ 1 & m \text{ odd} \\ A & m \text{ even} \end{cases} \qquad
l_2 = \begin{cases} l & m{=}0 \\ l-1 & m \text{ odd} \\ l-2 & m \text{ even} \end{cases}
$$

$$
l_4 = \begin{cases} l & m \text{ odd} \\ l-1 & m \text{ even and } 0 \end{cases} \qquad l_1 = \operatorname{mod}(l-m, 2) \qquad l_3 = 1 - \operatorname{mod}(l-m, 2)
$$

Other finite series for $\tau_l^{(m)}(\mu)$ takes the form:

$$
\pi_l^{(m)}(\mu) = \frac{(1+\mu)^{m/2}}{A} \sum_{k=m}^{l} (1-\mu)^{k-m/2} \frac{(l+k)!}{(l-k)!(k-m)!k!2^k} \qquad \text{(B.22)}
$$

which transforms to Eq.(B.16) at $l = m$. When $m = 1$, this series may be represented in a form suitable for calculation:

$$
\pi_l(\mu) = \frac{l(l+1)}{2} + \frac{l}{2}\sum_{k=2}^{l}(-1)^{k-1}\prod_{i=2}^{k}\frac{(l-i+1)(1-\mu)}{2(i-1)}\prod_{j=2}^{l}\frac{k+j}{j} \qquad \text{(B.23)}
$$

B.3 RECURSIONS

The general recursion over lower index l for $\pi_l^{(m)}$ is:

$$(2l+1)\mu\pi_l^{(m)} = (l-m+1)\pi_{l+1}^{(m)} + (l+m)\,\pi_{l-1}^{(m)} \tag{B.24}$$

Note, that the argument μ is omitted in the expressions below. When $m = 1$, the well known relation:

$$(2l+1)\mu\pi_l = l\pi_{l+1} + (l+1)\,\pi_{l-1} \tag{B.25}$$

is evident from Eq.(B.24). The general recursion over the upper index m for $\pi_l^{(m)}$ is:

$$\pi_l^{(m+2)} - 2(m+1)\frac{\mu}{A}\pi_l^{(m+1)} = -(l-m)(l+m+1)\pi_l^{(m)} \tag{B.26}$$

Recursions with simultaneous change of both indices:

$$\begin{aligned}
\pi_{l-1}^{(m)} - \mu\pi_l^{(m)} &= -(l-m+1)A\pi_l^{(m-1)} \\
\pi_{l+1}^{(m)} - \mu\pi_l^{(m)} &= (l+m)A\pi_l^{(m-1)} \\
\pi_{l-1}^{(m)} - \pi_{l+1}^{(m)} &= -(2l+1)A\pi_l^{(m-1)} \\
(2l+1)A\pi_l^{(m+1)} &= (l+m)(l+m+1)\pi_{l-1}^{(m)} - (l-m)(l-m+1)\pi_{l+1}^{(m)} \\
(l+m+1)\mu\pi_l^{(m)} - A\pi_l^{(m+1)} &= (l-m+1)\pi_{l+1}^{(m)} \\
(l-m)\mu\pi_l^{(m)} - (l+m)\pi_{l-1}^{(m)} &= -A\pi_l^{(m+1)}
\end{aligned} \tag{B.27}$$

Several additional expressions for $\pi_l^{(m)}$ functions are of some interest also:

$$\begin{aligned}
\pi_l^{(m)} &= \frac{A}{2m}\left[\pi_{l+1}^{(m+1)} - (l-m+1)(l-m+2)\pi_{l+1}^{(m-1)}\right] \\
\pi_{l+1}^{(m+1)} - \pi_{l-1}^{(m+1)} &= (l+m)(l+m-1)\pi_{l-1}^{(m-1)} \\
&\quad - (l-m+1)(l-m+2)\pi_{l+1}^{(m-1)} \\
(2l-1)(2l+1)(2l+3)\mu^2\pi_l^{(m)} &= \left[2l(l+1) - 2m^2 - 1\right](2l+1)\pi_l^{(m)} \\
&\quad + (2l-1)(l-m+1)(l-m+2)\pi_{l+2}^{(m)} \\
&\quad + (2l+3)(l+m)(l+m-1)\pi_{l-2}^{(m)}
\end{aligned} \tag{B.28}$$

$$
\begin{aligned}
(2l-1)(2l+1)(2l+3)A\mu\pi_l^{(m)} &= (2m+1)(2l+1)\pi_l^{(m+1)} \\
&+ (2l-1)(l-m+1)\pi_{l+2}^{(m+1)} \\
&- (2l+3)(l+m)\pi_{l-2}^{(m+1)} \\
(2l-1)(2l+1)(2l+3)A\mu\pi_l^{(m)} &= (2m-1)(2l+1)(l+m)(l-m+1)\pi_l^{(m-1)} \\
&- (2l-1)(l-m+1)(l-m+2)(l-m+3)\pi_{l+2}^{(m-1)} \\
&+ (2l+3)(l+m)(l+m-1)(l+m-2)\pi_{l-2}^{(m-1)} \\
(2l-1)(2l+1)(2l+3)A^2\pi_l^{(m)} &= -2(2l+1)\pi_l^{(m+2)} + (2l-1)\pi_{l+2}^{(m+2)} \\
&+ (2l+3)\pi_{l-2}^{(m+2)}
\end{aligned}
$$

$$
\begin{aligned}
&(2l-1)(2l+1)(2l+3)A^2\pi_l^{(m)} \\
&\quad = -2(2l+1)(l+m-1)(l+m)(l-m+1)(l-m+2)\pi_l^{(m-2)} \\
&\quad + (2l-1)(l-m+1)(l-m+2)(l-m+3)(l-m+4)\pi_{l+2}^{(m-2)} \\
&\quad + (2l+3)(l+m-3)(l+m-2)(l+m-1)(l+m)\pi_{l-2}^{(m-2)}
\end{aligned}
$$

The analogous recursions may be constructed for $\tau_l^{(m)}$ functions. But it is more suitable to obtain these functions through the known $\pi_l^{(m)}$ functions. Such expressions for constant m are:

$$
\begin{aligned}
\tau_l^{(m)} &= -\mu(l+1)\pi_l^{(m)} + (l-m+1)\pi_{l+1}^{(m)} \\
\tau_l^{(m)} &= l\mu\pi_l^{(m)} - (l+m)\pi_{l-1}^{(m)} \\
\tau_l^{(m)} &= \frac{1}{2l+1}[-(l+1)(l+m)\pi_{l-1}^{(m)} + l(l-m+1)\pi_{l+1}^{(m)}]
\end{aligned}
\tag{B.29}
$$

which transforms to the well known formulae at $m = 1$:

$$
\begin{aligned}
\tau_l &= -\mu(l+1)\pi_l + l\pi_{l+1} \\
\tau_l &= l\mu\pi_l - (l+1)\pi_{l-1} \\
\tau_l &= \frac{1}{2l+1}\left[l^2\pi_{l+1} - (l+1)^2\pi_{l-1}\right]
\end{aligned}
\tag{B.30}
$$

The expressions relating τ and π functions with constant l are as follows:

$$
\begin{aligned}
\tau_l^{(m)} &= -A\pi_l^{(m+1)} + m\mu\pi_l^{(m)} \\
\tau_l^{(m)} &= (l-m+1)(l+m)A\pi_l^{(m-1)} - m\mu\pi_l^{(m)} \\
\tau_l^{(m)} &= \frac{A}{2}[(l-m+1)(l+m)\pi_l^{(m-1)} - \pi_l^{(m+1)}]
\end{aligned}
\tag{B.31}
$$

Let us present for completeness some relations for $m = 1$ which follow immediately from the basic formulae:

$$
\begin{aligned}
\pi_{l-1} &= \frac{1}{l+1}\left[\mu(2l+1) - \frac{l(l+2)}{\mu(l+1)}\right]\pi_l - \frac{l}{\mu(l+1)^2}\tau_{l+1} \\
\tau_{l-1} &= \left[l-1-\frac{l(2l+1)}{(l+1)}\mu^2\right]\pi_l + \frac{l^2\mu}{(l+1)}\pi_{l+1} \\
\tau_l &= \frac{1}{2}[l\pi_{l+1} - \mu\pi_l - (l+1)\pi_{l-1}] \\
\tau_{l-1} &= \left[\frac{l^2-1}{\mu(2l+1)} - \mu l\right]\pi_{l-1} + \frac{l(l-1)}{\mu(2l+1)}\pi_{l+1} \\
\tau_{l+1} &= \left[\mu(l+1) - \frac{l(l+2)}{\mu(2l+1)}\right]\pi_{l+1} - \frac{(l+1)(l+2)}{\mu(2l+1)}\pi_{l-1} \\
\tau_{l-1} &= \left(-l\mu + \frac{l^2-1}{l\mu}\right)\pi_{l-1} + \frac{l-1}{l\mu}\tau_l \\
\tau_{l-1} &= \left(l-1-\frac{l^2}{l+1}\mu^2\right)\pi_l + \frac{l}{l+1}\mu\tau_l \\
\tau_{l-1} &= \left[\frac{l(l-1)}{(l+1)\mu} - \frac{l^3\mu}{(l+1)^2}\right]\pi_{l+1} + \left[\frac{l(2l+1)}{(l+1)^2}\mu - \frac{(l-1)}{(l+1)\mu}\right]\tau_l \\
\tau_{l+1} &= \left[\frac{(l+1)^3\mu}{l^2} - \frac{(l+1)(l+2)}{\mu l}\right]\pi_{l-1} \\
&\quad + \left[\frac{(l+1)(2l+1)}{l^2}\mu - \frac{(l-1)(l+2)}{l(l+1)\mu}\right]\tau_l, \\
\pi_l &= \frac{(l+1)^2\tau_{l-1} - l^2\tau_{l+1}}{(2l+1)\left[A^2 l(l+1) - 1\right]}, \\
\tau_l &= \frac{l(l^2A^2-1)^2\tau_{l+1} + (l+1)\left[-l(l+2)A^2 + \mu^2\right]\tau_{l-1}}{\mu(2l+1)\left[A^2 l(l+1) - 1\right]} \\
1 &= \frac{(l-1)^2(2l+3)\tau_{l+1} + (l+2)^2(2l-1)\tau_{l-1}}{(l+1)^2(l-1)^2\pi_{l+2} - l^2(l+2)^2\pi_{l-2}}
\end{aligned}
\tag{B.32}
$$

The expression of $\pi_l^{(m)}$ functions in terms of combinations of $\pi_l^{(0)}$ functions may be useful sometimes:

$$
\begin{aligned}
\pi_l^{(1)} &= \frac{l(l+1)}{(2l+1)A}\left[\pi_{l-1}^{(0)} - \pi_{l+1}^{(0)}\right] \\
\pi_l^{(2)} &= \frac{(l-1)l(l+1)(l+2)}{(2l+1)A^2}\left[\frac{\pi_{l-2}^{(0)}}{2l-1} - \frac{2(2l+1)}{(2l-1)(2l+3)}\pi_l^{(0)} + \frac{1}{2l+3}\pi_{l+2}^{(0)}\right]
\end{aligned}
$$

$$\pi_l^{(3)} = \frac{(l-2)....(l+3)}{(2l+1)A^3}\left[\frac{\pi_{l-3}^{(0)}}{(2l-3)(2l-1)} - \frac{3\pi_{l-1}^{(0)}}{(2l-3)(2l+3)}\right.$$
$$\left.+\frac{3\pi_{l+1}^{(0)}}{(2l-1)(2l+5)} - \frac{\pi_{l+3}^{(0)}}{(2l+3)(2l+5)}\right]$$
$$\pi_l^{(4)} = \frac{(l-3)....(l+4)}{(2l+1)A^4}\left[\frac{\pi_{l-4}^{(0)}}{(2l-5)(2l-3)(2l-1)} - \frac{4\pi_{l-2}^{(0)}}{(2l-5)(2l-1)(2l+3)}\right.$$
$$+\frac{6(2l+1)\pi_l^{(0)}}{(2l-3)(2l-1)(2l+3)(2l+5)} - \frac{4\pi_{l+2}^{(0)}}{(2l-1)(2l+3)(2l+7)}$$
$$\left.+\frac{\pi_{l+4}^{(0)}}{(2l+3)(2l+5)(2l+7)}\right] \tag{B.33}$$

B.4 INDEFINITE INTEGRALS FOR THE QUADRATIC COMBINATIONS OF THE ANGULAR FUNCTIONS

The list of the known closed analytical expressions for the indefinite integrals over the quadratic combinations of the angular functions, where the elements of combinations differ in the function type (π or τ) or in the value of the upper or lower index, includes three integrals (Bateman and Erdélyi, 1953*a*; Malushkov, 1974; Filippov, 1983).

It is apparent that the most general indefinite integral over a quadratic combination of angular functions having the closed analytical presentation is as follows:

$$\int\left[(l-k)(l+k+1)A^2 + (1-m^2)\right]\pi_l^{(m)}\pi_k^{(l)}d\mu = A^2\left[\pi_k^{(l)}\tau_l^{(m)} - \pi_l^{(m)}\tau_k^{(l)}\right] \tag{B.34}$$

Recall that $A = \sqrt{1-\mu^2}$. If set the equality of orders $m = l$ in Eq.(B.34) then:

$$\int A^2\pi_l^{(m)}\pi_k^{(m)}d\mu = \frac{A^2}{l(l+1)-k(k+1)}\left[\tau_l^{(m)}\pi_k^{(m)} - \pi_l^{(m)}\tau_k^{(m)}\right] \qquad l \neq k \tag{B.35}$$

If, on the contrary, $l = k$, then:

$$\int \pi_l^{(m)}\pi_l^{(k)}d\mu = \frac{A^2}{k^2-l^2}\left[\pi_l^{(k)}\tau_l^{(m)} - \pi_l^{(m)}\tau_l^{(k)}\right] \qquad k \neq m \tag{B.36}$$

There exists another indefinite integral having closed analytics:

$$\int A^{2(s-1)}\left[2(s-1)\mu\pi_l^{(m)}\pi_k^{(l)} + \pi_l^{(m)}\tau_k^{(l)} + \tau_l^{(m)}\pi_k^{(l)}\right]d\mu = -A^{2s}\pi_l^{(m)}\pi_k^{(l)} \tag{B.37}$$

If $s = 1$, then this integral is simplified to:

$$\int \left[\pi_j^{(m)}\tau_k^{(l)} + \tau_j^{(m)}\pi_k^{(l)}\right] d\mu = -A^2\pi_j^{(m)}\pi_k^{(l)} \tag{B.38}$$

The third indefinite integral is:

$$\begin{aligned}&\int \left[m^2\pi_l^{(m)}\pi_k^{(m)} + \tau_l^{(m)}\tau_k^{(m)}\right] d\mu \\ &\quad = \frac{A^2}{l(l+1)-k(k+1)}\left[k(k+1)\tau_l^{(m)}\pi_k^{(m)} - l(l+1)\pi_l^{(m)}\tau_k^{(m)}\right] \qquad l \neq k\end{aligned} \tag{B.39}$$

On the other hand, it immediately follows from Eq.(B.21) that the integrand of any indefinite integral of the types:

$$\int \pi_l^{(m)}\pi_k^{(m)}A^{2i}\mu^j d\mu \qquad \int \tau_l^{(m)}\tau_k^{(m)}A^{2i}\mu^j d\mu \qquad \int \pi_l^{(m)}\tau_k^{(m)}A^{2i}\mu^j d\mu$$

where i and j are the integers including zero, is the finite polynomial over powers of μ and, since, the corresponding integral is expressed in principle in the form of finite power series in μ. The sole exception is the integral $\int \tau_l^{(0)}\tau_k^{(0)}\mu^j d\mu$.

Equations (B.35) and (B.39) do not work when $l = k$. One can get out of this difficulty in the following way. Let us begin with notation:

$$\int \left[m^2\left(\pi_l^{(m)}\right)^2 + \left(\tau_l^{(m)}\right)^2\right] d\mu \equiv J_l^{(m)} \qquad \int A^2\left(\pi_l^{(m)}\right)^2 d\mu \equiv \frac{H_l^{(m)}}{(2l+1)} \tag{B.40}$$

It can readily be seen that:

$$J_l^{(m)} = \frac{l(l+1)}{2l+1}H_l^{(m)} - A^2\pi_l^{(m)}\tau_l^{(m)} \tag{B.41}$$

The following recursion over the lower index l is valid for $H_l^{(m)}$:

$$\begin{aligned}&(l+1-m)\,H_{l+1}^{(m)} = (l+m+1)\,H_l^{(m)} \\ &\quad + A^2\mu\left[(l+m+1)\left(\pi_l^{(m)}\right)^2 + (l-m+1)\left(\pi_{l+1}^{(m)}\right)^2\right] - 2(l+1)\pi_l^{(m)}\pi_{l+1}^{(m)}\end{aligned} \tag{B.42}$$

which stems from the known relation:

$$\begin{aligned}&(2l+3)(l-m+1)\left(P_{l+1}^{(m)}\right)^2 - (2l+1)(l+m+1)\left(P_l^{(m)}\right)^2 \\ &\quad = \frac{d}{d\mu}\left\{\mu\left[(l+m+1)\left(P_l^{(m)}\right)^2 + (l-m+1)\left(P_{l+1}^{(m)}\right)^2\right] - 2(l+1)P_l^{(m)}P_{l+1}^{(m)}\right\}\end{aligned}$$

The initial value of this recursion:

$$H_m^{(m)} = (2m+1)\int \left(\pi_l^{(m)}\right)^2 A^2 d\mu \tag{B.43}$$

may be calculated with help of Eq.(B.16):

$$H_m^{(m)} = (2m+1)\left[(2m-1)!!\right]^2 \int A^{2m} d\mu \equiv (2m+1)\left[(2m-1)!!\right]^2 I_m \tag{B.44}$$

It is simple to obtain the recursion for I_m:

$$I_m = \frac{1}{(2m+1)}\left[\mu\left(1-\mu^2\right)^m + 2mI_{m-1}\right] \tag{B.45}$$

whence the recursion for $H_m^{(m)}$ follows. If $m=1$ then immediately:

$$H_1^{(1)} = \mu(3-\mu^2) \tag{B.46}$$

B.5 DEFINITE INTEGRALS OVER THE QUADRATIC COMBINATIONS OF ANGULAR FUNCTIONS

B.5.1 Arbitrary order *m*

$$\begin{aligned}
&1. \int_{-1}^{1} \pi_j^{(m)} \pi_j^{(l)} d\mu = \delta_{ml} \frac{(j+m)!}{m(j-m)!} \qquad m \neq 0 \\
&2. \int_{-1}^{1} \pi_j^{(m)} \pi_l^{(m)} A^2 d\mu = \delta_{jl} \frac{2}{2j+1} \frac{(j+m)!}{(j-m)!} \\
&3. \int_{-1}^{1} \pi_j^{(m)} \pi_l^{(m)} \mu A^2 d\mu = \frac{2}{2j+1} \frac{(j+m)!}{(j-m)!} \left[\frac{j+m+1}{2j+3}\delta_{l,j+1} + \frac{j-m}{2j-1}\delta_{l,j-1}\right] \\
&4. \int_{-1}^{1} \tau_j^{(m)} \pi_l^{(m)} A^2 d\mu = \frac{2}{2j+1} \frac{(j+m)!}{(j-m)!} \left[\frac{j\,(j+m+1)}{2j+3}\delta_{l,j+1} - \frac{(j+1)\,(j-m)}{2j-1}\delta_{l,j-1}\right] \\
&5. \int_{-1}^{1} \left(\pi_j^{(m)} \tau_l^{(m)} + \pi_l^{(m)} \tau_j^{(m)}\right) d\mu = 0 \qquad m \neq 0 \\
&6. \int_{-1}^{1} \left(m^2 \pi_j^{(m)} \pi_l^{(m)} + \tau_l^{(m)} \tau_j^{(m)}\right) d\mu = \delta_{jl} \frac{2j(j+1)}{2j+1} \frac{(j+m)!}{(j-m)!} \\
&7. \int_{0}^{1} \left(\pi_j^{(m)}\right)^2 d\mu = \frac{1}{2m} \frac{(j+m)!}{(j-m)!}
\end{aligned} \tag{B.47}$$

B.5.2 Order *m=0*

$$\int_0^1 \pi_j^{(0)} \pi_l^{(0)} d\mu = \frac{\delta_{jl}}{2j+1} \qquad (j+l) \text{ even} \tag{B.48}$$

$$= \frac{(-1)^{(j+l-1)/2}}{l(l+1)-j(j+1)} \frac{(j-1)!!\, l!!}{j!!\, (l-1)!!} \qquad j \text{ even}, l \text{ odd}$$

B.5.3 Order *m=1*

$$1. \int_{0(-1)}^{1(0)} (\tau_j)^2 \, d\mu = \frac{j(j+1)(2j^2-1)}{2(2j+1)}$$

$$2. \int_{0(-1)}^{1(0)} \left[(\tau_j)^2 + (\pi_j)^2\right] d\mu = \frac{j^2(j+1)^2}{2j+1} \tag{B.49}$$

$$3. \int_{-1}^{1} \pi_j \pi_l \mu^2 A^2 d\mu = \frac{2j(j+1)}{2j+1} \left\{ \frac{1}{2j+1} \left[\frac{j(j+2)}{2j+3} + \frac{(j-1)(j+1)}{2j-1} \right] \delta_{lj} \right.$$
$$\left. + \frac{(j+2)\,(j+3)}{(2j+3)\,(2j+5)} \delta_{l,j+2} + \frac{(j-2)\,(j-1)}{(2j-3)\,(2j-1)} \delta_{l,j-2} \right\}$$

$$4. \int_{-1}^{1} \pi_j \pi_l \mu A^2 d\mu = \frac{2j(j+1)}{2j+1} \left(\frac{j+2}{2j+3} \delta_{l,j+1} + \frac{j-1}{2j-1} \delta_{l,j-1} \right)$$

$$5. \int_{-1}^{1} \pi_j \pi_l A^4 d\mu = \frac{2j(j+1)}{2j+1} \left\{ \frac{1}{2j+1} \left[\frac{(j+2)\,(j+3)}{2j+3} + \frac{(j-2)\,(j-1)}{2j-1} \right] \delta_{lj} \right.$$
$$\left. - \frac{(j+2)\,(j+3)}{(2j+3)\,(2j+5)} \delta_{l,j+2} - \frac{(j-2)\,(j-1)}{(2j-3)\,(2j-1)} \delta_{l,j-2} \right\}$$

$$6. \int_{-1}^{1} \tau_j \pi_l A^2 d\mu = \frac{2j(j+1)}{2j+1} \left[\frac{j\,(j+2)}{2j+3} \delta_{l,j+1} - \frac{(j-1)\,(j+1)}{2j-1} \delta_{l,j-1} \right]$$

$$7. \int_{-1}^{1} (\pi_j \tau_l + \tau_j \pi_l)\, A^2 d\mu = -\frac{4j(j+1)}{2j+1} \left(\frac{j+2}{2j+3} \delta_{l,j+1} + \frac{j-1}{2j-1} \delta_{l,j-1} \right)$$

$$8. \int_{-1}^{1} \tau_j \pi_l \mu A^2 d\mu = \frac{2j(j+1)}{2j+1} \left\{ \frac{1}{2j+1} \left[\frac{j^2(j+2)}{2j+3} - \frac{(j-1)(j+1)^2}{2j-1} \right] \delta_{lj} \right.$$
$$\left. + \frac{j\,(j+2)\,(j+3)}{(2j+3)\,(2j+5)} \delta_{l,j+2} - \frac{(j-2)\,(j-1)(j+1)}{(2j-3)\,(2j-1)} \delta_{l,j-2} \right\}$$

$$9. \int_{-1}^{1} \pi_j \tau_l \mu A^2 d\mu = \frac{2j(j+1)}{2j+1} \left\{ \frac{1}{2j+1} \left[\frac{j^2(j+2)}{2j+3} - \frac{(j-1)(j+1)^2}{2j-1} \right] \delta_{lj} \right.$$
$$\left. - \frac{(j+2)\,(j+3)^2}{(2j+3)\,(2j+5)} \delta_{l,j+2} + \frac{(j-2)^2\,(j-1)}{(2j-3)\,(2j-1)} \delta_{l,j-2} \right\}$$

$$10. \int_{-1}^{1} \tau_j \tau_l A^2 d\mu = \frac{2j(j+1)}{2j+1} \left\{ \frac{1}{2j+1} \left[\frac{j^3(j+2)}{2j+3} + \frac{(j-1)(j+1)^3}{2j-1} \right] \delta_{lj} \right.$$
$$\left. - \frac{j\,(j+2)\,(j+3)^2}{(2j+3)\,(2j+5)} \delta_{l,j+2} - \frac{(j-2)^2\,(j-1)(j+1)}{(2j-3)\,(2j-1)} \delta_{l,j-2} \right\}$$

The integrals 11-15 are the combinations of integrals 4 and 6:

$$11. \int_{-1}^{1} (\tau_j \pi_l - \mu \pi_j \pi_l)\, A^2 d\mu = \frac{2(j-1)j(j+1)(j+2)}{2j+1} \left(\frac{\delta_{l,j+1}}{2j+3} - \frac{\delta_{l,j-1}}{2j-1} \right)$$

$$12. \int_{-1}^{1} (3\tau_j \pi_l + \mu \pi_j \pi_l)\, A^2 d\mu$$
$$= \frac{2j(j+1)}{2j+1} \left[\frac{(j+2)(3j+1)}{2j+3} \delta_{l,j+1} - \frac{(j-1)(3j+2)}{2j-1} \delta_{l,j-1} \right]$$

$$13. \int_{-1}^{1} (\tau_j \pi_l + 3\mu \pi_j \pi_l)\, A^2 d\mu \qquad \text{(B.50)}$$
$$= \frac{2j(j+1)}{2j+1} \left[\frac{(j+2)(j+3)}{2j+3} \delta_{l,j+1} - \frac{(j-2)(j-1)}{2j-1} \delta_{l,j-1} \right]$$

$$14. \int_{-1}^{1} (\pi_j \tau_l + 3\mu \pi_j \pi_l)\, A^2 d\mu = \frac{2(j-1)j(j+1)(j+2)}{2j+1} \left(-\frac{\delta_{l,j+1}}{2j+3} + \frac{\delta_{l,j-1}}{2j-1} \right)$$

$$15. \int_{-1}^{1} (\pi_j \tau_l + \mu \pi_j \pi_l)\, A^2 d\mu = \frac{2j(j+1)}{2j+1} \left[-\frac{(j+1)(j+2)}{2j+3} \delta_{l,j+1} + \frac{(j-1)j}{2j-1} \delta_{l,j-1} \right]$$

$$16.\int_{-1}^{1}\left(\tau_j\tau_l+\pi_j\pi_l\right)\mu d\mu=\frac{2j(j+1)}{2j+1}\left[\frac{j(j+2)^2}{2j+3}\delta_{l,j+1}+\frac{(j-1)^2(j+1)}{2j-1}\delta_{l,j-1}\right]$$

$$17.\int_{-1}^{1}\left(\mu^2\tau_j\tau_l+\pi_j\pi_l\right)d\mu=$$

$$\frac{2j(j+1)}{2j+1}\left\{\frac{1}{2j+1}\left[j(j+1)(2j+1)-\frac{j^3(j+2)}{2j+3}-\frac{(j-1)(j+1)^3}{2j-1}\right]\delta_{lj}\right.$$
$$\left.+\frac{j\,(j+2)\,(j+3)^2}{(2j+3)\,(2j+5)}\delta_{l,j+2}+\frac{(j-2)^2\,(j-1)(j+1)}{(2j-3)\,(2j-1)}\delta_{l,j-2}\right\}$$

$$18.\int_{-1}^{1}\left(\tau_j\pi_l+\pi_j\tau_l\right)d\mu=\frac{2j(j+1)}{2j+1}\delta_{lj}$$

$$19.\int_{-1}^{1}\left(\tau_j\pi_l+\mu^2\pi_j\tau_l\right)d\mu=\frac{2j(j+1)}{2j+1}\left[\frac{(j+2)^2}{2j+3}\delta_{l,j+1}-\frac{(j-1)^2}{2j-1}\delta_{l,j-1}\right]$$

$$20.\int_{-1}^{1}\left(\tau_j\tau_l+\mu^2\pi_j\pi_l\right)d\mu=\frac{2j(j+1)}{2j+1}(j^2+j-1)\delta_{lj}$$

The integrals 21–23 are the combinations of integrals 4 and 6:

$$21.\int_{-1}^{1}\left(\tau_j\tau_l-\mu\tau_j\pi_l-\mu\pi_j\tau_l+\mu^2\pi_j\pi_l\right)d\mu=\frac{2(j-1)j(j+1)(j+2)}{2j+1}\delta_{lj}$$

$$22.\int_{-1}^{1}\left(3\tau_j\tau_l+\mu\tau_j\pi_l+\mu\pi_j\tau_l+3\mu^2\pi_j\pi_l\right)d\mu=\frac{2j(j+1)(3j^2+3j-2)}{2j+1}\delta_{lj}$$

$$23.\int_{-1}^{1}\left(\tau_j\tau_l+3\mu\tau_j\pi_l+3\mu\pi_j\tau_l+\mu^2\pi_j\pi_l\right)d\mu=\frac{2j(j+1)(j^2+j+2)}{2j+1}\delta_{lj}$$

$$24.\int_{-1}^{1}\left(\pi_j\pi_l+\tau_j\tau_l\right)d\mu=\frac{2j^2(j+1)^2}{2j+1}\delta_{lj}\qquad\text{(B.51)}$$

The integral 25 is a combination of integral 10 and the integral 2 in section B.5.1:

$$25. \int_{-1}^{1} (\pi_j \pi_l + \tau_j \tau_l) A^2 d\mu \tag{B.52}$$

$$= \frac{2j(j+1)}{2j+1} \left\{ \left[1 + \frac{1}{2j+1} \left(\frac{j^3(j+2)}{2j+3} + \frac{(j-1)(j+1)^3}{2j-1} \right) \right] \delta_{lj} - \frac{j(j+2)(j+3)^2}{(2j+3)(2j+5)} \delta_{l,j+2} - \frac{(j-2)^2(j-1)(j+1)}{(2j-1)(2j-3)} \delta_{l,j-2} \right\}$$

The integrals 26–32 are the combinations of integrals 16 and 19:

$$26. \int_{-1}^{1} \left(\mu \tau_j \tau_l - \tau_j \pi_l - \mu^2 \pi_j \tau_l + \mu \pi_j \pi_l \right) d\mu \tag{B.53}$$

$$= \frac{2(j-1)j(j+1)(j+2)}{2j+1} \left(\frac{j+2}{2j+3} \delta_{l,j+1} + \frac{j-1}{2j-1} \delta_{l,j-1} \right)$$

$$27. \int_{-1}^{1} \left(3\mu \tau_j \tau_l + \tau_j \pi_l + \mu^2 \pi_j \tau_l + 3\mu \pi_j \pi_l \right) d\mu$$

$$= \frac{2j(j+1)}{2j+1} \left[\frac{(j+2)^2(3j+1)}{2j+3} \delta_{l,j+1} + \frac{(j-1)^2(3j+2)}{2j-1} \delta_{l,j-1} \right]$$

$$28. \int_{-1}^{1} \left(\mu \tau_j \tau_l + 3\tau_j \pi_l + 3\mu^2 \pi_j \tau_l + \mu \pi_j \pi_l \right) d\mu$$

$$= \frac{2j(j+1)}{2j+1} \left[\frac{(j+2)^2(j+3)}{2j+3} \delta_{l,j+1} + \frac{(j-1)^2(j-2)}{2j-1} \delta_{l,j-1} \right]$$

$$29. \int_{-1}^{1} \left(\mu \tau_j \tau_l + \mu^2 \tau_j \pi_l + \pi_j \tau_l + \mu \pi_j \pi_l \right) d\mu$$

$$= \frac{2j^2(j+1)^2}{2j+1} \left(\frac{j+2}{2j+3} \delta_{l,j+1} + \frac{j-1}{2j-1} \delta_{l,j-1} \right)$$

30. $$\int_{-1}^{1} \left(\mu\tau_j\tau_l - \pi_j\tau_l - \mu^2\tau_j\pi_l + \mu\pi_j\pi_l\right) d\mu$$
$$= \frac{2j(j+1)}{2j+1}\left(\frac{j\,(j+2)\,(j+3)}{2j+3}\delta_{l,j+1} + \frac{(j-2)\,(j-1)\,(j+1)}{2j-1}\delta_{l,j-1}\right)$$

31. $$\int_{-1}^{1} \left(3\mu\tau_j\tau_l + \pi_j\tau_l + \mu^2\tau_j\pi_l + 3\mu\pi_j\pi_l\right) d\mu$$
$$= \frac{2j(j+1)}{2j+1}\left(\frac{j\,(j+2)\,(3j+5)}{2j+3}\delta_{l,j+1} + \frac{(j-1)\,(j+1)(3j-2)}{2j-1}\delta_{l,j-1}\right)$$

32. $$\int_{-1}^{1} \left(\mu\tau_j\tau_l + 3\pi_j\tau_l + 3\mu^2\tau_j\pi_l + \mu\pi_j\pi_l\right) d\mu$$
$$= \frac{2(j-1)j(j+1)(j+2)}{2j+1}\left(\frac{j}{2j+3}\delta_{l,j+1} + \frac{j+1}{2j-1}\delta_{l,j-1}\right)$$

The integrals 33–35 are the combinations of integrals 17 and 18:

33. $$\int_{-1}^{1} \left(3\mu^2\tau_j\tau_l + \mu\tau_j\pi_l + \mu\pi_j\tau_l + 3\pi_j\pi_l\right) d\mu \tag{B.54}$$
$$= \frac{2j(j+1)}{2j+1}\left\{\frac{1}{2j+1}\left[(2j+1)(3j^2+3j+1) - \frac{3j^3(j+2)}{2j+3} - \frac{3(j-1)(j+1)^3}{2j-1}\right]\delta_{lj}\right.$$
$$\left. + \frac{3j\,(j+2)\,(j+3)^2}{(2j+3)\,(2j+5)}\delta_{l,j+2} + \frac{3\,(j-2)^2\,(j-1)(j+1)}{(2j-3)\,(2j-1)}\delta_{l,j-2}\right\}$$

34. $$\int_{-1}^{1} \left(\mu^2\tau_j\tau_l + 3\mu\tau_j\pi_l + 3\mu\pi_j\tau_l + \pi_j\pi_l\right) d\mu$$
$$= \frac{2j(j+1)}{2j+1}\left\{\frac{1}{2j+1}\left[(2j+1)(j^2+j+3) - \frac{j^3(j+2)}{2j+3} - \frac{(j-1)(j+1)^3}{2j-1}\right]\delta_{lj}\right.$$
$$\left. + \frac{j\,(j+2)\,(j+3)^2}{(2j+3)\,(2j+5)}\delta_{l,j+2} + \frac{(j-2)^2\,(j-1)(j+1)}{(2j-3)\,(2j-1)}\delta_{l,j-2}\right\}$$

35. $$\int_{-1}^{1} \left(\mu^2\tau_j\tau_l - \mu\tau_j\pi_l - \mu\pi_j\tau_l + \pi_j\pi_l\right) d\mu$$
$$= \frac{2j(j+1)}{2j+1}\left\{\frac{1}{2j+1}\left[(2j+1)(j^2+j-1) - -\frac{j^3(j+2)}{2j+3} - \frac{(j-1)(j+1)^3}{2j-1}\right]\delta_{lj}\right.$$
$$\left. + \frac{j\,(j+2)\,(j+3)^2}{(2j+3)\,(2j+5)}\delta_{l,j+2} + \frac{(j-2)^2\,(j-1)(j+1)}{(2j-3)\,(2j-1)}\delta_{l,j-2}\right\}$$

B.5.4 Combination of *m=3* with *m=1*

$$1. \int_{-1}^{1} \pi_j^{(3)} \pi_l A^4 d\mu \tag{B.55}$$

$$= \frac{2(j-2)(j-1)j(j+1)(j+2)(j+3)}{2j+1} \left[-\frac{1}{2j+1} \left(\frac{1}{2j+3} + \frac{1}{2j-1} \right) \delta_{lj} + \frac{1}{(2j+3)(2j+5)} \delta_{l,j+2} + \frac{1}{(2j-3)(2j-1)} \delta_{l,j-2} \right]$$

$$2. \int_{-1}^{1} \left(\tau_j^{(3)} \pi_l + 3\pi_j^{(3)} \pi_l \right) A^2 d\mu$$

$$= \frac{2(j-2)(j-1)j(j+1)(j+2)(j+3)}{2j+1} \left[-\frac{\delta_{l,j+1}}{2j+3} + \frac{\delta_{l,j-1}}{2j-1} \right]$$

$$3. \int_{-1}^{1} \left(\pi_j^{(3)} \tau_l - \pi_j^{(3)} \pi_l \right) A^2 d\mu$$

$$= \frac{2(j-2)(j-1)j(j+1)(j+2)(j+3)}{2j+1} \left[\frac{\delta_{l,j+1}}{2j+3} - \frac{\delta_{l,j-1}}{2j-1} \right]$$

$$4. \int_{-1}^{1} \left(\mu\pi_j^{(3)} \tau_l - \pi_j^{(3)} \pi_l \right) A^2 d\mu = \frac{2(j-2)(j-1)j(j+1)(j+2)(j+3)}{2j+1}$$

$$\times \left[-\frac{3}{(2j-1)(2j+3)} \delta_{lj} + \frac{j+3}{(2j+3)(2j+5)} \delta_{l,j+2} - \frac{j-2}{(2j-3)(2j-1)} \delta_{l,j-2} \right]$$

$$5. \int_{-1}^{1} \left(\mu\tau_j^{(3)} \pi_l + 3\pi_j^{(3)} \pi_l \right) A^2 d\mu = \frac{2(j-2)(j-1)j(j+1)(j+2)(j+3)}{2j+1}$$

$$\times \left[-\frac{3}{(2j-1)(2j+3)} \delta_{lj} - \frac{j}{(2j+3)(2j+5)} \delta_{l,j+2} + \frac{j+1}{(2j-3)(2j-1)} \delta_{l,j-2} \right]$$

$$6. \int_{-1}^{1} \left(\tau_j^{(3)} \tau_l - \mu\tau_j^{(3)} \pi_l + 3\mu\pi_j^{(3)} \tau_l - 3\mu^2 \pi_j^{(3)} \pi_l \right) d\mu$$

$$= -\frac{2(j-2)(j-1)j(j+1)(j+2)(j+3)}{2j+1} \delta_{lj}$$

$$7. \int_{-1}^{1} \left(\mu \tau_j^{(3)} \tau_l - \tau_j^{(3)} \pi_l + 3\mu^2 \pi_j^{(3)} \tau_l - 3\mu \pi_j^{(3)} \pi_l \right) d\mu$$

$$= -\frac{2(j-2)(j-1)j(j+1)(j+2)(j+3)}{2j+1} \left(\frac{j+2}{2j+3} \delta_{l,j+1} + \frac{j-1}{2j-1} \delta_{l,j-1} \right)$$

$$8. \int_{-1}^{1} \left(\mu \tau_j^{(3)} \tau_l - \mu^2 \tau_j^{(3)} \pi_l + 3 \pi_j^{(3)} \tau_l - 3\mu \pi_j^{(3)} \pi_l \right) d\mu$$

$$= -\frac{2(j-2)(j-1)j(j+1)(j+2)(j+3)}{2j+1} \left(\frac{j}{2j+3} \delta_{l,j+1} + \frac{j+1}{2j-1} \delta_{l,j-1} \right)$$

$$9. \int_{-1}^{1} \left(\mu^2 \tau_j^{(3)} \tau_l - \mu \tau_j^{(3)} \pi_l + 3\mu \pi_j^{(3)} \tau_l - 3 \pi_j^{(3)} \pi_l \right) d\mu$$

$$= -\frac{2(j-2)(j-1)j(j+1)(j+2)(j+3)}{2j+1} \left[\frac{1}{2j+1} \left(\frac{j^2}{2j+3} + \frac{(j+1)^2}{2j-1} \right) \delta_{lj} \right.$$
$$\left. + \frac{j(j+3)}{(2j+3)(2j+5)} \delta_{l,j+2} - \frac{(j-2)(j+1)}{(2j-3)(2j-1)} \delta_{l,j-2} \right]$$

B.5.5 Combinations of *m=2* with *m=1*

$$1. \int_{-1}^{1} \pi_j^{(2)} \pi_l A^3 d\mu = \frac{2(j-1)j(j+1)(j+2)}{2j+1} \left(-\frac{\delta_{l,j+1}}{2j+3} + \frac{\delta_{l,j-1}}{2j-1} \right) \tag{B.56}$$

$$2. \int_{-1}^{1} \pi_j^{(2)} \pi_l \mu A^3 d\mu = \frac{2(j-1)j(j+1)(j+2)}{2j+1} \left[\frac{j+3}{2j+3} - \frac{j-2}{2j-1} \delta_{lj} \right.$$
$$\left. - \frac{j+3}{(2j+3)(2j+5)} \delta_{l,j+2} + \frac{j-2}{(2j-3)(2j-1)} \delta_{l,j-2} \right]$$

$$3. \int_{-1}^{1} \pi_j^{(2)} \tau_l A^3 d\mu = \frac{2(j-1)j(j+1)(j+2)}{2j+1} \left[\frac{1}{2j+1} \left(-\frac{j^2}{2j+3} - \frac{(j+1)^2}{2j-1} \right) \delta_{lj} \right.$$
$$\left. + \frac{(j+3)^2}{(2j+3)(2j+5)} \delta_{l,j+2} + \frac{(j-2)^2}{(2j-3)(2j-1)} \delta_{l,j-2} \right]$$

$$4. \int_{-1}^{1} \tau_j^{(2)} \pi_l A^3 d\mu = \frac{2(j-1)j(j+1)(j+2)}{2j+1} \left[\frac{1}{2j+1} \left(\frac{j(j+3)}{2j+3} + \frac{(j+1)(j+2)}{2j-1} \right) \delta_{lj} - \frac{j(j+3)}{(2j+3)(2j+5)} \delta_{l,j+2} - \frac{(j-2)(j+1)}{(2j-3)(2j-1)} \delta_{l,j-2} \right]$$

$$5. \int_{-1}^{1} \left(\tau_j^{(2)} \pi_l + 2\mu \pi_j^{(2)} \pi_l \right) A d\mu = \frac{2(j-1)j(j+1)(j+2)}{2j+1} \delta_{lj}$$

$$6. \int_{-1}^{1} \left(\pi_j^{(2)} \tau_l - \mu \pi_j^{(2)} \pi_l \right) A d\mu = -\frac{2(j-1)j(j+1)(j+2)}{2j+1} \delta_{lj}$$

$$7. \int_{-1}^{1} \left(\mu \pi_j^{(2)} \tau_l - \pi_j^{(2)} \pi_l \right) A d\mu = -\frac{2(j-1)j(j+1)(j+2)}{2j+1} \left(\frac{j+2}{2j+3} \delta_{l,j+1} + \frac{j-1}{2j-1} \delta_{l,j-1} \right)$$

$$8. \int_{-1}^{1} \left(\mu \tau_j^{(2)} \pi_l + 2\pi_j^{(2)} \pi_l \right) A d\mu = \frac{2(j-1)j(j+1)(j+2)}{2j+1} \left(\frac{j}{2j+3} \delta_{l,j+1} + \frac{j+1}{2j-1} \delta_{l,j-1} \right)$$

$$9. \int_{-1}^{1} \left(\mu \tau_j^{(2)} \pi_l + 2\mu^2 \pi_j^{(2)} \pi_l \right) A d\mu = \frac{2(j-1)j(j+1)(j+2)}{2j+1} \left(\frac{j+2}{2j+3} \delta_{l,j+1} + \frac{j-1}{2j-1} \delta_{l,j-1} \right)$$

$$10. \int_{-1}^{1} \left(\mu \pi_j^{(2)} \tau_l - \mu^2 \pi_j^{(2)} \pi_l \right) A d\mu = -\frac{2(j-1)j(j+1)(j+2)}{2j+1} \left(\frac{j+3}{2j+3} \delta_{l,j+1} + \frac{j-2}{2j-1} \delta_{l,j-1} \right)$$

$$11. \int_{-1}^{1} \left(\mu^2 \pi_j^{(2)} \tau_l - \mu \pi_j^{(2)} \pi_l \right) A d\mu = -\frac{2(j-1)j(j+1)(j+2)}{2j+1} \left[\frac{1}{2j+1} \left(\frac{j(j+3)}{2j+3} + \frac{(j-2)(j+1)}{2j-1} \right) \delta_{lj} + \frac{(j+3)^2}{(2j+3)(2j+5)} \delta_{l,j+2} + \frac{(j-2)^2}{(2j-3)(2j-1)} \delta_{l,j-2} \right]$$

12. $$\int_{-1}^{1}\left(\mu^2\tau_j^{(2)}\pi_l+2\mu\pi_j^{(2)}\pi_l\right)Ad\mu$$

$$=\frac{2(j-1)j(j+1)(j+2)}{2j+1}\left[\frac{1}{2j+1}\left(\frac{j^2}{2j+3}-\frac{(j+1)^2}{2j-1}\right)\delta_{lj}\right.$$
$$\left.+\frac{j\,(j+3)}{(2j+3)\,(2j+5)}\delta_{l,j+2}+\frac{(j-2)\,(j+1)}{(2j-3)\,(2j-1)}\delta_{l,j-2}\right]$$

13. $$\int_{-1}^{1}\left(\tau_j^{(2)}\tau_l-\mu\tau_j^{(2)}\pi_l\right)Ad\mu$$

$$=\frac{2(j-1)j(j+1)(j+2)}{2j+1}\left[-\frac{j\,(j+3)}{2j+3}\delta_{l,j+1}+\frac{(j-2)\,(j+1)}{2j-1}\delta_{l,j-1}\right]$$

14. $$\int_{-1}^{1}\left(\tau_j^{(2)}\tau_l+2\mu\pi_j^{(2)}\tau_l\right)Ad\mu$$

$$=\frac{2(j-1)j(j+1)(j+2)}{2j+1}\left[-\frac{(j+2)^2}{2j+3}\delta_{l,j+1}+\frac{(j-1)^2}{2j-1}\delta_{l,j-1}\right]$$

15. $$\int_{-1}^{1}\left(\mu\tau_j^{(2)}\tau_l-\tau_j^{(2)}\pi_l\right)Ad\mu$$

$$=\frac{2(j-1)j(j+1)(j+2)}{2j+1}\left[\frac{1}{2j+1}\left(-\frac{j^2\,(j+3)}{2j+3}+\frac{(j-2)\,(j+1)^2}{2j-1}\right)\delta_{lj}\right.$$
$$\left.-\frac{j\,(j+3)^2}{(2j+3)\,(2j+5)}\delta_{l,j+2}+\frac{(j-2)^2\,(j+1)}{(2j-3)\,(2j-1)}\delta_{l,j-2}\right]$$

16. $$\int_{-1}^{1}\left(\mu\tau_j^{(2)}\tau_l+2\pi_j^{(2)}\tau_l\right)Ad\mu$$

$$=\frac{2(j-1)j(j+1)(j+2)}{2j+1}\left[\frac{1}{2j+1}\left(\frac{j^3}{2j+3}-\frac{(j+1)^3}{2j-1}\right)\delta_{lj}\right.$$
$$\left.-\frac{j\,(j+3)^2}{(2j+3)\,(2j+5)}\delta_{l,j+2}+\frac{(j-2)^2\,(j+1)}{(2j-3)\,(2j-1)}\delta_{l,j-2}\right]$$

Appendix C

Vector spherical wave functions

C.1 DEFINITION OF VECTOR SPHERICAL WAVE FUNCTIONS

The vector spherical wave functions satisfying the vector wave equation in spherical coordinates are of the form:

$$\mathbf{M}^{(i)}_{\sigma ml}(k\mathbf{r}) = z_l^{(i)}(kr)\mathbf{m}_{\sigma ml} \tag{C.1}$$
$$\mathbf{N}^{(i)}_{\sigma ml}(k\mathbf{r}) = \frac{z_l^{(i)}(kr)}{kr}\mathbf{l}_{\sigma ml} + \frac{1}{kr}\frac{d}{dr}\left(rz_l^{(i)}(kr)\right)\mathbf{n}_{\sigma ml}$$
$$\mathbf{L}^{(i)}_{\sigma ml}(k\mathbf{r}) = \frac{1}{kl(l+1)}\frac{d}{dr}z_l^{(i)}(kr)\mathbf{l}_{\sigma ml} + \frac{z_l^{(i)}(kr)}{kr}\mathbf{n}_{\sigma ml}$$

where σ is the parity index taking the values $\sigma = e$ (even) and $\sigma = o$ (odd), $l = 1, 2, 3, ...$ and $m = 0, 1, 2, ..., l$. Instead of three indices σml one can write the generalized index (e.g., $\mathbf{M}_\nu$). The radial dependence in Eq.(C.1) is described by the spherical Bessel function $z_l^{(i)}(i = 1, 3)$ (see Eq.(A.4)), where index (1) and (3) correspond respectively to the function regular in the origin $z_l^{(1)}(u) = (\pi/2u)^{1/2}J_{l+1/2}(u)$ and the function $z_l^{(3)}(u) = (\pi/2u)^{1/2}H^{(1)}_{l+1/2}(u)$ which (for time-factor $\exp(-i\omega t)$ used by us) describes the outgoing spherical wave at a great distance from the particle. The dependence on angles θ and φ enters in Eq.(C.1) in the form of the so-called angular vector functions $\mathbf{m}, \mathbf{n}$, and $\mathbf{l}$ whose definition is below. If the more convenient Riccati–Bessel functions $\psi_l^{(i)}(u) = uz_l^{(i)}(u)$ are introduced into Eq.(C.1) instead of the spherical functions then:

$$\mathbf{M}^{(i)}_{\sigma ml}(k\mathbf{r}) = \frac{\psi_l^{(i)}(kr)}{kr}\mathbf{m}_{\sigma ml} \tag{C.2}$$
$$\mathbf{N}^{(i)}_{\sigma ml}(k\mathbf{r}) = \frac{\psi_l^{(i)}(kr)}{k^2r^2}\mathbf{l}_{\sigma ml} + \frac{\psi_l^{(i)\prime}(kr)}{kr}\mathbf{n}_{\sigma ml}$$
$$\mathbf{L}^{(i)}_{\sigma ml}(k\mathbf{r}) = \frac{1}{krl(l+1)}\left(\psi_l^{(i)\prime}(kr) - \frac{\psi_l^{(i)}(kr)}{kr}\right)\mathbf{l}_{\sigma ml} + \frac{\psi_l^{(i)}(kr)}{k^2r^2}\mathbf{n}_{\sigma ml}$$

The wave functions $\mathbf{M}_\nu$ and $\mathbf{N}_\nu$ describe the transverse waves:

$$\nabla \cdot \mathbf{M}_\nu = \nabla \cdot \mathbf{N}_\nu = 0 \tag{C.3}$$

and wave function $\mathbf{L}_\nu$ describes the longitudinal waves. The functions $\mathbf{M}_\nu$ and $\mathbf{N}_\nu$ are related as follows:

$$\nabla_r \times \mathbf{N}_\nu(k\mathbf{r}) = k\mathbf{M}_\nu(k\mathbf{r}) \qquad \nabla_r \times \mathbf{M}_\nu(k\mathbf{r}) = k\mathbf{N}_\nu(k\mathbf{r}) \tag{C.4}$$

The vector angular wave functions are defined as:

$$\begin{aligned}
\mathbf{m}_{\substack{e\\o}ml} &= \mp \mathbf{e}_\theta \frac{mP_l^{(m)}(\cos\theta)}{\sin\theta} \begin{matrix}\sin\\ \cos\end{matrix} m\varphi - \mathbf{e}_\varphi \frac{dP_l^{(m)}(\cos\theta)}{d\theta} \begin{matrix}\cos\\ \sin\end{matrix} m\varphi \\
\mathbf{n}_{\substack{e\\o}ml} &= \left[\mathbf{e}_r \times \mathbf{m}_{\substack{e\\o}ml}\right] = \mathbf{e}_\theta \frac{dP_l^{(m)}(\cos\theta)}{d\theta} \begin{matrix}\cos\\ \sin\end{matrix} m\varphi \mp \mathbf{e}_\varphi \frac{mP_l^{(m)}(\cos\theta)}{\sin\theta} \begin{matrix}\sin\\ \cos\end{matrix} m\varphi \\
\mathbf{l}_{\substack{e\\o}ml} &= \mathbf{e}_r l(l+1) P_l^{(m)}(\cos\theta) \begin{matrix}\cos\\ \sin\end{matrix} m\varphi
\end{aligned} \tag{C.5}$$

where $\mathbf{e}_r$, $\mathbf{e}_\theta$, and $\mathbf{e}_\varphi$ are the unit vectors along axes r, θ, and φ. If we introduce the angular functions $\pi_l^{(m)}(\mu)$ and $\tau_l^{(m)}(\mu)$ (see Appendix B), where $\mu = \cos\theta$, instead of the associated Legendre polynomials $P_l^{(m)}(\mu)$ then Eqs.(C.5) may be rewritten in a more convenient form:

$$\begin{aligned}
\mathbf{m}_{\substack{e\\o}ml} &= \mp \mathbf{e}_\theta m\pi_l^{(m)} \begin{matrix}\sin\\ \cos\end{matrix} m\varphi - \mathbf{e}_\varphi \tau_l^{(m)} \begin{matrix}\cos\\ \sin\end{matrix} m\varphi \\
\mathbf{n}_{\substack{e\\o}ml} &= \mathbf{e}_\theta \tau_l^{(m)} \begin{matrix}\cos\\ \sin\end{matrix} m\varphi \mp \mathbf{e}_\varphi m\pi_l^{(m)} \begin{matrix}\sin\\ \cos\end{matrix} m\varphi \\
\mathbf{l}_{\substack{e\\o}ml} &= \mathbf{e}_r l(l+1) \pi_l^{(m)} \sin\theta \begin{matrix}\cos\\ \sin\end{matrix} m\varphi
\end{aligned} \tag{C.6}$$

These functions obey the relations of orthogonality:

$$\mathbf{m}_{\sigma ml} \cdot \mathbf{l}_{\sigma ml} = \mathbf{e}_r \cdot \mathbf{m}_{\sigma ml} = 0 \tag{C.7}$$

Sometimes the following relations are useful:

$$\begin{aligned}
\left[\mathbf{M}^{(i)}_{\substack{e\\o}ml} \times \mathbf{e}_r\right] &= \frac{\psi_l^{(i)}(kr)}{kr} \left(\pm \mathbf{e}_\varphi m\pi_l^{(m)} \begin{matrix}\sin\\ \cos\end{matrix} m\varphi - \mathbf{e}_\theta \tau_l^{(m)} \begin{matrix}\cos\\ \sin\end{matrix} m\varphi\right) \\
\left[\mathbf{N}^{(i)}_{\substack{e\\o}ml} \times \mathbf{e}_r\right] &= \frac{\psi_l^{(i)\,\prime}(kr)}{kr} \left(-\mathbf{e}_\varphi \tau_l^{(m)} \begin{matrix}\cos\\ \sin\end{matrix} m\varphi \mp \mathbf{e}_\theta m\pi_l^{(m)} \begin{matrix}\sin\\ \cos\end{matrix} m\varphi\right)
\end{aligned} \tag{C.8}$$

C.2 INTEGRALS FOR THE VECTOR ANGULAR WAVE FUNCTIONS

The integral expressions for the vector angular functions follows from the integrals over the quadratic combinations of angular functions $\pi_l^{(m)}$ and $\tau_l^{(m)}$ in the range $(-1, 1)$ (see

Appendix B.5) and the known integrals over quadratic combinations of trigonometric functions in the range $(0, 2\pi)$:

$$\int_{(4\pi)} \mathbf{m}_{\sigma ml} \cdot \mathbf{m}_{\sigma' m' l'} d\Omega = \int_{(4\pi)} \mathbf{n}_{\sigma ml} \cdot \mathbf{n}_{\sigma' m' l'} d\Omega = \lambda_{ml}^{-1} \delta_{\sigma\sigma'} \delta_{mm'} \delta_{ll'}$$

$$\int_{(4\pi)} \mathbf{l}_{\sigma ml} \cdot \mathbf{l}_{\sigma' m' l'} d\Omega = l(l+1) \lambda_{ml}^{-1} \delta_{\sigma\sigma'} \delta_{mm'} \delta_{ll'} \tag{C.9}$$

where

$$\lambda_{ml} = \frac{(2l+1)(l-m)!}{2\pi l(l+1)(l+m)!(1+\delta_{0m})}$$

δ_{ij} is the Kronecker delta and the integration is done over the entire solid angle: $\int_{(4\pi)} d\Omega = \int_0^\pi \sin\theta d\theta \int_0^{2\pi} d\varphi$. Note also that:

$$\int_{(4\pi)} \mathbf{m}_{\sigma ml} \cdot \mathbf{l}_{\sigma' m' l'} d\Omega = 0 \tag{C.10}$$

C.3 INTEGRALS FOR THE VECTOR SPHERICAL WAVE FUNCTIONS

Upon integrating the scalar products $\mathbf{M} \cdot \mathbf{M}$ and $\mathbf{N} \cdot \mathbf{N}$ over all solid angle we have (see Eq.(C.9)):

$$\int_{(4\pi)} \mathbf{M}_{\sigma ml}^{(i)}(k\mathbf{r}) \cdot \mathbf{M}_{\sigma' m' l'}^{(j)}(k\mathbf{r})\, d\Omega = \frac{\psi_l^2(kr)}{k^2 r^2} \lambda_{ml}^{-1} \delta_{\sigma\sigma'} \delta_{mm'} \delta_{ll'}$$

$$\int_{(4\pi)} \mathbf{N}_{\sigma ml}^{(i)}(k\mathbf{r}) \cdot \mathbf{N}_{\sigma' m' l'}^{(j)}(k\mathbf{r})\, d\Omega \tag{C.11}$$

$$= \frac{\lambda_{ml}^{-1}}{(2l+1)\, k^2 r^2} \left[(l+1)\, \psi_{l-1}^2(kr) + l \psi_{l+1}^2(kr) \right] \delta_{\sigma\sigma'} \delta_{mm'} \delta_{ll'}$$

If Eqs.(A.29–A.30) are considered, we can obtain for the volume integrals ($d^3r = r^2 dr \sin\theta\, d\theta\, d\varphi$), where integration over r is indefinite:

$$\int_{(r)} \int_{(4\pi)} \mathbf{M}_{\sigma ml}^{(i)}(k_\nu \mathbf{r}) \cdot \mathbf{M}_{\sigma' m' l'}^{(j)}(k_p \mathbf{r})\, d^3r = \lambda_{ml}^{-1} \delta_{\sigma\sigma'} \delta_{mm'} \delta_{ll'}\, {}^{(1)}I_l^{ij}(k_\nu, k_p)$$

$$\int_{(r)} \int_{(4\pi)} \mathbf{N}_{\sigma ml}^{(i)}(k_\nu \mathbf{r}) \cdot \mathbf{N}_{\sigma' m' l'}^{(j)}(k_p \mathbf{r})\, d^3r = \lambda_{ml}^{-1} \delta_{\sigma\sigma'} \delta_{mm'} \delta_{ll'}\, {}^{(2)}I_l^{ij}(k_\nu, k_p) \tag{C.12}$$

where

$$ {}^{(1)}I_l^{ij}(k_\nu, k_p) \equiv \frac{1}{k_\nu k_p} \int \psi_l^{(i)}(k_\nu r)\psi_l^{(j)}(k_p r) = \tag{C.13} $$

$$ = \frac{1}{\left(k_\nu^2 - k_p^2\right) k_\nu k_p} \left[k_p \psi_l^{(i)}(k_\nu r)\psi_l^{(j)\prime}(k_p r) - k_\nu \psi_l^{(i)\prime}(k_\nu r)\psi_l^{(j)}(k_p r)\right] $$

$$ {}^{(2)}I_l^{ij}(k_\nu, k_p) \equiv \frac{1}{k_\nu k_p} \int \left[\psi_l^{(i)\prime}(k_\nu r)\psi_l^{(j)\prime}(k_p r) + \frac{l(l+1)}{r^2 k_\nu k_p}\psi_l^{(i)}(k_\nu r)\psi_l^{(j)}(k_p r)\right] dr \tag{C.14} $$

$$ = \frac{1}{\left(k_\nu^2 - k_p^2\right) k_\nu k_p} \left[k_\nu \psi_l^{(i)}(k_\nu r)\psi_l^{(j)\prime}(k_p r) - k_p \psi_l^{(i)\prime}(k_\nu r)\psi_l^{(j)}(k_p r)\right] $$

In a similar way one can obtain the integral:

$$ \int_{(r)} \int_{(4\pi)} \mathbf{N}_{\sigma ml}^{(i)}(k_\nu \mathbf{r}) \cdot \mathbf{L}_{\sigma' m' l'}^{(j)}(k_p \mathbf{r})\, d^3 r = \lambda_{ml}^{-1} \delta_{\sigma\sigma'} \delta_{mm'} \delta_{ll'} \frac{1}{k_\nu k_p}\, {}^{(3)}I_l^{ij}(k_\nu, k_p) \tag{C.15} $$

where

$$ {}^{(3)}I_l^{ij}(k_\nu, k_p) = \int \left[\frac{1}{k_\nu r}\psi_l^{(i)}(k_\nu r)\psi_l^{(j)\prime}(k_p r) \right. \tag{C.16} $$

$$ \left. + \frac{1}{k_p r}\psi_l^{(i)\prime}(k_\nu r)\psi_l^{(j)}(k_p r) - \frac{1}{k_\nu k_p r^2}\psi_l^{(i)}(k_\nu r)\psi_l^{(j)}(k_p r) dr\right] $$

$$ = \frac{1}{k_\nu k_p r}\psi_l^{(i)}(k_\nu r)\psi_l^{(j)}(k_p r) $$

Appendix D

Guide to our light scattering codes

During many years, numerous computer codes were written in our laboratory. They calculate the various characteristics of the scattered and internal fields for particles of various forms – homogeneous and inhomogeneous spheres, cylinders, and spheroids. A little part of this collection is now freely available on the website of V. Il'in (see Appendix E). All these codes are written on Fortran in double precision and supplemented by the detailed algorithms rigidly bound with code text. The list includes:

1. Program **DMAGMI** calculates integral and angular scattering characteristics (including the phase function asymmetry) for the homogeneous sphere with possible magnetic properties.
2. Program **DMAGPO** is the polydisperse variant of **DMAGMI**. The distribution function may be lognormal, gamma, or Junge type.
3. Program **DMAGMI2** calculates integral and angular scattering characteristics (including the phase function asymmetry) for the two-layered sphere with possible magnetic properties.
4. Program **DMAGMIP2** is the polydisperse variant of **DMAGMI2**. The distribution function may be lognormal, gamma, or Junge type. The constant ratio of core to shell radii is maintained.
5. Program **DMIINT** calculates the field inside the homogeneous sphere (Chapter 6).
6. Package of programs **KPOG** calculates the absorption integral characteristics for homogeneous sphere (see Chapter 5).
7. Program **LUCH** and its polydisperse variant **LUCHPO2** calculate integral and angular scattering characteristics for a two-layered spherical particle with homogeneous core and radially inhomogeneous shell with the power law of inhomogeneity (Section 8.4). The distribution function may be lognormal, gamma, or Junge type. The constant ratio of core to shell radii is maintained.
8. Program **SPHMUL** calculates integral scattering characteristics for the sphere with an arbitrary number of concentric layers (Section 3.5).

Note: The essential parts of the above programs are the subroutine **DLD** (the logarithmic derivatives of the Riccati–Bessel functions, Section 1.5.2.1), the subroutine **CB2** (the amplitude coefficients of the scattered field for multilayered sphere, Section 3.5) and the subroutine **DIFM** (the Bessel function with the complex index, Section 8.4).

This list of freely available codes will be updated and supplemented in the future. One of us (V.B.) also made available, for the light scattering community, his great database containing more than 7500 references to publications on various aspects of light interaction with dispersed media. This database is now on the same website of V.Il'in.

Appendix E

Light scattering codes on the Internet

A lot of codes and other information on light scattering are available via Internet. Some related websites are listed below. The first author of this book can be reached by the following e-mail: babenko@dragon.bus-net.by.

1. **M. Mishchenko** crmim@giss.nasa.gov http://www.giss.nasa.gov/~crmim/
This website contains a lot of useful information:

(a) the Fortran T-matrix codes for computing electromagnetic scattering by:

- randomly oriented and fixed-oriented non-spherical rotationally symmetric particles;
- randomly oriented two-sphere clusters with touching and separated components;
- multi-sphere clusters in fixed and random orientations;
- polydisperse spherical particles.

(b) the Fortran code for computing the bidirectional reflection function for flat particulate layers;
(c) the on-line list of publications of M. Mishchenko with freely available PDFs;
(d) the list of members of the Electromagnetic scattering Community with addresses, e-mails and URLs;
(e) newsletter on electromagnetic and light scattering;
(f) many links to other websites.

2. **T. Wriedt** thw@iwt.uni-bremen.de http://www.t-matrix.de
Dr. T. Wriedt created and now are maintaining and regularly updated the great library of the electromagnetic scattering codes written by many researchers and providing links to:

(a) many Mie-type codes for homogeneous and concentrically layered spheres, spheres with asymmetrically located spherical inclusions, particles on the surface;
(b) ray-tracing codes for non-spherical particles and polyhedral crystals;

(c) codes based on the method of moments, the method of finite elements, the finite difference time domain (FDTD) approach, the discrete dipole approximation (DDA), the general multipole technique (GMT), the point matching approach;
(d) T-matrix codes;
(e) bisphere and multi-sphere codes;
(f) the on-line list of publications of T. Wriedt with freely available PDFs;
(g) other websites.

3. **P.J. Flatau** flatau@ucsd.edu http://atol.ucsd.edu/~pflatau/scatlib/
This site contains the SCATTERLIB library of single light scattering codes:

(a) Mie-type codes (homogeneous and coated spheres, multispheres);
(b) codes for the infinite cylinder;
(c) T-matrix and Separation of variable (SVM) codes for spheroids;
(d) DDA codes;
(e) codes based on the anomalous diffraction approximation, and a collection of references to light scattering works.

Regrettably, this site has not been updated for the last several years.

4. **B. Draine** draine@astro.princeton.edu ftp://astro.princeton.edu/draine/scat/ddscat
This ftp contains the well documented version 5a.10 of the popular code DDSCAT (discrete dipole approximation).

5. **A. Macke** amacke@ifm.uni-kiel.de
http://www.ifm.uni-kiel.de/me/research/Projekte/RemSens/SourceCodes /source.html
This site contains codes in UNIX environments for:

(a) ray-tracing calculations of light scattering by ellipsoids, polyhedral shaped particles, circular cylinders, and nonspherical particles containing nonsymmetrical inclusions;
(b) MonteCarlo radiative transfer calculations of multiple scattering in the 3D inhomogeneous atmosphere;
(c) radiative transfer calculations for microwave frequencies.

6. **D. Mackowski** dmckwski@eng.auburn.edu
ftp://ftp.eng.auburn.edu/pub/dmskwski/scatcodes/index.html
This site provides access to two multi-sphere superposition T-matrix codes calculating scattering by the large-scale sphere cluster in fixed and chaotic orientations.

7. **V. Il'in** vi2087@VI2087.spb.edu http://www.astro.spbu.ru/staff/ilin2/ilin.html
This is an extremely large and rich site devoted to investigations of light scattering by particles of cosmic dust conducted in Astronomical Institute of S.-Petersburg University. In contains:

(a) SVM codes for homogeneous and core-mantle spheroids;

(b) SVM code for core-mantle infinitely long circular cylinder;
(c) T-matrix code for spheroids and special T-matrix-like code for axisymmetrical particles;
(d) Mie-type codes for homogeneous and multilayered spheres;
(e) code for quasistatic approximation for spheroids;
(f) Jena–Petersburg database on optical constants;
(g) a very great deal of links to other sites of astrophysical nature;
(h) the on-line list of publications of several researchers in Astronomical Institute of S.-Petersburg University with freely available PDFs.

It is evident that the above list of sites giving free access to some light scattering codes is far from being complete but the key sites are in it. We strongly recommend to study the content of these and analogous sites before creating your own code.

References

Aannestad, A. (1975) Absorptive properties of silicate core-mantle grains. *Astrophys. J.* **200**, 30–41.

Abashev, M. T. and Rozenberg, V. I. (1969) Scattering and extinction of radio-frequency radiation by water-containing hail. *Izvestiya AN USSR. Fizika Atmosfery i Okeana* **5**, 973–978.

Aden, A. L. and Kerker, M. (1951) Scattering of electromagnetic waves from two concentric spheres. *J. Appl. Phys.* **22**, 1242–1246.

Akulinin, A. A. (1988) Some special features of the algorithm to calculate scattering with Mie theory. *Optika Atmosfery* **1**, 127–129.

Albini, F. A. (1962) Scattering of a plane wave by an inhomogeneous sphere under the Born approximation. *J. Appl. Phys.* **33**, 3032– 3036.

Alexopoulos, N. G. (1969) Radar cross-section of perfectly conducting spheres coated with a certain class of radially inhomogeneous dielectrics. *IEEE Trans. Antenn. Propagat.* **AP–17**, 667–669.

Alexopoulos, N. G. and Barcilon, V. (1972) On the reflection of light from an inhomogeneous shell. *IEEE Trans. Antenn. Propagat.* **AP–20**, 390–392.

Alexopoulos, N. G. and Huebner, D. A. (1973) Pulse scattering from a sphere coated with an inhomogeneous sheath. *IEEE Trans. Antenn. Propagat.* **AP–21**, 727–728.

Alexopoulos, N. G. and Uslenghi, P. L. E. (1969) A special class of spherically inhomogeneous dielectrics. *Alta frequenza* **38** (Num. speciale), 65– 69.

Andreev, G. A., Bisyarin, V. P, Sokolov, A. V. and Strelkov, G. M. (1977) Propagation of laser radiation in Earth atmosphere. *Itogi Nauki i Tekhniki. Radiotekhnika* **11,** 94–148.

Andreev, S. D., Ivlev, L. S, Kabanov, M. V. and Pkhalagov, Ju. A. (1974) Influence of relative humidity on aerosol extinction of optical radiation in atmosphere. *Izvestiya VUZOV. Fizika* N 5, 54–57.

Aragon, S. R. and Elwenspoek, M. (1982) Mie scattering from thin spherical bubbles. *J. Chem. Phys.* **77**, 3406–3413.

Arnold, S. and Folan, L. M. (1989) Energy transfer and the photon lifetime within an aerosol particle. *Opt. Lett.* **14**, 387–389.

Arnush, D. (1964) Electromagnetic scattering from a spherical nonuniform medium. Part I: General theory. *IEEE Trans. Antenn. Propagat.* **AP–12**, 86–90.

Arsaev, I. E. (1982) Scattering of plane wave by axially symmetrical bodies with absorbing coatings. *Radiotehnika i Elektronika* **27**, 2101–2109.

Ashkin, A. (1970) Acceleration and trapping of particles by radiation pressure. *Phys. Rev. Lett.* **24**, 156–159.

Ashkin, A. and Dziedzic, J. M. (1977) Observations of resonances in the radiation pressure on dielectric spheres. *Phys. Rev. Lett.* 38, 1351–1354.

Ashkin, A. and Dziedzic, J. M. (1981) Observations of optical resonances of dielectric spheres by light scattering. *Appl. Opt.* **22**, 2302–2307.

Astafyeva, L. G. (1973) Energy distribution inside large weakly absorbing particles. *Jurnal Prikladnoi Spektroskopii* **18**, 470–2.

Astafyeva, L. G. and Lednyeva, G. P. (1999) Spatial structures at microparticles at resonance conditions. *Optika i Spektroskopiya* **86**, 810–814.

Astafyeva, L. G. and Prishivalko, A. P. (1978) On applicability of homogeneous particle model at optical remote sensing of atmospheric hazes. *Jurnal Prikladnoi Spektroskopii* **28**, 735–740.

Astafyeva, L. G. and Prishivalko, A. P. (1989) Absorption of IR-radiation and optical fields inside metallic particles with oxide films. *Optika Atmosfery* **9**, 1089–92.

Astafyeva, L. G. and Prishivalko, A. P. (1990*a*) The specific features of absorption of IR-radiation and optical fields inside the hollow metallic and metallized particles. *Doklady AN BSSR* **34**, 127–129.

Astafyeva, L. G. and Prishivalko, A. P. (1990*b*) The specific features of optical fields inside spherical particles with relative refractive index $\overline{n} < 1$. *Optika i Spectroskopiya* **69**, 398–401.

Astafyeva, L. G. and Prishivalko, A. P. (1991) Heating of solid aluminium particles by intense radiation with $\lambda = 0.63\ \mu m$. *Metallofizika* **13,** 122–126.

Astafyeva, L. G. and Prishivalko, A. P. (1995) Radiation absorption by metallic particles with oxide films. *Preprints of the 4-th Internat. Congress Optical Particle Sizing,* pp. 267–273. Nurnberg, Germany.

Astafyeva, L. G. , Prishivalko, A. P. and Katseva, I. R. (1981) Extinction and backscattering of radiation by systems of two- and three-layered particles of atmospheric aerosol. *Izvestiya AN BSSR. ser. fiz.-mat. nauk* N 2, 71–76.

Astafyeva, L. G., Lednyeva, H. P. and Prishivalko, A. P. (1986) Resonances in absorption of radiation by hollow spherical particles. *Doklady AN BSSR* **30**, 414–17.

Astafyeva, L. G., Prishivalko, A. P. and Leiko, S. T. (1989) About the movement of metallic particles by laser radiation. *Fizika i Khimiya Obrabotki Materialov* N 4, 35–9.

Astafyeva, L. G., Prishivalko, A. P. and Leiko, S. T. (1991) The heating of metallic particles by laser radiation. *Fizika i Khimiya Obrabotki Materialov* N 2, 64–9.

Astafyeva, L. G., Prishivalko, A. P. and Sorokina, E. A. (1992) Effect of the imaginary part of the refractive index on radiation attenuation by solid spherical particles. *Optika i Spectroskopiya* **72**, 502–505.

Babenko, V. A. (1975) On the theory of plane electromagnetic wave scattering by radially inhomogeneous sphere. *Izvestiya AN BSSR. ser. fiz.-mat. nauk* N 1, 80–91.

Babenko, V. A. (1991) Asymmetry of absorption by homogeneous sphere. *Optika Atmosfery* **4**, 803–808.

Babenko, V. A. (1996) Spectral dependencies of radiation specific extinction by aggregates of aluminum subparticles. *Jurnal Prikladnoi Spektroskopii* **63**, 486–493.

Babenko, V. A. (1999) Absorption of radiation in the restricted region of spherical particle. *Optika Atmosfery i Okeana **12**, 14–18.*

Babenko, V. A. and Astafyeva, L. G. (1985) Absorption of radiation in multilayered spherical particles. *Vesti AN BSSR. Ser. fiz.–mat. nauk.* n 1, 55–60.

Babenko, V. A. and Leiko, S. T. (1991) Method of calculation of light scattering characteristics for two-layered sphere with inhomogeneous shell. *Optika Atmosfery* **4**, 191–196.

Babenko, V. A. and Siniyk, A. F. (1993) Method of calculation the field inside two-layered sphere with inhomogeneous shell. *Optika Atmosfery i Okeana* **6**, 643–649.

Babenko, V. A. and Prishivalko, A. P. (1973) Light scattering by particles with given radial profile of refractive index inhomogeneity. *Doklady AN BSSR* **17**, 1093–1095.

Babenko, V. A. and Prishivalko, A. P. (1997) Similarity criterion for optical fields in weakly absorbing spherical aerosol particles. *Optika i Spektroskopiya* **83**, 372–376.

Babenko, V. A., Prishivalko, A. P. and Leiko, S. T. (1974) Light scattering by radially-inhomogeneous particles of atmospheric aerosol. *Doklady AN BSSR* **18**, 984–986.

Babenko, V. A., Prishivalko, A. P. and Leiko, S. T. (1975) Angular light scattering characteristics for radially-inhomogeneous particles of atmospheric aerosol. *Izvestiya AN USSR. Fizika Atmosfery i Okeana* **11**, 643–647.

Baev, S. Yu. and Belov, N. N. (1995) Heat release in the illuminated aerosol particle. *Izvestiya RAN. Fizika Atmosfery i Okeana* ***31****, 540–544.*

Baibulatov, F. Kh., Ivaniya, S. P. and Pas'ko, L. N. (1971) On calculations of light scattering by spherical particles. *Izvestiya Sibir. Otdeleniya AN USSR. ser. tehn. nauk* N1, 38–44.

Bakhanova, R. A., Solyanek, E. G. and Terzyev, F. S. (1971) Experiments on action on fog by transpiration of a queous suspensions of high-aliphatic alcohols. *Trudy UkrNIIGMI*, N. 106, 128–132.

Bambino, T. M. and Guimaraes, L. G. (1996) Resonance of a coated sphere. *Phys. Rev. E.* **53**, 2859–2863.

Barabanenkov, Yu. N. (1975) Multiple scattering of waves by ensemble of particles and theory of radiative transfer. *Uspekhi Fizicheskih Nauk* **117**, 49–78.

Barabanenkov, Yu. N. and Finkelberg,V. M. (1968) The optical theorem in the theory of multiple wave scattering. *Radiofizika* **11**, 719–725.

Barabanenkov, Yu. N. and Finkelberg, V. M. (1971) Green's functions method in the theory of multiple wave scattering. In: Stepanov, B. I and Ivanov, A .P. (eds.) *Theoretical and applied problems of light scattering,* pp.171–186. Nauka i tekhnika, Minsk, .

Barabanenkov, Yu. N., Kravtsov, Yu. A., Rytov, S. M. and Tatarskiy, V. I. (1970) A state of the theory of wave propagation in randomly non-uniform medium. *Uspekhi Fizicheskih Nauk* **102**, 3–42.

Barabanenkov, Yu. N., Vinogradov, A. G., Kravtsov, Yu. A. and Tatarskiy, V. I. (1975) Application of the multiple wave scattering theory to deriving the radiative transfer equation for statistically non-uniform medium. *Izvestiya VUZOV. Radiofizika* **15**, 1852–1860.

Baranov, V. G. (1966) Scattering of polarized light by great anisotropic particles. *Optika i Spektroskopiya* **21**, 610–612.

Barber, P. W. and Chang, R. K. (eds.) (1990) *Optical effects associated with small particles.* World Scientific, Singapore.

Barber, P. W. and Hill, S. C. (1990) *Light scattering by particles: computational methods.* World Scientific, Singapore.

Barber, P. and Yeh, C. (1975) Scattering of electromagnetic waves by arbitrary shaped dielectric bodies. *Appl. Opt.* **14**, 2864–2872.

Bateman, H. and Erdélyi, A. (1953*a*) *Higher transcendental functions. Vol.1.* McGraw Hill, New York.

Bateman, H. and Erdélyi, A. (1953*b*) *Higher transcendental functions. Vol.2.* McGraw Hill, New York.

Beker, B., Umashankar, K. R. and Taflove, A. (1989) Numerical analysis and validation of the combined field surface integral equations for electromagnetic scattering by arbitrary shaped two-dimensional anisotropic objects. *IEEE Trans. Antenn. Propagat.* **AP–37**, 1573–1581.

Belov, N. N. (1986) Inhomogeneity of electromagnetic field (10.6 μm) inside the particles of atmospheric aerosol. *Optika i Spektroskopiya* **61**, 1331–1336.

Belov, N. N. (1989) Complex refractive index of aluminium oxide. *Kristallografiya* **34**, 1185–1190.

Belov, N. N. (1992) Calculation of Riccati–Bessel functions with upward recursion. *Optika Atmosfery i Okeana* **5**, 165–8.

Benincasa, D. S., Barber, P. W, Zhang, J.-Z., Hsieh, W.-F. and Chang, R. K. (1987) Spatial distributions of the internal and near-field intensity of large cylindrical and spherical scatterers. *Appl. Opt.* **26**, 1348–1356.

Benner, R. E., Barber, P. W., Oven, J. F. and Chang, R. K. (1980) Observation of structure resonances in the fluorescence spectra from microspheres. *Phys. Rev. Lett.* **44**, 475–478.

Berry, M. V. and Percival, I. C. (1986) Optics of fractal cluster such as smoke. *Opt. Acta* **33**, 577–91.

Bhandari, R. (1985) Scattering coefficients for a multilayered sphere: analytic expressions and algorithms. *Appl. Opt.* **24**, 1960–1967.

Bhandari, R. (1986*a*) Tiny core or thin layer as a perturbation in scattering by a single-layered sphere. *J. Opt. Soc. Amer. A.* **3**, 319–328.

Bhandari, R. (1986*b*) Specific absorption of a tiny absorbing particle embedded within a non-absorbing particle. *Appl. Opt.* **25**, 3331–3333.

Bhardwaja, P. S., Herbert, J. and Charlson, R. J. (1974) Refractive index of atmospheric particulate matter: an in situ method for determination. *Appl. Opt.* **13**, 731–734.

Binenko, V. I. (1980) Definition of effective values of atmospheric aerosol refractive index. *Trudy GGO* N 434, 33–38.

Bisbing, P. E. (1966) Electromagnetic scattering by an exponentially inhomogeneous plasma sphere. *IEEE Trans. Antenn. Propagat.* **AP–14**, 216–226.

Bishop, G. C. (1986) Dipole scattering from a conducting sphere coated with a lossy dielectric. *J. Appl. Phys.* **60**, 2205–2217.

Blifford, I. H. and Gillette, D. A. (1971) Application of the lognormal frequency distribution to the chemical composition and size distribution of naturally occurring atmospheric aerosols. *Water, air and soil pollution* **1**, 106–114.

Blifford, I. H. and Ringer, L. D. (1969) The size and number distribution of aerosols in the continental troposphere. *J. Atm. Sci.* **26**, 716–726.

Blokh, A. G. (1967) *Thermal radiation in boilers.* Energiya, Moscow.

Bohren, C. F. (1975) Scattering of electromagnetic waves by an optically active spherical shell. *J. Chem. Phys.* **62**, 1566–1571.
Bohren, C. F. (1978) Scattering of electromagnetic waves by an optically active cylinder. *J. Coll. Interf. Sci.* **66**, 105–109.
Bohren, C. F. and Huffman, D. R. (1983) *Absorption and scattering of light by small particles*. Wiley, New York.
Borghese, F., Denti, P., Saija, R., Toscano, G. and Sindoni, O. I. (1987) Extinction coefficient for a random dispersion of small stratified spheres and random dispersion of their binary aggregates. *J. Opt. Soc. Amer. A.* **4**, 1984–1991.
Borghese, F., Denti, P. and Saija, R. (1994) Optical properties of spheres containing several spherical inclusions. *Appl. Opt.* **33**, 484–93.
Borisevich, N. A., Vereshchagin, V. G. and Validov, M. A. (1971) *Infrared filters*. Nauka i tekhnika, Minsk.
Born, M. and Wolf, E. (1999) *Principles of optics*. Cambridge University Press, Cambridge.
Bott, A. and and Zdunkowski, W. (1987) Electromagnetic energy within dielectric spheres. *J. Opt. Soc. Amer. A.* **4**, 1361–1365.
Bremmer, H. (1960) On the theory of wave propagation through a concentrically stratified troposphere with a smooth profile. 1. *J. Res. Nat. Bur Stand.* **D64**, 467–473.
Brenner, K.-H. and Singer, W. (1993) Light propagation through microlenses: a new simulation method. *Appl. Opt. 32, 4984–4988.*
Brown, M. J., Krauss, R. K. and Smith, R. M. (1968) Dust deposition and weather. *Weatherwise* **21**, 178–183.
Brunsting, A. and Mullaney, P. E. (1972*a*) Differential light scattering: a possible method of mammalian cell identification. *J. Coll. Interf. Sci.* **39**, 492–496.
Brunsting, A. and Mullaney, P. E. (1972*b*) Light scattering from coated spheres: Model for biological cells. *Appl. Opt.* **11**(3), 675–680.
Brunsting, A. and Mullaney, P. E. (1974) Differential light scattering from spherical mammalian cells. *Biophys. J.* **14**, 439–453.
Buley, E. R. (1967) Scattering from inhomogeneous plasma spheres in a collisionless approximation. *IEEE Trans. Antenn. Propagat.* **AP–15**, 677–681.
Bullrich, K. (1964) Scattered radiation in the atmosphere and the natural aerosol. *Advances in Geophysics* **10**, 99–260.
Bullrich, K., Eschelbach, J., Fischer, K. and Hänel, G. (1972) New aspects of scattering and absorbing properties of atmospheric aerosol particles. *J. Coll. Interf. Sci.* **39**, 546–550.
Burlak, G., Koshevaya, S., Sanchez-Mondragon, J. and Grimalsky, V. (2001) Electromagnetic eigenoscillations and fields in a dielectric microsphere with multilayer spherical stack. *Opt. Comm.* **187**, 91–105.
Cachorro, V. E. and Salsedo, L. L. (1991) New improvements for Mie scattering calculations. *J. Electromag. Waves. Applicat. 5, 913–926.*
Cachorro, V. E., Salsedo, L. L. and Casanova J. L. (1989) Programa LVEC–MIE para el calcudo de las magnitudes de la teoria de scattering de Mie. *Anales de Fisica B.* **85**, 198–211.

Cadle, R. D. (1976) Solid particles in atmosphere and cosmic space. In: *Chemistry of lower atmosphere.* pp. 90–154. Khimiya, Moscow.
Chandrasekhar, S. (1960) *Radiative transfer.* Dover Publications, New York.
Chen, H. C. and Cheng, D. K. (1964) Scattering of electromagnetic waves by an anisotropic plasma coated conducting cylinder. *IEEE Trans. Antenn. Propagat.* **AP–12**, 348–353.
Chen, H. C. and Cheng, D. K. (1965) Scattering electromagnetic and plasma waves by a conducting cylinder coated with a compressible plasma. *Appl. Sci. Res. B* **11**, 442–452.
Chen, H. C. and Cheng, D. K. (1966) Concerning lossy, compressible, magneto-ionic media: General formulation and equation decoupling. *IEEE Trans. Antenn. Propagat.* **AP–14**, 497–501.
Chen, H. G., Iskander M. F. and Penner, J. E. (1990) Light scattering and absorption by fractal agglomerates and coagulation of smoke aerosols. *J. Mod. Opt.* **37**, 171– 181.
Chen, T. W. (1991) Superposition principle of small-angle light scattering at high frequencies. *J. Appl. Phys.* **70**, 1031–1032.
Chen, Z., Sheng, P., Weitz, D. A., Lindsay, H. M. and Lin, M. J. (1988) Optical properties of aggregate clusters. *Phys. Rev. B.* **37**, 5232–5235.
Cherkasova, K. P. and Khijnyak, N. A. (1978) Scattering by ferrite sphere of plane electromagnetic wave, directed under arbitrary angle to magnetized field. *Ukrainskiy Fizicheskiy Jurna*l **23**, 1673–1681
Cherkasova, K. P., Lomonosov, M. I. and Khijnyak, N. A. (1977) Diffraction of electromagnetic wave on sphere with magnetic anisotropy. *Izvestiya VUZOV. Radiofizika* **20**, 913–923.
Chernov, L. A. (1975) *Waves in randomly non-uniform media.* Nauka, Moscow.
Chew, H. and Kerker, M. (1976) Abnormally low electromagnetic scattering cross sections. *J. Opt. Soc. Amer.* **66**, 445–449.
Chew, H., Sculley, M., Kerker, M., McNulty, P. J. and Cooke, D. D. (1978) Raman and fluorescent scattering by molecules embedded in small particles: Results for coherent optical processes. *J. Opt. Soc. Amer.* **68**, 1686–1689.
Ching, S. C., Lai, H. M. and Young, K. (1987) Dielectric microspheres as optical cavities: thermal spectrum and density of states. *J. Opt. Soc. Amer. B.* **4**, 1995–2003.
Choi, M. K. (2001) Numerical calculation of light scattering from a layered sphere by the boundary element method. *J. Opt. Soc. Amer. A.* **18**, 577–583.
Choi, M. K. and Brock J. R. (1997) Light scattering and absorption by radially inhomogeneous sphere: application of the hybrid numerical method. *J. Opt. Soc. Amer. B.* **14**, 620–626.
Choi, M. K. , Liebman, L. A. and Brock J. R. (1996) Finite element solution of the Maxwell equations for absorption and scattering of electromagnetic radiation by a coated dielectric particle. *Chem. Eng. Commun.* **151**, 5–17.
Chowdhury, D. Q., Barber, P. W. and Hill, S. C. (1992) Energy density distribution inside large non-absorbing spheres using by using Mie theory and geometrical optics. *Appl. Opt.* **31**, 3518–3523.
Chowdhury, D. Q., Hill, S. C. and Barber, P. W. (1991) Morphology dependent resonances in radially inhomogeneous spheres. *J. Opt. Soc. Amer. A.* ***8**, 1702–1705.*

Chýlek, P. (1973) Mie scattering into the backward hemisphere. *J. Opt. Soc. Amer.* **63**, 1467–1471.

Chýlek, P. (1976) Partial wave resonances and the ripple structure in the Mie normalized extinction cross section. *J. Opt. Soc. Amer.* **66**, 385–387.

Chýlek, P., Grams, G. W. and Pinnick , R. G. (1976) Light scattering by irregular randomly oriented particles. *Science* **193**, 480–482.

Chýlek, P., Kiehl, J. T. and Ko, M. K. W. (1978) Optical levitation and partial wave resonances. *Phys. Rev.* **A18**, 2229–2233.

Chýlek, P., Ramaswamy, V., Ashkin, A. and Dziedzič, J. M. (1983) Simultaneous determination of refractive index and size of spherical dielectric particles from light scattering data. *J. Opt. Soc. Am.* **58**, 2302–2307.

Chýlek, P., Ramaswamy, V. and Cheng, R. J. (1984) Effect of graphitic carbon on the albedo of clouds. *J. Atmosph. Sci.* ***41**, 3076–3084.*

Chýlek, P., Pendleton, J. D. and Pinnick, R. G. (1985) Internal and near–surface scattered field of a spherical particle at resonance condition. *Appl. Opt.* **24**, 3940–3942.

Chýlek, P., Srivastava, V., Pinnick, R. and Wang, R. T. (1988) Scattering of electromagnetic waves by composite spherical particles: experiment and effective medium approximations. *Appl. Opt.* **27**, 2396–2404.

Chýlek, P., Videen, G. and Ngo, D. (1998) Effect of air bubbles on absorption of solar radiation by water droplets. *J. Atm. Sci.* **55**, 340–3.

Chýlek, P., Videen, G., Geldart, D. J. W., Dobbie, J. S. and Tso, H. C. W. (2000) Effective medium approximations for heterogeneous particles. In: Mishchenko, M. I., Hovenier, J. W. and Travis, L. D., (eds.), *Light scattering by nonspherical particles.* pp.*273–308.* Academic Press, San Diego.

Clark, W. E. and Whitby, K. T. (1967) Concentration and size distribution measurements of atmospheric aerosols and a test of the theory of self-preserving size distributions. *J. Atm. Sci.* **24**, 677–687.

Cohoon, D. K. (1989) An exact solution of Mie type for scattering by a multilayer anisotropic sphere. *J. Electromagn. Waves Appl.* **3**, 421–448.

Cooray, M. F. R. and Ciric, I. R. (1993) Wave scattering by a chiral spheroid. *J. Opt. Soc. Amer. A.* ***10**, 1197–1203.*

d'Almeida, G. A., Koepke, P. and Shettle, E. P. (1991) *Atmospheric aerosols. Global climatology and radiative characteristics.* Deepak, Hampton.

Davies, C. N. (1974) Size distribution of atmospheric particles. *J. Aerosol Sci.* **5**, 293– 300.

Davis, V. A. and Schwartz, L. (1985) Electromagnetic propagation in close-packed disordered suspensions. *Phys. Rev. B.* **31**, 5155–5165.

Deirmendjian, D. (1969) *Electromagnetic scattering on spherical polydispersions.* Elsevier, New York.

Devaty, R. P. (1988) Extinction of electromagnetic waves by a small gyrotropic sphere. *Phys. Rev. B* **38**, 7972–7979.

Dobson, C. C. and Lewis, J. W. L. (1989) Survey of the Mie problem source function. *J. Opt. Soc. Amer. A.* **6**, 463–466.

Doicu, A. (2003) Null-field method to electromagnetic scattering from uniaxial anisoropic

particles. *Opt.Comm.* **218**, 11–17.
Dolginov, A. Z. and Silant'ev, N. A. (1976) Light scattering by anisotropic particles. *Astroph. Space Sci.* **43**, 319–336.
Dolginov, A. Z., Gnedin, Yu. N. and Silant'ev, N. A. (1995) *Propagation and polarization of radiation in cosmic media.* Gordon & Breach, Basel.
Dombrovskiy, L. A. (1974*a*) Light scattering and absorption by hollow spherical particles. *Izvestiya AN USSR. Fizika Atmosfery i Okeana* **10**, 720–727.
Dombrovskiy, L. A. (1974*b*) Radiation of plane-parallel layer from hollow spherical particles of aluminium oxide. *Teplofizika Vysokih Temperatur* **12**, 1316–1318.
Donn, B. and Swamy, K. S. (1969) Extinction by interstellar grains, Mie particles and polycyclic aromatic molecules. *Physica* **41**, 144–150.
Dotsenko, A. V., Efremov, A. M. and Sofronov, V. G. (1985) Scattering of light by two-layered particles of white pigments. *Jurnal Prikladnoi Spektroskopii 43, 281–285.*
Dotsenko, A. V. , Efremov, A. M. and Sofronov, V. G. (1988) Extinction of light by spherical two-layered particles with color pigment shell. *Optiko–Mekhanicheskaya Promyshlennost* n 9, 39–41.
Dotsenko, A. V. and Zakharov, V. K. (1974) Calculation of light extinction spectra for color centres in photochromic glass on the basis of argentum haloid. *Jurnal Prikladnoi Spektroskopii* **21**, 1052–1056.
Dotsenko, A. V. , Zakharov, V. K. and Loiko, V. A. (1984) Determination of sizes of light-sensitivity regions in photochromic glasses. *Jurnal Prikladnoi Spektroskopii **40**, 503–505.*
Dudo, N. I .(1973) Determination of refractive index of monodisperse particles from the characteristics of scattered light. *Jurnal Prikladnoi Spektroskopii* **19**, 904–910.
Dugin, V. P. , Golubitskiy, B. M. and Mirumyants, S. O. (1971) Experimental investigation of the optical characteristics of simulated ice clouds. *Izvestiya AN USSR. Fizika Atmosfery i Okeana* **7**, 871–877.
Dugin, V. P., Maksimiyk, V. S., Mirumyants, S. O. and Nikiforova, N. K. (1976) About anisotropy of light scattering by simulated crystalline cloudy formations (vertical irradiating of medium) . *Trudy IEM* **No. 13(58)**, 184–189.
Dugin, V. P., Volkovitskiy, O. A., Mirumyants, S. O. and Nikiforova, N. K. (1977) About light scattering anisotropy of simulated crystalline cloudy mediums. *Izvestiya USSR. Fizika Atmosfery i Okeana* **13**, 35–41.
Dusel, P. W., Kerker, M. and Cooke, D. D. (1979) Distribution of absorption centers within irradiated spheres. *J. Opt. Soc. Amer.* **69**, 5–59.
Eiden, R. (1966) The elliptical polarization of light scattered by a volume of atmospheric air. *Appl. Opt.* **5**, 569–575.
Emilenko, A. S. and Tolstobrov, V. G. (1981) *Light scattering by polydisperse sole.* Nauka, Moscow.
Eselson, M. P., Oshchepkov, S. L., Prishivalko, A. P. and Yanovskiy, V. Yu. (1982) Microphysical model of water-oil emulsions. *Kolloidnyi Jurnal* **44**, 676–683.
Espenscheid, W. F., Villis, E., Matijevic, E. and Kerker, M. (1965) Aerosol studies by light scattering. *J. Coll. Sci.* **20**, 501–521.
Fahlen, T. S. and Bryant, H. C. (1968) Optical back scattering from single water droplets.

J. Opt. Soc. Am. **58**, 304–310.

Farafonov, V. G., Voshchinnikov, N. V. and Somsikov, V. V. (1996) light scattering by a core-mantle spheroidal particle. *Appl. Opt.* **35**, 5412–5426.

Farone, W. A. (1965) The Rayleigh–Gans approximation for radially symmetric spheres. *J. Opt. Soc. Amer.* **55**, 737–738.

Fedorov, F. I. (1958) *Optics of anisotropic media.* Izdatelstvo AN BSSR, Minsk.

Felske, J. D., Chu, Z. Z. and Ku, J. C. (1983) Mie scattering subroutines (DBMIE and MIEVO): a comparison of computational times. *Appl.Opt.* **22**, 2240–2241.

Fenn, R. N. and Oser, H. (1965) Scattering properties of concentric soot-water spheres for visible and infrared light. *Appl. Opt.* **4**, 1504– 1509.

Fercher, A. F. (1975) Modellrechnungen zum bedeutungs bild biologische zeilen mit hilfe der Mieschen theorie. *Optik* **43**, 129–148.

Ferdinandov, E. S. (1967) Light diffraction from a spherical particle of thin concentric shells. *Bulg. J. Phys.* **3**, 323–333.

Fett, W. (1967) Sichtweite und atmospharishes aerosol in Berlin–Dahlem. *Beitr. Phys. Atmosph.* **40**, 262–278.

Figurovskiy, N. A. and Futran, M. F. (1947) On problem of nature and thickness of protecting emulsifier films on surface of emulsia droplets. *Kollodnyi Jurnal* **9**, 392–399.

Fikioris, J. G. and Uzunoglu, N. K. (1979) Scattering from eccentrically stratified dielectric spheres. *J. Opt. Soc. Amer.* **69**, 1359–1366.

Filippov, Yu. F. (1983) *Tables of indefinite integrals over higher transcendental functions.* Vishcha Shkola, Kharkov.

Fitzgerald, J. W. (1975) Approximation formulas for the equilibrium size of. an aerosol particle as a function of its dry size and composition and the ambient relative humidity. *J. Appl. Meteorol.* **14**, 1044–1049.

Foitzic, L. (1965) The spectral extinction of the atmosphere aerosol by Mie particles with different Gaussian distributions. *Gerlach Beitr. Geophys.* **73**, 199–206.

Folan, L. M., Arnold, S. and Druger, S. D. (1985) Enhanced energy transfer within a microparticle. *Chem. Phys. Lett.* **118**, 322–327.

Ford, G. W. and Werner, S. A. (1978) Scattering and absorption of electromagnetic waves by a gyrotropic sphere. *Phys. Rev. B* **18**, 6752–6769.

Forrest, S. R. and Witten, T. A. (1979) Long-range correlations in smoke particle aggregates. *J. Phys. A.* **12**, L109–L117.

Fucile, E., Borghese, F., Denti, P. and Saija, R. (1993) Theoretical description of dynamic light scattering from an assembly of large axially symmetric particles. *J. Opt. Soc. Amer. A.* **10**, 2611–17.

Fuller, K. A. (1993) Scattering of light by coated spheres. *Opt. Lett.* **18**, 257–259.

Fuller, K. A. and Mackowski, D. W. (2000) Electromagnetic scattering by compounded spherical particles. *In:* Mishchenko, M. I., Hovenier, J. W. and Travis, L. D., (eds.), *Light scattering by nonspherical particles.* pp. *225–272.* Academic Press, San Diego.

Garbacz, R. I. (1962) Electromagnetic scattering from radially inhomogeneous spheres. *Proc. IRE* **50**, 1837–1838.

Gashevskaya, O. S., Loskutov, V. S. and Strelkov, G. M. (1990) Some peculiarities of optical fields for spherical aerosol particles. *Optika i Spektroskopiya* **69**, 144–149.

Gaunaurd, G. C. and Überall, H. (1981) Solution of the inverse electromagnetic scattering problem in the resonance case. *IEEE Trans. Antenn. Propagat.* **AP–29**, 293–297.

Gaunaurd, G. C., Überall, H. and Moser, P. J. (1980) Resonances of dielectrically-coated conducting spheres and the inverse scattering problem. *J. Appl. Phvs.* **52**, 35–43.

Georgievskiy, Yu. S. (1969) About spectral transparence of hazes in spectral range from 0,37 up to 1,0 micron. *Izvestiya AN USSR. Fizika Atmosfery i Okeana* **5**, 388–394.

Georgievskiy, Yu. S. and Rozenberg, G. V. (1973) Humidity as the factor of aerosol variability. *Izvestiya AN USSR. Fizika Atmosfery i Okeana* **9**, 126–137.

Gillette, D. A. and Blifford, I. H. (1972) Measurements of aerosol size distribution and vertical fluxes of aerosols on land subject to wind erosion. *J. Appl. Meteorol.* **11**, 977–987.

Goldberger, M. L. and Watson, K. M. (1973) *Collision theory.* Mir, Moscow.

Goldstein, M. (1956) Influence of anisotropy of spherulites on their light scattering. *J. Polym. Sci.* **20**, 413–415.

Golubyatnikov, A. P. and Katsenelenbaum, B. Z. (1997) Scattering by gradient dielectric sphere (Luneberg lens). *Radioteknika i Elektronika* **42**, 1429–46.

Gorbunov, N. I. (1963) *High-dispersive minerals and methods of their investigation.* Izdatelstvo AN USSR, Moscow.

Gorchakov, G. I. (1966) Light scattering matrixes for surface air. *Izvestiya AN USSR. Fizika Atmosfery i Okeana* **2**, 595–605.

Gorchakov, G. I. and Emilenko, A. S. (1974) Modeling of atmospheric aerosol microstructure and light scattering matrix. *Izvestiya AN USSR. Fizika Atmosfery i Okean*a **10**, 350–359.

Gorchakov, G. I. and Sidorov, V. N. (1980) On the spectral dependence of light scattering matrix components for atmospheric haze. *Izvestiya AN USSR. Fizika Atmosfery i Okeana* **16**, 974–978.

Gould, R. N. and Burman, R. (1964) Some electromagnetic wave functions for propagation in stratified media. *J. Atm. Terr. Phys.* **26**, 335–340.

Gradshteyn, I. S. and Ryzhik, I. W. (1963) *Tables of integrals, series and products.* Academic, New York.

Graglia, R. D. and Uslenghi, P. L. E. (1984) Electromagnetic scattering from anisotropic materials. Part I: General theory. *IEEE Trans. Antenn. Propagat.* **AP–32**, 867–869.

Graglia, R. D. and Uslenghi, P. L. E. (1987) Electromagnetic scattering from anisotropic materials. Part II: Computer code and numerical results in two dimensions. *IEEE Trans. Antenn. Propagat.* **AP–35**, *225–232.*

Graglia, R. D., Uslenghi, P. L. E. and Zich, R. S. (1989) Moment method with isoparametric elements for three dimensional anisotropic scatterers. *Proc. IEEE* **77**, 750–760.

Grams, G. W., Blifford, I. H., Schuster, B. G. and De Luisi, J. J. (1972) Complex index of refraction of airborne fly ash determined by laser radar and collection of particles at 13 km. *J. Aim. Sci.* **29**, 900–905.

Grams, C. W., Blifford, I. H., Gillette, D. A. and Russel, P. B. *(*1974) Complex index of refraction of airborne soil particles. *J. Appl. Meteorol.* **13**, 459–471.

Green, H. L. and Lane, W. P. (1972) *Particulate Clouds: Dust, Smokes and Mists.* Khimiya, Leningrad.

Greenberg, J. M. (1968) Interstellar grains. In: Middlehurst, B. M. and Aller, L. H. (eds.) *Stars and stellar systems.* V.7, ch.6. Univ. Chicago Press, Chicago.

Greene, W. M., Spjut, R. E., Bar-Ziv, E., Sarofim, A. F. and Longwell, J. P. (1985) Photophoresis of irradiated sphere: absorption centers. *J. Opt. Soc. Amer. B.* **2**, 998–1004.

Grehan, G. and Gouesbet, G. (1979) Mie theory calculations: new progress with emphasis on particle sizing. *Appl. Opt.* **18**, 3489–3493.

Gurwich, I., Shiloah, N. and Kleimann, M. (2001) Calculations of the Mie scattering coefficients for multilayered particles with large size parameters. *J. Quant. Spectrosc. Rad. Transf.* **70**, 433–440.

Gustafson, B. Å. S., Kolokolova, L. and Videen, G. (eds.) (2002) *Electromagnetic and light scattering by nonspherical particles.* Army Res. Lab., Adelphi.

Gutman, A. L. (1965) Application of cross sections method to problem on electromagnetic wave diffraction by non-uniform sphere. *Radiotekhnika i Elektronika* **10**, 1583–1593.

Gutman, A. L. and Chesnokov, V. A. (1969) Comparison of ray approximation with exact solution of problem on scattering of plane wave by non-uniform plasma sphere. *Radiotekhnika i Elektronika* **14**, 335–336.

Güttler, A. (1952) Die Miesche theorie der beugung durch dielektrische kugeln mit absorbierendem kern und ihre bedeutung fur probleme der interstellaren materie und des atmospharischen aerosols. *Ann. der Phys.* **11**, 65–98.

Hale, G. M. and Querry, M. R. (1973) Optical constants of water in the 200 nm to 200 μm wavelength region. *Appl. Opt.* **12**, 555–563.

Hänel, G. (1968) The real part of mean complex refractive index and the mean density of samples of atmospheric aerosol particles. *Tellus* **20**, 371–379.

Hänel, G. (1970) The size of atmospheric aerosol particles as a function of the relative humidity. *Beitr. Phys. Atmosph.* **43**, 119–132.

Hänel, G. (1971) New results concerning the dependence of visibility on relative humidity and their significance in a model for visibility forecast. *Beitr Phys. Atmosph.* **44**, 137–167.

Hänel, G. (1972) Computation of the extinction of visible radiation by atmospheric aerosol particles as a function of the relative humidity, based upon measured properties. *J. Aerosol Sci.* **3**, 337–386.

Hänel, G. (1975) Physikalisch-chemishe und optische eigenschaften von aero-solteilchen. *Meteorol. Fortbildung* N 3, 11–14.

Hänel, G. (1976) The properties of atmospheric aerosol particles as functions of the relative humidity at thermodyriamic equilibrium with the surrounding moist air. *Advances in Geophysics* **19**, 73–188.

Hänel, G. and Bullrich, K. (1978) Physico-chemical property models of tropospheric aerosol particles. *Beitr. Phys. Atmosph.* **51**, 129–138.

Hänel, G. and Lehman, M. (1980) Aerosolteilchen und relative feuchte: neue messergebnisse und meteorologische konsequenzen. *Ann. Meteorol.* N 15, 37–38.

Hänel, G. and Lehman, M. (1981) Equilibrium size of aerosol particles and relative humidity: New experimental data from various aerosol types and their treatment for cloud physics application. *Beitr. Phys. Atmosph.* **54**, 57–71.

Hänel, G. and Zankl, B. (1979) Aerosol size and relative humidity: water uptake by mixtures of salts. *Tellus* **31**, 478–486.

Hardaker, P. J. and Holt, A. R. (1994) Electromagnetic scattering from the spherical Luneberg lens. *J. Electrom. Waves Applicat. 8, 391–406.*

Harrison, H., Herbert, J. and Waggoner, A. P. (1972) Mie-theory computation of lidar and nephelometric scattering parameters for power law aerosols. *Appl. Opt.* **11**, 2880–2885.

Helaly, A. and Shafai, L. (1989) Scattering by two concentric spheres using the impedance boundary condition on the inner surface. *Can. J. Electr. Comput. Eng.* **14**, 122–126.

Heller, W. (1965) Remarks on refractive index mixture rule. *J. Phys. Chem.* **69**, 1128–1129.

Herman, B. M. and Battan, L. J. (1961) Calculation of Mie backscattering from melting ice spheres. *J. Meteorol.* **18**, 468–478.

Hidy, G. M., Mueller, P. K., and Wange, H. H. (1974) Observation of aerosol over southern California coastal waters. *J. Appl. Meteorol.* **13**, 96–107.

Hightower, R. L. and Richardson, C. B. (1988) Resonant Mie scattering from a layered sphere. *Appl. Opt.* **27**, 4850–4855.

Hill, S. C. and Benner, R. E. (1988) Morphologically dependent resonances. In: Barber, P. W. and Chang, R. K. (eds.) *Optical effects associated with small particles*, pp. 3–61. World Scientific, Singapore.

Hill, S. C., Benner, R. E., Rushforth, C. K. and Conwell, P. R. (1984) Structural resonances observed in the fluorescence emission from small spheres on substrates. *Appl. Opt.* **23**, 1680–1683.

Hill, S. C., Rushforth, C. K., Benner, R. E., and Conwell, P. R. (1985) Sizing dielectric spheres and cylinders by aligning measured and computed structural resonance locations: algorithm for multiple orders. *Appl. Opt.* **24**, 2380–2390.

Hizal, A. and Tosun, H. (1973), State-space formulation of scattering with application to spherically symmetrical objects. *Can. J. Phys.* **51**, 549– 556.

Hoekstra, A. , Rahola, J. and Sloot, P. (1998) Accuracy of internal fields in volume integral equation simulations of light scattering. *Appl. Opt.* **37**, 8482–8497.

Hoidale, G. B. and Blanco, A. J. (1968) An infrared spectroscopic view of the nature of giant and large particle atmospheric dust. *J. Rech. Atmosph.* **3**, 561–567.

Hoidale, G. B. and Smith, S. M. (1968) Analysis of the giant particle component of the atmosphere over an interior desert basin. *Tellus* **20**, 251–268.

Holland, A. C. and Draper, J. S. (1967) Analytical and experimental investigations of light scattering from polydispersions of Mie particles. *Appl. Opt.* **6**, 511–518.

Holland, A. C. and Gagne, G. (1970) The scattering of polarized light by polydisperse systems of irregular particles. *Appl. Opt.* **9**, 1113–1121.

Holland, A. C. and Gagne, G. (1971) Comment on: The scattering of polarized light by polydisperse systems of irregular particles. *Appl. Opt.*, **10**, 1173–1174.

Holoubek, J. (1972) A simple representation of small angle light scattering from an anisotropic sphere. *J. Polym. Sci.* **10**, 1461–1465.

Holoubek, J. (1991) Small-angle light-scattering from an anisotropic sphere in the Rayleigh–Gans–Debye approximation: the Mueller matrix formalism. *Appl. Opt.* **30**, 4987–4992.

Holt, A. R., Uzunoglu, N. K. and Evans, B. G. (1978) An integral equation solution to the scattering of electromagnetic radiation by dielectric spheroids and ellipsoids. *IEEE*

Trans. Antenn. Propagat. **AP–26**, 706–712.
Hui, P. M. and Stroud, D. (1986) Complex dielectric response of metal-particle clusters. *Phys. Rev. B.* **33**, 2163–2169.
Huibers, P. D. T. (1997) Models for the wavelength dependence of the index of refraction of water. *Appl. Opt.* **36**, 3785–3787.
Inagaki, T., Yamamoto, Y. and Yabushita, S. (1991) The ultraviolet extinction by hollow spherical particles of graphite. *Astrophys. Space Sci.* **182**, 75–80.
Ioannidou, M. P. and Chrissoulidis, D. P. (2002) Electromagnetic-wave scattering by a sphere with multiple spherical inclusions. *J. Opt. Soc. Amer. A* ***19***, *505–512.*
Ippolitov, I. I., Krekov, G. M. and Lopasova, T. A. (1987) Optical properties of clouds: the effect of non-uniform internal structure of droplets. In: Zuev V.E. (ed.), *Optical and meteorological investigations of Earth atmosphere*, pp.53–63, Nauka, Novosibirsk.
Irvine, W. M. (1965) Light scattering by spherical particles: radiation pressure, asymmetry factor, and extinction cross section. *J. Opt. Soc. Amer.* **55**, 16–21.
Irvine, W. M. and Pollack, J. B. (1968) Infrared optical properties of water and ice spheres. *Icarus* **8**, 324–360.
Isakov, A. I., Krupinina, L. A., and Leonov, Yu. S (1976) Method of automated selection of polymer shells from foam particles. *Kratkie Soobshcheniya po Fizike* N 5, 8–12.
Ishimaru, A. (1981) *Wave propagation and scattering. in random media.* Academic, New York.
Ivanov, A. P. (1969) *Optics of scattering media.* Nauka i tehnika, Minsk.
Ivanov, E. A. (1968) *Diffraction of electromagnetic waves on two bodies.* Nauka i tehnika, Minsk.
Ivlev, L. S. and Andreev, S. D. (1986) *Optical properties of atmospheric aerosols.* Leningrad Univ. Press, Leningrad.
Ivlev, L. S. and Popova, S. I. (1972) Optical constants of atmospheric aerosol substances. *Izvestiya VUZOV. Fizika* N 5, 91–97.
Ivlev, L. S. and Popova, S. I. (1973) Complex refractive index of disperse phase of atmospheric aerosol. *Izvestiya AN USSR. Fizika Atmosfery i Okeana* **9**, 1034–1042.
Ivlev, L. S., Spajakina, N. K., Grigorov, N. O. and Yanchenko, E. L. (1980) Influence of humidity change on the optical and microphysical characteristics of atmospheric aerosol. *Problemy Fiziki Atmosfery* N 8, 112–123.
Jaaskelainen, T. and Ruuskanen, J. (1981) Note on Lentz's algorithm. *Appl. Opt.* **20**, 3289–3290.
Jaenicke, R. and Davies, C. N. (1976) The mathematical expression of the size distribution of atmospheric aerosols. *J. Aerosol Sci.* **7**, 265–269.
Jakobi, W., Junge, C. E. and Lippert, W. (1952) Reihenunter suchungen des naturlichen aerosols mittels electronenmicroskop. *Arch. Meteorol. Geophys. Biokl.* **A5**, 166–170.
Jarzembski, M. A. and Srivastava, V. (1989) Electromagnetic fields enhancement in small liquid droplets using geometrical optics. *Appl. Opt.* **28**, 4962–4965.
Jazbi, B. and Wickramasinghe, N. C. (1991) Scattering by low refractive index dielectric spheres and cylinders compared with rigirous calculations for hollow particles. *Astrophys. Space Sci.* **179**, 303–311.
Johnson, B. R. (1996) Light scattering by multilayer sphere. *Appl. Opt.* **35**, 3286–3296.
Johnson, B. R. (1999) Exact theory of electromagnetic scattering by a heterogenous mul-

tilayer sphere in the infinite-layer limit: effective media approximation. *J. Opt. Soc. Amer. A.* **16**, 845–852.

Jonhson, P. B. and Christy, R. W. (1972) Optical constants of the noble metals. *Phys. Rev. B.* **6**, 4370–4379.

Jones, A. P. (1988) Modelling interstellar extinction. 1. Porous grains. *Mon. Not. Royal Astron. Soc.* **234**, 209–218.

Jones, A. R. (1983) Calculations of the ratios of complex Riccati–Bessel functions for Mie scattering. *J. Phys. D: Appl. Phys.* ***16,*** L49–L52.

Jullien, R. and Botet, R. (1987) *Aggregation and fractal aggregates.* World Scientific, Singapore.

Jullien, R. and Botet, R. (1989) Light scattering by fractal aggregates: a Monte Carlo method in the geometrical optics limit. *J. Phys. (France)* **50**, 1983–1994.

Junge, C. (1955) The size distribution and aging of natural aerosols as determined from electrical and optical data on the atmosphere. *J. Meteorol.* **12**, 13–25.

Junge, C. E. (1957) Atmospheric chemistry. *Advances in Geophysics* **4**, 1–81.

Junge, C. E. (1963) *Air chemistry and radioactivity*. Academic, New York.

Kafalas, P. and Ferdinand, A. P. (1973) Fog droplet vaporization and fragmentation by a 10.6 μm laser pulse. *Appl. Opt.* **12**, 29–39.

Kai, L. and D'Alessio, A. (1995) Electromagnetic radiation at the interface of two concentric spheres. *J. Mod. Optics* **42**, 1603–1610.

Kai, L. and Massoli, P. (1994) Scattering of electromagnetic plane wave by radially-inhomogeneous sphere: a finely-stratified sphere model. *Appl. Opt.* **33**, 501–511.

Kaiser, T., Lange, S. and Schweiger, G. (1994) Structural resonances in a coated sphere: investigation of the volume-averaged source function and resonance positions. *Appl. Opt.* **33**, 7789–7797.

Kaiser, T. and Schweiger, G. (1993) Stable algorithm for the computation of Mie coefficients for scattered and transmitted fields of a coated sphere. *Comput. Phys.* **7**, 782–786.

Kapustin, V. N. and Lyubovtseva ,Yu. S. (1975) Relative humidity and parameters of natural aerosol. *Izvestiya AN USSR. Fizika Atmosfery i Okeana* **11**, 908–915.

Karam, M. A. and Fung, A. K. (1993) Electromagnetic energy absorbed within a Mie sphere. *J. Electrom. Waves Applicat.* **7**, 1379–1388.

Karonis, G. J., Kaklamani, D. I. and Uzunoglu, N. K. (1999) Scattering of electromagnetic waves by a chiral object of ellipsoidal shape. *J. Electromagn. Waves and Appl.* **13**, 817–846.

Kasten, F. (1969) Visibility forecast in the phase of precondensation. *Tellus* **21**, 631– 635.

Kattawar, G. W. and Eisner, M. (1970) Radiation from a homogeneous isothermal sphere. *Appl. Opt.* **9**, 2685–2690.

Kattawar, C. W. and Hood, D. A. (1976) Electromagnetic scattering from a spherical polydispersion of coated spheres. *Appl. Opt.* **15**, 1996–1999.

Kaufman, Y. J., Tanre, D., Gordon, H. R., Nakajima, T. (eds.) (1997) Special issue on passive remote sensing of tropospheric aerosols and atmospheric corrections of the aerosol effects. *J. Geophys. Res.* **102**, 16815–17217.

Kerker, M. (1969) *The scattering of light and other electromagnetic radiation.* Academic, New York.

Kerker, M. (1975) Invisible bodies. *J. Opt. Soc. Amer.* **65**, 376–379.
Kerker, M. (1977) Some recent reflections on light scattering. *J. Coll. Interf. Sci.* **58**, 100–112.
Kerker, M. (1985) The optics of colloidal silver. Something old and something new. *J. Coll. Interf. Sci.* **105**, 297–314.
Kerker, M. and Blatchford, C. G. (1982) Elastic scattering, absorption and SERS by concentric spheres comprised of a metallic and a dielectric region. *Phys. Rev. B.* **26**, 4052–4063.
Kerker, M. and Cooke, D. D. (1972) Comments on: Effect of dispersing agents on the angular dependence of light scattered from the polystyrene sphere/water sols. *Appl. Opt.* **11**, 689–690.
Kerker, M. and Cooke, D. D. (1973) Radiation pressure on absorbing spheres and photophoresis. *Appl. Opt.* **12**, 1378–1379.
Kerker, M., Kratohvil, J. P. and Matijevič, E. (1962) Light scattering functions for concentric spheres. Total scattering coefficient, m_1=2.1050, m_2=1.4821. *J. Opt. Soc. Amer.* **52**, 551–561.
Kerker, M., Kauffman, L. H. and Farone, W. A. (1966) Scattering of electromagnetic waves from two concentric spheres when the outer shell has a variable refractive index. Numerical results. *J. Opt. Soc. Amer.* **56**, 1053–1056.
Kerker, M., Cooke, D. D. and Ross, W. D. (1975) Pigmented microvoid coatings: theoretical study of three models. *J. Paint Technology* **47**, 33–42.
Kerker, M., Cooke, D. D, Chew, H. and McNulty, P. J. (1978) Light scattering by structured spheres. *J. Opt. Soc. Amer.* **68**, 592–601.
Khlebtsov, N. G. (1979a) Scattering matrix for anisotropic ellipsoids comparable to light wavelength. *Optika i Spektroskopiya* **46**, 529–531.
Khlebtsov, N. G. (1979b) The analysis and numerical calculation of scattering matrix for soft spherulites comparable to light wavelength. *Optika i Spektroskopiya* **46**, 341–349.
Khlebtsov, N. G. (1996) Spectroturbidimetry of fractal clusters: test of density correlation function cutoff. *Appl. Opt.* **35**, 4261–4270.
Khlebtsov, N. G. and Melnikov, A. G. (1990) The theory of dichroism of oriented dispersions of anisotropic particles. *Kolloidnyi Jurnal 52, 928–934.*
Khlebtsov, N. G. and Melnikov, A. G. (1994) Structure factor and exponent of scattering by polydisperse fractal colloidal aggregates. *J. Coll. Interf. Sci.* **163**, 145–151.
Khlebtsov, N. G., Dykman, L. A., Krasnov, Ya. M. and Melnikov, A. G. (2000) Light absorption by clusters of gold and silver particles forming in the regimes of fast and slow aggregation. *Kolloidnyi Jurnal* **62,** 765–779.
Khrgian, A. Kh. (1969) *Physics of atmosphere.* Gidrometeoizdat, Moscow.
Kiselev, A. D., Reshetnyak, V. Yu. and Sluckin, T. J. (2002) Light scattering by optically anisotropic scatterers: T–matrix theory for radiaal and uniform anisotropies. *Phys. Rev. E* **65**, 056609.
Klyatskin, V. I. (1973) The statistical theory of light propagation in randomly non-uniform media. *Izvestiya VUZOV. Radiofizika* **16**, 1629–1641.
Klyatskin, V. I. (1975) *Statistical description of dynamic systems with fluctuating parameters.* Nauka, Moscow.

Klyatskin, V. I. and Tatarskiy, V. I . (1972) The statistical theory of light propagation in turbulent media. *Izvestiya VUZOV. Radiofizika* **15**, 1433–1455.

Klyatskin, V. I. and Tatarskiy, V. I. (1973) Approximation of diffusion process in some non-stationary statistical problems of physics. *Uspekhi Fizicheskih Nauk* **110**, 499–536.

Klyuchnik, A. V., Lozovik, Yu. E. and Nersesyan, O. N. (1986) Anomalies in scattering of electromagnetic waves by particles with magnetic coating. *Pisma v Jurnal Technicheskoi Fiziki* **12**, 348–356.

Kokhanovsky, A. A. (2001) *Optics of light scattering media. Problems and solutions.* Wiley–Praxis, London.

Kokorin, A. M. and Shifrin, K. S. (1989) Use of model of enlighten sphere for calculation of light scattering by particles of ocean suspension. *Izvestiya AN SSSR. Fizika Atmosfery i Okeana* **25**, 193–201.

Kokorin, A. M. and Shifrin, K. S. (1998) Influence of internal structure of radially inhomogeneous particles of marine suspension on the phase function and polarization of scattered light. *Optiko–mekhanicheskaya Promyshlennost.* **65**, 26–31.

Kondratiev, K. Ya. (ed.) (1978) *Atmospheric aerosol and its influence on radiative transfer.* Gidrometeoizdat, Leningrad.

Kondratiev, K. Ya. and Pozdnyakov, D. V. (1981) *Aerosol models of atmosphere.* Nauka, Moscow.

Kondratiev, K. Ya. and Ter-Markaryants, N. E., (eds.) (1976) *Complete radiative experiment.* Gidrometeoizdat, Leningrad.

Kondratiev, K. Ya., Vasil'ev, O. B., and Ivlev, L. S. (1973) *Influence of aerosol on radiative transfer: possible climatic impact.* Leningrad Univ. Press, Leningrad.

Kopilovich, L. E. (1980) Method of calculation of Mie coefficients. *Radiotechnika i Elektronika* **25**, 858–860.

Kopytin, Yu. D., Belov, N. N., Bukatyi, V. I., Skripkin, A. M. and Sorokin, Yu. M. (1990) *Optical discharge in aerosols.* Nauka, Novosibirsk.

Korostina, O. M. and Rozanov, V. V. (1988) Sensitivity of optical characteristics of two-layered spherical particles to variations of refractive indexes of core and shell. *Optika Atmosfery* **1**, 25–31.

Kotomtseva, L. A., Lednyeva, G. P. and Astafyeva, L. G. (1998) Microparticles as laser sources and bistable elements. *Laser Physics* **8**, 34–37.

Kotov, Yu. N. and Novikov, A. S. (1967) On scattering of a polarized light by polymer films: Rayleigh–Gans scattering by radial spherulite. *Jurnal Prikladnoi Spektroskopii* **7**, 726–730.

Kou, L., Labril, D. and Chýlek, P. (1993) Refractive index of water and ice in the $0.65 \div 2.5$ micron spectral range. *Appl. Opt.* **32**, 3531–3540.

Kravtsov,Yu. A., Rytov, S. M. and Tatarskiy, V. I. (1975) Statistical problems in the diffraction theory. *Uspekhi Fizicheskih Nauk* **115**, 239–262.

Kravtsov, Yu. A. (1993) New effects in wave propagation and scattering in random media (mini-review). *Appl. Opt.* **32**, 2681–2691.

Kravtsov, Yu. A. and Tatarskiy, V. I. (1975) *Statistical phenomena at wave diffraction.* Ryazan.

Kreibig, U. and Vollmer, M. (1995) *Optics of metal clusters.* Springer, Berlin.

Kreibig, U. and von Fragstein, C. (1969) The limitation of electron mean free path in small silver particles. *Z. Phys.* **224**, 307–323.

Krekov, G. M. and Rakhimov, R. F. (1973) On calculation of the radiative characteristics of polydisperse concentric spheres. *Izvestiya VUZOV. Fizika* N 6, 30–35.

Krekov, G. M and Rakhimov, R. F. (1975*a*) About algorithm of calculation the optical characteristics of two-layered particles . In: *Physics of mesosphere and mesospheric clouds*, pp. 78–83. Nauka, Moscow.

Krekov, G. M. and Rakhimov, R. F. (1975*b*) The optical characteristics of some types of mesospheric aerosol with two-layered structure. In: *Physics of mesosphere and mesospheric clouds*, pp. 84–91. Nauka, Moscow.

Krekov, G. M. and Rakhimov, R. F. (1982) *Optical remote sensing model of continental aerosol.* Nauka (Siberian branch), Novosibirsk.

Krekov, G. M., Rakhimov, R. F., Savel'ev, B. A. and Fadeev, V. Ya. (1976) Influence of condensation centres on optical properties of atmospheric water aerosol. *Izvestiya VUSOV. Fizika* N 1, 128–131.

Krepak, V. N., Nazarenko, L. A. and Yakimenko, I. P. (1973) On the account of transition region in problems of electromagnetic wave scattering by the radially-inhomogeneous plasma objects. *Radiotekhnika i Elektronika* **18**, 2225–2228.

Kudinov, V. V. and Ivanov, V. M. (1981) *Deposition by Plasma of Refractory Coatings.* Nauka, Moscow.

Kudinov, V. V., Puzanov, A. A. and Zambrzhitsky, A. P. (1981) *Optics of Plasma Coatings.* Nauka, Moscow.

Kutuzov, Yu. I. and Klenin, V. I. (1983) Calculation of integral functions for light scattering by two-layered spherical particles. *Optika i Spektroskopiya* 55, 383–389.

Kuzmin, V. N. (1979) Matrix of electromagnetic wave scattering by spherical particle in coordinate and pulsing representations. *Izvestiya AN BSSR. Ser. fiz.–mat. nauk* N 1, 92–101.

Kuzmin, V. N. (1980) Electromagnetic wave scattering matrix for spherical anisotropic particles. *Izvestiya AN BSSR. Ser. fiz.–mat. nauk* N 1, 70–75.

Kuzmin, V. N. (1981*a*) Linear dichroism of scattering systems from anisotropic quartz particles. *Jurnal Prikladnoi Spektroskopii* **35**, 929–931.

Kuzmin, V. N. (1981*b*) Spectral transparence of systems of anisotropic quartz particles. *Jurnal Prikladnoi Spektroskopii* **35**, 1024–1028.

Kuzmin, V. N. (1992) Influence of atmospheric soot aggregation on its optical properties in the model of fractal cluster. *Izvestiya RAN. Fizika Atmosfery i Okeana* **28**, 953–957.

Kuzmin, V. N. and Babenko, V. A. (1981) Light scattering by weakly anisotropic spherical particle. *Optika i Spektroskopiya* **50**, 498–505.

Kuzmin, V. N. and Babenko, V. A. (1982) Some features of angular light scattering characteristics for crystalline particles. *Jurnal Prikladnoi Spektroskopii* **36**, 305–309.

Kuzmin, V. N. and Moroz, O. V. (1992) Plasmon extinction spectra of the volume fractal clusters. *Jurnal Prikladnoi Spektroskopii* **56**, 446–450.

Kuzmin, V. N. and Prishivalko, A. P. (1980) On quasi-homogeneity of scattering mediums. *Izvestiya AN BSSR. Ser. fiz.–mat. nauk* N 1, 70–76.

Kuzmin, V. N., Moroz, O. V. and Prishivalko, A. P. (1988) Asymptotic model for describing the light scattering properties of fractal clusters. *Doklady AN SSSR.* **302**, 332–334.
Lagendijk, A. and van Tiggelen, B. A. (1996) Resonant multiple scattering of light. *Physics Reports* **270**, 143–215.
Lai, H. M., Leung, P. T., Poon, K. L. and Young, K. (1991) Characterization of the internal energy density in Mie scattering. *J. Opt. Soc. Amer. A.* **8**, 1553–1558.
Laikin, M. (1961) The refractive index of quartz in the visible region. *J. Opt. Soc. Amer.* **51**, 238–242.
Lakhtakia, A. (1992) General theory of the Purcell–Pennypaker scattering approach and its extension to bianisotropic scatterers. *Astroph. J.* **394**, 494–499.
Lakhtakia, A., Varadan, V. K. and Varadan V. V. (1985) Scattering and absorption characteristics of lossy dielectric, chiral, nonspherical objects. *Appl. Opt.* **24**, 4146–4154.
Laktionov, A. G. (1972) The content of water soluble substances in atmospheric aerosol particles. *Izvestiya AN USSR. Fizika Atmosfery i Okeana* **8**, 389–395.
Laktionov, A. G. and Bogomolov, Yu. P. (1972) Dependence of the particle sizes of natural aerosol on air humidity. *Izvestiya AN USSR. Fizika Atmosfery i Okeana* **8**, 291–297.
Lange, S. and Schweiger, G. (1994) Thermal radiation from spherical microparticles: a new dipole model. *J. Opt. Soc. Amer. B.* **11**, 2444–2451.
Latham, R. W. (1968) Electromagnetic scattering from cylindrically and spherically stratified bodies. *Can. J. Phys.* **46**, 1463–1468.
Laxpati, S. R. and Uslenghi, P. L. E. (1973) Numerical and asymptotic scattering from two inhomogeneous lenses. *Appl. Phys.* **2**, 11–17.
Lee, S.-I., Noh, T. W., Golben, J. and Gaines, J. R. (1986) Effect of an oxide coating on the infrared and visible absorption of small silver particles. *Phys. Rev. B.* **33**, 5844–5847.
Lentz, W. J. (1976) Generating Bessel functions in Mie scattering calculations using continued fractions. *Appl. Opt.* **15**, 668–671.
Lettiery, T. R. and Preston, R. L. (1985) Observation of Raman spectrum of single optically levitated microdroplet. *Opt. Comm.* **54**, 349–352.
Levine, S. and Kerker, M. (1963) The scattering of electromagnetic waves from two spheres, when the outer sphere has a variable index of refraction. In: Kerker, M. (ed.), *Electromagnetic scattering,* pp. 37–46. Pergamon, New York.
Li, A. and Greenberg, J. M. (1997) A unified model of interstellar dust. *Astron. Astrophys.* **323**, 566–584.
Lindberg, J. D. and Gillespie, J. B. (1977) Relationship between particle size and imaginary refractive index in atmospheric dust. *Appl. Opt.* **16**, 2628–2630.
Lingart, Yu. K., Petrov, V. A. and Tichonova, L. A. (1982) The optical properties of leucosapphire for high temperatures. 1. Region of semitransparency. *Teplofizika Vysokih Temperatur* **20**, 872–880.
Liou, K. N. (1992) *Radiation and cloud processes in the atmosphere.* Oxford Univ. Press, Oxford.
Liou, K. N., Takano, Y. and Yang, P. (2000) Light scattering and radiative transfer in ice crystals clouds: application to climate research. In: Mishchenko, M. I., Hovenier, J. W. and Travis, L. D. (eds.), *Light scattering by nonspherical particles*, pp. 417–450. Academic Press, San Diego.

Liu, L. H., Tan, H. P. and Tong, T. W. (2002) Internal distribution of radiation absorption in a semitransparent spherical particles. *J. Quantit. Spectrosc. Rad. Transf.* **72**, 747–756.

Livshits, G. Sh., Solov'eva, A. I. and Travina, T. B. (1974) On relation of scattering factor of atmospheric haze with relative humidity. *Izvestiya AN USSR. Fizika Atmosfery i Okeana* **10**, 991–992.

Lock, J. A. (1990) Interference enhancement of the internal fields at structural resonances of a coated sphere. *Appl. Opt.* **29**, 3180–3187.

Lock, J. A. (2001) Semiclassical scattering of an electric dipole inside a spherical particle. *J. Opt. Soc. Amer. A.* **18**, 3085–3097.

Logan, N. A. (1965) Survey of some early studies of the scattering of plane waves by a sphere. *Proc. IEEE* **53**, 773–785.

Lopatin, V. N. and Sid'ko, F. Ya. (1977) Influence of shell on light extinction and scattering by spherical three-layered particles – cell models. *Optika i Spektroskopiya* **43**, 930–935.

Lopatin, V. N. and Sid'ko, F. Ya. (1978) Influence of shell and core on phase function and degree of polarization of radiation, scattered by "soft" small biological particles. *Optika i Spektroskopiya* **45**, 704–709.

Lopatin, V. N. and Sid'ko, F. Ya. (1988) *Introduction to optics of cell suspensions.* Nauka, Novosibirsk.

Loskutov, V. S. and Strelkov, G. M. (1984) Internal field of a water droplet on the wavelength of laser radiation in the range 3.39–10.6 μm. *Optika i Spektroskopiya* **56**, 128–131.

Ludlow, I. K. and Everitt, J. (1996) Systematic behavior of the Mie scattering coefficients of spheres as a function of order. *Phys. Rev. E.* **53**, 2909–2924.

Ludlow, I. K. and Everitt, J. (2000) Inverse Mie scattering sceme. *J. Opt. Soc. Amer. A.*, **17**, 2229–2235.

Lumme, K. (2000) Scattering properties of interplanetary dust particles. In: Mishchenko, M. I., Hovenier, J. W. and Travis, L. D, (eds), *Light scattering by nonspherical particles.* pp.*555–583.* Academic Press, San Diego.

Lundgren, D. A. and Cooper, D. W. (1969) Effect of humidity on light-scattering methods of measuring particle concentration. *J. Air Pollut. Assoc.* **19**, 243–247.

Lynch, R. J. (1963) Backscatter from inhomogeneous media. *Phys. Rev.* **130**, 1235–1243.

Mackowski, D. W., Altenkirch, R. A. and Mengus, M. P. (1990) Internal absorption cross-sections in a stratified sphere. *Appl. Opt.* **29**, 1551–1559.

Malushkov, G. D. (1974) Integral expressions for spherical harmonics. *Trudy MIREA.* No 4, 153–157.

Mandelbrott, B. B. (1982) *The fractal geometry of nature.* Freeman, New York.

Margulies, R. S. and Scarf, F. L. (1964) Electromagnetic scattering from a spherical nonuniform medium. Part II: The radar cross-section of a flare. *IEEE Trans. Anten. Propagat.* **AP–12**, 91–96.

Markel, V. A. and George, Th. F. (eds.) (2000) *Optics of nanostructured materials.* Wiley, New York.

Marston, P. L., Bilette, S. C. and Dean, C. E. (1988) Scattering of light by a coated bubble in water near the critical and Brewster scattering angles. *Proc. SPIE* **925**, 308–316.

Mason, B. J. (1957) *The physics of clouds.* Oxford. Univ. Press, Oxford.

Mathai, C. V. and Harrison, A. W. (1980) Estimation of atmospheric aerosol refractive index. *Atmosph. Environ.* **14**, 1131–1135.

Matijevič, E., Kerker, M. and Schulz, K. F. (1960) Light scattering by coated aerosols. Part I – Scattering by the AgCl cores. *Disc. Faraday Soc.* **n 3**, 178–189.

Matijevič, E. , Schulz, K. F. and Kerker, M. (1962) Light scattering by coated aerosols. II: Scattering by linolenic acid aerosols. *J. Colloid Sci.* **17**, 26–38.

Mayhon, R. J. and Schultz F. V. (1967) Electromagnetic plane wave scattering from a plasma coated conducting cylinder. *Radio Sci.* **2**, 853–865.

McCartney, E. J. (1977) *Optics of the atmosphere.* Wiley, New York.

McNulty, P. J., Chew, H. W. and Kerker, M. (1980) Inelastic light scattering. In: Marlow, W. H. (ed.) *Aerosol Microphysics I,* pp. 89–116. Springer, New York.

Meeten, G. H. (1980*a*) The intrinsic optical anisotropy of colloidal particles in the anomalous diffraction approximation. *J. Coll. Interf. Sci.* **74**, 181–185.

Meeten, G. H. (1980*b*) The birefringence of colloidal dispersions in the Rayleigh and anomalous diffraction approximations. *J. Coll. Interf. Sci.* **73**, 38–44.

Meeten, G. H. (1980*c*) Refractive index of colloidal dispersions of spheroidal particles. *J. Coll. Interf. Sci.* **77**, 1–5.

Meeten, G. H. (1984) Computation of $S_1 - S_2$ in Mie scattering theory. *J. Phys. D.: Appl. Phys.* **18**, L89–L91.

Mehta, N. K., Grimison, A. and Block, A. (1971) Effect of dispersing agents on the angular dependence of light scattering from polystyrene sphere/water sols. *Appl. Opt.* **10**, 2031–2034.

Mei, K. K. (1971) On the perturbation solution to the dyadic Green's functions of Maxwell equations in anisotropic media. *IEEE Trans. Antenn. Propagat.* **AP–19**, 665–696.

Meszaroš, A. (1971) On the variation of the size distributions of large and giant atmospheric particles as a function of the relative humidity. *Tellus* **23**, 436–440.

Meszaroš, A. (1977) An attempt to explain the relation between visual range and relative humidity on the basis of aerosol measurements. *J. Aerosol Sci.* **8**, 31–38.

Meszaroš, E. (1966) Some data on the chemical composition of atmospheric particles in the submicroscopic range of sizes. *Idöjárás* **70**, 257– 261.

Meszaroš, E. (1968) On the size distribution of water soluble particles in the atmosphere. *Tellus* **20**, 443–448.

Meszaroš, E. (1975) On the thermodynamics of the equilibrium size of an aerosol particle as a function of its dry size and composition and the ambient relative humidity. *J. Appl. Meteorol.* **14**, 1044–1049.

Meyer, R. A. and Brunsting, A. (1975) Light scattering from nucleated biological cells. *Biophys.J.* **15**, 191–204.

Mikulski, J. J. and Murphy, E. L. (1963) The computation of electromagnetic scattering from concentric spherical structures. *IEEE Trans. Antenn. Propagat.* **AP–11**, 169–177.

Mishchenko, M. I. (2002) Vector radiative transfer equation for arbitrarily shaped and arbitrarily oriented particles: a microphysical derivation from statistical electomagnetics. *Appl. Opt.* **41**, 7114–7135.

Mishchenko, M. I., Mackowski, D. W., and Travis, L. D. (1995) Scattering of light by bispheres with touching and separated components. *Appl. Opt.* **34**, 4589–4599.

Mishchenko, M. I., Travis, L. D., Kahn, R. A. and West, R. A. (1997) Modeling phase functions for dustlike tropospheric aerosols using a shape mixture of randomly oriented polydisperse spheroids. *J. Geophys. Res.* **102**, 16831–16847.

Mishchenko, M. I., Hovenier, J. W. and Travis, L. D. (eds.) (2000*a*) *Light scattering by nonspherical particles.* Academic Press, San Diego.

Mishchenko, M. I., Hovenier, J. W. and Travis, L. D. (2000*b*) Concepts, terms, notation. In: Mishchenko, M. I., Hovenier, J. W. and Travis, L. D, (eds.), *Light scattering by nonspherical particles.* pp. 3–27. Academic Press, San Diego.

Mishchenko, M. I., Wiscombe, W. J., Hovenier, J. W. and Travis, L. D. (2000*c*) Overview of scattering by nonspherical particles. In: Mishchenko, M. I., Hovenier, J. W. and Travis, L. D, (eds.), *Light scattering by nonspherical particles.* pp. 29–60. Academic Press, San Diego.

Mishchenko, M. I., Travis, L. D and Lacis, A. A. (2002*a*) *Scattering, absorption, and emission of light by small particles.* Cambridge Univ. Press, Cambridge.

Mishchenko, M. I., Penner, J. and Anderson, D. (eds.) (2002*b*). Special issue on global aerosol climatology project. *J. Atmos. Sci.* **59**, 249–783.

Misra, A. and Stein, R. S. (1972) Light scattering studies of the early stage of the crystallization of polyethylene. *J. Polym. Sci.*, **B6**, 473–477.

Monzon, J. C. (1987) Three-dimensional scattering by an infinite homogeneous anisotropic circular cylinder: a spectral approach. *IEEE Trans. Antenn. Propagat.* **35**, 670–682.

Morriss, R. H. and Collins, L. F. (1964) Optical properties of multilayer colloids. *J. Chem. Phys.* **41**, 3357–3363.

Moser, P. J., Murphy, J. D., Nagl, A. and Überall, H. (1981) Resonances and surface waves on conducting spheres with a dielectric coating. *Radio Sci.* **16**, 279–288.

Morse, P. M. and Feshbach, H. (1953) *Methods of theoretical physics.* McGraw–Hill, New York.

Mullaney, P. F. and Fiel, R. J. (1976) Cellular structure as revealed by visible light scattering: studies on suspensions of red blood cell ghost. *Appl. Opt.* **15**, 310–312.

Naumenko, E. K. and Prishivalko, A. P. (1981) Interference effects in two-layered absorbing particles and their influence on the extinction coefficient of aerosol. *Jurnal Prikladnoi Spektroskopii* **35**, 156–162.

Naumenko, E. K., Prishivalko, A. P. and Katseva, I. R. (1975) Optimization of computing algorithm of radiation scattering by polydisperse media. *Izvestiya AN BSSR. Ser. fiz.–mat. nauk* N 1, 71–79.

Newton, R. (1982) *Scattering theory of waves and particles.* Springer, New York.

Nelson, J. (1989) Test of mean field theory for the optics of fractal clusters. *J. Mod. Opt.* **36**, 1031–1057.

Ngo, D., Videen, G. and Chýlek, P. (1996) A FORTRAN code for the scattering of EM waves by a sphere with a nonconcentric spherical inclusion. *Comp. Phys. Comm.* **1077**, 94–112.

Nikol'skiy, V. V. (1977) Projection methods in electrodynamics (screened and open systems). *Prikladnaya Elektrodinamika* **1**(1), 4–50.

Nomura, Y. and Takaku, K. (1955) On the propagation of the electromagnetic waves in an inhomogeneous atmosphere. 1. *Res. Inst. Electr. Comm. Tohoku Univ.* N 7B, 107–114.

Novak, M. and Deway, T. G. (eds.) (1999) *Fractal frontiers*. World Scientific, Singapore.

Oganessyan, S. S. (1982) Diffraction of plane electromagnetic wave on an anisotropic spheroid. *Izvestiya VUZOV. Radiofizika 25, 971–976.*

Oganessyan, S. S. and Baregamyan, V. A. (1981) Diffraction of plane electromagnetic wave on an anisotropic sphere. *Izvestiya AN Armenian SSR. Fizika* **16**, 37–43.

Ohba, V. (1963) Diffraction by a conducting circular cylinder slab with an anisotropic plasma sheath. *Can. J. Phys.* **41**, 881– 89.

Ordal, M. A., Bell, R. J., Alexander, R. W., Long, L. L. and Querry, M. R. (1985) Optical properties of fourteen metals in the infrared and far infrared: Al, Co, Cu, Au, Fe, Pb, Mo, Ni, Pd, Pt, Ag, Ti, V, and W. *Appl. Opt.* **24**, 4493–4499.

Oshchepkov, S. L. and Sorokina, E. A. (1986) Optical manifestation of adsorption-solvate layers in water-oil emulsions of low concentration. *Jurnal Prikladnoi Spektroskopii* **44**, 628–633.

Oshchepkov, S. L. and Sorokina, E. A. (1989) Generalized parameters of marine suspension microstructure at solution of restricted inverse problem of scattering. *Doklady AN SSSR. Fizika Atmosfery i Okeana* **25**, 870–877.

Palik, E. D. (ed) (1985) *Handbook of optical constants of solids*. Academic, Orlando.

Papadakis, S. N, Usunoglu N. K. and Capsalis, C. N. (1990) Scattering of a plane wave by a general anisotropic dielectric ellipsoids. *J. Opt. Soc. Amer. A* **7**, 991–997.

Pasceri, R. E. and Fridlander, S. K. (1965) Measurements on the particle size distribution of the atmospheric aerosol. II. Experimental results and discussion. *J. Atm. Sci.* **22**, 577–584.

Patterson, E. M. (1977) Atmospheric extinction between 0.55μm and 10.6μm due to soil-derived aerosols. *Appl. Opt.* **16**, 2414–2418.

Patterson, E. M. and Gillette, D. A. (1977) Commonalities in measured size distribution for aerosols having a soil-derived component. *J. Geophys. Res.* **82**, 2074–2082.

Patterson, E. M., Gillette, D. A. and Stockton B. H. (1977) Complex index of refraction between 300–700 nm for Saharan aerosols. *J. Geophys. Res.* **82**, 3153–3160.

Pendleton, J. D., Rosen, D. L. (1998) Light scattering from an optically active sphere into a circular aperture. *Appl. Opt.* **37**, 7897–7905.

Penndorf, R. B. (1962) Scattering and extinction coefficients for small absorbing and non-absorbing aerosols. *J. Opt. Soc. Amer.* **52**, 896–904.

Perelman, A. Y. (1979) Scattering in spherically symmetric media. *Appl. Opt.* **18**, 2307–2314.

Perelman, A. Y. (1994) Improvement of convergence of series for absorption cross section of soft sphere. *Optika i Spektroskopiya* **77**, 572–576.

Perelman, A. Y. (1995) Diffraction by spherically symmetric nonhomogeneous particles. *Optika i Spektroskopiya* **78**, 741–750.

Perelman, A. Y. (1996) Scattering by particles with radially variable refractive indices. *Appl. Opt.* **35**, 5452–5460.

Perelman, A. Y. and Zinov'eva, T. V. (2002) Light scattering by sphere with variable optical properties of intermediate layer. *Optika i spektroskopiya* **92**, 221–226.
Perina, Ya. *Light coherence. (1974)*. Mir, Moscow.
Permyakov, V. A. (1968) Diffraction of electromagnetic waves by radially inhomogeneous plasma sphere and cylinder. *Izvestiya VUZOV. Radiofizika* **11**, 531–537.
Petrov, Yu. I. (1986) *Clusters and small particles.* Nauka, Moscow.
Piazza, R., Stavans, J., Bellini, T. and Degiorgio V. (1989) Light scattering study of crystalline latex particles. *Opt. Comm.* **73**, 263–267.
Picot, C., Weill, G. and Benoit, H. (1968) Light scattering by anisotropic discs: Application to polymer single crystals. *J. Coll. Interf. Sci.* **27**, 360–376.
Pilat, M. J. (1967) Optical efficiency factors for concentric spheres. *Appl. Opt.* **6**, 1555–1558.
Piller, N. B. and Martin, O. J. F. (1998) Extension of the generalized multipole technique to anisotropic medias. *Opt. Comm.* **150**, 9–14.
Pinnick, R. G., Biswas, A., Pendleton, J. D. and Armstrong, R. L. (1992) Aerosol induced laser breakdown thresholds: effect of resonant particles. *Appl. Opt.* **31**, 311–318.
Plass, G. N. (1964) Mie scattering and absorption cross sections for aluminium oxide and magnesium oxide. *Appl. Opt.* **3**, 867–872.
Plass, G. N. and Kattawar, G. W. (1971) Comment on: The scattering of polarized light by polydisperse systems of irregular particles. *Appl. Opt.* **10**, 1172–1173.
Platzman, M. and Ozaki, H. T. (1960) Scattering of electromagnetic wave in infinitely long magnetized cylindrical plasma. *J. Appl. Phys.* **31**, 1597–1601.
Pluchino, A. B. (1981) Surface waves and the radiative properties of micron-sized particles. *Appl. Opt.* **20**, 2986–2992.
Pochil, P. F. (1972) *Combustion of powered metals in active media.* Nauka, Moscow.
Pollack, J. B., Toon, O. B. and Khare, B. W. (1973) Optical properties of some terrestrial rocks and glasses. *Icarus* **19**, 372–389.
Pontier, L. and Dechambenoy, C. (1966) Determination des constantes optiques de l'eau liquide entre 1 et 40μ : Application au calcul de son pouvoir reflecteur et de son emissivite. *Ann. Geophys.* **22**, 633–641.
Pontikis, C. (1968) Retro–diffusion d'une onde electromagnetique sur des obstacles spherique composes de plusieurs spheres concentriques. *Meteorologie* N7, 271–283.
Press, W. H., Teukolsky, S. A., Vetterling, W. T. and Flannery, B. P. (1997) *Numerical recipes in Fortran 77. The art of scientific computing.* Cambridge Univ. Press, Cambridge.
Prishivalko, A. P. (1977) Relative humidity and light scattering by systems of homogeneous and two-layered particles of atmospheric aerosol. *Izvestiya AN USSR. Fizika Atmosfery i Okeana* **13**, 828–837.
Prishivalko, A. P. (1978) Influence of air relative humidity on elements of light scattering matrix for systems of homogeneous and nonuniform particles of atmospheric aerosol. *Trudy IEM*, N 18 (71), 128–140.
Prishivalko, A. P. (1981) Investigation of opportunity to increase the absorption efficiency for radiation λ= 10.6 micron by water droplets at presence of surface films. *Izvestiya AN BSSR. Ser. fiz.–mat. nauk* N 6, 84–87.

Prishivalko, A. P. (1983) *Optical and thermal fields inside the light scattering particles.* Nauka i tekhnika, Minsk.

Prishivalko, A. P. (1984) Laser heating of large water droplets with surface films. *Jurnal Prikladnoi Spektroskopii* ***41****, 33–39.*

Prishivalko, A. P. (1986) On the influence of the imaginary part of refractive index of spherical particles on absorption of radiation and structure of internal field at resonance condition. *Jurnal Prikladnoi Spektroskopii* ***45****, 64–69.*

Prishivalko, A. P. and Astafyeva, L. G. (1972*a*) On energy distribution inside homogeneous absorbing particles illuminated by parallel light beam. *Jurnal Prikladnoi Spektroskopii* **16**, 344–350.

Prishivalko, A. P. and Astafyeva L. G. (1972*b*) Influence of optical constants on the energy distribution inside homogeneous particles, illuminated by parallel light beam, *Doklady AN BSSR* **16,** 305–7.

Prishivalko, A. P. and Astafyeva, L. G. (1974) Light absorption, scattering and extinction by atmospheric particles with aqueous shell. *Izvestiya AN USSR. Fizika Atmosfery i Okeana* **10**, 1322–1327.

Prishivalko, A. P. and Astafyeva, L. G. (1975) Investigation of light absorption by two-layered particles. *Jurnal Prikladnoi Spektroskopii* **22**, 908–915

Prishivalko, A. P. and Astafyeva, L. G. (1991) On the distribution of the energy of laser beam in small metal particles with $n < 1$. *Izvestiya AN SSSR. Ser. Metally.* n *3, 192–195.*

Prishivalko, A. P. and Babenko, V. A. (1975) Integral characteristics of light scattering by radially inhomogeneous particles with continuously varying refractive index. *Optika i Spektroskopiya* **29**, 162–169.

Prishivalko, A. P. and Babenko, V. A. (1979) Light scattering by systems of radially-inhomogeneous atmospheric particles. *Izvestiya AN USSR. Fizika Atmosfery i Okeana* **15**, 710–716.

Prishivalko, A. P. and Dudo, N. I. (1975) On opportunity to detect the layers of the adsorbed substances on surface of light scattering particles by optical method. *Jurnal Prikladnoi Spektroskopii* **22**, 512–517.

Prishivalko, A. P. and Dudo, N. I. (1983) Patent 807824 (USSR). Dispersion light filter.

Prishivalko, A. P., Astafyeva, L. G. and Gladkaya, S. V. (1975) Investigation of light scattering and extinction by two-layered particles . *Izvestiya AN BSSR. Ser. fiz.–mat. nauk* N 3, 88–95.

Prishivalko, A. P., Kuzmin, V. N., Naumenko, E. K. and Dudo, N. I. (1980) Quasi-homogeneity of scattering mediums and opportunity to monochromatize the radiation. *Jurnal Prikladnoi Spektroskopii* **33**, 493–498.

Prishivalko, A. P., Naumenko, E. K. and Dudo, N. I. (1981) Patent 683598 (USSR). The filter for IR wavelength range.

Prishivalko, A. P., Naumenko, E. K. and Dudo, N. I. (1983) Patent 841224 (USSR). Method to manufacture the dispersion filters.

Prishivalko, A. P., Astafyeva, L. G., Veremchuk, M. S. and Lednyeva, G. P. (1984*a*) On influence of water droplet size on the absorption efficiency factor and structure of the internal field at resonance condition. *Jurnal Prikladnoi Spektroskopii* ***41****, 846–849.*

Prishivalko, A. P. Astafyeva, L. G., Veremchuk, M. S. and Lednyeva, G. P. (1984*b*) Optical fields inside and near particle at resonance. *Jurnal Prikladnoi Spektroskopii **41**, 641–647.*

Prishivalko, A. P., Astafyeva, L. G. and Lednyeva, G. P. (1984*c*) About influence of water droplets sizes on the radiation absorption factors and on the structure of internal field in the resonant conditions. *Jurnal Prikladnoi Spektroskopii* **41**, 846–849.

Prishivalko, A. P., Astafyeva, L. G. and Lednyeva, G. P. (1988) On influence of low-absorbing solid aerosol particles on lowering of threshold for air optical breakdown. *Injenerno–Fizicheskiy Jurnal* **54**, 582–585.

Prishivalko, A. P., Semenov, L. P., Astafyeva, L. G. and Leiko, S. T. (1988) Thermal destruction of spherical ice particles under the action of radiation with $\lambda = 10.6\ \mu m$. *Injenerno–Fizicheskiy Jurnal* **54**, 103–108.

Pueschel, R. F. and Noll, K. E. (1967) Visibility and aerosol size frequency distribution. *J Appl. Meteorol.* **6**, 1045–1052.

Quiney, R. G. and Carswell, A. I. (1972) Laboratory measurements of light scattering by simulated atmospheric aerosols. *Appl. Opt.* **11**, 1611–1618.

Raes, F. and Bates, T. (2000) Special issue on the $2^{\rm nd}$ aerosol characterization experiment (ACE–2). *Tellus B* **52**, 111–908.

Rakhimov, R. F. (1998) Non-additive absorption of light by particles of atmospheric haze in the liquid-phase clouds. *Optika Atmosfery i Okeana* **11**, 659–668.

Reagan, J. A., Byrne, D. M., King, M. D., Spinhirne, J. D. and Herman, B. M. (1980) Determination of the complex refractive index and size distribution of atmospheric particulates from bistatic-monostatic lidar and solar radiometer measurements. *J. Geophys. Res.* **85**, 1591–1599.

Reagan, J. A., Spiegel, R. J. and Herman, B. M. (1972) Determination of aerosol size distribution parameters by optimization technique. In: *4-th Conf. Laser Atmospheric Studies: Conf. Absts.* pp. 92–93. Univ. of Arizona, Tucson.

Reilly, E. D. (1973) Resonant scattering from inhomogeneous nonspherical targets. *J. Comput. Phys.* **11**, 463–492.

Rheinstein, J. (1964) Scattering of electromagnetic waves from dielectric coated conducting spheres. *IEEE Trans. Antenn. Propagat.* **AP–12**, 334–340.

Richmond, J. H. (1987) Scattering by a ferrite-coated conducting sphere. *IEEE Trans. Antenn. Propag.* **AP–35**, 73–79.

Ricka, J., Borkovec, M. and Hofmeier, U. (1991) Coated droplet model of microemulsions: optical matching and polydispersity. *J. Chem. Phys.* **94**, 8503–8509.

Rojas, R. G. (1992) Integral equation for the scattering by three dimensional inhomogeneous chiral bodies. *J. Electromagn. Waves Appl.* **6**, 733–750.

Roll, G., Kaizer, T. and Schweiger, G. (1998*a*) Controlled modification of the expansion order as a tool in Mie computations. *Appl. Opt.* **37**, 2483–2492.

Roll, G., Kaizer, T., Lange, S. and Schweiger, G. (1998*b*) Ray interpretation of the multipole fields in spherical dielectric cavities. *J. Opt. Soc. Amer. A.* **15**, 2879–2891.

Ross, W. D. (1972) Computation of Bessel functions in light scattering studies. *Appl. Opt.* **11**, 1919–1923.

Roth, J. and Dignam, M. J. (1973) Scattering extinction cross section for spherical particle coated with an oriented molecular layer. *J. Opt. Soc. Amer.* **63**, 308–311.

Rozenberg, G. V. (1958) *Optics of thin coatings*. GIFML, Moscow.

Rozenberg, G. V. (1967) Properties of atmospheric aerosol on data of optical investigations. *Izvestiya AN USSR. Fizika Atmosfery i Okeana*, **3**, 936–949.

Rozenberg, G. V. (1968) Optical investigations of atmospheric aerosol. *Uspekhi Fizicheskih Nauk*, **95**, 159–208.

Rozenberg, G. V. (1977) The light ray (on the theory of light field). *Uspekhi Fizicheskih Nauk*, **121**, 97-133.

Rozenberg, G. V. (1982) The spectral theory of light field. In : Ivanov, A.P. (ed.), *Light propagation in dispersive medium*, pp. 22–36. Nauka and tehnika, Minsk.

Rozenberg, V. I. (1972) *Scattering and extinction of electromagnetic radiation by atmospheric particles*. Gidrometeoizdat, Leningrad.

Rozenberg, V. I. (1970) Scattering of microradiowaves by layered hailstone. *Izvestiya AN USSR. Fizika Atmosfery i Okeana* **6**, 168–177.

Ruck, G. T. (ed) (1970) *Radar cross-section handbook*. vol. 1. Plenum, New York.

Rusch, W. V. T. and Yeh, C. (1967) Scattering by infinite cylinder coated with inhomogeneous and anisotropic plasma sheath. *IEEE Trans. Antenn. Propagat.* **AP–15**, 453–455.

Rysakov, W. and Ston, M. (2001) The light scattering by core-mantle sphere. *J. Quant. Spectr. Rad. Transf. 69, 121–129.*

Rytov, S. M., Kravtsov, Yu. A. and Tatarskiy, V. I. (1978) *Introduction into statistical physics*. vol.2. Nauka, Moscow.

Ryym, R. I. (1974) Features of light absorption and scattering by two-layered Mie particles with absorbing core. *Izvestiya AN USSR. Fizika Atmosfery i Okeana* **10**, 434–439.

Sano, Y. (1989) The state of polarization of light scattered by anisotropic spheroids. *Bull. Chem. Soc. Jap.* **62**, 697–700.

Sassen, K. (2000) Lidar backscatter depolarization technique for cloud and aerosol research. In: Mishchenko, M. I., Hovenier, J. W. and Travis, L. D. (eds.), *Light scattering by nonspherical particles*. pp. 393–416. Academic Press, San Diego.

Savel'ev, B. A. and Fadeev, V. Ya. (1975) Complex investigation of optical properties of polydisperse water aerosol. In: Zuev, V.E. (ed.) *Propagation of optical waves in atmosphere.* pp. 47–52. Nauka (Siberian branch), Novosibirsk.

Schaaf, J. M. and Williams, D. (1973) Optical constants of ice in the infrared. *J. Opt. Soc. Amer.* **63**, 726–732.

Scharfman, H. (1954) Scattering from dielectric coated spheres in the region of the first resonance. *J. Appl. Phys.* **25**, 1352–1356.

Schneider, M., Hirleman, E. D., Saleheen, H. I., Chowdhury, D. Q. and Hill, S. C. (1993) Light scattering by radially inhomogeneous fuel droplets evaporating in a high-temperature environment. *Proc. SPIE* **1862**, 269–286.

Schulz, L. (1954) Optical constants of silver, gold, copper and aluminum. I. The absorption coefficient k. *J. Opt. Soc. Am.* **44**, 357–362.

Schulz, L. and Tangherlini, M. (1954) The optical constants of silver, gold, copper, and aluminum. II. The index of refraction n. *J. Opt. Soc. Am.* **44**, 362–368.

Schweiger, G. (1990) Raman scattering on single aerosol particles and on flowing aerosols: a review. *J. Aerosol Sci.* **21**, 483–509.

Sedunov, Yu. S. (1972) *Physics of liquid droplet phase formation in atmosphere.* Gidrometeoizdat, Leningrad.
Selezneva, E. S. (1966) *Atmospheric aerosols.* Gidrometeoizdat, Leningrad.
Seltman, J. (1985) Erfahrungen mit einem neuen Program zur Mie-streung. *Geofyz. Veroff. Karl–Marx–Univ. Leipzig* **3**, 171–185.
Shafai, L. (1972) Scattering by spherically symmetrical objects. *Can. J. Phys.* **50**, 749–753.
Shah, G. A. (1985) A new representation of Güttler's theory of electromagnetic scattering by a composite sphere. *Astron. Nachr.* **306**, 63–66.
Shalaev, V. M. (2000) *Nonlinear optics of random media: fractal composites and metal dielectric films.* Springer–Verlag, Berlin.
Sheng, P. (1980) Theory for the dielectric function of granular composite media. *Phys. Rev. Lett.* **45**, 60–63.
Shiff, L. I. (1962) Scattering of waves and particles by inhomogeneous regions. *J. Opt. Soc. Amer.* **52**, 140–146.
Shifrin, K. S. (1951) *Scattering of light in a turbid medium.* Gostechteorizdat, Moscow. [English translation: *NASA Techn. Transl.* **NASA.TT F–447** (1968)].
Shifrin, K. S. (1952) Light scattering by two-layered particles. *Izvestiya AN USSR. Ser. geophysics* N 2, 15–21.
Shifrin, K. S., Perelman, A. Ya. and Kokorin, A. M. (1985) Light scattering by two-layer dielectric particles with continuous optical properties. *Optika i Spektroskopiya* **59**, 361–364.
Shigapov, A. B. (1990) Errors in calculation of radiative properties of polydisperse system of particles. *Teplofizika Vysokih Temperatur* **28**, 553–557.
Shybanov, E. B. and Haltrin, V.I. (2002) Mie scattering by very large spherical particles. *Appl. Opt.* (submitted).
Singh, P. R. and Singh, S. S. (1973) The studies on the continuous phasic near by the droplets of emulsion. *J. Indian Chem. Soc.* **50**, 125–128.
Singham, S. B. (1987) Form and intrinsic optical activity in light scattering by chiral particles. *J. Chem. Phys.* **87**, 1873–1881.
Singham, S. B. (1988) Coupled dipoles in light scattering by randomly oriented chiral particles. *J. Chem. Phys.* ***88**, 1522–1527.*
Singham, S. B. (1989) Theoretical factors in modeling polarized light scattering by arbitrary particles. *Appl. Opt.* **28**, 5058–5064.
Sinzig, J. and Quinten, M. (1994) Scattering and absorption by spherical multilayer particle. *Appl. Phys. A.* **58**, 157–162.
Sitarski, M. (1987*a*) Internal heating of multilayered aerosol particles by electromagnetic radiation. *Langmuir* **3**, 85–93.
Sitarski, M. (1987*b*) Absorption of IR radiation inside an explosively boiling file coal-water particles. *Part. Sci. Technol.* **5**, 193–205.
Skaropoulos, N. C., Ioannidou, M. P. and Chrissoulidis, D. P. (1996) Induced electromagnetic field in a layered eccentric spheres model of the head: plane wave and localized source exposure. *IEEE Trans. Microw. Theor. Techn.* **MTT–44**, 1963–73.

Slonim, I. Ya., Fodiman, N. M. and Ustinova, Z. M. (1960) Definition of the particle size on light scattering. *Optika i Spektroskopiya* **8**, 243–248.

Smart, C. and Wills, E. (1967) Determination of refractive indices of polystyrene latexes by light scattering. *J. Coll. Interf. Sci.* **25**, 577–583.

Smirnov, B. M. (1986) Fractal clusters. *Uspekhi fizicheskih nauk* **149**, 177–219.

Smirnov, V. I. (1973) About approximation of empirical size distributions of cloudy droplets and others aerosol particles. *Izvestiya AN USSR. Fizika Atmosfery i Okeana* **9**, 54–65.

Sorensen, C. M. (2001) Light scattering by fractal aggregates: a review. *Aerosol Sci. Technol.* **36**, 648–687.

Sosman, R. B. (1927) *The properties of silica.* Chemical Press, New York.

Speranskaya, T. A. and Tarutina, Z. I. (1976) *Optical Properties of Polymers.* Khimiya, Leningrad.

Starojilova, A. I. (1969) On passivation of water-absorbing condensation cores by an adsorption os SAS from gas phase. *Fizika Aerodispesnykh Sistem* N. 1, 40–45.

Stephens, G. L. (1994) *Remote sensing of the lower atmosphere.* Oxford Univ. Press, Oxford.

Stein, R. S., Rhodes, M. B. and Parter, R. S. (1968) Light scattering by liquid crystals. *J. Coll. Interf. Sci.* **27**, 336–354.

Stoimenova, M., Labaki, L. and Stoylov, S. (1980) Orientationally induced conservative dichroism in dilute disperse systems. *J. Coll. Interf. Sci.* **77**, 53–56.

Stratton, J. A. (1941) *Electromagnetic. theory.* McGraw Hill, New York.

Strawbridge, K. B. and Hallett, F. R. (1992) Polydispersed Mie theory applied to hollow latex spheres: an integrated light scattering study. *Can. J. Phys.* **70**, 401–6.

Su, C. C. (1989) Electromagnetic scattering by a dielectric body with arbitrary inhomogeneity and anisotropy. *IEEE Trans. Antenn. Propagat.* **AP–37,** 384–389.

Sundararajan, P. R. and Marchessault, K. H. (1970) Light scattering by anisotropic shells. *J. Appl. Phys.* **41**, 4256–4260.

Svatoš, S. and Secanina, Z. (1968) The efficiency factors Q_{ext} and Q_{scat} of the core mantle grains in the infrared region 2–3.4 μm. *Acta Univ. Carolinae Math. et Phys.* N 1, 3–52.

Sveshnikov, A. G. and Eremin, Yu. A. (1980) Numerical investigation of diffraction problems for axially symmetrical body by method of non-orthogonal series. *Izvestiya VUZOV. Radiofizika* **23**, 1006–1008.

Sveshnikov, A. G. and Il'inskiy, A. S. (1971) Direct method for problems of diffraction on local non-uniform body. *Jurnal Vychislitelnoi Matematiki i Matematicheskoi Fiziki* **11**, 960–968.

Swamy, K. S. and O'Dell, C. R. (1967) Distinction between models of interstellar grains. *Astrophys. J.* **147**, 937–942.

Swarner, W. J. and Peters, L. (1963) Radar cross-section of dielectric or plasma coated spheres and circular cylinders. *IEEE Trans. Antenn. Propagat.* **AP–11**, 558–569.

Syd'ko, F. Ya, Lopatin, V. N. and Paramonov, L. E. (1990) *Polarization characteristics of suspensions of biological particles.* Nauka, Novosibirsk.

Tai, C.-T. (1958) An electromagnetic theory of the spherical Luneberg lens. *Appl. Sci. Res.* **B7**, 113–130.

Tai, C.-T. (1963) Some wave functions and potentials functions pertaining to spherically stratified media. *J. Res. Nat. Bur. Stand.* **D67**, 199–202.

Tai, C.-T. (1973) On the eigenfunction expansion of dyadic Green's functions. *Proc. IEEE* **61**, 480–481.

Tai, C.-T. (1987) Some essential formulas in dyadic analysis and their applications. *Radio Sci.* **22**, 1283–1288.

Tai, C.-T. (1993) *Dyadic Green functions in electromagnetic theory.* IEEE Press, New York.

Tai, C.-T. and Rosenfeld, P. (1976) Different representations of dyadic Green's functions for a rectangular cavity. *IEEE Trans. Microwave Theory Techn.* **MTT–24**, 597–601

Tatarskiy, V. I. (1967) *Wave propagation in turbulent atmosphere.* Nauka, Moscow.

Terskov, I. A., Lopatin, V. N. and Sid'ko, F. Ya. (1977) Influence of three-layered spherical particles structure (model of cells) on light absorption and scattering. *Doklady AN USSR* **233**, 712–715.

Thurn, R. and Kiefer, W. (1984) Raman-microsampling technique applying optical levitation by radiation pressure. *Appl. Spectrosc.* **38**, 78–83.

Thurn, R. and Kiefer, W. (1985) Structural resonances observed in the Raman spectra of optically levitated liquid droplets. *Appl. Opt.* **24,** 1515–1519.

Toon, O. B. and Ackerman, T. P. (1981) Algorithms for the calculation of scattering by stratified spheres. *Appl. Opt.* **20**, 3657–3660.

Tuomi, T. J. (1976) The influence of relative humidity and its stratification on lidar backscatter. *Beitr. Phys. Atmosph.* **49**, 98–113.

Tuomi, T. J. (1980) Light scattering by aerosols with layered humidity-dependent structure. *J. Aerosol Sci.* **11**, 367–375.

Turner, L. (1973) Rayleigh–Gans–Born light scattering by ensembles of randomly oriented anisotropic particles. *Appl. Opt.* **12**, 1085–1090.

Uslenghi, P. L. E. (1969) Electromagnetic and optical behavior of two classes of dielectric lenses. *IEEE Trans. Antenn. Propagat.* **AP–17**, 235–236.

Uslenghi, P. L. E. (1990) Scattering by impedance sphere coated with a chiral layer. *Electromagnetics* **10**, 201–211.

Uslenghi, P. L. E. and Weston, V. H. (1970) High-frequency scattering from a metal-like dielectric lens. *Appl. Sci. Res.* **23**, 147–163.

Urgiris, E. E. and Costaschuk, F. M. (1973) Investigation of colloid stability in polyelertrolyte solutions by laser Doppler spectroscopy. *Nature* **242**, 77–79.

van Bladel, J. (1961) Some remarks on Green's dyadic for infinite space. *IRE Trans. Antenn. Propagat.* **9**, 563–566.

van de Hulst, H. C. (1981) *Light scattering by small particles.* Dover, New York.

Varadan, V. V., Lakhtakia, A. and Varadan, V. K. (1989) Scattering by three-dimensional anisotropic scatterers. *IEEE Trans. Antenn. Propagat.* **AP–37**, 800–802.

Varshalovich, D. A., Moskalev, A. N. and Khersonskij, V. K. (1988) *Quantum theory of angular momemtum.* World Scientific, Singapore.

Velesco, N., Kaiser, T. and Schweiger, G. (1997) Computation of the internal field of a large spherical particle by use of the geometrical optics approximation. *Appl. Opt.* **36**, 8724–8728.

Volkovitskiy, O. A., Nikiforova, N. K. and Pavlova, L. N. (1974*a*) Preliminary results of extinction and scattering measurement for 10.6 micron radiation in droplet and crystalline mists. *Trudy IEM* N. 5 (43), 145–155.

Volkovitskiy, O. A., Nikiforova, N. K., Pavlova, L. N and Petrushin, A. G. (1974*b*) Some data on CO_2 laser radiation propagation in crystalline cloudy medium. *Izvestiya AN USSR. Fizika Atmosfery i Okeana* **10**, 1157–1162.

Volkovitskiy, O. A., Pavlova, L. N. and Snykov, V. P. (1975) On the problem of scattering properties asymmetry for crystalline cloudy media. *Izvestiya AN USSR. Fizika Atmosfery i Okeana* **11**, 757–760.

Volkovitskiy, O. A., Sedunov, Yu. S. and Semenov, L. P. (1982) *Propagation of intense laser radiation in clouds.* Gidrometeoizdat, Leningrad.

Volten, H., Muñoz, O., and Rol, E. Scattering matrices of mineral aerosol particles at 441.6 and 632.8 nm. *J. Geophys. Res.* **106**, 17375–17402.

Volz, F. (1972) Infrared refractive index of atmospheric aerosol substances. *Appl. Opt.* **11**, 755–759.

von Roos, O. (1971) Method for the solution of electromagnetic scattering problems for inhomogeneous dielectrics as a power series in the ratio (dimension of scatterer) / wavelength. *J. Appl. Phys.* **42**, 4197–4201.

Voronkova, E. M., Grechushnikov, B. M., Distler, G. I. and Petrov, I. P. (1965) *Optical materials for infrared technique.* Nauka, Moscow.

Vorontsov, A. A. and Mirovitskaya, S. D. (1993) Diffraction of plane wave on two-layered sphere. *Radiotekhnika* **7**, 70–72.

Voshchinnikov, N. V. (2002) Optics of cosmic dust. Part I. *Astrophys. Space Phys.* **12**, 1–169.

Voshchinnikov, N. V. and Farafonov, V. G. (1993) Optical properties of spheroidal particles. *Astrophys. Space Sci.* **204**, 19–86.

Voshchinnikov, N. V. and Mathis, J. S. (1999) Calculating cross-sections of cosmic interstellar grains. *Astrophys. J.* **526**, 257–264.

Waggoner, A. P. , Ahlquist, N. C. and Charlson, R. J. (1972) Measurement of the aerosol total scatter-backscatter ratio. *Appl. Opt.* **11**, 2856– 2889.

Wait, J. R. (1961) Some boundary value problems involving plasma media. *J. Res. Nat. Bur. Stand.* **65**, 137–150.

Wait, J. R. (1963) Electromagnetic scattering from a radially inhomogeneous sphere. *Appl. Sci. Res.* **B10**, 441–450.

Ward, G., Cushing, K. M., McPeters, R. D. and Green, A. E. S. (1973) Atmospheric aerosol index of refraction and size-altitude distribution from bistatic laser scattering and solar aureole measurements. *Appl. Opt.* **12**, 2585–2592.

Warren, S. G. (1984) Optical constants of ice from the ultraviolet to the microwave. *Appl. Opt.* **23**, 1206–1225.

Watson, K. M. (1969) Multiple scattering of electromagnetic waves in an under-dense plasma. *J. Math. Phys.* **10**, 688–703.

Weiser, P., Stolzenburg, M. and Baumueller, J. (1977) Aerosol particles: remarks on the optical properties. *Appl. Opt.* **16**, 2651–2653.

Werner, C. (1972) Lidar measurements of atmospheric aerosol as a function of relative humidity. *Opto–Electronics* **4**, 125–132.
Westcott, B. S. (1968*a*) Exact solution for electromagnetic wave propagation .in spherically stratified isotropic media. *Proc. Cambr. Phil. Soc.* **64**, 227–235.
Westcott, B. S. (1968*b*) Electromagnetic wave propagation in spherically stratified isotropic media. *Electronics Lett.* **4**, 572–573.
Whitby, K. T. (1978) The physical characteristics of sulfur aerosols. *Atmosph. Environ.* **12**, 135–159.
Whitby, K. T. (1981) Determination of aerosol growth rates in the atmosphere using lumped mode aerosol dynamics. *J. Aerosol Sci.* **12**, 173–178.
Whitby, K. T., Husar, R. B. and Lin, B. Y. H. (1972) The aerosol size distribution of Los Angeles smog. *J. Coll. Interf. Sci.* **39**, 177–204.
Wickramasinghe, N. C. (1965) On the growth and destruction of ice mantles on interstellar graphite grains. *Month. Not. Roy. Astron. Soc.* **131**, 177–190.
Wickramasinghe, N. C. (1967) *Interstellar grains.* Chapman and Hall, London.
Wickramasinghe, N. C. and Nandy, K. (1968) Extinction curves for graphite particles coated with solid hydrogen. *Nature* **219**, 1347–1349.
Wickramasinghe, N. C. and Swamy, K. S. (1968) Comments on the possibility of interstellar quartz grains. *Astrophys. J.* **154**, 397–400.
Wickramasinghe, N. C., Dharmawardhana M. N. C. and Wyld C. (1966a) Light scattering by graphite core-ice mantle grains. *Month. Not. Roy. Astron. Soc.* **134**, 25–36.
Wickramasinghe, N. C., Donn, B., Stecher, T. P. and Williams, D. A. (1966b) Wavelength dependence of interstellar polarization by graphite grains. *Nature* **212**, 167–168.
Winkler, P. (1969) Unter suchungen iiber das grossen wachstum naturlicher aerosolteilchen mit der relativen feuchte nach einer wagemethode. *Ann. Meteorol.* N4, 134–137.
Winkler, P. (1973) The growth of atmospheric aerosol particles as a function of the relative humidity. II. Improved concept of mixed nuclei. *J. Aerosol Sci.* **4**, 373–387.
Winkler, P. and Junge, C. (1972) The growth of atmospheric aerosol particles as a function of relative humidity. Part I: Method and measurements at different locations. *J. Rech. Atmosph.* **6**, 617–638.
Wiscombe, W. J. (1980) Improved Mie scattering algorithms. *Appl. Opt.* **19**, 1505–1509.
Wiscombe, W. J. and Chýlek, P. (1977) Mie scattering between any two angles. *J. Opt. Soc. Amer.* **67**, 572–573.
Witt, G. (1968) Optical characteristics of mesospheric aerosol distributions in relation to noctilucent clouds. *Tellus* **20**, 98–114.
Witten, T. A. and Sander, L. M. (1981) Diffusion-limited aggregation, a kinetic critical phenomenon. *Phys. Rev. Lett.* **47**, 1400–1403.
Wu, Z. S. and Wang, Y. P. (1991) Electromagnetic scattering for multilayered sphere: recursive algorithms. *Radio Sci.* **26**, 1393–1401.
Wu, Z. S., Guo, L. X., Ren K. F., Gouesbet, G. and Grehan, G. (1997) Improved algorithm for electromagnetic scattering of plane waves and shaped beams by multilayered sphere. *Appl. Opt.* **36**, 5188–5198.
Wyatt, P. J. (1962) Scattering of electromagnetic plane waves from inhomogeneous spherically symmetric objects. *Phys. Rev.* **127**, 1837–1843.

Wyatt, P. J. (1963) Interdependence of plasma frequency and surface diffusivity from electromagnetic scattering analysis. *J. Appl. Phys.* **34**, 2078–2083.

Wyatt, P. J. (1965) Electromagnetic scattering by finite dense plasmas. *J. Appl. Phys.*, **36**(12), 3875–3882.

Wyatt, J. and Phillips, D. T. (1972) Structure of single bacteria from light scattering. *J. Theor. Biol.* **37**, 493–501.

Yakovlev, D. A. and Afonin, O. A. (1997) The method of calculation of amplitude scattering matrix for inhomogeneous anisotropic particles in the anomalous diffraction approximation. *Optika i Spektroskopiya* **82**, 86–92.

Yamamoto, G. and Tanaka, M. (1969) Determination of aerosol size distribution from spectral attenuation measurements. *Appl. Opt.* **8**, 447– 453.

Yeh, C. (1964) Electromagnetic radiation from an arbitrary slot on conducting cylinder coated with a uniform cold plasma sheath with an axial static magnetic field. *Can. J. Phys.* **42**, 1369–1381.

Zardecki, A. and Pendelton, J. D. (1989) Hydrodfynamics of water droplets irradiated by a pulsed CO_2 laser. *Appl. Opt.* **28**, 638–640.

Zeldovich, Ya. B. and Sokolov, D. D. (1985) Fractals, scaling, intermediate asymptotes. *Uspekhi Fizicheskih Nauk **146**, 493–506.*

Zich, R. (1969) The use of dyadic Green's functions in problems involving inhomogeneous media. *Alta frequenza* **38** (Num. Speciale), 30–34.

Zuev, V.E. (1970) *Propagation of visual and infrared waves in atmosphere.* Sovetskoe radio, Moscow.

Zuev, V. E., Ivlev, L. S. and Kondratiev, K. Ya. (1973) New results of atmospheric aerosol investigations. *Izvestiya AN USSR. Fizika atmosfery i okeana* **9**, 371–385.

Zuev, V. E., Kopytin, Yu. D. and Kuzikovsky, A. V. (1980) *Non-linear optical effects in aerosols.* Nauka, Novosibirsk.

Zumer, S. (1988) Light scattering from nematic droplets: the anomalous diffraction approximation. *Phys. Rev. A* **37**, 4006–4015.

Zumer, S. and Doane, J. W. (1986) Light scattering from a small nematic droplets. *Phys. Rev. A* **34**, 3373–3386.

Zumer, S., Golumme, A. and Doane, J. W. (1989) Light extinction in a dispersion of small nematic droplets. *J. Opt. Soc. Amer. A.* **6,** 403–411.

Index

MIX
Papier aus verantwortungsvollen Quellen
Paper from responsible sources
FSC® C105338

If you have any concerns about our products,
you can contact us on
ProductSafety@springernature.com

In case Publisher is established outside the EU,
the EU authorized representative is:
Springer Nature Customer Service Center GmbH
Europaplatz 3, 69115 Heidelberg, Germany

Printed by Libri Plureos GmbH
in Hamburg, Germany